T5-CAO-659

ASTRONOMICAL DATA ANALYSIS SOFTWARE AND SYSTEMS VII

A SERIES OF BOOKS ON RECENT DEVELOPMENTS IN ASTRONOMY AND ASTROPHYSICS

Managing Editor, D. Harold McNamara
Production Manager, Elizabeth S. Holloman

Printed by BookCrafters, Inc.

First published 1998

ISBN: 1-886733-65-1 **ISSN: 1080-7926**

Please contact proper address for information on:

PUBLISHING:
Managing Editor
PO Box 24463
Brigham Young University
Provo, UT 84602-4463
801-378-2298

pasp@astro.byu.edu
Fax: 801-378-2265

ORDERING BOOKS:
Astronomical Society of the Pacific
CONFERENCE SERIES
390 Ashton Avenue
San Francisco, CA 94112 - 1722 USA
415-337-2624

catalog@aspsky.org
Fax: 415-337-5205

A SERIES OF BOOKS ON RECENT DEVELOPMENTS IN ASTRONOMY AND ASTROPHYSICS

Vol. 1-Progress and Opportunities in Southern Hemisphere Optical Astronomy: CTIO 25th Anniversary Symposium
ed. V. M. Blanco and M. M. Phillips ISBN 0-937707-18-X

Vol. 2-Proceedings of a Workshop on Optical Surveys for Quasars
ed. P. S. Osmer, A. C. Porter, R. F. Green, and C. B. Foltz ISBN 0-937707-19-8

Vol. 3-Fiber Optics in Astronomy
ed. S. C. Barden ISBN 0-937707-20-1

Vol. 4-The Extragalactic Distance Scale: Proceedings of the ASP 100th Anniversary Symposium
ed. S. van den Bergh and C. J. Pritchet ISBN 0-937707-21-X

Vol. 5-The Minnesota Lectures on Clusters of Galaxies and Large-Scale Structure
ed. J. M. Dickey ISBN 0-937707-22-8

Vol. 6-Synthesis Imaging in Radio Astronomy: A Collection of Lectures from the Third NRAO Synthesis Imaging Summer School
ed. R. A. Perley, F. R. Schwab, and A. H. Bridle ISBN 0-937707-23-6

Vol. 7-Properties of Hot Luminous Stars: Boulder-Munich Workshop
ed. C. D. Garmany ISBN 0-937707-24-4

Vol. 8-CCDs in Astronomy
ed. G. H. Jacoby ISBN 0-937707-25-2

Vol. 9-Cool Stars, Stellar Systems, and the Sun. Sixth Cambridge Workshop
ed. G. Wallerstein ISBN 0-937707-27-9

Vol. 10-Evolution of the Universe of Galaxies: Edwin Hubble Centennial Symposium
ed. R. G. Kron ISBN 0-937707-28-7

Vol. 11-Confrontation Between Stellar Pulsation and Evolution
ed. C. Cacciari and G. Clementini ISBN 0-937707-30-9

Vol. 12-The Evolution of the Interstellar Medium
ed. L. Blitz ISBN 0-937707-31-7

Vol. 13-The Formation and Evolution of Star Clusters
ed. K. Janes ISBN 0-937707-32-5

Vol. 14-Astrophysics with Infrared Arrays
ed. R. Elston ISBN 0-937707-33-3

Vol. 15-Large-Scale Structures and Peculiar Motions in the Universe
ed. D. W. Latham and L. A. N. da Costa ISBN 0-937707-34-1

Vol. 16-Proceedings of the 3rd Haystack Observatory Conference on Atoms, Ions and Molecules: New Results in Spectral Line Astrophysics
ed. A. D. Haschick and P. T. P. Ho ISBN 0-937707-35-X

Vol. 17-Light Pollution, Radio Interference, and Space Debris
ed. D. L. Crawford ISBN 0-937707-36-8

Vol. 18-The Interpretation of Modern Synthesis Observations of Spiral Galaxies
ed. N. Duric and P. C. Crane ISBN 0-937707-37-6

Vol. 19-Radio Interferometry: Theory, Techniques, and Applications, IAU Colloquium 131
ed. T. J. Cornwell and R. A. Perley ISBN 0-937707-38-4

Vol. 20-Frontiers of Stellar Evolution: 50th Anniversary McDonald Observatory (1939-1989)
ed. D. L. Lambert ISBN 0-937707-39-2

Vol. 21-The Space Distribution of Quasars
ed. D. Crampton ISBN 0-937707-40-6

Vol. 22-Nonisotropic and Variable Outflows from Stars
ed. L. Drissen, C. Leitherer, and A. Nota ISBN 0-937707-41-4

Vol. 23-Astronomical CCD Observing and Reduction Techniques
ed. S. B. Howell ISBN 0-937707-42-4

Vol. 24-Cosmology and Large-Scale Structure in the Universe
ed. R. R. de Carvalho ISBN 0-937707-43-0

Vol. 25-Astronomical Data Analysis Software and Systems I
ed. D. M. Worrall, C. Biemesderfer, and J. Barnes ISBN 0-937707-44-9

Vol. 26-Cool Stars, Stellar Systems, and the Sun, Seventh Cambridge Workshop
ed. M. S. Giampapa and J. A. Bookbinder ISBN 0-937707-45-7

Vol. 27-The Solar Cycle: Proceedings of the National Solar Observatory/Sacramento Peak 12th Summer Workshop
ed. K. L. Harvey ISBN 0-937707-46-5

Vol. 28-Automated Telescopes for Photometry and Imaging
ed. S. J. Adelman, R. J. Dukes, Jr., and C. J. Adelman ISBN 0-937707-47-3

Vol. 29-Viña Del Mar Workshop on Catacysmic Variable Stars
ed. N. Vogt ISBN 0-937707-48-1

Vol. 30-Variable Stars and Galaxies
ed. B. Warner ISBN 0-937707-49-X

Vol. 31-Relationships Between Active Galactic Nuclei and Starburst Galaxies
ed. A. V. Filippenko ISBN 0-937707-50-3

Vol. 32-Complementary Approaches to Double and Multiple Star Research, IAU Colloquium 135
ed. H. A. McAlister and W. I. Hartkopf ISBN 0-937707-51-1

Vol. 33-Research Amateur Astronomy
ed. S. J. Edberg ISBN 0-937707-52-X

Vol. 34-Robotic Telescopes in the 1990s
ed. A. V. Filippenko ISBN 0-937707-53-8

Vol. 35-Massive Stars: Their Lives in the Interstellar Medium
ed. J. P. Cassinelli and E. B. Churchwell ISBN 0-937707-54-6

Vol. 36-Planets Around Pulsars
ed. J. A. Phillips, S. E. Thorsett, and S. R. Kulkarni ISBN 0-937707-55-4

Vol. 37-Fiber Optics in Astronomy II
ed. P. M. Gray ISBN 0-937707-56-2

Vol. 38-New Frontiers in Binary Star Research: Pacific Rim Colloquium
ed. K. C. Leung and I.-S. Nha ISBN 0-937707-57-0

Vol. 39-The Minnesota Lectures on the Structure and Dynamics of the Milky Way
ed. Roberta M. Humphreys ISBN 0-937707-58-9

Vol. 40-Inside the Stars, IAU Colloquium 137
ed. Werner W. Weiss and Annie Baglin ISBN 0-937707-59-7

Vol. 41-Astronomical Infrared Spectroscopy: Future Observational Directions
ed. Sun Kwok ISBN 0-937707-60-0

Vol. 42-GONG 1992: Seismic Investigation of the Sun and Stars
ed. Timothy M. Brown ISBN 0-937707-61-9

Vol. 43-Sky Surveys: Protostars to Protogalaxies
ed. B. T. Soifer ISBN 0-937707-62-7

Vol. 44-Peculiar Versus Normal Phenomena in A-Type and Related Stars, IAU Colloquium 138
ed. M. M. Dworetsky, F. Castelli, and R. Faraggiana ISBN 0-937707-63-5

Vol. 45-Luminous High-Latitude Stars
ed. D. D. Sasselov ISBN 0-937707-64-3

Vol. 46-The Magnetic and Velocity Fields of Solar Active Regions, IAU Colloquium 141
ed. H. Zirin, G. Ai, and H. Wang ISBN 0-937707-65-1

Vol. 47-Third Decennial US-USSR Conference on SETI
ed. G. Seth Shostak ISBN 0-937707-66-X

Vol. 48-The Globular Cluster-Galaxy Connection
ed. Graeme H. Smith and Jean P. Brodie ISBN 0-937707-67-8

Vol. 49-Galaxy Evolution: The Milky Way Perspective
ed. Steven R. Majewski ISBN 0-937707-68-6

Vol. 50-Structure and Dynamics of Globular Clusters
ed. S. G. Djorgovski and G. Meylan ISBN 0-937707-69-4

Vol. 51-Observational Cosmology
ed. G. Chincarini, A. Iovino, T. Maccacaro, and D. Maccagni ISBN 0-937707-70-8

Vol. 52-Astronomical Data Analysis Software and Systems II
ed. R. J. Hanisch, R. J. V. Brissenden, and Jeannette Barnes ISBN 0-937707-71-6

Vol. 53-Blue Stragglers
ed. Rex A. Saffer ISBN 0-937707-72-4

Vol. 54-The First Stromlo Symposium: The Physics of Active Galaxies
ed. Geoffrey V. Bicknell, Michael A. Dopita, and Peter J. Quinn ISBN 0-937707-73-2

Vol. 55-Optical Astronomy from the Earth and Moon
ed. Diane M. Pyper and Ronald J. Angione ISBN 0-937707-74-0

Vol. 56-Interacting Binary Stars
ed. Allen W. Shafter ISBN 0-937707-75-9

Vol. 57-Stellar and Circumstellar Astrophysics
ed. George Wallerstein and Alberto Noriega-Crespo ISBN 0-937707-76-7

Vol. 58-The First Symposium on the Infrared Cirrus and Diffuse Interstellar Clouds
ed. Roc M. Cutri and William B. Latter ISBN 0-937707-77-5

Vol. 59-Astronomy with Millimeter and Submillimeter Wave Interferometry, IAU Colloquium 140
ed. M. Ishiguro and Wm. J. Welch ISBN 0-937707-78-3

Vol. 60-The MK Process at 50 Years: A Powerful Tool for Astrophysical Insight: A Workshop of the Vatican Observatory
ed. C. J. Corbally, R. O. Gray, and R. F. Garrison ISBN 0-937707-79-1

Vol. 61-Astronomical Data Analysis Software and Systems III
ed. Dennis R. Crabtree, R. J. Hanisch, and Jeannette Barnes ISBN 0-937707-80-5

Vol. 62-The Nature and Evolutionary Status of Herbig Ae / Be Stars
ed. P. S. Thé, M. R. Pérez, and E. P. J. van den Heuvel ISBN 0-937707-81-3

Vol. 63-Seventy-Five Years of Hirayama Asteroid Families: The role of Collisions in the Solar System History
ed. R. P. Binzel, Y. Kozai, and T. Hirayama ISBN 0-937707-82-1

Vol. 64-Cool Stars, Stellar Systems, and the Sun, Eighth Cambridge Workshop
ed. Jean-Pierre Caillault ISBN 0-937707-83-X

Vol. 65-Clouds, Cores, and Low Mass Stars
ed. Dan P. Clemens and Richard Barvainis ISBN 0-937707-84-8

Vol. 66- Physics of the Gaseous and Stellar Disks of the Galaxy
ed. Ivan R. King ISBN 0-937707-85-6

Vol. 67-Unveiling Large-Scale Structures Behind the Milky Way
ed. C. Balkowski and R. C. Kraan-Korteweg ISBN 0-937707-86-4

Vol. 68-Solar Active Region Evolution: Comparing Models with Observations
ed. K. S. Balasubramaniam and George W. Simon ISBN 0-937707-87-2

Vol. 69-Reverberation Mapping of the Broad-Line Region in Active Galactic Nuclei
ed. P. M. Gondhalekar, K. Horne, and B. M. Peterson ISBN 0-937707-88-0

Vol. 70-Groups of Galaxies
ed. Otto G. Richter and Kirk Borne ISBN 0-937707-89-9

Vol. 71-Tridimensional Optical Spectroscopic Methods in Astrophysics, IAU Colloquium 149
ed. G. Comte and M. Marcelin ISBN 0-937707-90-2

Vol. 72-Millisecond Pulsars: A Decade of Surprise,
ed. A. A. Fruchter, M. Tavani, and D. C. Backer ISBN 0-937707-91-0

Vol. 73-Airborne Astronomy Symposium on the Galactic Ecosystem: From Gas to Stars to Dust
ed. M. R. Haas, J. A. Davidson, and E. F. Erickson ISBN 0-937707-92-9

Vol. 74-Progress in the Search for Extraterrestrial Life: 1993 Bioastronomy Symposium
ed. G. Seth Shostak ISBN 0-937707-93-7

Vol. 75-Multi-Feed Systems for Radio Telescopes
ed. D. T. Emerson and J. M. Payne ISBN 0-937707-94-5

Vol. 76-GONG '94: Helio- and Astero-Seismology from the Earth and Space
ed. Roger K. Ulrich, Edward J. Rhodes, Jr., and Werner Däppen ISBN 0-937707-95-3

Vol. 77-Astronomical Data Analysis Software and Systems IV
ed. R. A. Shaw, H. E. Payne, and J. J. E. Hayes ISBN 0-937707-96-1

Vol. 78-Astrophysical Applications of Powerful New Databases: Joint Discussion No. 16 of the 22nd General Assembly of the IAU
ed. S. J. Adelman and W. L. Wiese ISBN 0-937707-97-X

Vol. 79-Robotic Telescopes: Current Capabilities, Present Developments, and Future Prospects for Automated Astronomy
ed. Gregory W. Henry and Joel A. Eaton ISBN 0-937707-98-8

Vol. 80-The Physics of the Interstellar Medium and Intergalactic Medium
ed. A. Ferrara, C. F. McKee, C. Heiles, and P. R. Shapiro ISBN 0-937707-99-6

Vol. 81-Laboratory and Astronomical High Resolution Spectra
ed. A. J. Sauval, R. Blomme, and N. Grevesse ISBN 1-886733-01-5

Vol. 82-Very Long Baseline Interferometry and the VLBA
ed. J. A. Zensus, P. J. Diamond, and P. J. Napier ISBN 1-886733-02-3

Vol. 83-Astrophysical Applications of Stellar Pulsation, IAU Colloquium 155
ed. R. S. Stobie and P. A. Whitelock ISBN 1-886733-03-1

Vol. 84-The Future Utilisation of Schmidt Telescopes, IAU Colloquium 148
ed. Jessica Chapman, Russell Cannon, Sandra Harrison, and Bambang Hidayat ISBN 1-886733-05-8

Vol. 85-Cape Workshop on Magnetic Cataclysmic Variables
ed. D. A. H. Buckley and B. Warner ISBN 1-886733-06-6

Vol. 86-Fresh Views of Elliptical Galaxies
ed. Alberto Buzzoni, Alvio Renzini, and Alfonso Serrano ISBN 1-886733-07-4

Vol. 87-New Observing Modes for the Next Century
ed. Todd Boroson, John Davies, and Ian Robson ISBN 1-886733-08-2

Vol. 88-Clusters, Lensing, and the Future of the Universe
ed. Virginia Trimble and Andreas Reisenegger ISBN 1-886733-09-0

Vol. 89-Astronomy Education: Current Developments, Future Coordination
ed. John R. Percy ISBN 1-886733-10-4

Vol. 90-The Origins, Evolution, and Destinies of Binary Stars in Clusters
ed. E. F. Milone and J. -C. Mermilliod ISBN 1-886733-11-2

Vol. 91-Barred Galaxies, IAU Colloquium 157
ed. R. Buta, D. A. Crocker, and B. G. Elmegreen ISBN 1-886733-12-0

Vol. 92-Formation of the Galactic Halo--Inside and Out
ed. H. L. Morrison and A. Sarajedini ISBN 1-886733-13-9

Vol. 93-Radio Emission from the Stars and the Sun
ed. A. R. Taylor and J. M. Paredes ISBN 1-886733-14-7

Vol. 94-Mapping, Measuring, and Modelling the Universe
ed. Peter Coles, Vicent Martinez, and Maria-Jesus Pons-Borderia ISBN 1-886733-15-5

Vol. 95-Solar Drivers of Interplanetary and Terrestrial Disturbances: Proceedings of 16th International Workshop, National Solar Observatory/Sacramento Peak
ed. K.S. Balasubramaniam, S. L. Keil, and R. N. Smartt ISBN 1-886733-16-3

Vol. 96-Hydrogen-Deficient Stars
ed. C. S. Jeffery and U. Heber ISBN 1-886733-17-1

Vol. 97-Polarimetry of the Interstellar Medium
ed. W. G. Roberge and D. C. B. Whittet ISBN 1-886733-18-X

Vol. 98-From Stars to Galaxies: The Impact of Stellar Physics on Galaxy Evolution
ed. Claus Leitherer, Uta Fritze-von Alvensleben, and John Huchra ISBN 1-886733-19-8

Vol. 99-Cosmic Abundances: Proceedings of the 6th Annual October Astrophysics Conference
ed. Stephen S. Holt and George Sonneborn ISBN 1-886733-20-1

Vol. 100-Energy Transport in Radio Galaxies and Quasars
ed. P. E. Hardee, A. H. Bridle, and J. A. Zensus ISBN 1-886733-21-X

Vol. 101-Astronomical Data Analysis Software and Systems V
ed. George H. Jacoby and Jeannette Barnes ISSN 1080-7926

Vol. 102-The Galactic Center, 4th ESO/CTIO Workshop
ed. Roland Gredel ISBN 1-886733-22-8

Vol. 103-The Physics of Liners in View of Recent Observations
ed. M. Eracleous, A. Koratkar, C. Leitherer, and L. Ho ISBN 1-886733-23-6

Vol. 104-Physics, Chemistry, and Dynamics of Interplanetary Dust, IAU Colloquium 150
ed. Bo A. S. Gustafson and Martha S. Hanner ISBN 1-886733-24-4

Vol. 105-Pulsars: Problems and Progress, IAU Colloquium 160
ed. M. Bailes, S. Johnston, and M.A. Walker ISBN 1-886733-25-2

Vol. 106-Minnesota Lectures on Extragalactic Neutral Hydrogen
ed. Evan D. Skillman ISBN 1-886733-26-0

Vol. 107-Completing the Inventory of the Solar System: A Symposium held in conjunction with the 106th Annual Meeting of the ASP
ed. Terrence W. Rettig and Joseph M. Hahn ISBN 1-886733-27-9

Vol. 108-M. A. S. S. Model Atmospheres and Spectrum Synthesis: 5th Vienna Workshop
ed. S. J. Adelman, F. Kupka, and W. W. Weiss ISBN 1-886733-28-7

Vol. 109-Cool Stars, Stellar Systems, and the Sun, Ninth Cambridge Workshop
ed. Roberto Pallavicini and Andrea K. Dupree ISBN 1-886733-29-5

Vol. 110-Blazar Continuum Variability
ed. H. R. Miller, J. R. Webb, and J. C. Noble ISBN 1-886733-30-9

Vol. 111-Magnetic Reconnection in the Solar Atmosphere: Proceedings of a Yohkoh Conference
ed. R. D. Bentley and J. T. Mariska ISBN 1-886733-31-7

Vol. 112- The History of the Milky Way and Its Satellite System
ed. A. Burkert, D. H. Hartmann, and S. R. Majewski ISBN 1-886733-32-5

Vol. 113-Emission Lines in Active Galaxies: New Methods and Techniques, IAU Colloquium 159
ed. B. M. Peterson, F.-Z. Cheng, and A. S. Wilson ISBN 1-886733-33-3

Vol. 114-Young Galaxies and QSO Absorption-Line Systems
ed. Sueli M. Viegas, Ruth Gruenwald, and Reinaldo R. de Carvalho ISBN 1-886733-34-1

Vol. 115-Galactic and Cluster Cooling Flows
ed. Noam Soker ISBN 1-886733-35-X

Vol. 116-The Second Stromlo Symposium: The Nature of Elliptical Galaxies
ed. M. Arnaboldi, G. S. Da Costa, and P. Saha ISBN 1-886733-36-8

Vol. 117-Dark and Visible Matter in Galaxies
ed. Massimo Persic and Paolo Salucci ISBN 1-866733-37-6

Vol. 118-First Advances in Solar Physics Euroconference: Advances in the Physics of Sunspots
ed. B. Schmieder, J. C. del Toro Iniesta, and M. Vázquez ISBN 1-886733-38-4

Vol. 119-Planets Beyond the Solar System and the Next Generation of Space Missions
ed. David R. Soderblom ISBN 1-886733-39-2

Vol. 120-Luminous Blue Variables: Massive Stars in Transition
ed. Antonella Nota and Henny J. G. L. M. Lamers ISBN 1-886733-40-6

Vol. 121-Accretion Phenomena and Related Outflows, IAU Colloquium 163
ed. D. T. Wickramasinghe, G. V. Bicknell and L. Ferrario ISBN 1-886733-41-4

Vol. 122-From Stardust to Planetesimals: Symposium held as part of the 108th Annual Meeting of the ASP
ed. Yvonne J. Pendleton and A. G. G. M. Tielens ISBN 1-886733-42-2

Vol. 123-The 12th 'Kingston Meeting': Computational Astrophysics
ed. David A. Clarke and Michael J. West ISBN 1-886733-43-0

Vol. 124-Diffuse Infrared Radiation and the IRTS
ed. Haruyuki Okuda, Toshio Matsumoto, and Thomas L. Roellig ISBN 1-886733-44-9

Vol. 125-Astronomical Data Analysis Software and Systems VI
ed. Gareth Hunt and H. E. Payne ISBN 1-886733-45-7

Vol. 126-From Quantum Fluctuations to Cosmological Structures
ed. D. Valls-Gabaud, M. A. Hendry, P. Molaro, and K. Chamcham ISBN 1-886733-46-5

Vol. 127-Proper Motions and Galactic Astronomy
ed. Roberta M. Humphreys ISBN 1-886733-47-3

Vol. 128-Mass Ejection from AGN (Active Galactic Nuclei)
ed. N. Arav, I. Shlosman, and R. J. Weymann ISBN 1-886733-48-1

Vol. 129-The George Gamow Symposium
ed. E. Harper, W. C. Parke, and G. D. Anderson ISBN 1-886733-49-X

Vol. 130–The Third Pacific Rim Conference on Recent Development on Binary Star Research
ed. Kam-Ching Leung ISBN 1-886733-50-3

Vol. 131–Boulder-Munich II: Properties of Hot, Luminous Stars
ed. Ian D. Howarth ISBN 1-886733-51-1

Vol. 132–Star Formation with the Infrared Space Observatory (ISO)
ed. João L. Yun and René Liseau ISBN 1-886733-52-X

Vol. 133–Science with the NGST
ed. Eric P. Smith and Anuradha Koratkar ISBN 1-886733-53-8

Vol. 134–Brown Dwarfs and Extrasolar Planets
ed. Rafael Rebolo, Eduardo L. Martin,
and Maria Rosa Zapatero Osorio ISBN 1-886733-54-6

Vol. 135–A Half Century of Stellar Pulsation Interpretations: A Tribute to Arthur N. Cox
ed. P. A. Bradley and J. A. Guzik ISBN 1-886733-55-4

Vol. 136–Galactic Halos: A UC Santa Cruz Workshop
ed. Dennis Zaritsky ISBN 1-886733-56-2

Vol. 137–Wild Stars in the Old West: Proceedings of the 13th North American Workshop
on Cataclysmic Variables and Related Objects
ed. S. Howell, K. Kuulkers, and C. Woodward ISBN 1-886733-57-0

Vol. 138–1997 Pacific Rim Conference on Stellar Astrophysics
ed. Kwing L. Chan, K. S. Cheng, and Harinder P. Singh ISBN 1-886733-58-9

Vol. 139–Preserving the Astronomical Windows
ed. Syuzo Isobe and Tomohiro Hirayama ISBN 1-886733-59-7

Vol. 140–Synoptic Solar Physics
ed. K. S. Balasubramaniam, J. W. Harvey, and D. M. Rabin ISBN 1-886733-60-0

Vol. 141–Astrophysics from Antarctica
ed. Giles Novak and Randall H. Landsberg ISBN 1-886733-61-9

Vol. 142–The Stellar Initial Mass Function, 38th Herstmonceux Conference
ed. Gerry Gilmore and Debbie Howell ISBN 1-886733-62-7

Vol. 143–The Scientific Impact of the Goddard High Resolution Spectrograph
ed. John C. Brandt, Thomas B. Ake III, and Carolyn Collins Petersen ISBN 1-886733-63-5

Vol. 144–Radio Emission from Galactic and Extragalactic Compact Sources
ed. J. Anton Zensus, G. B. Taylor, and J. M. Wrobel ISBN 1-886733-64-3

Vol. 145-Astronomical Data Analysis Software and Systems VII
ed. Rudolf Albrecht, Richard N. Hook, and Howard A. Bushouse ISBN 1-886733-65-1

Inquiries concerning these volumes should be directed to the:
Astronomical Society of the Pacific
CONFERENCE SERIES
390 Ashton Avenue
San Francisco, CA 94112-1722 USA
415-337-2126
catalog@aspsky.org
Fax: 415-337-5205

ASTRONOMICAL SOCIETY OF THE PACIFIC
CONFERENCE SERIES

Volume 145

ASTRONOMICAL DATA ANALYSIS SOFTWARE AND SYSTEMS VII

Proceedings from a meeting held in Sonthofen, Germany 14-17 September 1997

Edited by
Rudolf Albrecht, Richard N. Hook,
and Howard A. Bushouse

Contents

Part 8. Astrostatistics and Databases

Preface

The 1997 conference of the Astronomical Data Analysis Software and Systems (ADASS) Conference series was organized by the Space Telescope European Coordinating Facility and the European Southern Observatory from 14–17 September, 1997. It was the first time this conference had been held in Europe.

The conference took place in the Allgäu Stern Hotel and Conference Centre in Sonthofen, Bavaria. There were 222 registered participants from 20 different countries; the distribution was 38 participants from Germany (including ESO), 100 from the US and Canada, and 84 others. Through the financial aid program we were able to support 15 participants from 9 countries. We made special efforts to support scientists from eastern Europe to help our colleagues there during difficult times.

The program of the conference consisted of 6 invited papers and 39 oral contributions. 92 posters were on display and 11 computer demonstrations were presented. Seven birds-of-a-feather (BOF) sessions were held on the following topics: Astrobrowse, Linux, IRAF, IDL, AIPS/AIPS++, FITS, and Scheduling/Pipelines. 119 contributions were submitted for inclusion in the proceedings, including 4 of the invited papers, 30 of the contributed papers, and 85 of the poster papers.

Because of infrastructure limitations there was no pre-conference tutorial and no IRAF Developers Workshop. However, there was a tag-along meeting on the topic of Converging Computing Methodologies in Astronomy (CCMA).

The Bavarian mountains and the beautiful weather made the conference quite enjoyable for the participants. The high point of the social program was the Conference Banquet which was held in a Bavarian brewery and which featured enormous quantities of food and, upholding an ancient Bavarian custom, free beer. As an added bonus there was a total lunar eclipse during the banquet.

The conference sponsors included the European Southern Observatory (ESO), the European Space Agency (ESA), the Space Telescope European Coordinating Facility (ST-ECF), the Space Telescope Science Institute (STScI), the National Radio Astronomy Observatory (NRAO), Sun Microsystems Deutschland, CREASO (IDL Germany), Sybase, and Pink Aviation Services. We want to thank the sponsors for their generous support.

The Program Organizing Committee of the conference consisted of the following members: Rudi Albrecht (ST-ECF/ESO), Roger Brissenden (SAO), Tim Cornwell (NRAO), R. Crutcher (University of Illinois), F. Rick Harnden - Chair (SAO), George Jacoby (NOAO), Jonathan McDowell (SAO), Jan Noordam (NFRA), Dick Shaw (STScI), Richard Simon (NRAO), Britt Sjöberg (ST-ECF), Doug Tody (NOAO), and Dave Van Buren (IPAC).

The Local Organizing Committee members were Miguel Albrecht (ESO), Rudi Albrecht - Chair (ST-ECF), Piero Benvenuti (ST-ECF), Preben Grosbøl (ESO), Richard Hook (ST-ECF), Benoît Pirenne (ST-ECF), Peter Quinn (ESO), Britt Sjöberg (ST-ECF), Christine Telander (ESO), and Eline Tolstoy (ST-ECF).

We also had the help of numerous other people. Setting up a LAN and an Internet connection on short notice in a hotel environment turned out to be a non-trivial effort and was only possible due to the heroic efforts of ST-ECF staff members and of staff of SERCo, the ESO computer support contractor.

ADASS VIII will be hosted by the National Center for Supercomputer Applications (NCSA) and the University of Illinois Astronomy Department. It will be held November 1–4, 1998 at the University of Illinois in Urbana, Illinois, USA. Information can be found at http://www.ncsa.uiuc.edu/ADASS98/.

Rudi Albrecht
Richard Hook

Space Telescope — European Coordinating Facility

Howard Bushouse

Space Telescope Science Institute

April 1998

Cover Illustration: This figure, pertaining to the paper by Markus Dolensky, Alberto Micol, Benoit Pirenne and Michael Rosa (page 304), shows the "HST Jitter Ball" visualized by a Java applet. The deviations of the HST line of sight from the nominal position during a 1000 second WFPC2 exposure are plotted on a scale of milliarcseconds. HST jitter information is of obvious importance for assessing the quality of an individual observation. It has been found that the center of gravity of the jitter ball moves on a subpixel scale between revisits of the same target. This information can be used to combine images in an optimal way and with minimum loss of resolution using the "drizzling" technique described at the last ADASS and used for combining the images of the Hubble Deep Field. Another application of drizzling is given in the paper by Hook et al. in these proceedings (page 320).

Participant List

Timothy Abbott, Canada France Hawaii Telescope Corporation, Astronomy Dept., 65-1238 Mamalahoa Hwy., Kamuela, HI 96743, USA (tmca@cfht.hawaii.edu)

Alberto Accomazzi, Harvard-Smithsonian Center for Astrophysics, High Energy Astrophysics Division, 60 Garden Street, Cambridge, MA 02138, USA (aaccomazzi@cfa.harvard.edu)

Michail Agafonov, Radiophysical Research Institute (NIRFI), B. Pecherskaya St. 25, 603600 Nizhny Novgorod, Russia (agfn@nirfi.nnov.su)

Miguel Albrecht, European Southern Observatory, DMD, Karl-Schwarzschild-Str. 2, D-85748 Garching, Germany (malbrech@eso.org)

Rudolf Albrecht, ST-ECF, Karl-Schwarzschild-Str. 2, D-85748 Garching, Germany (ralbrech@eso.org)

Anastasia Alexov, Smithsonian Astrophysical Observatory, Data Systems/ HEAD, 60 Garden Street, Cambridge, MA 02138, USA (asquared@head-cfa.harvard.edu)

Alex (Sandy) Antunes, NASA / GSFC, Astro-E, Code 664, Greenbelt, MD 20771, USA (antunes@lheamail.gsfc.nasa.gov)

Charlie Backus, Jet Propulsion Laboratory, Dept. of Astrophysics, 4800 Oak Grove Dr., Pasadena, CA 91109, USA (Charles.R.Backus@jpl.nasa.gov)

Pascal Ballester, European Southern Observatory, DMD, Karl-Schwarzschild-Str. 2, D-85748 Garching, Germany (pballest@eso.org)

Klaus Banse, European Southern Observatory, Data Management Division, Karl-Schwarzschild-Str. 2, D-85748 Garching, Germany (kbanse@eso.org)

David Barnes, The University of Melbourne, School of Physics, Grattan St., Parkville, VIC 3052, Australia (dbarnes@physics.unimelb.edu.au)

Andrea Baruffolo, Osservatorio Astronomico, Vicolo dell'Osservatorio 5, I-35122 Padova, Italy (baruffolo@astrpd.pd.astro.it)

Michael Barylak, ESA Villafranca del Castillo Satellite Tracking Station, SSD, Aptdo. 50727, E-28080 Madrid, Spain (mb@iuearc.vilspa.esa.es)

Ugo Becciani, Astrophysical Observatory of Catania, Viale A. Doria, 6, I-95125 Catania, Italy (ube@sunct.ct.astro.it)

Steffi Beckhaus, GMD - Forschungszentrum Informationstechnik GmbH, IMK.MAT, Schloß Birlinghoven, D-53754 Sankt Augustin, Germany (steffi.beckhaus@gmd.de)

David Bell, National Optical Astronomy Observatories, Central Computer Services, 950 N. Cherry Ave., P.O. Box 26732, Tucson, AZ 85726, USA (dbell@noao.edu)

Leopoldo Benacchio, Osservatorio Astronomico, Vicolo dell'Osservatorio 5, I-35122 Padova, Italy (benacchio@astrpd.pd.astro.it)

Piero Benvenuti, ST-ECF, Karl-Schwarzschild-Str. 2, D-85748 Garching, Germany (pbenvenu@eso.org)

Martin Bly, Rutherford Appleton Laboratory, Chilton, Didcot OX11 0QX, United Kingdom (bly@star.rl.ac.uk)

Carlo Boarotto, European Southern Observatory, DMD/Serco, Karl-Schwarzschild-Str.2, D-85748 Garching, Germany (cboarott@eso.org)

Bruce Bohannan, Kitt Peak National Observatory, P.O. Box 26732, Tucson, AZ 85726, USA (bruce@noao.edu)

David Bohlender, Herzberg Institute of Astrophysics, Canadian Astronomy Data Centre, 5071 W. Saanich Road, Victoria, V8X 4M6, Canada (David.Bohlender@hia.nrc.ca)

Jerzy Borkowski, Integral Science Data Centre, Chemin d'Ecogia 16, CH-1290 Versoix, Switzerland (Jerzy.Borkowski@obs.unige.ch)

Andrei Boulatov, Institut Solnechno-Zemnoi Fiziki, Dept. of Solar Physics, P.O. Box 4026, UL. Lermontova, 126, 664033 Irkutsk-33, Russia (bulat@iszf.irk.ru)

Sylvie Brau-Nogue, Observatoire Midi-Pyrenées, B.P. 136; 9, rue du Pont de la Moulette, F-65201 Bagnères de Bigorre Cedex, France (brau@obs-mip.fr)

Alan Bridger, Royal Observatory, Technology Dept., Blackford Hill, Edinburgh EH9 3HJ, United Kingdom (ab@roe.ac.uk)

Melania Brolis, Osservatorio Astronomico, Vicolo dell'Osservatorio 5, I-35122 Padova, Italy (brolis@astrpd.pd.astro.it)

Loredana Bruca, Nuova Telespazio, Control Centre Division, Via Corcolle 19, I-00131 Rome, Italy (bruca@saxnet.sdc.asi.it)

Peter Bunclark, Royal Greenwich Observatory, Madingley Road, Cambridge CB3 0HJ, United Kingdom (psb@ast.cam.ac.uk)

Howard Bushouse, Space Telescope Science Institute, Science Software Group, 3700 San Martin Drive, Baltimore, MD 21218, USA (bushouse@stsci.edu)

Sepideh Chakavek, GMD - Forschungszentrum Informationstechnik GmbH, IMK.MAT, Schloß Birlinghoven, D-53754 Sankt Augustin, Germany (chakaveh@gmd.de)

Josephine Chan, University of Cambridge, Institute of Astronomy, Madingley Road, Cambridge CB3 0HA, United Kingdom (sjchan@ast.cam.ac.uk)

Sumitra Chary, Smithsonian Astrophysical Observatory, High Energy Astrophysics Division, 60 Garden Street, Cambridge, MA 02138, USA (schary@cfa.harvard.edu)

Alberto Maurizio Chavan, European Southern Observatory, Karl-Schwarzschild-Str.2, D-85748 Garching, Germany (amchavan@eso.org)

Yoshihiro Chikada, National Astronomical Observatory, Osawa, 2-21-1, Mitaka, Tokyo 181, Japan (chikada@optik.mtk.nao.ac.jp)

Carol Christian, Space Telescope Science Institute, Office of Public Outreach, 3700 San Martin Drive, Baltimore, MD 21218, USA (carolc@stsci.edu)

Norbert Christlieb, Hamburger Sternwarte, Gojenbergsweg 112, D-21029 Hamburg, Germany (nchristlieb@hs.uni-hamburg.de)

Alessandro Coletta, Nuova Telespazio, Control Centre Division, Via Corcolle 19, I-00131 Rome, Italy (coletta@saxnet.sdc.asi.it)

Maureen Conroy, Smithsonian Astrophysical Observatory, MS-19, 60 Garden Street, Cambridge, MA 02138, USA (mo@cfa.harvard.edu)

Tim Cornwell, National Radio Astronomy Observatory, P.O. Box 0, Socorro, NM 87801, USA (tcornwel@nrao.edu)

Dennis Crabtree, Canada-France-Hawaii Telescope, P.O. Box 1597, Kamuela, HI 96743, USA (crabtree@cfht.hawaii.edu)

Richard Crutcher, University of Illinois, Dept. of Astronomy, 1002 West Green Street, Urbana, IL 61801, USA (crutcher@uiuc.edu)

Pierre Cruzalebes, Observatoire de la Côte d'Azur, FRESNEL, Avenue Copernic, F-06130 Grasse, France (cruzalebes@obs-azur.fr)

Malcolm Currie, Rutherford Appleton Laboratory, Starlink, Chilton, Didcot OX11 0QX, United Kingdeom (mjc@star.rl.ac.uk)

Jadwiga Daszynska, Wroclaw University, Physics and Astronomy, Kopernika 11, PL-51-622 Wroclaw, Poland (daszynska@astro.uni.wroc.pl)

Lindsey Davis, National Optical Astronomy Observatories, 950 N. Cherry Ave., P.O. Box 26732, Tucson, AZ 85726, USA (ldavis@noao.edu)

Jean-Pierre De Cuyper, Royal Observatory of Belgium, Ringlaan 3, B-1180 Ukkel, Belgium (Jean-Pierre.DeCuyper@ksb-orb.oma.be)

Michele De La Pena, Space Telescope Science Institute, Science Software Group (STSDAS), 3700 San Martin Drive, Baltimore, MD 21218, USA (delapena@stsci.edu)

Susana Delgado, Instituto de Astrofisica de Canarias, Computer Centre, c/ Via Lactea s/n, E-38200 La Laguna, Tenerife, Spain (sdm@iac.es)

Janet DePonte, Smithsonian Astrophysical Observatory, ASC, 60 Garden Street, Cambridge, MA 02138, USA (janet@cfa.harvard.edu)

Stephen Doe, Smithsonian Astrophysical Observatory, High Energy Astrophysics Division, MS 81, 60 Garden Street, Cambridge, MA 02139, USA (sdoe@head-cfa.harvard.edu)

Markus Dolensky, ST-ECF, Karl-Schwarzschild-Str. 2, D-85748 Garching, Germany (mdolensk@eso.org)

Daniel Durand, Herzberg Institute of Astrophysics, Canadian Astronomy Data Centre, 5071 W. Saanich Road, Victoria, V8X 4M6, Canada (Daniel.Durand@hia.nrc.ca)

Frossie Economou, Joint Astronomy Centre, 660 North A'ohoku Place, Hilo, HI 96720, USA (frossie@jach.hawaii.edu)

Daniel Egret, Observatoire Astronomique de Strasbourg, 11, rue de l'Université, F-67000 Strasbourg, France (Daniel.Egret@astro.u-strasbg.fr)

Guenther Eichhorn, Smithsonian Astrophysical Observatory, High Energy Dept., MS-83, 60 Garden Street, Cambridge, MA 02138, USA (gei@cfa.harvard.edu)

Pierre Fernique, Observatoire Astronomique de Strasbourg, 11, rue de l'Université, F-67000 Strasbourg, France (Pierre.Fernique@astro.u-strasbg.fr)

Anthony Ferro, University of Arizona, Steward Observatory, 933 N. Cherry Ave., Tucson, AZ 85721-0065, USA (aferro@as.arizona.edu)

Richard Fisher, National Radio Astronomy Observatory, P.O. Box 2, Green Bank, WV 24944, USA (rfisher@hyades.gb.nrao.edu)

Mike Fitzpatrick, National Optical Astronomy Observatories, IRAF Group, 950 N. Cherry Ave., P.O Box 26732, Tucson, AZ 85726, USA (fitz@noao.edu)

Wolfram Freudling, ST-ECF, Karl-Schwarzschild-Str. 2, D-85748 Garching, Germany (wfreudli@eso.org)

Carlos Gabriel, ESA Villafranca del Castillo Satellite Tracking Station, SAI, Aptdo. 50727, E-28080 Madrid, Spain (cgabriel@iso.vilspa.esa.es)

Niall Gaffney, University of Texas, McDonald Observatory, RLM 15.308, Austin, TX 78712, USA (niall@rhea.as.utexas.edu)

Kevin Gamiel, National Center for Supercomputing Applications, ITECH, 152 Computing Applications Building, MC-476, Champaign, IL 61820, USA (kgamiel@ncsa.uiuc.edu)

Robert Garwood, National Radio Astronomy Observatory, 520 Edgemont Road, Charlottesville, VA 22911, USA (bgarwood@nrao.edu) René Gastaud, DAPNIA, SEI, F-91191 Gif-sur-Yvette, France (gastaud@sapi01.saclay.cea.fr)

Severin Gaudet, Herzberg Institute of Astrophysics, Canadian Astronomy Data Center, 5071 W. Saanich Road, Victoria, V8X 4M6, Canada (Severin.Gaudet@hia.nrc.ca)

Françoise Genova, Observatoire astronomique de Strasbourg, 11, rue de l'Université, F-67000 Strasbourg, France (genova@astro.u-strasbg.fr)

Gino Giannone, European Southern Observatory, DMD/Serco, Karl-Schwarzschild-Str.2, D-85748 Garching, Germany (ggiannon@eso.org)

Mark Giuliano, Space Telescope Science Institute, PRESTO, 3700 San Martin Drive, Baltimore, MD 21218, USA (giuliano@stsci.edu)

Brian Glendenning, National Radio Astronomy Observatory, AIPS++, PO Box 0, Socorro, NM 87801, USA (bglenden@nrao.edu)

Perry Greenfield, Space Telescope Science Institute, Science Software Group, 3700 San Martin Drive, Baltimore, MD 21218, USA (perry@stsci.edu)

Eric Greisen, National Radio Astronomy Observatory, 520 Edgemont Road, Charlottesville, VA 22903-2475, USA (egreisen@primate.cv.nrao.edu)

Elena Grocheva, Pulkovo Observatory, Photographic Astrometry, 65 k 1 shosse Pulkovskoe, 196140 St. Petersburg, Russia (gl@spb.iomail.lek.ru)

Ted Groner, Smithsonian Astrophysical Observatory, FLWO, 670 Mount Hopkins Road, Tucson, AZ 85645, USA (ted@sparky.sao.arizona.edu)

Preben Grosbøl, European Southern Observatory, DMD, Karl-Schwarzschild-Str. 2, D-85748 Garching, Germany (pgrosbol@eso.org)

Rainer Gruber, Max-Planck-Institut für Extraterrestrische Physik, Giessenbachstr. 2, D-85748 Garching, Germany (gru@rosat.mpe-garching.mpg.de)

Bernt Grundseth, Canada France Hawaii Telescope Corporation, Software Group, 65-1238 Mamalahoa Hwy., Kamuela, HI 96743, USA (bernt@cfht.hawaii.edu)

Ranjan Gupta, Inter University Centre for Astronomy and Astrophysics, Post Bag 4, Ganeshkhind, Pune 411007, India (rag@iucaa.ernet.in)

Warren Hack, Space Telescope Science Institute, Software Support Group, 3700 San Martin Drive, Baltimore, MD 21218, USA (hack@stsci.edu)

Robert Hanisch, Space Telescope Science Institute, 3700 San Martin Drive, Baltimore, MD 21218, USA (hanisch@stsci.edu)

F. Rick Harnden, Jr., Smithsonian Astrophysical Observatory, High Energy Astrophysics Division, 60 Garden St. MS-2, Cambridge, MA 02138, USA (frh@cfa.harvard.edu)

R. Lee Hawkins, Wellesley College, Department of Astronomy, Wellesley, MA 02181, USA (lhawkins@wellesley.edu)

Jeffrey Hayes, Space Telescope Science Institute, STIS, 3700 San Martin Drive, Baltimore, MD 21218, USA (hayes@stsci.edu)

Helen He, Smithsonian Astrophysical Observatory, High Energy Astrophysics Division, 60 Garden Street, Cambridge, MA 02173, USA (hhe@cfa.harvard.edu)

Sara Heap, NASA Goddard Space Flight Center, Code 681, Greenbelt, MD 20771, USA (hrsheap@stars.gsfc.nasa.gov)

André Heck, Observatoire Astronomique, 11, rue de l'Université, F-67000 Strasbourg, France (heck@astro.u-strasbg.fr)

Herman Hensberge, Royal Observatory of Belgium, Astrophysics, Ringlaan 3, B-1180 Brussels, Belgium (Herman.Hensberge@oma.be)

Paul Hilton, Institute of Space and Astronautical Science, High Energy X-ray, 3-1-1 Yoshinodai, Sagamihara, 229 Kanagawa, Japan (paul@astro.isas.ac.jp)

Philip Hodge, Space Telescope Science Institute, 3700 San Martin Drive, Baltimore, MD 21218, USA (hodge@stsci.edu)

Andras Holl, Konkoly Observatory, P.O. Box 67, H-1525 Budapest, Hungary (holl@ogyalla.konkoly.hu)

Richard Hook, ST-ECF, Karl-Schwarzschild-Str. 2, D-85748 Garching, Germany (rhook@eso.org)

Wolfgang Hovest, Ruhr-Univeristät Bochum, Astronomisches Institut, D-44780 Bochum, Germany (hovest@astro.ruhr-uni-bochum.de)

Jin-Chung Hsu, Space Telescope Science Institute, 3700 San Martin Drive, Baltimore, MD 21218, USA (hsu@stsci.edu)

Edwin Huizinga, Space Telescope Science Institute, DSD, 3700 San Martin Drive, Baltimore, MD 21218, USA (huizinga@stsci.edu)

Stephen Hulbert, Space Telescope Science Institute, SSD/STIS, 3700 San Martin Drive, Baltimore, MD 21218, USA (hulbert@stsci.edu)

Gareth Hunt, National Radio Astronomy Observatory, 520 Edgemont Road, Charlottesville, VA 22903, USA (ghunt@nrao.edu)

Steffen Huth, ESA Villafranca del Castillo Satellite Tracking Station, ISO Science Operations Centre, Aptdo. 50727, E-28080 Madrid, Spain (shuth@iso.vilspa.esa.es)

Sidik Isani, Canada France Hawaii Telescope Corporation, Software Group, 65-1238 Mamalahoa Hwy., Kamuela, HI 96743, USA (isani@cfht.hawaii.edu)

Yasuhide Ishihara, Fujitsu Limited, Earth Science Systems, Nakase 1-9- 3, Mihama-ku, Chiba 261, Japan (ishi@ssd.se.fujitsu.co.jp)

George Jacoby, National Optical Astronomy Observatories, Kitt Peak National Observatory, 950 N. Cherry Avenue; P.O. Box 26732, Tucson, AZ 85726-6732, USA (jacoby@noao.edu)

Tim Jenness, Joint Astronomy Centre, 660 North A'ohoku Place, Hilo, HI 96720, USA (timj@jach.hawaii.edu)

Donald Jennings, Integral Science Data Centre, Chemin d'Ecogia 16, CH- 1290 Versoix, Switzerland (Don.Jennings@obs.unige.ch)

William Joye, Smithsonian Astrophysical Observatory, High Energy Astrophysics Division, 60 Garden Street, Cambridge, MA 02138, USA (wjoye@cfa.harvard.edu)

Athol Kemball, National Radio Astronomy Observatory, AIPS++ Project, P.O. Box 0, Socorro, NM 87801, USA (akemball@nrao.edu)

Ellyne Kinney, Space Telescope Science Institute, SSD/STIS, 3700 San Martin Drive, Baltimore, MD 21218, USA (ekinney@stsci.edu)

George Kosugi, National Astronomical Observatory of Japan, Hawaii Observatory, Osawa 2-21-1, Mitaka, Tokyo 181, Japan (george@optik.mtk.nao.ac.jp)

Reinhold Kroll, Instituto de Astrofisica de Canarias, Computer Centre, c/ Via Lactea s/n, E-38200 La Laguna, Tenerife, Spain (kroll@iac.es)

Anthony Krueger, Space Telescope Science Institute, PRESTO, 3700 San Martin Drive, Baltimore, MD 21218, USA (krueger@stsci.edu)

Michael Kurtz, Harvard-Smithsonian Center for Astrophysics, 60 Garden Street, Cambridge, MA 02138, USA (kurtz@cfa.harvard.edu)

Fred Lahuis, ESA Villafranca del Castillo Satellite Tracking Station, ISO Science Operations Centre, Aptdo. 50727, E-28080 Madrid, Spain (flahuis@iso.vilspa.esa.es)

Robert Lamontagne, Université de Montréal, Département de Physique, C.P. 6128, Succ. Centre-Ville, Montreal, H3C 3J7, Canada (lamont@astro.umontreal.ca)

Wayne Landsman, NASA Goddard Space Flight Center, Code 681, Greenbelt, MD 20771, USA (landsman@mpb.gsfc.nasa.gov)

Soizick Lesteven, Observatoire Astronomique de Strasbourg, CDS, 11, rue de l'Université, F-67000 Strasbourg, France (lesteven@astro.u-strasbg.fr)

James Lewis, Royal Greenwich Observatory, Madingley Road, Cambridge CB3 0HA, United Kingdom (jrl@ast.cam.ac.uk)

Peter Linde, Lund Observatory, Box 43, S-22100 Lund, Sweden (peter@astro.lu.se)

Don Lindler, Advanced Computer Concepts, Inc, 11518 Gainsborough Road, Potomac, MD 20854, USA (lindler@rockit.gsfc.nasa.gov)

Antoine Llebaria, Laboratoire d'Astronomie Spatiale, Traitement des Images, Traverse du Siphon; B.P. 8, F-13376 Marseille Cedex 12, France (antoine@astrsp-mrs.fr)

Dyer Lytle, University of Arizona, NICMOS Project, Steward Observatory, 950 N. Cherry Ave., Tucson, AZ 85721, USA (dlytle@as.arizona.edu)

Maria Concetta Maccarone, Consiglio Nazionale delle Ricerche, Ist. Fisica Cosmica e Appliazioni Informatica, Via Ugo La Malfa 153, I-90146 Palermo, Italy (cettina@ifcai.pa.cnr.it)

Eric Mandel, Smithsonian Astrophysical Observatory, High Energy Astrophysics Division, 60 Garden Street, Cambridge, MA 02138, USA (eric@cfa.harvard.edu)

Ralph Martin, Royal Greenwich Observatory, Computing & Information Services, Madingley Road, Cambridge CB3 0EZ, United Kingdom (ralf@ast.cam.ac.uk)

Thomas McGlynn, Laboratory for High Energy Astrophysics, HEASARC, Code 660.2, Greenbelt, MD 20771, USA (tam@silk.gsfc.nasa.gov)

Nuria McKay, University of Manchester, Nuffield Radio Astronomy Laboratories, Jodrell Bank, Cheshire SK11 9DL, United Kingdom (nm@jb.man.ac.uk)

Warren McLaughlin, Smithsonian Astrophysical Observatory, ASC, 60 Garden Street, Cambridge, MA 02138, USA (wmclaugh@head-cfa.harvard.edu)

Brian McLean, Space Telescope Science Institute, 3700 San Martin Drive, Baltimore, MD 21218, USA (mclean@stsci.edu)

Alberto Micol, ST-ECF, Karl-Schwarzschild-Str. 2, D-85748 Garching, Germany (amicol@eso.org)

Manfred Miebach, Space Telescope Science Institute, 3700 San Martin Drive, Baltimore, MD 21218, USA (miebach@stsci.edu)

Glenn Miller, Space Telescope Science Institute, 3700 San Martin Drive, Baltimore, MD 21218, USA (miller@stsci.edu)

Douglas Mink, Smithsonian Astrophysical Observatory, 60 Garden Street, Cambridge, MA 02138, USA (dmink@cfa.harvard.edu)

Gilles Missonnier, Institut d'Astrophysique de Paris, 98 bis Boulevard Arago, F-75014 Paris, France (missonni@iap.fr)

Yoshihiko Mizumoto, National Astronomical Observatory of Japan, Osawa 2-21-1, Mitaka, Tokyo 181, Japan (mizumoto@optik.mtk.nao.ac.jp)

Gianluca Morgante, Osservatorio Astronomico di Trieste, Technology Division, Via G.B. Tiepolo 11, I-34131 Trieste, Italy (morgante@ts.astro.it)

Fionn Murtagh, University of Ulster, Magee College, Faculty of Informatics, Londonderry BT48 8JW, United Kingdom (fd.murtagh@ulst.ac.uk)

Dan Nguyen, Smithsonian Astrophysical Observatory, High Energy Astrophysics Division, 60 Garden Street, Cambridge, MA 02138, USA (dtn@head-cfa.harvard.edu)

Eiji Nishihara, National Astronomical Observatory of Japan, Okayama Astrophysical Observatory, Kamogata-cho, Asakuchi-gun, Okayama 719-02, Japan (eiji@nao.ac.jp)

Jan Noordam, Netherlands Foundation for Research in Astronomy (NFRA), R&D, P.O. Box 2, NL-7990 AA Dwingeloo, The Netherlands (jnoordam@nfra.nl)

Michael Norman, Max-Planck-Institute für Astrophysik, Karl-Schwarzschild-Str. 1, D-85716 Garching, Germany (norman@mpa-garching.mpg.de)

François Ochsenbein, Observatoire Astronomique de Strasbourg, CDS, 11 , rue de l'Université, F-67000 Strasbourg, France (francois@simbad.u-strasbg.fr)

Victor Oknyanskij, Sternberg State Astronomical Institute, Universitetskij Prospekt 13, 119899 Moscow, Russia (oknyan@sai.msu.su)

Friso Olnon, Netherlands Foundation for Research in Astronomy, Postbus 2, NL-7990 AA Dwingeloo, The Netherlands (folnon@nfra.nl)

Patrizio Ortiz, University of Chile, Dept. of Astronomy, Casilla 36-D, Santiago, Chile (ortiz@das.uchile.cl)

Stephan Ott, ESA Satellite Tracking Station, ISO Science Operations Centre, P.O. Box 50727, E-28080 Madrid, Spain (sott@iso.vilspa.esa.es)

Clive Page, University of Leicester, Department of Physics and Astronomy, University Road, Leicester LE1 7RH, United Kingdom (cgp@star.le.ac.uk)

Fabio Pasian, Osservatorio Astronomico, Technology Division, Via G.B. Tiepolo 11, I-34131 Trieste, Italy (pasian@ts.astro.it)

Scott Paswaters, US Naval Research Laboratory, Code 7660, 4555 Overlook Ave. SW, Washington, DC 20375, USA (scott@argus.nrl.navy.mil)

Harry Payne, Space Telescope Science Institute, 3700 San Martin Drive, Baltimore, MD 21218, USA (payne@stsci.edu)

William Pence, NASA Goddard Space Flight Center, Mail Code 662, Greenbelt, MD 20771, USA (pence@tetra.gsfc.nasa.gov)

Jeffery Percival, University of Wisconsin, Dept. of Astronomy, 1150 University Avenue, Madison, WI 53706, USA (jwp@sal.wisc.edu)

Matt Phelps, Smithsonian Astrophysical Observatory, Computation Facility, 60 Garden Street, Cambridge, MA 02138, USA (mphelps@cfa.harvard.edu)

Benoît Pirenne, ST-ECF, Karl-Schwarzschild-Str. 2, D-85748 Garching, Germany (bpirenne@eso.org)

Norbert Pirzkal, ST-ECF, Karl-Schwarzschild-Str. 2, D-85748 Garching, Germany (npirzkal@eso.org)

Raymond Plante, University of Illinois, National Computational Science Alliance, 405 N. Mathews Avenue, Urbana, IL 61801, USA (rplante@ncsa.uiuc.edu)

Borut Podlipnik, Max-Planck-Institut für Aeronomie, LASCO, Max-Planck Str. 2, D-37191 Katlenburg-Lindau, Germany (borut@corona.mpae.gwdg.de)

Francesco Polcaro, National Research Council, Institute of Space Astrophysics, via Enrico Fermi 21-23, I-00044 Frascati, Italy (polcaro@astrma.rm.astro.it)

Joseph Pollizzi, III, Space Telescope Science Institute, Science and Engineering Systems Division, 3700 San Martin Drive, Baltimore, MD 21218, USA (pollizzi@stsci.edu)

Grzegorz Polubek, Wroclaw University, Physics and Astronomy, Kopernika 11, PL-51-622 Wroclaw, Poland (polubek@astro.uni.wroc.pl)

Daniel Ponz, ESA Villafranca del Castillo Satellite Tracking Station, GSED, Aptdo. 50727, E-28080 Madrid, Spain (jdp@vilspa.esa.es)

Matthias Pruksch, Optische und elektronische Systeme GmbH, Dr.Neumeyer-Str. 240, D-91349 Egloffstein, Germany (Matthias_Pruksch@fue.maus.de)

Peter Quinn, European Southern Observatory, DMD, Karl-Schwarzschild- Str. 2, D-85748 Garching, Germany (pjq@eso.org)

Ernst Raimond, Netherlands Foundation for Research in Astronomy, Postbus 2, NL-7990 AA Dwingeloo, The Netherlands (exr@nfra.nl)

Walter Rauh, Max-Planck-Institut für Astronomie, EDV, Königstuhl 17, D-69117 Heidelberg, Germany (rauh@mpia-hd.mpg.de)

Gotthard Richter, Astrophysical Institute Potsdam, An der Sternwarte 16, D-14482 Potsdam, Germany (gmrichter@aip.de)

Reiner Rohlfs, Integral Science Data Centre, Chemin d'Ecogia 16, CH- 1290 Versoix, Switzerland (Reiner.Rohlfs@obs.unige.ch)

Michael Rosa, Space Telescope - European Coordinating Facility, Karl-Schwarzschild-Str. 2, D-85748 Garching, Germany (mrosa@eso.org)

Jim Rose, Space Telescope Science Institute, OPUS/DPT/SESD, 3700 San Martin Drive, Baltimore, MD 21218, USA (rose@stsci.edu)

Sean Ryan, Royal Greenwich Observatory, Madingley Road, Cambridge CB3 0HJ, United Kingdom (sgr@ast.cam.ac.uk)

Ivo Saviane, Osservatorio Astronomico, Vicolo dell'Osservatorio 5, I- 35122 Padova, Italy (saviane@astrpd.pd.astro.it)

Richard Saxton, ESA Satellite Tracking Station, ISO Operations Centre, Apartado 50727, E-28080 Madrid, Spain (rsaxton@iso.vilspa.esa.es)

David Schade, Herzberg Institute of Astrophysics, Canadian Astronomy Data Center, 5071 W. Saanich Road, Victoria, V8X 4M6, Canada (David.Schade@hia.nrc.ca)

Skip Schaller, University of Arizona, Steward Observatory, 933 N. Cherry Ave., Tucson, AZ 85721, USA (skip@as.arizona.edu)

Jürgen Schmidt, Max-Planck-Institut für Radioastronomie, Rechnerabteilung, Auf dem Hugel 69, D-53121 Bonn, Germany (jsm@mpifr-bonn.mpg.de)

Markus Schöller, European Southern Observatory, ODG, Karl-Schwarzschild-Str. 2, D-85748 Garching, Germany (mschoell@eso.org)

Ethan Schreier, Space Telescope Science Institute, 3700 San Martin Drive, Baltimore, MD 21218, USA (schreier@stsci.edu)

Joseph Schwarz, European Southern Observatory, Karl-Schwarzschild-Str. 2, D-85748 Garching, Germany (jschwarz@eso.org)

Steve Scott, Caltech/OVRO, P.O. Box 968, Big Pine, CA 93513, USA (scott@ovro.caltech.edu)

Robert Seaman, National Optical Astronomy Observatories, IRAF Group, 950 N. Cherry Ave., P.O. Box 26732, Tucson, AZ 85726, USA (rseaman@noao.edu)

Richard Shaw, Space Telescope Science Institute, Science Support Division, 3700 San Martin Drive, Baltimore, MD 21218, USA (shaw@stsci.edu)

William Sherwood, Max-Planck-Institut für Radioastronomie, Auf dem Hugel 69, D-53121 Bonn, Germany (p166she@mpifr-bonn.mpg.de)

Patrick Shopbell, California Institute of Technology, Department of Astronomy, Mail Code 105-24, 1200 E. California Blvd., Pasadena, CA 91125, USA (pls@astro.caltech.edu)

David Silva, European Southern Observatory, DMD, Karl-Schwarzschild- Str. 2, D-85748 Garching, Germany (dsilva@eso.org)

Luc Simard, University of California - Santa Cruz, Lick Observatory, Kerr Hall, 1156 High Street, Santa Cruz, CA 95062, USA (simard@ucolick.org)

Richard Simon, National Radio Astronomy Observatory, 520 Edgemont Road, Charlottesville, VA 22903-2475, USA (rsimon@nrao.edu)

Jean-Luc Starck, CEA, DSM/DAPNIA, F-91191 Gif-sur-Yvette Cedex, France (jstarck@cea.fr)

David Stern, Research Systems, Inc., 2995 Wilderness Place, Boulder, CO 80301, USA (dave@rsinc.com)

Julian Sternberg, ESTEC, Space Science Department, Postbus 299, NL-2200 AG Noordwijk, The Netherlands (jsternbe@astro.estec.esa.nl)

Elizabeth Stobie, University of Arizona, Steward Observatory, 933 N. Cherry Avenue, Tucson, AZ 85721, USA (bstobie@as.arizona.edu)

Eckhard Sturm, Max-Planck-Institut für Extraterrestrische Physik, Giessenbachstr. 1, D-85740 Garching, Germany (sturm@mpe-garching.mpg.de)

Daryl Swade, Space Telescope Science Institute, DPT/SESD, 3700 San Martin Drive, Baltimore, MD 21218, USA (swade@stsci.edu)

Ian Taylor, University of Wales, Department of Physics/Astronomy, P.O. Box 913, Cardiff CF2 3YB, United Kingdom (ian.taylor@astro.cf.ac.uk)

Peter Teuben, University of Maryland, Astronomy Department, College Park, MD 20742, USA (teuben@astro.umd.edu)

Oleg Titov, Institute of Applied Astronomy, 8, Zhadanovskaya st., 197110 St. Petersburg, Russia (titov@ipa.rssi.ru)

Doug Tody, National Optical Astronomy Observatories, IRAF Group, 950 N. Cherry Ave., P.O. Box 26732, Tucson, AZ 85726, USA (dtody@noao.edu)

Susan Tokarz, Smithsonian Astrophysical Observatory, Optical and Infrared Dept., 60 Garden Street, Cambridge, MA 02138, USA (tokarz@cfa.harvard.edu)

Ralph Tremmel, Max-Planck-Institut für Astronomie, EDV, Königstuhl 17, D-69117 Heidelberg, Germany (tremmel@mpia-hd.mpg.de)

Luc Turbide, Université de Montréal, Département de Physique, C.P. 6128, Succ. "Centre-Ville", Montréal, H3C 3J7, Canada (turbide@astro.umontreal.ca)

Frank Valdes, National Optical Astronomy Observatories, IRAF Group, 950 N. Cherry Ave., P.O. Box 26732, Tucson, AZ 85726, USA (valdes@noao.edu)

Robert Vallance, The University of Birmingham, School of Physics and Astronomy, Edgbaston Park Road, Birmingham B15 2TT, United Kingdom (rjv@star.sr.bham.ac.uk)

Ger Van Diepen, Netherlands Foundation for Research in Astronomy (NFRA), R&D, P.O. Box 2, NL-7990 AA Dwingeloo, The Netherlands (gvandiep@nfra.nl)

Gustaaf Van Moorsel, National Radio Astronomy Observatory, Computing Dept., 1003 Lopezville Road, Socorro, NM 87801, USA (gvanmoor@nrao.edu)

Serhei Vasilyev, Solar-Environmental Research Centre, P. O. Box 30, 310052 Kharkov, Ukraine (vvs@land.kharkov.ua)

Christian Veillet, Canada France Hawaii Telescope Corporation, Astronomy Dept., 65-1238 Mamalahoa Hwy., Kamuela, HI 96743, USA (veillet@cfht.hawaii.edu)

Patrick Wallace, Council for the Central Laboratory of the Research Council, Rutherford Appleton Laboratory, Chilton, Didcot, OX11 0QX, United Kingdom (ptw@star.rl.ac.uk)

Roland Walter, INTEGRAL Science Data Centre, Chemin d'Ecogia 16, CH-1290 Versoix, Switzerland (Roland.Walter@obs.unige.ch)

Zhong Wang, Smithsonian Astrophysical Observatory, OIR, 60 Garden Street, Cambridge, MA 02138, USA (zwang@cfa.harvard.edu)

Rein Warmels, European Southern Observatory, DMD, Karl-Schwarzschild-Str. 2, D-85748 Garching, Germany (rwarmels@eso.org)

Rodney Warren-Smith, Council for the Central Laboratory of the Research Council, Rutherford Appleton Laboratory, Chilton, Didcot OX11 0QX, United Kingdom (rfws@star.rl.ac.uk)

Don Wells, National Radio Astronomy Observatory, 520 Edgemont Road, Charlottesville, VA 22903-2475, USA (dwells@nrao.edu)

Marc Wenger, Observatoire Astronomique de Strasbourg, Centre de Données Astronomiques, 11, rue de l'Université, F-67000 Strasbourg, France (wenger@astro.u-strasbg.fr)

Michael Werger, ESA/ESTEC, Astrophysics Division, Keplerlaan 1, NL-2200 AG Noordwijk, The Netherlands (mwerger@estec.esa.nl)

Richard White, Space Telescope Science Institute, 3700 San Martin Drive, Baltimore, MD 21218, USA (rlw@stsci.edu)

Andreas Wicenec, European Southern Observatory, DMD, Karl-Schwarzschild-Str. 2, D-85748 Garching, Germany (awicenec@eso.org)

Ekkehard Wieprecht, Max-Planck-Institut für Extraterrestrische Physik, Giessenbachstr. 1, D-85740 Garching, Germany (ewieprec@iso.vilspa.esa.es)

Nailong Wu, York University, Department of Physics and Astronomy, 4700 Keele St., North York M3J 1P3, Canada (nwu@sal.phys.yorku.ca)

Hiroshi Yanaka, Fujitsu Limited, Earth Science Systems, Nakase 1-9-3, Mihama-ku, Chiba 261, Japan (yanaka@ssd.se.fujitsu.co.jp)

Peter Young, Australian National University, Mount Stromlo Observatory, via Cotter Road, Weston Creek, Canberra 2611, Australia (pjy@mso.anu.edu.au)

Nelson Zarate, National Optical Astronomy Observatories, IRAF Group, 950 N. Cherry Ave., P.O. Box 26732, Tucson, AZ 85726, USA (nzarate@noao.edu)

Werner Zeilinger, University of Vienna, Institute for Astronomy, Türkenschanzstr. 17, A-1180 Vienna, Austria (wzeil@doradus.ast.univie.ac.at)

Hans-Ulrich Zimmermann, Max-Planck-Institut für extraterrestrische Physik, Giessenbachstr. 1, D-85740 Garching, Germany (zim@rosat.mpe-garching.mpg.de)

Panagoula Zografou, Smithsonian Aastrophysical Observatory, High Energy Astrophysics Division, 60 Garden Street, Cambridge, MA 02138, USA (pzografou@head-cfa.harvard.edu)

Rob Zondag, CARA Software and Services International Limited, Meidoomlaan 5, NL-2231 XG Rijnsburg, The Netherlands (rzondag@astro.estec.esa.nl)

Ed Zuiderwijk, Royal Greenwich Observatory, Madingley Road, Cambridge CB3 0HJ, United Kingdom (ejz@ast.cam.ac.uk)

Part 1. Computational Astrophysics

Astronomical Data Analysis Software and Systems VII
ASP Conference Series, Vol. 145, 1998
R. Albrecht, R. N. Hook and H. A. Bushouse, eds.

VRML and Collaborative Environments: New Tools for Networked Visualization

R. M. Crutcher,[1] R. L. Plante and P. Rajlich

National Center for Supercomputing Applications (NCSA), University of Illinois, Urbana, IL 61801, Email: crutcher@uiuc.edu

Abstract. We present two new applications that engage the network as a tool for astronomical research and/or education. The first is a VRML server which allows users over the Web to interactively create three-dimensional visualizations of FITS images contained in the NCSA Astronomy Digital Image Library (ADIL). The server's Web interface allows users to select images from the ADIL, fill in processing parameters, and create renderings featuring isosurfaces, slices, contours, and annotations; the often extensive computations are carried out on an NCSA SGI supercomputer server without the user having an individual account on the system. The user can then download the 3D visualizations as VRML files, which may be rotated and manipulated locally on virtually any class of computer. The second application is the ADILBrowser, a part of the NCSA Horizon Image Data Browser Java package. ADILBrowser allows a group of participants to browse images from the ADIL within a collaborative session. The collaborative environment is provided by the NCSA Habanero package which includes text and audio chat tools and a white board. The ADILBrowser is just an example of a collaborative tool that can be built with the Horizon and Habanero packages. The classes provided by these packages can be assembled to create custom collaborative applications that visualize data either from local disk or from anywhere on the network.

1. Introduction

Network software applications are becoming increasingly important in astronomy. In this paper we report on two such applications which we have developed. The first is a VRML (virtual reality modeling language) server application that makes it possible for users to create three-dimensional (3D) visualizations of astronomy image data sets which reside in the NCSA Astronomy Digital Image Library (ADIL)[2]. The second is ADILBrowser, a part of the NCSA Horizon Image Data Browser Java package. ADILBrowser allows a group of participants to browse images from the ADIL within a collaborative session.

---

[1] Astronomy Department, University of Illinois, Urbana, IL 61801

[2] http://imagelib.ncsa.uiuc.edu/imagelib

2. VRML Server

The VRML Server tool allows a user to interactively create 3D visualizations of astronomical images found in the ADIL. The visualizations are returned to the user as VRML files. To create the visualizations, you first tell the server which images you wish to visualize by providing their ADIL image codenames. The server then fetches the requested images from the Library archive. You then set up the visualization parameters by filling out an HTML form; when you submit the form, the VRML visualization is returned to you. The visualization can be altered by returning to the form and further editing the parameters.

To view the VRML visualizations you either need a Web browser that is VRML-capable (such as Netscape with a VRML plug-in attached) or a VRML viewer. In the latter case, interaction with the VRML server works best when your browser is configured to start up your VRML viewer as a helper application. For a free VRML viewer available on a wide variety of platforms, we recommend VRweb[3] (supports VRML 1.0 only).

The VRML server consists of two components. The first is an executable that takes command line arguments and produces VRML output. The second component is a set of CGI scripts that provide the user interface through HTML forms.

2.1. The VRML Server Executable

The VRML Server executable, *vrserver*, takes input from the command line and generates VRML output. The *vrserver* is completely independent of the Web interface. Locally, the executable can be used directly. All of the information necessary can be specified with the command-line arguments. The *vrserver* itself is written in C++. It makes use of SGI's OpenInventor library and our own Sv library. The Sv library is an extension of the OpenInventor library that facilitates building visualization applications. Using the Sv library, the *vrserver* can generate isosurfaces, slices, contours, and surface (height field) displays. When the *vrserver* is run, it builds an internal representation of the desired visualization. It then goes through that internal representation and generates the corresponding VRML output.

2.2. The VRML Server Interface

The VRML Server interface appears to the user as an HTML forms page driven by a set of CGI (Common Gateway Interface, a standard for connecting server applications to the Web) scripts written in Perl. The CGI scripts generate HTML forms, process user inputs, manage a disk cache for storing the FITS and VRML files, and executes the *vrserver* executable.

The initial form in the interface allows the user to select which FITS files from ADIL will be used. In subsequent forms, these file names appear in pull-down menus, making it easier to select a datafile for a specific visualization. The initial form is also used to select whether the output will be VRML 1.0 or VRML 2.0. When this form is submitted the server first looks for the images

---

[3]http://www.iicm.edu/vrweb

in its local cache; if they are not found there, they are retrieved automatically from the ADIL.

After the initial form, the user is presented with a second form in which to provide the parameters of the visualization. Multiple datasets may be visualized in the same VRML file. For example, one might represent HCN data as an isosurface and $H^{13}CN$ data as a raster or contour slice. Initially, the parameters form has default values already filled in. The user may modify these or submit the form unchanged. The parameters form also gives the user access to digests of the FITS headers for the images requested (which include the minimum and maximum values of each image) as well as access to help documentation for the form. The VRML Server tutorial found on the VRML server Web page contains complete information about how to use the forms page to generate a variety of visualizations, combinations of visualizations, labels, etc. When the form is submitted, the *vrserver* executable is run with the appropriate arguments. The VRML that is generated is stored in a temporary file.

After the VRML is generated and stored in a file, the last form is generated. It tells the user the size of the VRML output. At this point, the user can choose to transmit the VRML or to go back and change some of the parameters. Sometimes, the VRML output may be too large for the user's VRML browser to handle. For most workstations currently in use, VRML files larger than 400 to 600 kilobytes are too big to manipulate smoothly and easily; thus one should try to choose visualization parameters that produce files smaller than this limit. If it is determined that this will be the case, then the user can go back and make modifications, such as increasing the rendering stride. When the user decides to transmit the VRML data, the contents of the temporary VRML file are sent across the Web to the user's local machine which will then start up its VRML browser or plugin with that VRML data.

3. Collaboration with NCSA Habanero and Horizon

Two Java packages under development at NCSA will make it possible for astronomers to interact with astronomical images in real time and in a truly collaborative way. The first package is NCSA Habanero[4], a collaborative environment that allows a group of participants located on different machines to interact with a single tool as a group. The second package is the NCSA Horizon Image Data Browser[5] (Plante et al. 1997, 1998); this Java package includes a variety of platform-independent applications for browsing scientific images and includes readers for FITS, HDF, GIF, and JPEG formats. It is also a toolkit for creating new applications. The capabilities of these two packages are combined by running Horizon applications within the Habanero environment.

Figure 1 shows an example of running a Horizon application, the ADIL-Browser, within the Habanero environment. When a person starts a Habanero session, he or she gets a session management window (upper left) which can be used to contact collaborators by e-mail. Anyone that joins the session can start

---

[4]http://www.ncsa.uiuc.edu/SDG/Software/Habanero

[5]http://imagelib.ncsa.uiuc.edu/Horizon

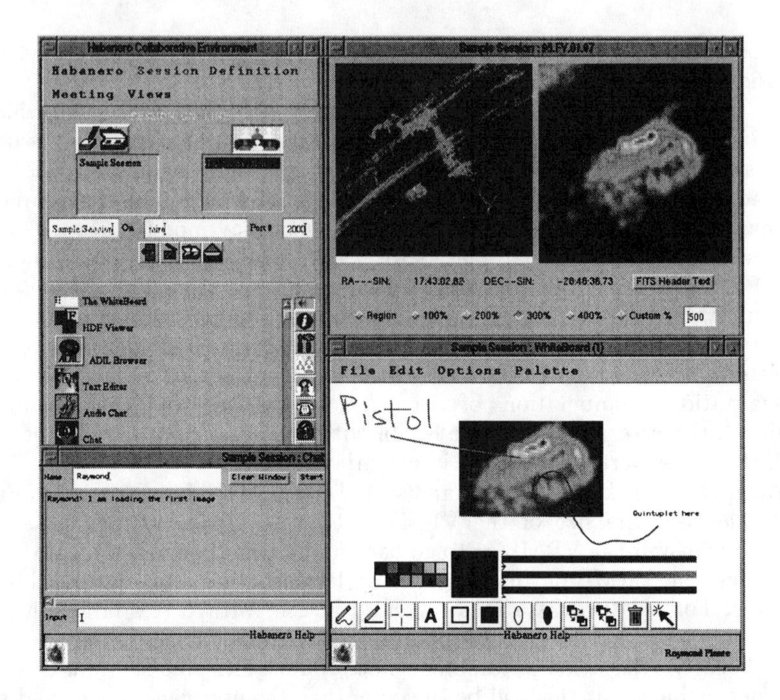

Figure 1. The ADILBrowser running within a Habanero session.

up various tools, copies of which will appear on the screens of all the participants. The window in the upper right corner of the figure shows the ADILBrowser, a Horizon applet for browsing images in the ADIL. When one of the participants clicks in the left image, a zoomed image appears on the right which all of the participants can see. The remaining windows are extra tools provided by Habanero, which include a chat window for discussion, and a white board for pasting images and scribbling. Habanero comes with a number of other tools, including audio chat, a collaborative text editor, and other more specialized applets from a variety of scientific disciplines.

Acknowledgments. This work is supported by the National Center for Supercomputing Applications, the National Science Foundation under grant NSF ASC 9217384, and the National Aeronautics and Space Administration under grant NASA NCC5-106.

References

Plante, R., Goscha, G., Crutcher, R., Plutchak, J., McGrath, R., Lu, X., & Folk, M. 1997, in ASP Conf. Ser., Vol. 125, Astronomical Data Analysis Software and Systems VI, ed. Gareth Hunt & H. E. Payne (San Francisco: ASP), 341.

Plante, R., Xie, W., Plutchak, J., McGrath, R., & Lu, X. 1998, this volume

Astronomical Data Analysis Software and Systems VII
ASP Conference Series, Vol. 145, 1998
R. Albrecht, R. N. Hook and H. A. Bushouse, eds.

Parallel Tree N-body Code: Data Distribution and DLB on the CRAY T3D for Large Simulations

U. Becciani, V. Antonuccio-Delogu, M. Gambera and A. Pagliaro

Osservatorio Astrofisico di Catania, Città Universitaria, Viale A. Doria, 6 – I-95125 Catania - Italy

R. Ansaloni

Silicon Graphics S.p.A. St.6 Pal. N.3, Milanofiori I-20089 Rozzano (MI) - Italy

G. Erbacci

Cineca, Via Magnanelli, 6/3 I-40033 Casalecchio di Reno (BO) - Italy

Abstract. We describe a strategy for optimal memory and work distribution. We have performed a series of tests to find an *optimal data distribution* in the Cray T3D memory, and to identify a strategy for the *Dynamic Load Balance* (DLB). The results of tests show that the step duration depends on two main factors: the data locality and the network contention. In a very large simulation, due to network contention, an unbalanced load arises. To remedy this we have devised an automatic work redistribution mechanism which provided a good DLB.

1. Introduction

N-body simulations are one of the most important tools in contemporary theoretical cosmology, however the number of particles required to reach a significant mass resolution is more larger than those allowed even by present-day state-of-the-art massively parallel (Gouhing 1995; Romeel 1997 & Salmon 1997) supercomputers (hereafter MPP). The most popular algorithms are generally based on grid methods like the P^3M. The main problem with this method lies in the fact that the grid has typically a fixed mesh size, while the cosmological problem is inherently highly irregular. On the other hand the Barnes & Hut (1986, hereafter BH) oct-tree recursive method is inherently adaptive, and allows one to achieve a higher mass resolution. Because of these features, the computational problem can easily become unbalanced and cause performance degradation. For this reason, we have undertaken a study of the optimal work- and data-sharing distribution for our parallel treecode.

Our Work- and Data-Sharing Parallel Tree-code (hereafter WDSH-PTc) is based on this algorithm tree scheme, which we have modified to run on a shared-memory MPPs (Becciani, 1996). The BH-Tree algorithm is a $NlogN$ procedure to compute the gravitational force, a more detailed discussion on the BH tree method can be found in Barnes & Hut (1986). For our purposes, we can distinguish three main phases in each timestep: TREE_FORMATION (TF), FORCE_COMPUTE (FC), UPDATE_POSITION. Besides, in the FC phase we can distinguish two important subphase: TREE_INSPECTION (TI) and ACCEL_COMPONENTS (AC).

2. The WDSH-PTc

During the FC phase each PE computes the acceleration components for each body in asynchronous mode and only at the end of the phase an explicit barrier statement is set. Our results show that the most time-consuming phases (TF, TI and AC) are executed in a parallel regime. Tests were carried out, fixing the constraint that each PE executes the FC phase only for all bodies residing in the local memory. A bodies data distribution ranging from contiguous blocks (coarse grain: CDIR$ SHARED POS(:BLOCK,:)) to a fine grain distribution (tf) (CDIR$ SHARED POS(:BLOCK(1),:)) was adopted. We studied different tree data distributions ranging from assigning to contiguous blocks (tc) a number of cells equal to the expected number of internal cells (N_{Tcell}), [] (coarse grain: CDIR$ SHARED POS_CELL(BLOCK(:N_{Tcell}/N$PEs),:)), to a simple fine grain distribution (CDIR$ SHARED POS_CELL:BLOCK(1),:)), (tf).

All the tests were performed for two different set of initial conditions, namely uniform and clustered distribution having 2^{20} particles each and they were carried out using from 16 to 128 PEs. In Tab. 1 we report only the most significant results obtained with 128 PEs. Our results show that the best tree data distribution is obtained using a block factor equal to 1, then we can to conclude that a (tf) should be used for the kind of codes that we deal in this paper.

	PE#	p/sec	FC phase	T-step	UF
1Mun_tf_bf	128	4129	230.05	249.5	4.22
1Mcl_tf_bf	128	3832	250.32	268.81	4.57
1Mun_tf_bm	128	3547	270.51	290.45	5.90
1Mcl_tf_bm	128	3308	291.63	312.26	6.32
1Mun_tf_bc	128	4875	186.31	205.32	4.14
1Mcl_tf_bc	128	4490	203.37	222.72	4.38
1Mun_tc_bc	128	837	1051.93	1230.0	16.33
1Mcl_tc_bc	128	750	1173.24	1373.4	17.62

Table 1. Results of our tests for 10^6 of particles both in uniform initial conditions (1Mun) and in clustered (1Mcl); being UF the Unbalance Factor. The times in FC phase and in T-step are in second.

The fine grain bodies data distribution (bf) is obtained using a block factor $N = 1$; i.e., bodies are shared among the PE but there is no spatial relation in the body set residing in the same PE local memory. The medium grain bodies data distribution (bm) is obtained using a block factor $N = N_{bod}/2 * N\$PEs$; i.e., each PE has two data block of bodies properties residing in the local memory, each block having a close bodies set. At the end the coarse grain bodies

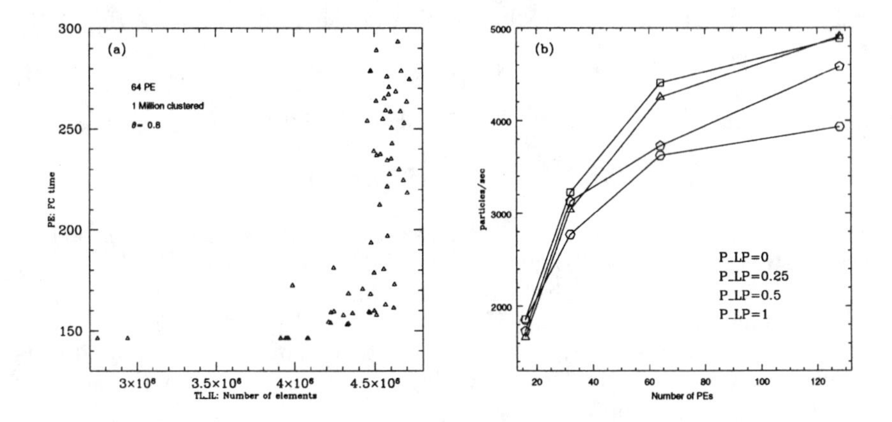

Figure 1. **a)** 64 PE run; **b)** 1 Million of particles: homogeneous configuration

data distribution (bc) is obtained using a block factor $N = N_{bod}/N\$PEs$; i.e., each PE has one close data set block of bodies residing in the local memory. The results reported in Tab. 1 show that the best bodies data distribution, having the highest code performance in terms of particles per second, is obtained using the block factor $N = N_{bod}/N\$PEs$ as expected, due to the data locality effect.

3. Dynamic Load Balance

Here, we present the results of a new DLB strategy, that allows us to avoid any large overhead. The total time spent in a parallel region T_{tot}, can be considered as the sum of the following terms

$$T_{tot} = T_s + KT_p/p + T_o(p) \tag{1}$$

where p is the number of processors executing the job, T_s is the time spent in the serial portion of the code, T_p is the time spent by a single processor ($p = 1$) to execute the parallel region, $T_o(p)$ the overhead time due to the remote data access and to the synchronization points, and K is a constant.

In the FC phase, there are no serial regions, so $T_p \propto$ to the length of the interaction list (IL). Using a coarse grain subdivision, each PE has a block of close bodies in the local memory ($Np = N_{bod}/N\$PEs$); in a uniform distribution initial condition, the PEs having extreme numeration in the pool of available PEs, have a lower load at each timestep. This kind of effect may be enhanced, if a clustered initial condition is used. If the number of PEs involved in the simulation increases we note that the data dispersion on the T3D torus increases. Our results do not show (see Figure 1a) the existence of a relationship between the time spent in the FC phase and the total length of the IL. The adopted technique is to perform a load redistribution among the PEs so that all PEs have the same load in the FC phase. We force each PE to execute this phase *only* for a fixed portion of the bodies residing in the local memory NB_{lp}. NB_{lp} is given by

$$NB_{lp} = (N_{bod}/N\$PEs) * P_{lp} \quad (2)$$

being $P_{lp} = \text{const.}$ $(0 \leq P_{lp} \leq 1)$. The FC phase for all the remaining bodies

$$N_f = N\$PEs * (N_{bod}/N\$PEs) * (1 - P_{lp}) \quad (3)$$

is executed from all the PEs that have finished the FC phase for the NB_{lp} bodies. No correlation between the PE memory and the PE, executing the FC phase for it, is found. If P_{lp}=1 all PEs execute the FC phase only for bodies residing in the local memory, on the contrary if $P_{lp} = 0$, $N_f = N_{bod}$ $(NB_{lp} = 0)$, the PEs execute only N_f bodies and the locality is not taken into account. Several tests were performed with different values of P_{lp}. We report in Figure 1b, only, the case with 10^6 particles uniform.

The results obtained lead us that it is possible to fix a P_{lp} value allowing the best code performances. In particular, we note that is convenient to fix the P_{lp} value near to 0, that is maximize the load balance. The data show that, fixing the PEs number and the particles number, the same P_{lp} value gives the best performance both in uniform and clustered conditions. This means that it is not necessary to recompute the P_{lp} value to have good performances.

4. Final Considerations

The results obtained using the WDSH-PTc code, at present, give performances comparable to those obtained with different approaches such as Local Essential Tree (LET) (Dubinski 1996), with the advantage of avoiding the LET and an excessive demand for memory. Besides, a strategy for the *automatic* DLB has been described, which does not introduce a significant overhead.

The results of this work will allow us to obtain, in the next future, a WDSH-PTc version for the CRAY T3E system, using the HPF-CRAFT and the shmem library. The new version will include an enhanced grouping strategy and periodic boundary conditions (Gambera & Becciani 1997).

References

Gouhing, X., 1995, ApJ. Supp., 97, 884

Romeel, D., Dubinski, J., & Hernquist, L., 1997, New Astronomy, 2, 277

Salmon, J., 1997, Proc. of the 8th SIAM Conf.

Barnes, J., & Hut, P., 1986, Nature, 324, 446

Becciani, U., Antonuccio-Delogu, V., & Pagliaro, A., 1996, Comp. Phys. Com., 99, 9

Salmon, J., 1990, Ph.D. Thesis, Caltech

Dubinski, J., 1996, New Astronomy, 1, 133

Gambera, M., & Becciani, U., 1997, in preparation

Astronomical Data Analysis Software and Systems VII
ASP Conference Series, Vol. 145, 1998
R. Albrecht, R. N. Hook and H. A. Bushouse, eds.

Modelling Spectro-photometric Characteristics of Nonradially Pulsating Stars

Jadwiga Daszynska and Henryk Cugier

Astronomical Institute of the Wroclaw University PL-51-622 Wroclaw, ul. Kopernika 11, Poland

Abstract.
We present a computing code for modelling energy flux distributions, photometric indices and spectral line profiles of (non-)radially pulsating Main-Sequence stars. The model is based on the perturbation-expansion formalism taking into account geometrical and nonadiabatic effects.

1. Introduction

For a given mode of oscillation the harmonic time dependence, $\exp(\mathrm{i}\omega_{nlm}t)$, and spherical harmonic horizontal dependence, $Y_l^m(\theta,\phi)$, are assumed for the first order perturbed quantities. The mass displacement for the spheroidal modes is described by y- and z-eigenfunctions and for toroidal modes by τ-eigenfunctions, cf. Dziembowski & Goode (1992). In the case of slowly rotating stars one can use the zero-rotation approximation to describe stellar pulsations. Such a model was used already by Cugier, Dziembowski & Pamyatnykh (1994) to study nonadiabatic observables of β Cephei stars. Apart from $y_{nlm}(r)$ and $z_{nlm}(r)$ it is desirable to use the eigenfunction $p_{nlm}(r)$, connected with the Lagrangian perturbation of pressure, and the $f_{nlm}(r)-$eigenfunction, which describes the variations of the local luminosity. In the nonadiabatic theory of pulsation the eigenvalues ω_{nlm} and the eigenfunctions are complex (cf. e.g., Dziembowski 1977) and $\psi_{nlm} = \arg(f_{nlm}/y_{nlm})$ is the phase lag between the light and radius variations.

2. Continuum Flux Behaviour

2.1. Numerical Integration

The monochromatic flux of radiation is given by

$$\mathcal{F}_\lambda = \int I_\lambda(r,\theta,\phi,\vec{o}\cdot\vec{n})\vec{o}\cdot\vec{n}\frac{dS}{R^2}. \qquad (1)$$

where $\vec{o}\cdot\vec{n}$ is the scalar product of the observer's direction, $\vec{o}$, and the normal vector, $\vec{n}$, and dS- the area of the surface element.

In the program the specific intensity data for the new generation line-blanketed model atmospheres of Kurucz (1996) were used in order to study the continuum flux behaviour and photometric indices. Kurucz's (1994) data contain monochromatic fluxes for 1221 wavelengths and monochromatic intensities at 17 points of $\tilde{\mu} = \vec{o} \cdot \vec{n}$. Using these data one can interpolate the monochromatic intensities for the local values of T_{eff}, $\log g$ and $\tilde{\mu}$. We can also introduce the linear or quadratic shape for the limb-darkening law as defined by Wade & Rucinski (1985).

2.2. Semi-analytical Method

Integrating Eq.1 over the surface in the linear approach we can obtain the semi-analytical solution, cf. Daszynska & Cugier (1997) for details,

$$\frac{\Delta \mathcal{F}_\lambda}{\mathcal{F}^0_\lambda} = \varepsilon d_{lm0} N^0_l \Big[(T_1 + T_2) \cos((\omega_{nlm} - m\Omega)t + \tilde{\psi}_{nlm}) + (T_3 + T_4 + T_5) \cos(\omega_{nlm} - m\Omega)t \Big]. \tag{2}$$

In this formula the temperature effects are described by two terms T_1 and T_2, whereas the effects of the pressure changes during the pulsation cycle are included in T_4 and T_5. The T_2 and T_5 terms reflect the sensitivity of the limb-darkening parameters to temperature and gravity variations, respectively. T_3 corresponds to the geometrical effects. N^0_l is a normalizing factor.

3. Accuracy of the Model Calculations

We examined how the results are influenced by different methods of integration over the stellar surface. The following cases were considered:

- Model 1: the semi-analytical method (Eq.2) with the quadratic form for the limb-darkening law,

- Model 2: the numerical integration of Eq.1 with the quadratic form for the limb-darkening law; constant limb-darkening coefficients corresponding to the equilibrium model were assumed,

- Model 3: the same as Model 2, but the limb-darkening coefficients were interpolated for local values of T_{eff} and $\log g$,

- Model 4: numerical integration over stellar surface with specific intensities interpolated for the local values of T_{eff}, $\log g$ and $\tilde{\mu}$.

As an example we consider the energy flux distribution and nonadiabatic observables for a β Cep model. We chose the stellar model ($\log T^0_{\text{eff}} = 4.33668$, $\log g^0 = 4.07842$) calculated with OPAL opacities. This model shows unstable $l = 0$, 1 and 2 modes of oscillations. We calculated theoretical fluxes and the corresponding Strömgren photometric indices at pulsating phases $\varphi = 0.05\ n$ (n=0,...,20). Subsequently amplitudes and phases of the light curves were computed by the least-square method. The accuracy of these calculations can be estimated from Table 1, which gives the results for the Models 1 - 4. The calculations were made on Sun Ultra 1 (192 MB RAM, 166 MHz) computer. The CPU time per 1 pulsating stellar model is from about 2 seconds (for Model 1) to about 10 hours (for Model 4).

Table 1. Nonadiabatic observables.

	l	A_y*	φ_y	A_u/A_y	$\varphi_u - \varphi_y$	$\frac{A_{u-y}}{A_y}$	$\varphi_{u-y} - \varphi_y$
Model 1	0	0.0211	3.3166	2.0024	-0.0381	0.8241	-0.0701
Model 2	0	0.0211	3.3167	2.0000	-0.0381	0.8220	-0.0715
Model 3	0	0.0213	3.3167	2.0000	-0.0381	0.8220	-0.0715
Model 4	0	0.0211	3.3168	2.0000	-0.0381	0.8217	-0.0718
Model 1	1	0.0207	3.1916	1.5958	0.0004	0.4876	0.0038
Model 2	1	0.0268	3.1929	1.6119	0.0002	0.4975	0.0022
Model 3	1	0.0268	3.1929	1.6112	0.0002	0.4975	0.0022
Model 4	1	0.0222	3.1910	1.5526	0.0009	0.4535	0.0048
Model 1	2	0.0204	3.2077	1.3476	0.0164	0.2560	0.0790
Model 2	2	0.0195	3.2083	1.3457	0.0161	0.2568	0.0801
Model 3	2	0.0195	3.2083	1.3457	0.0161	0.2568	0.0801
Model 4	2	0.0077	3.2084	1.3170	0.0160	0.2459	0.0804

*assumed

4. Line Profiles

The velocity field of pulsating stars may be found by calculating the time derivative of the Lagrangian displacement. Including the first order effect, the radial component v_p as seen by a distant observer is:

$$v_p = \vec{v}_{puls} \cdot (-\mathbf{e}_z) = \mathrm{Re}\{i\omega_{nlm}[\cos\theta\delta r(R,\theta,\phi,t) - r\sin\theta\delta\theta(R,\theta,\phi,t)]\}$$

$$= \omega_{nlm}\Big[\cos\theta r[y_{nlm}(r) + \frac{2m\Omega}{\omega^0_{nlm}}\tilde{y}_{nlm}(r)]\sum_{k=-l}^{l} d_{lmk}(i)N_l^k P_l^k(\theta)\sin((\omega_{nlm} - m\Omega)t + k\phi)$$

$$-r\sin\theta\Big([z_{nlm}(r) + \frac{2m\Omega}{\omega^0_{nlm}}\tilde{z}_{nlm}]\sum_{k=-l}^{l} d_{lmk}(i)N_l^k\frac{\partial P_l^k(\theta)}{\partial\theta}\sin((\omega_{nlm} - m\Omega)t + k\phi)$$

$$+\frac{\tau'_{l+1,m}}{\sin\theta}\sum_{k=-(l+1)}^{l+1} d_{l+1,m,k}(i)kN_{l+1}^k P_{l+1}^k(\theta,\phi)\cos((\omega_{nlm} - m\Omega)t + k\phi)$$

$$+\frac{\tau'_{l-1,m}}{\sin\theta}\sum_{k=-(l-1)}^{l-1} d_{l-1,m,k}(i)kN_{l-1}^k P_{l-1}^k(\theta,\phi)\cos((\omega_{nlm} - m\Omega)t + k\phi)\Big)\Big]. \quad (3)$$

The radial velocity due to pulsation and rotation is then

$$v_\mathrm{r} = v_\mathrm{p} - v_\mathrm{e}\sin i\sin\theta\sin\phi, \quad (4)$$

where v_e corresponds to the equatorial velocity of rotation and i is the angle between the rotation axis and the direction to the observer.

We illustrate the predicted behaviour of Si III 455.262 nm line profiles for stellar model given in Sect.3. We considered Kurucz's (1994) model atmospheres

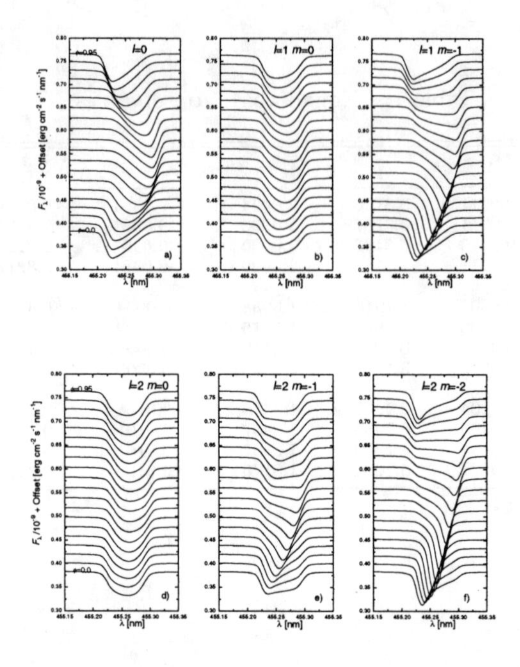

Figure 1. Theoretical SiIII 455.262 nm line profiles for various modes.

with the solar chemical composition and the microturbulent velocity $v_t = 0$. All calculations were made for the amplitude of the stellar radius variations $\varepsilon = 0.01$ and rigid rotation. Figures $1a - f$ show the theoretical line profiles for different phases of pulsation for $i = 77°$ and the equatorial velocity $V_e = 25$ km s^{-1}. The spectra are given in absolute units. In order to avoid overlap, vertical offsets were added to each spectrum using the relationship: $\mathcal{F}_\lambda + n \cdot 0.02 \cdot 10^{-9}$.

Acknowledgments. This work was supported by the research grant No.2 P03D00108 from the Polish Scientific Research Committee (KBN).

References

Cugier, H., Dziembowski W. A., & Pamyatnykh A. A. 1994, A&A, 291, 143

Daszynska J., & Cugier H. 1997. submitted for publication

Dziembowski W. A. 1977, Acta Astron. 27, 95

Dziembowski W. A., & Goode P. R. 1992, ApJ, 394, 670

Kurucz R. L. 1994, CD-ROM No.19

Kurucz R. L. 1996, private communication

Wade R. A., & Rucinski S. M. 1985, A&AS, 60, 471

Astronomical Data Analysis Software and Systems VII
ASP Conference Series, Vol. 145, 1998
R. Albrecht, R. N. Hook and H. A. Bushouse, eds.

Identification and Analysis of Binary Star Systems using Probability Theory

E. Grocheva and A. Kiselev

Photographic Astrometry Department, Pulkovo Observatory, S.Petersburg, 196140, Russia, Email: gl@spb.iomail.lek.ru

Abstract. Accurate identification of physical binary star systems has become of increasing interest in recent years. The frequency of occurrence of binary and multiple star systems remains uncertain at the present time, and is a stumbling block to our understanding of the formation of stars, stellar systems, and planetary systems. Identification of physical binary star systems, with an estimated reliability of such identifications based on the theory of probability, is also important for current and future efforts to find binary star systems that are candidates for observations with the aim of determining their orbits and component masses.

We shall present methods which allow the identification of physical binary stars in a probabilistic sense. The application of probability theory provides a more complete picture of the frequency of stellar binarity than simple methods based only on proximity or proper motion. We will also present preliminary results from the application of these methods to Pulkovo's observation program of binary star systems, and outline how such methods might be applied to present and future high precision astrometric catalogues.

1. Introduction

It is usually difficult to identify physical binary star systems without long-term observations. A simple comparison of proper motions does not provide certain verification of the connection between the components. In many cases we do not know the distances to stars or stellar space velocities. Therefore two stars seen in the same direction and having similar proper motions may have different distances and space velocities. Estimating the component masses is interesting from the point of view of searching for hidden masses. It is possible to make this identification on the basis of probability theory. We propose to estimate the probability **P** of the random disposition of two or more stars, having similar proper motions, on small angular distance ρ. True physical systems will have the lowest such probabilities.

2. Estimation of Probability.

We propose to use the real distribution of proper motions for estimation of **P**. Let the probability of finding the primary component in the area $\sigma = \pi\rho^2$ equal

1 as a probability of a reliable event. The probability **P** of finding n components in a small area σ may be represented as:

$$P = P_{n-1}(\rho) P_{\mu}(\mu_A, \mu_B, \varepsilon) \tag{1}$$

where

$P_n(\rho)$ - probability of finding n stars in small area limited with angular distance ρ for the case of a random distribution, (Deutsch,1962):

$$P_n(\rho) = \frac{1}{n!} \left(\frac{\sigma N}{\Sigma} \right) e^{-\frac{\sigma N}{\Sigma}} \tag{2}$$

n - is multiplicity of a star system;

Σ - is area where N stars are randomly disposed;

$P_{\mu}(\mu_A, \mu_B, \varepsilon)$ - probability of the proximity of proper motions μ_A and μ_B;

ε - error of determination of proper motions.

Let $\vec{\mu}$ be a random vector, $\mu = (\mu_x, \mu_y)$, where $\mu_x = 15\mu_{\alpha} \cos\delta$, $\mu_y = \mu_{\delta}$. If $\vec{\mu}$ takes on a value from the cell $g_{ij} = (\mu_{xj}, \mu_{xj} + \Delta\mu_x) \times (\mu_{yi}, \mu_{yi} + \Delta\mu_y)$, then we say that random vector $\vec{\mu}$ takes the value μ_{ij}. Hence, vector μ may take the following values:

$$\mu_{11}, \mu_{12}, .., \mu_{nn}.$$

Let the probabilities that μ takes one or another value be equal correspondingly to

$$p_{11}, p_{12}, .., p_{nn}.$$

and

$$\sum_{ij}^{n} p_{ij} = 1$$

i.e., the density of probability of the random vector is known. In the case of statistical estimation of probabilities the quantities p_{ij} are the relative frequencies $\frac{n_{ij}}{N}$, then $P_{\mu}(\mu_A, \mu_B, \varepsilon)$ is equal to:

$$P_{\mu} = \sum_{\mu_{ij} \in G^{\star}} \frac{n_{ij}}{N} = \frac{S}{N} \tag{3}$$

where $G^{\star}$ is the Borel's set, where the tips of vectors are situated at:

$$G^{\star} = (\mu_{x min} - \varepsilon, \mu_{x max} + \varepsilon) \times (\mu_{y_{min}} - \varepsilon, \mu_{y_{max}} + \varepsilon) \tag{4}$$

Now we will describe a procedure for the solution of this problem without a concrete realisation.

- Using any astrometric catalogue with stellar positions and proper motions we obtain the differential law of distribution (density of probability) of proper motions. We construct the matrix M whose elements will be n_{ij} (3). Let us consider the stars whose α and δ satisfy certain specified conditions, for example, $\delta > 70$. Properly speaking, selection conditions can be very different, but for the present we solve this problem from an astrometric point of view only. For every star we calculate the position of corresponding element in matrix M. We add 1 to the element thus found. The elements of resulting matrix will yield the probability density of proper motion distribution.

- Let us consider the binary star whose components have proper motions μ_A,μ_B and angular spacing ρ. The required probability is equal to

$$P = \frac{\sigma N}{\Sigma}\frac{S}{N} = \frac{\pi \rho^2 S}{\Sigma}$$

according to (1) -(3).

- We choose assumed binaries from some catalogue. The sample must be restricted to pairs whose ρ are limited by some quantity. For example it is possible to use the Aitken's criterion(Aitken,1932). Then we determine the probability of random distribution for every pair. The sample must include few known optical and physical pairs to obtain a definite criterion for the identification.

3. Results of Analysis of Stars from Pulkovo's Observation Program.

The distribution of proper motions was derived from the PPM catalogue for stars of the North-polar area. The parameters of the distributions are in Table 1. The predominance of negative motions, especially of μ_y, is readily observable. This is due to the location of the North-polar area relative to the Solar apex. We chose 76 double stars from Pulkovo's observation program and calculated the probabilities of a random distribution for these pairs. There are 8 physical and 12 optical systems among them (Grocheva,1996 & Catalogue of relative positions and motions of 200 visual double stars,1988). The proper motions of these pairs was obtained by using the catalogue *"Carte du Ciel"* and modern observations with the 26"refractor. The precision of μ is 0".005 /yr.

Table 1. Parameters of the proper motions distribution

	Mean	Variance	max	min
μ_x	-0".0006	0.047	1".545	-2".97
μ_y	-0".0059	0.041	0".818	-1".74

Analysing the resulting probabilities we conclude that only the probability of random proximity of proper motions $P_\mu = S/N$ can be used to identify true physical binaries. Multiplication P_μ by $\sigma N/\Sigma$ corrupts the probability pattern. Figure 1 shows probabilities P_μ on a logarithmic scale (we numbered the pairs of this sample from 1 to 76 and used these numbers as x-coordinates). We see that probabilities for physical pairs are less than 0.01, whilst those for optical ones are large. Hence, the quantity P_μ may be used to identify physical binaries and the limit of probability of proper motions proximity S/N is 0.01 for physical pairs. It turned out that only 27 physical binaries were among the sample of 76 double stars.

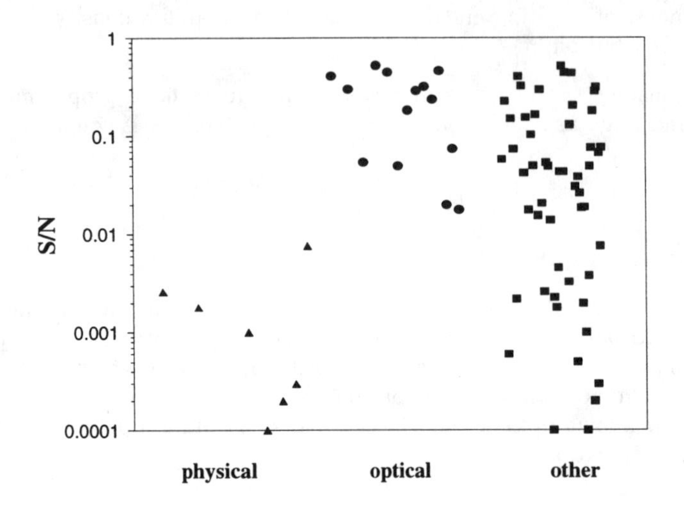

Figure 1. The pattern of probabilities $P_\mu = S/N$ shown on a logarithmic scale.

4. Conclusion.

The technique presented identifies physical binaries. This method has a simple algorithm and can be used for the automatic treatment of large stellar catalogues. We are pleased to also note that this method requires only minimal data such as positions and proper motions. This method was used to correct the Pulkovo's program of binary star observations.

References

Deutsch, A. N., 1962, "The visual double stars". in The course of astrophysics and stellar astronomy, p.60, Moscow, (in Russian).

Catalogue of relative positions and motions of 200 visual double stars.,1988, Saint-Petersburg, (in Russian).

Grocheva, E. A., 1996, "Physical and optical double stars...", Workshop The Visual Double Stars.., Spain.

Aitken, R. G., 1932, New General Catalogue of Double Stars, Edinburgh

Astronomical Data Analysis Software and Systems VII
ASP Conference Series, Vol. 145, 1998
R. Albrecht, R. N. Hook and H. A. Bushouse, eds.

Modelling Spectro-photometric Characteristics of Eclipsing Binaries

Grzegorz Połubek

Astronomical Institute of the Wrocław University PL-51-622 Wrocław, ul. Kopernika 11, Poland

Abstract. We present a computer program for modelling energy flux distribution and light curves of eclipsing binaries from far ultraviolet to infrared regions. The Roche model is assumed. Proximity effects of components (reflection of radiation and surface gravity darkening) are taken into account. The calculations are made in the absolute flux units and the newest Kurucz's (1996) models of stellar atmospheres are used. As an example, we consider the Algol (β Persei) system, where an eclipsing pair (Algol A-B) is accompanied by the third component (Algol C).

1. Model Description

The photospheres of eclipsing components are represented by a grid of surface elements, which are treated as plane-parallel and homogeneous ones. In order to calculate the synthetic fluxes of radiation an integration over all surface elements visible at a given orbital phase, ϕ, should be done. For this purpose geometrical parameters (i.e., area of elements and their positions) as well as physical ones (temperatures and surface gravities) for all elements have to be known. To describe the system's geometry we assumed the Roche model presented by Kopal (1959) and Limber (1963) for synchronous and nonsynchronous rotation, respectively. Assuming synchronous rotation of the components, the total potential Ψ can be written as:

$$\Psi = G\frac{M_A}{r_1} + G\frac{M_B}{r_2} + \frac{\omega^2}{2}\left[(x - \frac{aM_B}{M_A + M_B})^2 + y^2\right], \tag{1}$$

where M_A and M_B are stellar masses, a – orbital separation of components, ω – angular velocity of the system about the z-axis, G – the gravitational constant, ${r_1}^2 = x^2+y^2+z^2$, and ${r_2}^2 = (a-x)^2+y^2+z^2$. Using spherical polar coordinates (r, θ, φ) equation (1) can be rewritten as:

$$\Omega = \frac{1}{r} + q\left[\frac{1}{(1-2lr+r^2)^{\frac{1}{2}}} - lr\right] + \frac{1}{2}(1+q)(1-n^2)r^2, \tag{2}$$

where $q = M_B/M_A$, Ω is the dimensionless Roche potential

$$\Omega = \frac{a\Psi}{GM} - \frac{1}{2}\frac{q^2}{1+q}, \tag{3}$$

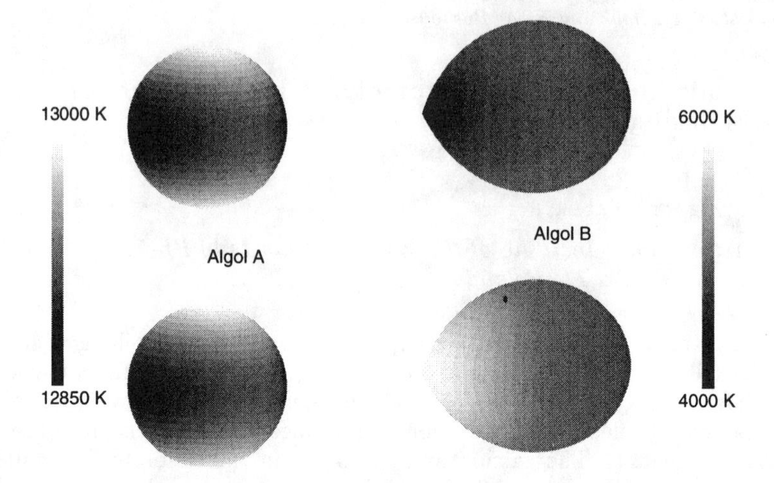

Figure 1. The surface temperature distribution for Algol A-B system before (upper plot) and after (lower plot) correction for temperature due to the irradiation effect. It is clearly seen that the irradiation effect affects mainly the B component's temperature distribution. The binary system is displayed at orbital phase $\varphi = 0.3$.

and l, m, n are direction cosines of the radius vector, i.e. $l = \sin\theta\cos\varphi$, $m = \sin\theta\sin\varphi$, and $n = \cos\theta$. If we assume that the photospheres of primary and secondary components are equipotential surfaces with potentials Ω_1 and Ω_2, respectively, then the equation (2) allows us to determine the total surfaces of both components. We solve this equation by the iterative Newton-Raphson method.

The local surface gravity follows from the equation

$$\vec{g} = -\mathrm{grad}\Psi = -\left(\frac{\partial\Psi}{\partial x}, \frac{\partial\Psi}{\partial y}, \frac{\partial\Psi}{\partial z}\right) . \tag{4}$$

If we neglect the effect of irradiation by the companion, the temperature at (i, j)-point of the surface grid depends only on the local value of gravity

$$T_{\mathrm{eff}}(i, j, 0) = T_{\mathrm{eff}}(\mathrm{pole}, 0)\left(\frac{g(i, j)}{g(\mathrm{pole})}\right)^{\beta} , \tag{5}$$

where β is the gravity-darkening exponent. Index 0 refers to effective temperatures in case of neglecting of the irradiation effect (hereafter referred to as intrinsic ones). For photospheres in radiative equilibrium the integral flux is proportional to the local value of gravity (von Zeipel 1924) which implies $\beta = 0.25$. The gravity-darkening exponent is smaller for photospheres with convection and in case of thick convective envelope $\beta = 0.08$ (Lucy 1967). If we take into account the irradiation effect, the local temperatures have to be modified. For this purpose an iterative procedure based on the Chen & Rhein (1969) approach was used. In the first step, the intrinsic effective temperatures of both

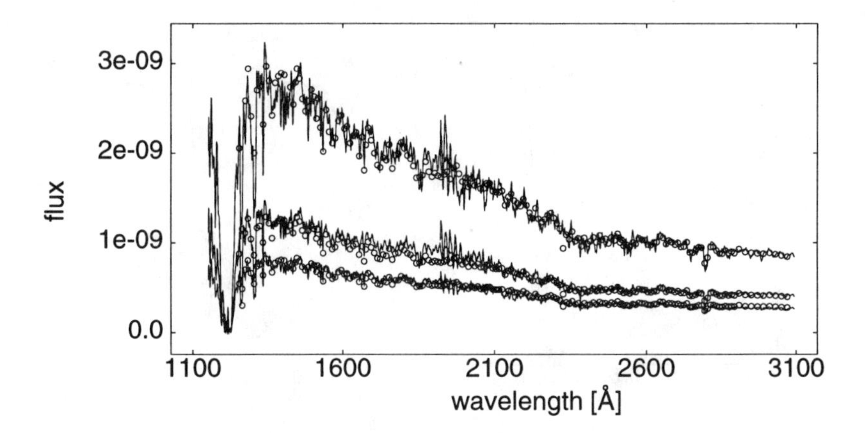

Figure 2. Examples of IUE spectra of Algol (solid lines) in $\mathrm{erg\,cm^{-2}\,s^{-1}\,\AA^{-1}}$, in comparison with the best-fit model calculations (circles) for orbital phases $\phi = 0.925, 0.979, 0.996$.

components were set. Then, for each surface element (i, j) of the given component the flux $F(i, j)$ received from the other component was evaluated. The corrected temperature $T_{\rm eff}(i, j, c)$ follows from the equation

$$T^4_{\rm eff}(i, j, c) = T^4_{\rm eff}(i, j, 0) + A\frac{F(i, j)}{\sigma}, \tag{6}$$

where A is bolometric albedo, σ - Stefan-Boltzmann constant. A was assumed to be constant over the stellar surfaces. In the case of radiative equilibrium all absorbed energy must be re-emitted; this means that A equals 1.0. For the convective models the bolometric albedo can be smaller than 1.0 and we assume $A = 0.5$ (cf. Rucinski 1969). The influence of the irradiation effect for the Algol eclipsing pair is shown in Figure 1. Next, we calculate the radiation flux received from the system:

$$F_\nu = \int I_\nu(\mu)\mu\, dS\,, \tag{7}$$

where integration extends over visible parts of all components, $\mu = \cos\gamma$ (γ is the angle between the surface normal and the line of sight). In this paper we use the second-order limb-darkening law for the evaluation of the specific intensity

$$I_\nu(\mu) = I_\nu(1)[1 - u_1(1-\mu) - u_2(1-\mu)^2]. \tag{8}$$

The coefficients u_1 and u_2, defined by Wade & Rucinski (1985), were calculated for Kurucz's (1996) unpublished models of stellar atmospheres.

2. An Example

We analyse UV spectra of Algol collected by the IUE satellite. We use 15 pairs of IUE spectra from short wavelength primary (SWP) and long wavelength primary

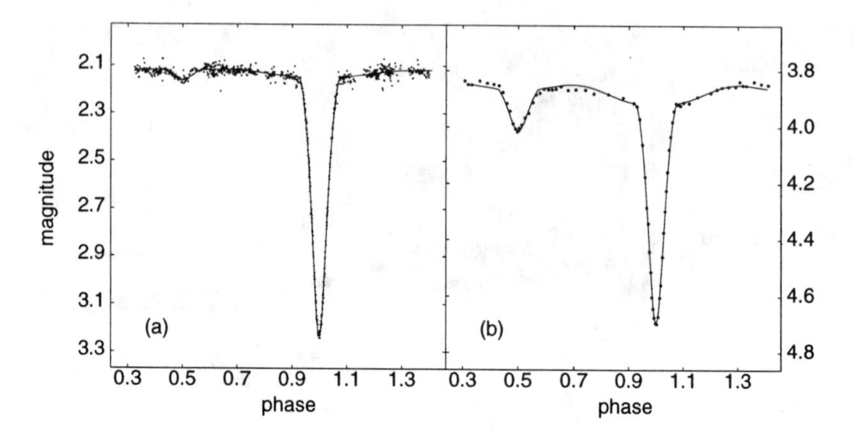

Figure 3. Comparison of the theoretical light curves of Algol (solid lines) with observations (dots) in Johnson V band (a) and in the infrared at 1.6 μm (b).

(LWP) cameras. We found T_{eff}^{pole}(Algol A) = 13000 K, T_{eff}^{pole}(Algol B) = 5000 K, T_{eff}^{pole}(Algol C) = 7000 K, $i_{A-B} = 82.^{\circ}2$ and other parameters similar to those given by Tomkin & Lambert (1978). The best-fit solution for the three pairs of spectra is displayed in Figure 2. This model was further verified by *UBV* observations of Algol taken from Wilson et al. (1972) and by the infrared observations of Chen & Reuning (1966). The comparison of these observations with the theoretical light curves are shown in Figure 3.

Acknowledgments. This work was supported by the research grant No. 2 P03D 001 08 from the Polish Scientific Research Committee (KBN).

References

Chen, K.-Y., & Reuning, E. G., 1966, AJ, 71, 283

Chen, K. -Y., & Rhein, W. J., 1969, PASP, 81, 387

Kopal, Z., 1959, Close Binary Systems, International Astrophysics Series Vol. 5

Kurucz, R.L., 1996, CD-ROM No.19

Limber, D.N., 1963, ApJ, 138, 1112

Lucy, L.B., 1967, Z. Astrophys., 65, 89

Rucinski, S.M., 1969, Acta Astron., **19**, 245

Tomkin, J., & Lambert, D. L., 1978, ApJ, 222, L119

Wade, R. A., & Rucinski, S. M., 1985, A&AS, 60, 471

Wilson, R. E., DeLuccia, M. R., Johnson, K., & Mango, S. A., 1972, ApJ, 177, 191

von Zeipel, H., 1924, MNRAS, 84, 702

Part 2. Applications

Astronomical Data Analysis Software and Systems VII
ASP Conference Series, Vol. 145, 1998
R. Albrecht, R. N. Hook and H. A. Bushouse, eds.

Object-Oriented Experiences with GBT Monitor and Control

J. R. Fisher

National Radio Astronomy Observatory, Green Bank, WV 24944, USA

Abstract. The Green Bank Telescope Monitor and Control software group adopted object-oriented design techniques as implemented in C++. The OO approach has led to a fairly coherent software system and a fair amount of module (class) reuse. Many devices (front-ends, spectrometers, LO's, etc.) share the same software structure, and implementing new devices in the latter part of the project has been relatively easy, as is to be hoped with an OO design. One disadvantage of a long design phase is that it is hard to evaluate progress and to have much sense for how the design satisfies the real user needs. The OO process is only as good at the requirement specifications, and the process has had to deal with continually emerging requirements all though the analysis, design, and implementation phases. Large and medium scale tests of the system in the midst of the implementation phase have required quite a bit of time and coordination effort. This has tended to inhibit progress evaluations.

1. Introduction

I should make clear from the outset that the Green Bank Telescope (GBT) monitor and control system is the work of a team of software designers of which I am not a principal member. This paper contains the thoughts and impressions of a part-time contributor, local consultant, and design philosopher to the project. Members of the group may not necessarily agree with all of my assessments, so the only justification for my giving this presentation at all is to offer a somewhat detached perspective. Any reference to this work should be to the primary documentation (Clark, Brandt, & Ford, http://www.gb.nrao.edu), and please see the acknowledgments section for the names of the designers.

Figure 1 shows the basic structure of the GBT monitor and control system. The object-oriented design referred to in this talk mainly concerns the parts of the software below the dotted line in this figure. Other software shown here does use OO design, particularly AIPS++, but this is not part of monitor and control. The operating systems to which this software has been ported are Solaris and VxWorks, and a port is planned to Windows 95/NT.

The design method used for the GBT monitor and control is the "Object Model" in Rumbaugh et al. (1991).

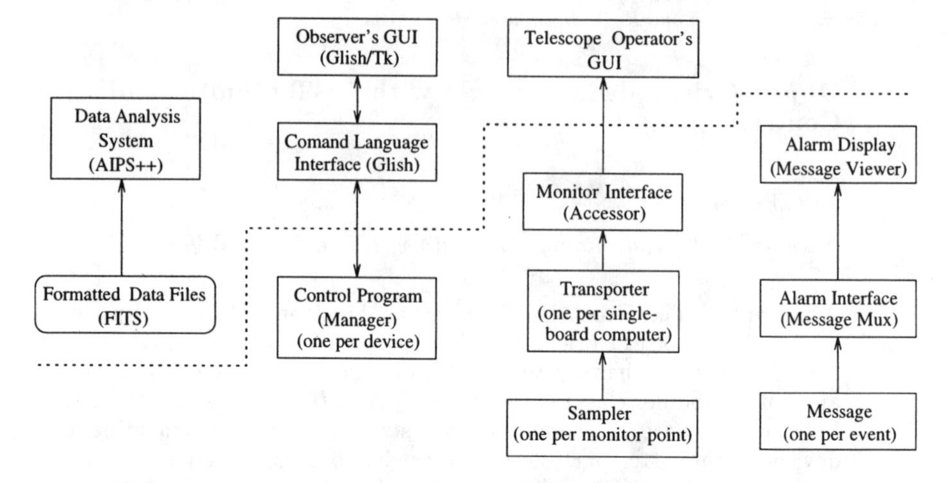

Figure 1. The basic software components of GBT monitor and control. The parts below the dotted line are the main subjects of this presentation.

2. Long Analysis Phase

The most evident feature of the object-oriented approach has been the long analysis phase. In a rigorous design this is as it should be, but it demands that the external requirements for the software be defined early. This has been a weakness in the GBT project, so the designers have been forced to proceed with rough requirements and modify and add to them as the project has progressed. The advantage of early requirements is a more coherent design, and considerable effort has been expended trying to reap this benefit.

Comprehensive requirements are very hard for future users to write. Formal requirements demand a certain amount of abstraction that is quite different from how software is used. It seems fair to say that object-oriented design adds another layer of abstraction and, hence, another language barrier to the communication between user and designer. One of the roles that I have played in this project is translator between the user and software domains. Still, most of the effort of ferreting out requirements has fallen on the designers.

A common method for converging on users' requirements is by building prototypes that the user can comment on and then iterate a number of times until the prototypes are acceptable. On the GBT project this hasn't been very effective because most of the design has remained in software abstractions. It is not clear to me whether this is an inherent practical drawback of the object-oriented approach or whether it is just a weakness in our implementation. It is likely a bit of both. The tension between getting something working and doing a careful design has existed throughout the project. In retrospect, the project might have been divided into smaller units with tangible products at early and mid stages of the project.

Another aspect of the GBT project that has worked against an early design phase is that the software and hardware were built concurrently. In principle, the designs could evolve together, but the difference in design languages was again a barrier. The parts of the GBT that were experimental for most of the construction were quite difficult to fold into the software design. To a certain extent, one could only hope that the design that fitted the known part of the hardware would apply without too much modification to the experimental parts when they became production items.

None of this is to say that we would not use an object-oriented design again. Many of the benefits that come with this approach were, in fact, realized.

3. Modularity

Software modularity was rule number one in the GBT design. This comes from bitter experience with system interdependencies that have been extremely hard to debug in earlier telescope control implementations. The attraction of object-oriented modules is that a few generic modules can be developed and thoroughly debugged, and this robustness can be carried through to specific instantiations of these modules. This has had the added benefit that different designers of various software/hardware modules have been required to use common code libraries (in our case C++ classes) which encourage much similarity between the devices. Certainly, this similarity can be circumvented, but it is generally easier to adopt the common theme.

The adherence to modularity and code reuse has had an interesting effect on the operational aspects of the GBT system. Large closed loops are contrary to this design philosophy because they impose module interdependence. A typical example is having the start time of the data acquisition modules depend on the position of the telescope. Functionally, this make perfect sense, and it does not really introduce a great deal of complexity, but a combination of these interdependencies may have quite serious consequences.

The goal of the GBT design has been that the failure or removal of any subsystem, even the antenna, will not affect the continued operation of the rest of the system. It may not make sense from the user's point of view to continue, but the intention is system robustness. How well this turns out to work remains to be seen.

This independence goal has dictated that all servo loops be closed at the lowest level possible and almost always within the same CPU. All synchronization of operation is done with a common clock which each CPU reads to the accuracy necessary for its function. Each module is responsible for executing its commands as given, and any negotiation of actions is done with a common coordinator module.

Action sequences are grouped by "scans" where every action of each module is fully specified at the beginning of the scan. This could be as simple as one integration on a fixed sky position or as complex as a fast raster map of a moderately large area around a radio source. Doppler tracking, for example, is done on the basis of commanded antenna pointing, not actual pointing. Failure of a module to live up to expectations can be anything from a flagged condition to a fault error, but it affects independent modules only if a common action is decided upon at the top control level.

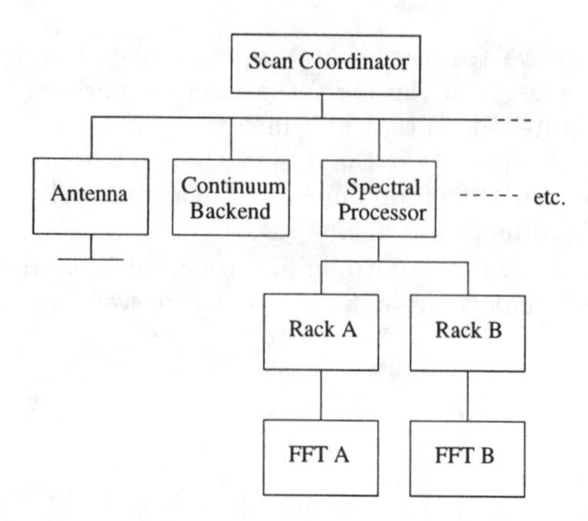

Figure 2. The hierarchy of managers in the GBT system. Each box is a manager which shares common interface, parameter control, and scan sequencing mechanisms with all others.

There is a module hierarchy in the system, but communication is strictly between parent and child.

4. Common Interfaces to Hardware Devices and Systems

A very important aspect of a common software module design is that all interfaces to modules derived from this design are identical. This is not unique to object-oriented design, but it is a valuable byproduct of strong adherence to the method.

The most ubiquitous code unit in the GBT is something called a "manager." In its use at the lowest levels of the system there is a one-to-one correspondence between a manager and a piece of hardware, a spectrometer, for example. The manager contains the mechanisms for receiving and computing parameter setups and for sequencing a scan on the basis of clock time. Managers can be controlled by another manager all the way up to the highest level where the scan coordinator resides. A piece of this hierarchy is shown in Figure 2

The only differences between specific instances of managers are the parameters they contain and, in the case of the lowest level managers, the hardware drivers. Each manager contains a state machine shown in Figure 3.

Upon receiving an activate command the managers that are connected directly to hardware automatically sequence themselves through the Ready, Activating, Committed, Running, Stopping, and Ready states. Parent managers echo the most advanced state of any of their children.

At the beginning of a scan the scan coordinator queries each of its children to find out who has the longest delay to the beginning of the next scan, given the specified scan parameters. The antenna is usually the slowest. If the user has said, "Start as soon as possible." the scan coordinator then issues the same earliest feasible start time to all sub-managers and then leaves them to go about their business until all have returned to the ready state. If the slew time to

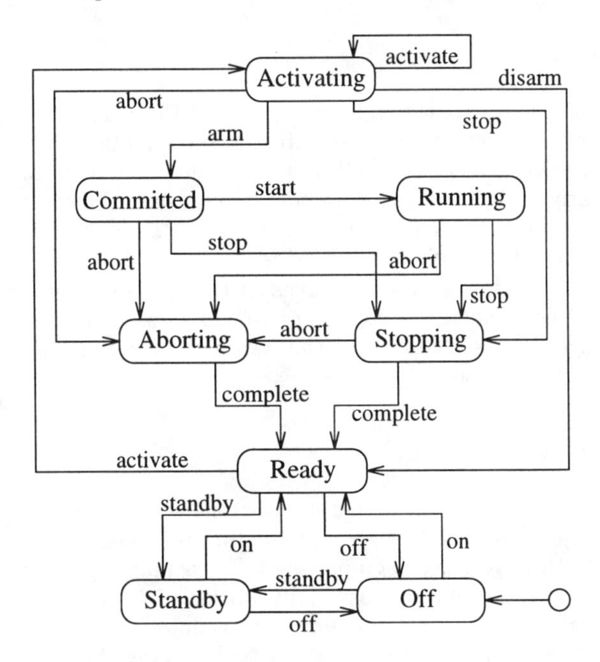

Figure 3. The manager state machine. Each state is shown in a rounded box, and the actions that cause transitions between states are the arrow labels. A typical scan sequence is Ready, Activating, Committed, Running, Stopping, Ready. Data are acquired during the Running state.

the next object is quite long, the scan coordinator might query the antenna for an update on the time-to-source estimate and issue an amended start time to everyone.

In the more complex hardware devices there is a complicated mapping of setup parameters that make sense to the user to values that are required by the hardware. In many cases the order in which the input parameters are used in the calculation of hardware values is important. The generic manager's parameter control mechanism was specifically designed to enforce the correct order of dependent parameter calculations. In most cases the user changes only a few inputs, but the current values of all the rest are still crucial. Keeping the calculation order straight while avoiding a complete recalculation of all unaffected hardware values is a very difficult coding problem. This was solved by a mechanism similar to the compiling *make* rules, plus a few coding rules largely enforced by the parameter class.

All managers inherit a common set of control commands such as 'set a parameter', 'start', and 'abort'. This makes building a user interface to them a very consistent bit of code construction, and it makes connecting one manager to another quite straightforward.

5. Code Reuse

Object-oriented design extends code reuse beyond the sharing of common functions and header parameters. In small software projects this advantage is small, but in a large project, such as the GBT, the benefits can be considerable. However, a lot of early design analysis is required to uncover the common properties of apparently disparate parts of a system. It doesn't happen automatically with the adoption of an object-oriented language.

I have already mentioned a few areas of code reuse, or more importantly design reuse. Others include the error reporting message system, interprocess communication (RPC++ class), and the hardware sampling and monitor system.

6. Generality vs Efficiency

There are two areas where our attempt at extensive software generality, as encouraged by object-oriented design and modularity, might come into conflict with efficiency. However, neither appears to be a major problem. One is the distribution of GBT processes across many workstations and single-board computers. The really time-critical functions are isolated in hardware and closely connected processors. The whole system is tied together by the scan coordinator which communicates with other device managers over a dedicated Ethernet. The Ethernet protocol does not guarantee a maximum transmission time so we need to be conservative about the time latency that can be tolerated by the start-of-scan coordination process.

The other area where efficiency and generality are at odds is in monitoring a large number of hardware diagnostic points. The sampler software nearest the hardware points transfers data continuously into ring buffers at the highest rate expected to be requested by the user of the monitor system for each hardware test point. This allows considerable flexibility for the user to connect to and disconnect from any test point without affecting the hardware and low level software configuration. It does mean that, at any given moment, most data are either ignored or greatly decimated in time before being looked at. Some tuning of the system will be required to avoid a significant processor load at the hardware connections.

7. A Virtual Telescope

We have all heard discussions about reusing software effort from one telescope to the next, but this has worked only in the very few cases where the goals, time scales, and communication proximity of two or more groups have been well matched. Too many factors that deter from system reuse are at work to assign much of the blame to the not-invented-here syndrome. We all have visions of the "virtual telescope" to which all software is designed and every user can use with only one learning curve. Let me finish up with a few thoughts on this from our design experience.

Every astronomical telescope is different for very good reasons. Attempts to hide these differences will usually hide the strengths of an instrument and make it more difficult for the observer to understand the telescope system. A good user interface should be informative, not an abstraction to some universal instrument.

We are pleased with the design and code structure that has emerged from the GBT effort. Our sense of a virtual telescope is one where a common connection between a user interface and the hardware control software can be defined (the dotted line in Figure 1). This connection is defined in terms of message definitions (start, set_parameter, get_parameter_value, etc.) that are independent of the specifics of the instrument. Within the GBT effort this has made building user interfaces to individual parts of the system much easier because the connections are all the same. Only the setup parameters are different.

From the programmer's point of view, the common functional elements, like scan sequencing, parameter control, and message handling, could be reused from one design to the next. To the extent that the operation of these embedded features is correct, it makes sense to carry them to other telescope implementations. However, this certainly will not happen unless the code system is documented to the extent that a knowledgeable programmer can pick up a manual like the one for, say, Tcl/Tk and start using the code system. This is an extremely tall order for any design group faced with the immediate demands from their home institution. The only alternative would be to carry a design from one project to the next in the heads of key personnel. Even then it is not clear that any of us would ever program a system the same way twice. There always seems to be a better way.

Acknowledgments. Full credit for the analysis and design of the GBT monitor and control system must go to the software engineers, particularly Mark Clark (project leader), Joe Brandt, John Ford, and Aron Benett of the NRAO in Green Bank, WV.

References

Rumbaugh, J., Blaha, M., Premerlani, W., Eddy, F., Lorensen, W. 1991, Object-Oriented Modeling and Design, (Prentice Hall)

Astronomical Data Analysis Software and Systems VII
ASP Conference Series, Vol. 145, 1998
R. Albrecht, R. N. Hook and H. A. Bushouse, eds.

Realtime, Object-oriented Reduction of Parkes Multibeam Data using AIPS++

D. G. Barnes

School of Physics, The University of Melbourne, Parkville, VIC 3052, Australia; Email: dbarnes@physics.unimelb.edu.au

Abstract. An overview of the Australia Telescope National Facility (ATNF) Parkes Multibeam Software is presented. The new thirteen-beam Parkes 21 cm Multibeam Receiver is being used for the neutral hydrogen (HI) Parkes All Sky Survey (HIPASS). This survey will search the entire southern sky for HI in the redshift range -1200 km s^{-1} to $+12600$ km s^{-1}; with a limiting column density of $N_{\rm HI} \simeq 5 \times 10^{17}$ cm^{-2}. Observations for the survey began in late February, 1997, and will continue through to the year 2000.

A complete reduction package for the HIPASS survey has been developed, based on the AIPS++ library. The major software component is realtime, and uses advanced inter-process communication coupled to a graphical user interface, provided by AIPS++, to apply bandpass removal, flux calibration, velocity frame conversion and spectral smoothing to 26 spectra of 1024 channels each, every five seconds. AIPS++ connections have been added to ATNF-developed visualization software to provide on-line visual monitoring of the data quality. The non-realtime component of the software is responsible for gridding the spectra into position-velocity cubes; typically 200000 spectra are gridded into an $8^\circ \times 8^\circ$ cube.

1. Introduction

1.1. The Parkes 21 cm Multibeam Receiver

The Parkes 21 cm Multibeam Receiver (hereafter "Multibeam") is described in detail by Staveley-Smith et al. (1996). The Multibeam is a thirteen-beam, cooled 21 cm receiver designed and built by the Australia Telescope National Facility (ATNF) with assistance from the Australian Commonwealth Scientific and Industrial Research Organisation (CSIRO) Division of Telecommunications and Industrial Physics. The Multibeam was installed at the prime focus of the ATNF-operated Parkes 64 m radio telescope on January 21, 1997. The Multibeam was funded by the ATNF and a large Australian Research Council grant, to undertake the ambitious HI (neutral hydrogen) Parkes All Sky Survey (HIPASS) project. The collaborating institutions are the Universities of Melbourne, Western Sydney, Sydney and Wales Cardiff; Mount Stromlo Observatory, Jodrell Bank and the ATNF.

The thirteen circular feed horns of the Multibeam are positioned in a hexagonal arrangement on the focal plane, with a single central feed, and inner and outer rings of six receivers each. The feed horns are all identical, have diameters at the focal plane of 240 mm, and are sensitive to orthogonal linear polarisations of radiation in the frequency range 1.27–1.47 GHz. The instantaneous bandwidth is 64 MHz for each of the 26 channels.

1.2. The HIPASS Project

The Multibeam was funded to undertake the HIPASS project whose primary aim is to make deep, large-area surveys for neutral hydrogen emission from external galaxies. HIPASS consists of two major surveys:

1. a survey for 21 cm emission from redshift -1200 km s^{-1} to $+12600$ km s^{-1}, over the entire southern sky south of declination $+2°$, with an effective integration time of 430 s per pointing; and

2. a deeper (by a factor of two) survey of a part of the "Zone of Avoidance" (ZOA) region—specifically $l = 213°$ to $33°$ with $|b| < 5°$, with velocity range the same as for the all sky survey.

The scientific potential of these two major surveys is great. The HI mass and luminosity functions for the nearby Universe will be determined better than ever before. The HIPASS project will provide completely new information on the distribution of galaxies, the density parameter, the space density of optically rare and invisible galaxies, and on group and supercluster dynamics. In particular, the ZOA survey will search a part of the sky in which typically 10 magnitudes of extinction at optical wavelengths conceal the southern crossing of the Local Supercluster and the likely connection between the Hydra-Centaurus and Pavo-Indus-Telescopium superclusters.

Survey parameters. Staveley-Smith (1997) investigated observational techniques for optimising the sensitivity, and uniformity of sensitivity, of the all sky and ZOA surveys. Subsequently, an active scanning approach was selected for both surveys, whereby the telescope is driven across the sky at a rate of 1° min^{-1} and the Multibeam correlator is programmed to cycle every five seconds. For the all sky survey, 8.6° scans are made along lines of equal right ascension, spaced by 7.0 arcmin; ZOA survey scans of length 8.6° are taken along lines of equal galactic latitude, spaced by 1.4 arcmin.[1]

The total observing time for the all sky survey will be close to 3000 hr, during which time 17000 scans will be acquired, each scan file containing approximately 2700 spectra. For the ZOA survey, 10000 scans will be collected over a total observing time of 1700 hr. Given the large predicted investment of telescope, observer and analyst time, it became clear during the planning phase for the project that realtime calibration of the survey data was desirable, if not necessary. This would enable realtime visualization of the data, thereby

[1]During the commissioning of the Multibeam it became clear that any movement within the focus cabin led to horrendous baseline ripple in bandpass corrected spectra, thus parallactification is not enabled for HIPASS scanning. As a result, the scans made by the twelve non-central beams are slightly curved on the sky. In practice this does not matter, since both surveys have tremendous redundancy built into the scan scheduling.

providing immediate feedback to observers on system integrity and survey data quality. It was also evident early on that radio frequency interference (RFI) had the potential to seriously contaminate a significant fraction of the survey data. Thus any automated data handling algorithms had to be robust to a wide variety of RFI conditions, yet operate in realtime.

When development of a realtime software solution for the HIPASS project commenced, there were no systems in existence, to our knowledge, which could realise the dual objectives of realtime and robust data processing of radio astronomical data. It was necessary to develop a dedicated software package for the HIPASS project.

1.3. AIPS++

The AIPS++ (Astronomical Information Processing System) project (Croes, 1993; Glendenning, 1996) is a new, modern, object-oriented C++ software package designed to process (principally) radio astronomy data. AIPS++ is being developed by an international consortium of seven partners who amongst them operate most of the world's synthesis arrays, and many of the best single dish telescopes. At the core of AIPS++ is an extensive set of classes for storing and manipulating (mathematically and logically) large multi-dimensional arrays of data—the bread and butter of radio astronomy.

The decision to base the Multibeam Software on AIPS++ was taken for several strong reasons. First and foremost, AIPS++ promised to provide (and indeed has provided) what is essentially a *realtime environment* in which to access, manipulate and store or pipe (to further processes) radio telescope data. Secondly, through its complex but rich MeasurementSet storage paradigm, AIPS++ gives the programmer and user *direct access to their data* in the AIPS++ shell ('glish'). This was seen as a particular advantage for the trialling of new algorithms prior to implementing binary clients, and thus improved the speed of development and debugging of the Multibeam Software. Furthermore, at the time development of the Multibeam Software commenced, AIPS++ already provided *advanced, object-oriented techniques for inter-process communication*, and simple methods for displaying and controlling a *graphical user interface* (GUI).

2. The Multibeam Software–an Overview

The Multibeam correlator is programmed to write data into RPFITS (Radiophysics FITS) format files. These files were originally designed to store complex, cross-correlation, visibility data from the Australia Telescope Compact Array, and have been modified to store real auto-correlation data from the Multibeam receiver. In HIPASS mode, the correlator cycle time is 5 s, and spectra are written for each beam and polarisation at the end of each cycle. There are two polarisations per beam, so a total of 26 spectra are written each cycle, each having 1024 channels. Each channel value is stored as a single precision floating point number, which takes four bytes of storage. Thus the HIPASS raw data rate is 104 kb/cycle, or 1.2 Mb/min. The HIPASS RPFITS file is referred to as an HPF (HIPASS FITS) file, and is closed and reopened each cycle by the correlator to ensure that the very latest data can be read from the end of the file.

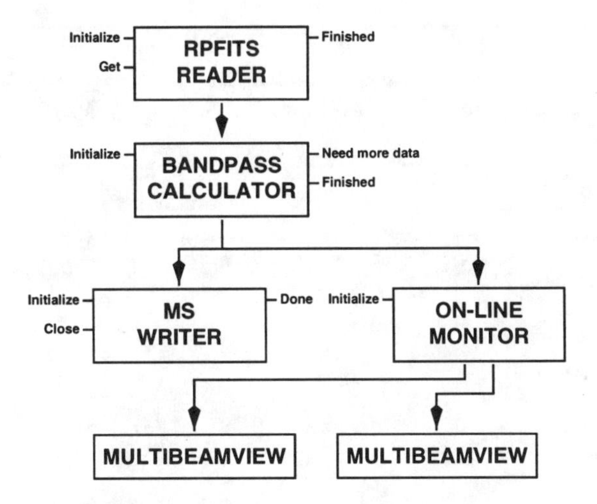

Figure 1. Flow diagram for the realtime Multibeam Software. The data is transported along the solid arrows as a glish record, which can contain several cycles of data, but normally contains only one. Left tags indicate the important commands that can be sent to each client; right tags indicate status events emitted by the clients, to which the LiveData script must respond.

The overall flow diagram for the realtime component of the Multibeam Software is shown in Figure 1. The Multibeam Software reads the HPF files in realtime, or as close to realtime as possible, and applies robust bandpass and baseline removal, velocity frame conversion, and spectral smoothing to all 26 spectra every cycle. The calibrated data is then written to an AIPS++ MeasurementSet. At this stage, realtime operations cease.[2] A complex, event-based glish script is used to control realtime processing of data by a set of binary glish clients; this script must recognise when new HPF files are available, queue these files, and set up control structures to process the next queued file whenever the processing system becomes idle. The glish control script, "LiveData", can be operated from the glish command line, or via an informative GUI (Figure 2). The GUI is written in glish. The script also has hooks so that the observer or data reducer can monitor the data on-screen using adapted visualisation tools from the ATNF "karma" library. In general, two on-line monitors are run concurrently to display the data for both polarizations of a selected beam. Bundles of spectra, usually 26 at a time, are transported in LiveData between the binary clients by means of a "Multibeam Glish Record", structured specifically for the Multibeam Software.

Off-line, a number of MeasurementSets, typically of order 100 covering an area of sky of order 70 square degrees, are collected to be gridded into a position-velocity cube. The gridding software is relatively straightforward, although in

[2]However, we have plans for pipelining the data into a pseudo-realtime gridding client running on a remote machine.

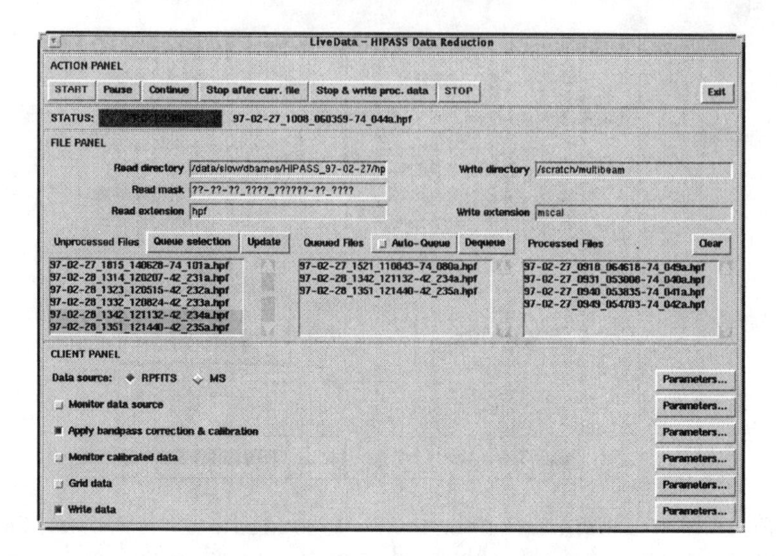

Figure 2. The LiveData graphical user interface—the user can dequeue files, including those which have been automatically queued, or change the order of queued files if the reduction system is falling behind the observing system.

lieu of a machine with a vast amount of memory (say, at least 1 Gb), the cubes need to be built in a number of velocity "slabs", and "glued" together afterwards.

3. Reading Realtime Telescope Data

A binary client was designed to read HPF files in realtime, and prepare a glish record structure (ie. a Multibeam Glish Record) containing an entire cycle's worth of data from the HPF file ("RPFITS READER" in Figure 1). The glish record is emitted from the client, and appropriate control structures in the LiveData glish script forward the record on to other clients. The RPFITS READER is able to package more than one cycle of data into a single Multibeam Glish Record. This can be useful when the reduction system is being operated in "catch-up" mode, and speed can be gained from the lower overhead of passing fewer, but larger, glish records to and from clients.

An alternative to the RPFITS READER was the combination of a client which converts HPF files to MeasurementSets, and a client which extracts Multibeam Glish Records from MeasurementSets. Whilst this approach proved too slow because of the intermediate writing and reading of a MeasurementSet, this client is a useful utility which enables inspection of the contents of HPF files using the AIPS++ TableBrowser application.

4. Realtime Data Manipulation

A binary glish client (the "BANDPASS CALCULATOR") has been developed for the principal purpose of robustly removing the gross bandpass from the spectra in realtime, and calibrating the spectra using the system temperature values written to the HPF file by the correlator. This client also provides further data manipulation capabilities, specifically robust zeroth order baseline removal, spectral smoothing and velocity frame conversion. In cases where the system temperatures written by the correlator are known to be wrong, the client can prescale the raw spectra such that an approximate calibration is made. The BANDPASS CALCULATOR is the most sophisticated component of the Multibeam Software, and is also the most time consuming in the realtime pipeline. A brief description of the components of the BANDPASS CALCULATOR is given below; for further details refer to Barnes et al. (1998).

The BANDPASS CALCULATOR keeps spectral data in a four-dimensional buffer—the four axes are cycle, beam, polarization and channel number. For approximately two minutes from the commencement of a particular scan, the client fills this buffer without actually making any calculations, since at least two minutes of data are required for generation of a bandpass estimate. After this filling period, the behaviour of the client changes, and instead of simply emitting a request for more data, it begins to calculate and apply bandpass removal, baseline removal, spectral smoothing and velocity frame conversion; culminating in the emission of corrected spectra in a Multibeam Glish Record.

4.1. Bandpass and Baseline Removal

The spectra written by the correlator suffer from a number of effects which must be removed during processing to obtain useful spectra of the sky. The most important of these effects is the presence of a "bandpass" in the correlator output spectra. The bandpass reflects the combined shape of all the filtering applied to the data by the receiver and correlator sub-systems, and may also include artifacts from strong continuum sources in or even out of the beams of the telescope. Whilst the bandpass for spectra from different beams of the Multibeam may exhibit large scale similarities, they are expected to differ at smaller scales, since each of the 26 signals pass through independent receivers, amplifiers, down-converters, filters and correlators.

For the HIPASS project, where the telescope is actively driven across the sky, traditional bandpass removal techniques (eg. signal–reference subtraction) are not suitable. Instead, a statistical estimate of the bandpass at the position and time that a particular spectrum—the *target* spectrum—was acquired is made from a set of earlier and later spectra observed by the same beam—these are the *reference* spectra. The reference spectra are selected individually for each target spectrum, and must be independent (with respect to the telescope beam) measures of the HI sky to that of the target spectrum, but taken with timestamps near that of the target spectrum, so that time variations of the bandpass are not significant. Furthermore, valid reference spectra must be from the same right ascension (or galactic latitude) scan series as the target spectrum, since movements of the receiver with respect to the telescope surface introduce gross phase changes in the bandpass.

Dividing the target spectrum by the statistical bandpass estimate yields the bandpass-corrected spectrum. This spectrum is then scaled using system temperature information, yielding a calibrated spectrum. The overall shape of the resultant spectrum is referred to as the *baseline* of the calibrated spectrum, and ideally should be flat. It is often the case, however, that an offset exists such that the mean (or median) of all the channel values for a calibrated spectrum is non-zero; and it is occasionally the case that low order curvature is present in bandpass-corrected spectra. Such curvature may be due to very long wavelength ripple of which only a portion of the cycle is visible in the spectra, or it may be some more complicated artifact of hardware or software processing. The final stage in calibration of the correlator spectra involves the removal of any flux density offset by subtracting from each channel value the median of all channel values for a particular spectrum.

4.2. Spectral Smoothing

Some form of smoothing is required to lessen the ringing that is associated with strong Galactic HI emission, and is caused by the sidelobes in the spectral response of the system. In some cases, ringing can be seen throughout the entire 21 cm spectrum from the correlator. Several smoothing techniques were applied to the data, in order to assess the best trade-off between loss of velocity resolution, and suppression of the ringing. A Tukey 25% filter was selected, and implemented in a C++ class where the filter is applied to the spectra in the Fourier domain. Although the suppression of the spectral sidelobes by this filter is not as strong as by a Hanning filter, this filter still gives good sidelobe suppression, and the loss in spectral resolution is only 15 per cent.

4.3. Velocity Tracking

Observations for the HIPASS project are made in topocentric mode, whereby the actual observing frequencies for each channel remain fixed throughout the survey. Consequently, the velocity range observed varies over the course of a year, and some correction must be made to the spectra. A C++ class was implemented to address this problem, and does so in the Fourier domain. The AIPS++ Measures classes were used to calculate the required velocity shift which was converted to a phase gradient; a subsequent inverse Fourier transform yields a spectrum in a fixed, barycentric reference frame. Note that the velocity frame conversion is applied *after* bandpass removal, since many components of the bandpass are internally generated in a fixed frame.

4.4. Realtime Monitoring Software

As one of the primary objectives of the development of the Multibeam Software was to enable realtime visualization of both raw and calibrated data, some time was spent investigating the best way to do this. The two options available at the time were AipsView, a young visualization tool which can be operated from glish; and the ATNF visualization software, specifically "kview". Whilst AipsView could already accept glish events, it was necessary to introduce a client between AipsView and LiveData to translate the Multibeam Glish Records into simple three-dimensional glish arrays (having channel, cycle and beam number as axes) that AipsView could handle. This client (the "ON-LINE MONITOR"), can be used to mask out a particular channel range, or set of beams, or even apply averaging to the data to reduce the CPU utilization of the client which actually writes the spectra to the display.

Unfortunately, we found that AipsView was not intuitive to use, and it lacked some features that the ATNF in-house visualization software already had.[3] Furthermore, at the time, AipsView was not supported on the DEC Alpha architecture. It turned out that we were able to add a simple C++ wrapper to one of the main ATNF applications, kview, so that it could handle glish arrays of the format being emitted by the ON-LINE MONITOR. This modification was very straightforward, and provided us with the visualization client ("MULTIBEAMVIEW") that is currently in use at Parkes for on-line monitoring of HIPASS data, and which retains all of the features and the same user interface as kview.

5. Data Storage

The calibrated, bandpass-corrected spectra are piped by LiveData to the final client in the realtime system, the MS WRITER. This client accepts incoming Multibeam Glish Records, and writes them to an AIPS++ MeasurementSet. We refer to the MeasurementSets written by the on-line reduction system as MSCAL files. The MS WRITER has been designed to write and read AIPS++ MeasurementSets which contain HIPASS data. The MSCAL files can be inspected directly using the AIPS++ TableBrowser application. Prior to archiving on CD-ROM media, the MSCAL files are converted to SDFITS (Single Dish FITS) files; this conversion does not take place in realtime.

6. Pseudo-realtime and Non-realtime Gridding

For the all-sky survey, 75 scans are needed to produce full survey coverage cubes of the sky; for the ZOA survey, 375 scans are needed. Since a complete set of scans for a given cube are acquired over a number of months, gridding of spectra into final survey cubes is not done at the telescope. Indeed, as of September 1997, only two fully sampled all sky survey cubes had been observed and generated.

The gridding is done by a single glish client (the "LARGE SCALE GRIDDER") designed specifically to generate HIPASS cubes; although it can in principle grid spectral line data from any single dish radio telescope. Because of the large data bandwidth needed by the LARGE SCALE GRIDDER (the MSCAL files for a ZOA cube occupy 4.3 Gb), it was necessary to design this client to read MeasurementSets and write position-velocity FITS images (cubes) itself, ie. without transferring data through glish. The gridding algorithm at present is very simple, and will probably be altered prior to publication of the survey. For a particular pixel in the cube, the value for the ith channel is calculated by taking the median of all the ith channel fluxes from spectra having position stamps that are within six minutes of arc of the fixed pixel position. Whilst this gridding technique is robust to even quite strong interference, it unfortunately introduces a downward

---

[3] Indeed, AipsView is now frozen at its current release, and will be superceded by a number of AIPS++ applications based on the AIPS++ Display Library.

bias in the gridded fluxes, since this algorithm fails to take account of the beam shape.

7. Conclusion

AIPS++ was adopted by the HIPASS project in late 1995 as the platform for the planned realtime, object-oriented data reduction pipeline. The resultant Multibeam Software was successfully operating at the Parkes telescope from day one of the HIPASS project, in February 1997. As of November 1997, approximately 45 Gb of data have been robustly bandpass corrected, smoothed, and velocity corrected, *mostly in real time,* using a combination of AIPS++ infrastructure, and binary clients and glish scripts developed by us but based on the AIPS++ library. The HIPASS project members now look forward to extracting astronomy from calibrated HI cubes, rather than manually grinding their way through gigabytes of unprocessed spectra!

Acknowledgments. I am sincerely grateful to my collaborators on the development of the Multibeam Software for their ideas, time and direction: Lister Staveley-Smith, Taisheng Ye, Tom Oosterloo, Mike Kesteven, Warwick Wilson and Richard Gooch. We are grateful to the Australia Telescope National Facility, the Universities of Melbourne, Western Sydney, Sydney, and Wales Cardiff; Mount Stromlo Observatory, Jodrell Bank and the Australian Research Council for supporting the development of the Parkes Multibeam Receiver. We acknowledge the large investment of time and personnel towards the design and construction of the receiver by the CSIRO Division of Telecommunications and Industrial Physics. We were delighted with the frequent and high-quality help provided by the many AIPS++ programmers during the planning and programming of the Parkes Multibeam Software. DGB acknowledges the support of an Australian Postgraduate Award, and is grateful for the financial assistance to attend ADASS '97 provided by the ADASS POC and a Melbourne Abroad Scholarship.

References

Barnes, D. G., Staveley-Smith, L., Ye, T., & Oosterloo, T. 1998, this volume

Croes, G. A. 1993, in ASP Conf. Ser., Vol. 52, Astronomical Data Analysis Software and Systems II, ed. R. J. Hanisch, R. J. V. Brissenden & Jeannette Barnes (San Francisco: ASP), 156

Glendenning, B. E. 1996, in ASP Conf. Ser., Vol. 101, Astronomical Data Analysis Software and Systems V, ed. George H. Jacoby & Jeannette Barnes (San Francisco: ASP), 271

Staveley-Smith, L. 1997, PASA, 14, 111

Staveley-Smith, L., Wilson, W. E., Bird, T. S., Disney, M. J., Ekers, R. D., Freeman, K. C., Haynes, R. F., Sinclair, M. W., Vaile, R. A., Webster, R. L., & Wright, A. E. 1996, PASA, 13, 243

Astronomical Data Analysis Software and Systems VII
ASP Conference Series, Vol. 145, 1998
R. Albrecht, R. N. Hook and H. A. Bushouse, eds.

World Coordinate Systems as Objects

R.F. Warren-Smith

Starlink, Rutherford Appleton Laboratory, Chilton, DIDCOT, Oxon, OX11 0QX, UK

D.S. Berry

Starlink, Department of Astronomy, University of Manchester, Oxford Road, MANCHESTER, M13 9PL, UK

Abstract. We describe a new library (AST) which provides a flexible high-level programming interface for handling world coordinate systems in astronomy and for producing graphical output. It includes, but is not limited to, a wide range of celestial coordinate systems and supports the Digitised Sky Survey plate solutions and the draft FITS WCS proposals amongst other possibilities. AST is portable and environment independent.

1. Introduction

Writing applications which handle non-linear world coordinate systems (WCS), such as celestial coordinates, in a general way currently presents significant difficulties. Although good algorithms exist to transform between celestial coordinate systems, understanding the relationship between the many different systems in use requires considerable expertise. Storing and retrieving WCS information in datasets also demands familiarity with complicated and changing conventions, such as the many variants of FITS in use. Presenting WCS information graphically (e.g. as coordinate grids) is also algorithmically complex, especially if all-sky plots which include the polar regions must be accommodated.

To address these problems, we have developed a library, AST, which provides a high-level model and programming interface for manipulating WCS data in astronomy. AST stands for 'ASTrometry Library', although astrometry is, in fact, only a small part of its function.

Our primary objective has been to insulate programmers from the problems described above by delivering 'best practice' solutions in an accessible and flexible form.

2. Design Criteria

AST is designed to be useful in a wide range of software projects and, to this end, dependencies on other software have been minimised (only the widely available SLALIB positional astronomy library is required). AST is implemented in

ANSI C for portability. It makes extensive use of object-oriented techniques, but conventional C and FORTRAN77 interfaces are provided — the latter by an additional C layer (so that only a C compiler is required for building). Provision has been made for new language bindings if needed in future.

Graphical output is via a small group of functions which may easily be implemented over most graphics systems (a PGPLOT implementation is provided). A similar mechanism is used for delivering error messages. This, together with an ability to perform I/O via text and FITS headers, ensures independence of any particular programming environment.

Currently, AST is implemented on PC Linux, Solaris and DEC Unix.

3. Inter-Relating Coordinate Systems (Mappings)

Relationships between coordinate systems are represented within AST by objects called *Mappings.* A Mapping, like any AST object, is created by a constructor function which returns a pointer (an integer in FORTRAN) through which the object is manipulated.

A Mapping does not describe a coordinate system, but merely the inter-relationship between two (unspecified) coordinate systems. It is a 'black box' to which coordinate values may be given in return for a set of transformed coordinates. This operation may, in principle, be performed in either direction (the forward and inverse transformations). A Mapping may use any number of input and output coordinates so as to match the, possibly different, dimensionalities of the coordinate systems it inter-relates.

AST provides a selection of different Mappings to support a wide range of celestial coordinate transformations and sky projections. It also provides a range of utility Mappings, such as linear transformations, look-up tables, etc.

An important feature is that any pair of Mappings may be combined together to form a compound Mapping, or *CmpMap.* A CmpMap is itself a Mapping, so this process may be repeated. In this way, Mappings of arbitrary complexity may be built, giving AST great flexibility in the coordinate transformations it can represent.

4. Representing Coordinate Systems (Frames)

While Mappings represent the relationships between coordinate systems, the coordinate systems themselves are represented by objects called *Frames.* An AST Frame is similar in concept to the frame one might draw around a graph. It contains information about the labels which appear on the axes, the axis units, a title, knowledge of how to format the coordinate values on each axis, etc. A Frame is not, however, restricted to two dimensions and may have any number of axes.

A basic Frame may be used to represent a Cartesian coordinate system by setting values for its *attributes* (all AST objects have attributes which may be set and enquired). Usually, this would involve setting appropriate axis labels

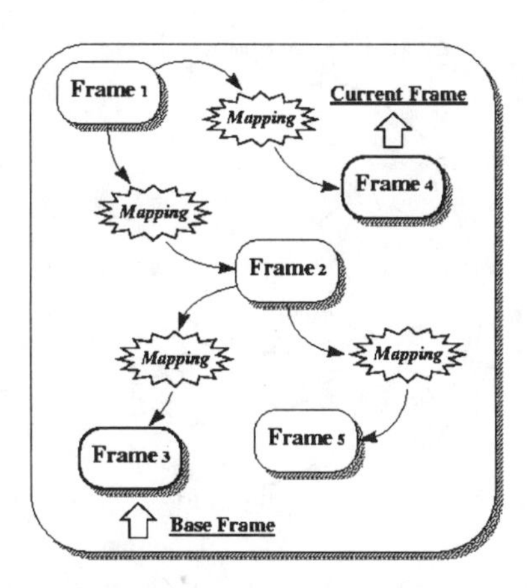

Figure 1. A FrameSet containing Frames inter-connected by Mappings.

and units, for example. Like all objects, a Frame also provides methods.[1] These perform operations such as formatting coordinate values as text, calculating distances between points, interchanging axes, etc.

A derived class, the *SkyFrame,* is provided to represent celestial coordinate systems, of which a wide range are supported and may be selected by setting appropriate SkyFrame attributes. A SkyFrame provides the additional functionality required when handling celestial coordinates — such as sexagesimal formatting and great circle distances. It also encapsulates knowledge of how to convert between any pair of celestial coordinate systems, making this available through a method.

As with Mappings (§3), it is possible to merge two Frames together into a compound Frame, or *CmpFrame,* in which both sets of axes are combined. One could, for example, have celestial coordinates on two axes and an unrelated coordinate (wavelength, perhaps) on a third. Knowledge of the relationship between the axes is preserved internally by the process of constructing the CmpFrame which represents them.

5. Coordinate Networks (FrameSets)

Mappings and Frames may be connected together to form networks called *FrameSets* (Figure 1). Such a network is extended by adding a new Frame and an associated Mapping which relates the new coordinate system to one already present. This ensures that there is always exactly one path, via Mappings, between any pair of Frames. A method is provided for identifying this path and returning the complete Mapping.

One of the Frames in a FrameSet is termed the *base* Frame. This underlies the FrameSet's purpose, which is to calibrate datasets and other entities by attaching coordinate systems to them. In this context, the base Frame represents

---

[1] In AST, methods are functions which take an object pointer as their first argument.

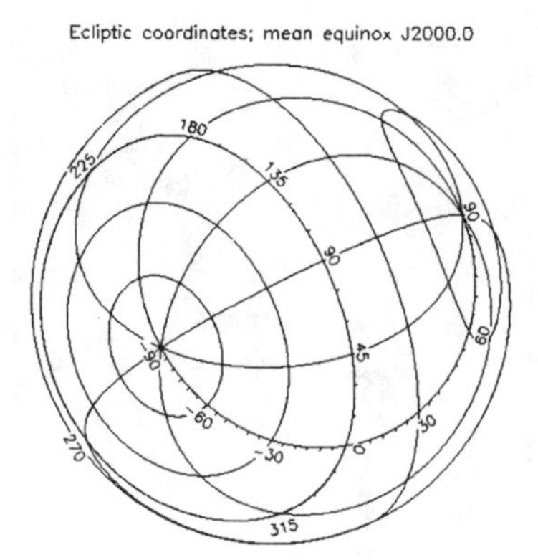

Figure 2. A labelled coordinate grid for an all-sky zenithal equal area projection in ecliptic coordinates, composed and plotted using a single function call.

the 'native' coordinate system (for example, the pixel coordinates of an image). Similarly, one Frame is termed the *current* Frame and represents the 'currently-selected' coordinates. It might, typically, be a celestial coordinate system and would be used during interactions with a user (as when plotting axes on a graph or producing a table of results). Other Frames within the FrameSet represent a library of alternative coordinate systems which a software user can select by making them current.

6. Graphical Output (Plots)

Graphical output is supported by a specialised class of FrameSet called a *Plot.* A Plot's base Frame corresponds with the native coordinates of the underlying graphics system. Plotting operations are specified, using AST Plot methods, in *physical* coordinates which correspond with the Plot's current Frame (typically this might be a celestial coordinate system).

Operations, such as drawing lines, are automatically transformed from physical to graphical coordinates before plotting, using an adaptive algorithm which ensures smooth curves (the transformation is usually non-linear). 'Missing' coordinates (e.g. graphical coordinates which do not project on to the celestial sphere), discontinuities and generalised clipping are all consistently handled. It is possible, for example, to plot in equatorial coordinates and clip in galactic coordinates. The usual plotting operations are provided (text, markers), but a geodesic curve replaces the primitive straight line element. There is also a method for drawing axis lines, which are normally not geodesics.

Perhaps the most useful Plot method is for drawing fully annotated coordinate grids (Figure 2). This uses a general algorithm which does not depend on knowledge of the coordinates being represented, so can also handle programmer-defined coordinate systems. Grids for all-sky projections, including polar regions, can be drawn and most aspects of the output (colour, line style, etc.) can be adjusted by setting appropriate Plot attributes.

Astronomical Data Analysis Software and Systems VII
ASP Conference Series, Vol. 145, 1998
R. Albrecht, R. N. Hook and H. A. Bushouse, eds.

Constructing and Reducing Sets of HST Observations Using Accurate Spacecraft Pointing Information

A. Micol and B. Pirenne

Space Telescope – European Coordinating Facility, ESA/ESO, Garching D-85748, Germany

P. Bristow[1]

Space Telescope – European Coordinating Facility

Abstract. The implementation of "On-The-Fly" Re-Calibration at the ST-ECF and CADC goes some way towards alleviating the problem of obtaining good and timely calibration of HST exposures. However, the data access paradigm is still to consider each exposure individually, re-calibrate them and offer the results to users, who subsequently process the data further.

We describe here techniques to automatically group together HST WFPC2 exposures for cosmic ray removal, co-addition and combination into mosaics with minimal resolution loss. We show that the execution of these tasks has been made essentially automatic.

The ST-ECF archive now offers the possibility to select *"associations"* of datasets and the automatically combined final products. A further spin off of this project is that more reliable pointing information for all exposures is provided.

1. Introduction

In 1995 the ST–ECF introduced, along with the CADC, the re-calibration on-the-fly (OTF) of archived HST exposures. By making use of the best current calibration software and reference files, OTF guarantees that archive users always retrieve from ST–ECF and CADC archives the current best products (i.e., calibrated data).

Pushing the concept further, ST–ECF and CADC are working in the direction of providing the astronomers with an *active* archive system, able to relieve the astronomers of time consuming and boring tasks, therefore increasing the productivity of archive researchers. In this framework the ST–ECF embarked upon a project aimed at building associations of WFPC2 exposures, recovering the otherwise lost information of which observation strategy (e.g., CR-SPLIT, dithering) a WFPC2 proposer decided to employ.

[1]Now at NATO/SACLANT Undersea Research Centre

When looking at an association the archive users immediately see what the shifts between all the exposures were without having to compute them manually. Upon request the archive system can produce a cosmic ray-free mosaic out of the association. Indeed a reliable automatic pipeline has been put in place.

All this is possible thanks to the reliable HST pointing information provided by the jitter files which are a product of the Observatory Monitoring System software at STScI.

2. Jitter Files as a Source of Reliable Pointing Information

After some tests it was clear that:

- the world coordinate system (WCS) keywords available in the header of the exposure FITS files are unreliable. For example, cases where two exposures with identical WCS keywords were found to be clearly shifted.
- Cross correlation to compute shifts among exposures cannot be easily automated, since not all the exposures have enough features or a good signal to noise ratio, and because of the presence of cosmic rays.

Since October 20th 1994 the Observatory Monitoring System subsection of the HST pipeline (STScI) has generated the so called jitter files. Jitter files are normally used to monitor the telescope pointing stability and the trends in the telescope and instrument performance as the orbital environment changes. By correlating the HST Mission Schedule and the time-tagged engineering telemetry data stream downloaded from HST it is possible to reconstruct with great accuracy the pointing sequence of any given scientific observation.

Since the initial production of jitter files, different formats have been adopted. The current format jitter files contain:

- a table of pointing and environment measurements, *e.g.,* right ascension, declination, magnetometer readouts, etc., sampled every 3 seconds.
- a 2D histogram showing the number of times HST was pointing in each element of a 64x64 grid with each pixel 2 milliarcsec in size.

2.1. Jitter Pointing Accuracy

Due to relative errors in Guide Stars coordinates (about 0.3 to 0.4 arcsec), the Fine Guide Stars alignment uncertainty (up to 50 milli-arcseconds [mas]), thermal breathing of the telescope (around 15 mas) and other minor effects, the absolute accuracy in the pointing ranges between 0.5 and 2 seconds of arc. On the other hand the relative accuracy is better than 10 mas within the same HST visit, that is, if the telescope didn't have to re-acquire a guide star during the sequence of observations.

Problems arise if the observation is taken in parallel mode. While the velocity aberration is corrected for the primary instrument, the differential velocity aberration causes the secondary instrument aperture to experience a drift on the sky. The effect can be as high as 50 mas for a full orbit observation, depending on the relative position on the focal plane of the primary and secondary instruments.

3. WFPC2 Associations: Computing the Shifts

Shifts among exposures can hence be computed with good accuracy using the jitter information in the case when all the observations were made during the same visit and WFPC2 was the primary instrument. The procedure consists of:

- Computing the right ascension, declination and roll angle averages and standard deviations from the jitter table, along with some telemetry keywords.
- Assigning a jitter quality flag to each exposure depending on the telemetry keywords and on the standard deviations. The flag can assume the value:
  - 'P' for exposures with small standard deviations and no suspect keyword values,
  - 'G' for exposures with not so small standard deviations or with bad keyword values like GUIDEACT not equal to 'FINE LOCK' etc.
  - 'B' for exposures with missing jitter information or with standard deviations too high or having the SLEWING flag on, etc.
- Grouping together all the 'P' and 'G' exposures
  - taken within the same HST visit
  - in the same filter
  - by the same PI
  - where the distance between each pair is less than 25 PC pixels
  - and the difference between the roll angle of two exposures cannot account for a shift bigger than a tenth of a PC pixel on the full image.
- Computing the shifts Δx and Δy in PC pixels
- Identifying those exposures in the association which are well registered within a tenth of a PC pixel and placing them in sub groups for cosmic ray removal.

4. WFPC2 Association Pipeline

Once we can compute the shifts we may recover the observation strategy (CR-SPLIT, POS-TARG) adopted by a PI. By making use of the computed shifts and rejecting all the exposures not flagged as 'P' (see above), it is possible to run an automatic pipeline which provides not only on-the-fly re-calibrated observations but also offers cosmic ray-free images from all the exposures found to be well aligned within the association.

The overall mosaic of the association can also be requested. In this case the 'drizzle' software (Hook & Fruchter, 1997) is used to build the mosaic of all the cosmic ray-free images preserving photometry, minimising loss of resolution (for sub-pixel shifts) and correcting for geometric distortion using the polynomial model of Trauger et al.

We plan to use a simpler, less CPU demanding, 'shift-and-add' technique for those associations where the shifts are simple integer numbers of pixels.

5. Conclusions

The great reliability of the jitter information has allowed the ST–ECF to:

- Reconstruct the WFPC2 PI's observation strategy (CR-SPLIT, POS-TARG)
- Build associations of WFPC2 exposures
- Compute the shifts among exposures within an association
- Build an automatic pipeline able to clean cosmic rays from aligned exposures and to co-add them with shift-and-add or drizzle techniques.

The ST–ECF archive users can now concentrate on their science while the bulk work of re-calibration, removal of cosmic rays, co-addition of WFPC2 frames is taken care of by the ST–ECF archive system.

The same system will be installed at the Canadian Astronomy Data Centre (CADC).

More information and access to the associations can be found at the following URLs: `http://archive.eso.org/archive/hst/wfpc2_asn/` (ST–ECF) and `http://cadcwww.dao.nrc.ca/` (CADC).

Acknowledgments. We would like to thank M. Lallo and J. Baum (STScI) who helped a lot in understanding the jitter file secrets; D. Durand and S. Gaudet (CADC) for the always fruitful collaboration; D. Shade (CADC) and D. Durand who helped in tuning the WFPC2 pipeline; R. Hook (ST–ECF) and A. Fruchter (STScI) for providing us with details of the drizzle method.

References

Lupie, O., Toth, B.A., & Lallo, M., "Observation Logs", STScI, 1997

Bely, P.Y., & Toth, B.A., "Line of sight jitter reconstruction from guide stars motion", STScI Report, TR-88-01, 22 January 1988

Micol, A., Bristow, P., & Pirenne, B., "Association of WFPC2 exposures", 1997 HST Calibration Workshop, STScI, S. Casertano, et al., eds.

Hook, R., & Fruchter, A., 1997, in ASP Conf. Ser., Vol. 125, Astronomical Data Analysis Software and Systems VI, ed. Gareth Hunt & H. E. Payne (San Francisco: ASP), 147

Micol, A., Dolensky, M., & Pirenne, B., ST–ECF Newsletter No 25, in prep.

Dolensky, M., Micol, A., & Pirenne, B., "Browsing the HST archive with Java-enriched Database Access", this volume

Dolensky, M., Micol, A., & Pirenne, B., Rosa, M., "Enhanced HST Pointing and Calibration Accuracy: Generating HST Jitter Files at ST-ECF", this volume

Astronomical Data Analysis Software and Systems VII
ASP Conference Series, Vol. 145, 1998
R. Albrecht, R. N. Hook and H. A. Bushouse, eds.

The New User Interface for the OVRO Millimeter Array

Steve Scott and Ray Finch

Owens Valley Radio Observatory, PO Box 986, Big Pine, CA 93513

Abstract. A new user interface for the OVRO Millimeter Array is in the early phase of implementation. A rich interface has been developed that combines the use of color highlights, graphical representation of data, and audio. Java and Internet protocols are used to extend the interface across the World Wide Web. Compression is used to enable presentation of the interface over low bandwidth links.

1. Introduction

Modern astronomical instruments share the challenge of presenting a complex and changing system to observers, engineers and technicians. Caltech's six element millimeter wave aperture synthesis array at the Owens Valley Radio Observatory (OVRO) is in the process of implementing a new user interface to facilitate observing and troubleshooting of an increasingly complex instrument. Around the clock operation of the array is done by astronomers and students from a variety of institutions without the help of telescope operators. This style of operation works well when there is easy access to expert observers, instrumentation designers and maintenance personnel and when these people in turn have transparent access to the current state of the array. By porting the array interface to the Web, the instrument is available to everyone with an Internet capable system, independent of operating system or location. Extra steps have been taken to ensure that a modem connection provides a satisfying response for the user. Our current implementation has been done in the tradition of the Web, as we have released our software before completion. This premature release concentrates on the monitoring aspect of some of the most critical parts of our system and serves as proof of concept. The instant utility of this new capability indicates that this was a good choice.

2. Features

The design utilizes a client server architecture with a Java client providing the user interface on an arbitrary computer on the Internet. The Unix machine that runs the array is the host for the server programs which are written in C++. A user initially sees a menu[1] of 16 different monitoring windows that can be

[1] http://www.ovro.caltech.edu/java/cma/menu.html

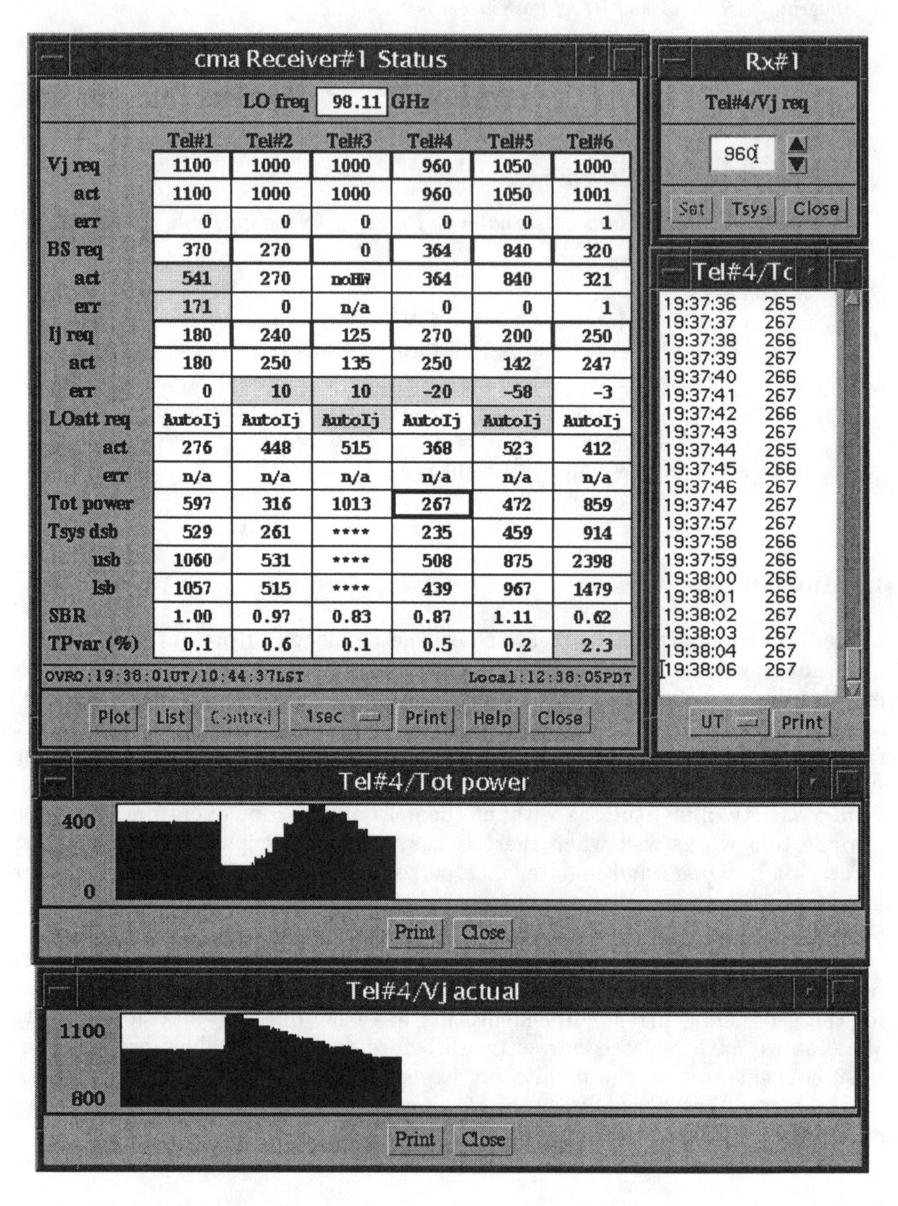

Figure 1. A single realtime window surrounded by its children.

launched onto their screen. Each of these windows is independent, and several of them are typically run simultaneously on the user's screen. The local window manager can be used to resize and place the windows to customize the layout as is illustrated in Figure 1.

The rectangles containing text on a light background are instances of the realtime data cell (RDC) that are the critical building blocks of the windows. The values within the cell are updated on a user selectable time scale that can be set from one second to thirty minutes using the choice menu at the bottom of the window. An RDC can optionally have the cell background color change as a function of the value in the cell. Typical color assignments are yellow for warning and red for alert. An audio chime can also be associated with the cell to emphasize the alert. When the window is initiated, each RDC begins to cache a history that can contain the most recent 200 samples. This history can then be accessed by other parts of the program. The RDCs are also used as part of a compact menu for selecting the advanced features. When an RDC is clicked with the mouse, its border is enhanced and it becomes the "selected" cell. In Figure 1 the "Tot Power" for telescope #4 is the selected cell. Clicking on the plot or list button at the bottom of the window brings up the plot or list window for that cell.

The plot and list features coupled with the history are an important part of the user interface for the RDCs and are shown in Figure 1. When a plot or list window is initiated it is seeded with the values in the RDC history. This allows the user to see an event occur in the main window and then to initiate a plot to look back in time before the event. Both plot and list contain print options. Additionally, the contents of the list window can be written to disk in a two column ASCII format. The plotting is intentionally simple. The ordinate is auto scaled but the abscissa simply uses one pixel per sample to allow a maximal number of samples to be plotted and avoid unnecessary rescaling. The ability of the list window to write to disk allows more complex analysis using a spread sheet or other tools. Because the histories for all of the RDCs within a window are synchronized, plots and listings are time aligned for cross comparison of cells. Each plot or listing contains an extension of the history buffer that can contain 1000 points.

Control functionality can also be added to RDCs. Using the same cell selection metaphor and the control button at the bottom of the window, a control widget can be launched. Cells with control capability are indicated by a purple outline rather than the standard black. An example of a control widget can be seen in the upper right hand corner of Figure 1. Security for control features is currently based on the Internet address of the client but will be changed to a login based mechanism using our Unix password files. If control is not authorized then the control button does not appear on the window.

Many of the RDCs are arranged in tabular form to provide a dense display and to emphasize the relationship of the data. It is also possible to lay out the RDCs in a freer format where there is a label associated with each RDC. These two styles can then be stacked vertically to provide a window with multiple sections.

3. Implementation

To eliminate the need for synchronized programming in the client and the server, the server programs send the client complete configuration information on startup and the client then configures itself accordingly. The result is clients with significantly different appearances even though they are running copies of the same Java code.

The life cycle of a new window begins when the client opens a TCP/IP socket connection to a pre-assigned port on the array control computer with a request for a specific type of realtime window. The Unix system then runs a simple "Internet service" program attached to this port that forks a copy of the specific server program for the type of window requested. This establishes a one-to-one mapping between the client and a window specific server program. The client now requests its layout from the server, and after reacting to the reply it makes itself visible. It then begins a loop of requesting an update of the live data from the server, displaying the data, and then sleeping for the update interval. The multi-threaded nature of the Java environment allows reaction to user initiated events, such as plotting, while in this loop. Driving the update cycle from the client is appropriate for monitoring data as it is then very robust to unforeseen delays. These delays only cause a gradual degradation of response rather than total failure.

An efficient programming environment for the C++ servers is based on shared memory containing all of the data of interest about the array. This shared memory is fed by eight different realtime microcomputers embedded in the array hardware that send UDP datagrams twice a second. The server programs then simply map the shared memory into their address space to have access to the live data. The fundamental classes for the RDCs and layouts that are used to craft a window contain about 1500 lines of code. By using these classes a new server program can be written in a few pages, making the creation of a new window almost trivial.

4. Resources

Resource utilization is important in our choice of C++ for the server program because it is currently much more efficient than Java. Our midrange Unix processor can serve 50 realtime windows without substantial impact. At the Java client end, an average PC can easily display six realtime windows. The server to client data rate for three windows at a two second update is about 1KB/sec, which is a large enough fraction of the 2-3KB/sec available over a modem to make the response sluggish. To improve throughput, the data stream is compressed by transmitting only changes to the existing screen and escape codes that encode the number of unchanged characters to skip. This algorithm matches the data quite well when the screen appearance often shows just the twinkling of the least significant digits. Typical compression rates of 5 to 10 allow four or more windows to be displayed over a modem link with good response time.

Astronomical Data Analysis Software and Systems VII
ASP Conference Series, Vol. 145, 1998
R. Albrecht, R. N. Hook and H. A. Bushouse, eds.

The IRAF Mosaic Data Reduction Package

Francisco G. Valdes

IRAF Group, NOAO[1], PO Box 26732, Tucson, AZ 85726

Abstract. The IRAF Mosaic Data Reduction Package, **mscred**, processes data from mosaics of CCDs such as the NOAO CCD Mosaic Camera[2]. A brief design of the package was presented earlier by Valdes (1997a). Since then a first implementation of the software has been completed. This paper summarizes the current status of the software and our plans for future developments.

1. Current Status

Most of the components of the original design (Valdes 1997a) for an IRAF Mosaic Data Reduction Package have been implemented in the first version of the **mscred** package. This has allowed complete, end-to-end reductions of data from the NOAO CCD Mosaic Camera. The package operates on multiextension FITS format (MEF) files consisting of a global header and individual image extensions for each amplifier (Valdes 1997b). The raw data files are processed so that each image extension is flux and coordinate system calibrated. Mosaic cameras have gaps and misalignments between the CCD elements so, when a complete image of the sky is desired, multiple "dithered" exposures are taken. The **mscred** package provides tasks to resample the data into a final image with the geometric effects (gaps, alignments, and optical distortions) removed. This process can also improve the signal-to-noise and eliminate cosmic rays and cosmetic defects. Observations have been obtained with the NOAO CCD Mosaic Camera and processed with the **mscred** package which produce images of high scientific and aesthetic quality despite the engineering grade CCDs used in the current instrument.

The functionality of the **mscred** package can be broken down into the following categories: (1) display, (2) basic CCD calibrations, (3) coordinate registration, (4) mosaicing, (5) taping, and (6) miscellaneous. The package is used both at the telescope and after the observing run for quick-look or full reductions and for data analysis. Some of the categories apply to both uses as, for example, the display of mosaic data. In this paper the package tasks are identified with their names in bold font.

---

[1]National Optical Astronomy Observatories, operated by the Association of Universities for Research in Astronomy, Inc. (AURA) under cooperative agreement with the National Science Foundation.

[2]http://www.noao.edu/kpno/mosaic/

Being able to display an exposure as an approximation of a complete (mosaiced) image of the sky is a prime requirement both at the telescope and during data reduction. We are developing a new display capability for this purpose as part of the NOAO Mosaic Data Handling System (Tody 1997). The **mscred** package provides an interim task (**mscdisplay**) to display multiextension data as an image in a standard display server such as Ximtool. **Mscdisplay** includes real-time capabilities to display the data while a readout is in progress. Related tools allow users to interact with the displayed mosaic exposure (even during readout) to evaluate focus (**mscfocus**) and to do quick-look analysis (**mscexamine**) including PSF fitting, statistics, graphics, and celestial coordinate measurements.

Basic CCD calibration provides for combining sequences of calibration exposures (**zerocombine**, **darkcombine**, and **flatcombine**) and the standard CCD calibration operations of overscan subtraction, trimming, bad pixel replacement, zero level subtraction, dark count subtraction, and flat-fielding (**ccdproc**). The input and output of these operations are MEF files.

The **mscred** package places emphasis on having an accurate celestial coordinate system (called the world coordinate system or WCS). Using prototype astrometry tools in the **mscred** package (**msctpeak**) an accurate WCS consisting of independent solutions for each CCD relative to a common reference pointing including all optical distortion and alignment terms has been derived for the NOAO Mosaic at the Kitt Peak National Observatory Mayall 4-meter and 0.9-meter telescopes (Davis, 1998). This was done using exposures of astrometric fields. This WCS is part of the raw data produced by the Mosaic Data Capture Agent (DCA) (Tody & Valdes 1998) with the coordinate system reference point set to the telescope pointing coordinates.

The zero point of the WCS, which is initially set by the telescope pointing, can be adjusted to a precise absolute coordinate or to common coordinates in a dithered set of exposures by displaying the exposures and identifying one or more reference stars (**msczero**). Given that the raw data already have relatively good coordinates there is a task (still evolving) that takes a set of overlapping exposures and either a set of coordinates or random regions and registers their WCS using cross-correlation (**mscregister**).

Using the WCS, the multiple images from each amplifier in a mosaic can be resampled to make a single image on a uniform pixel grid having a standard WCS, such as a tangent plane projection (**mscimage**). By using one exposure as a reference, multiple dithered exposures can all be resampled to the same pixel grid system (the same tangent point and pixel scale) so that the images may be stacked (**mscstack**) to make a final image without further resampling. The stacking process excludes the gaps and may include use of bad pixel masks and various scaling and pixel rejection algorithms found in the standard IRAF **imcombine** routine. Since combining dithered exposures is a common operation a higher level task (**mscdither**) combines **mscregister**, **mscimage**, and **mscstack** to directly produce a final image.

Tasks for taping of data are only included as an interim measure until generic IRAF tasks include direct support for disk FITS files in MEF format. For MEF data the duties of the taping tasks (**mscwfits** and **mscrfits**) are simply to transfer the FITS file to and from tape with the appropriate FITS blocking, efficient listing of the contents of tapes with multiextension files, and recording the disk filenames and restoring the files to disk with their original filenames. The tape is a valid FITS tape.

The **mscred** software is packaged as a standard IRAF external package for IRAF version 2.11 and later. Although this is an early version of the software it has been made available to users of the NOAO Mosaic Camera and other interested parties developing mosaic cameras. Releases of the software will be made periodically as new features are added.

2. Future Work

There are many things which still need to be added. These range from minor improvements to a few major research and development items. The major items are discussed in the following sections. The minor items consist of an improved syntax to interface MEF data to existing IRAF tasks that operate on lists of images, expanding the CCD processing task **ccdproc** to provide for incremental reductions, using a better WCS representation, and a task to restore flux conservation in flat-fielded data.

A wildcard syntax is needed to easily select a set of image extensions from an MEF file rather than the current requirement that each extension be listed explicitly. It takes special care to produce a good flat-field for a wide-field mosaic so for quick-look and initial reductions it is desired to apply archival calibration data, such as a high quality master sky flat, and then continue with incremental calibration using data acquired during the course of observing. **Ccdproc** needs to be modified to easily support incremental calibration.

The WCS representation for a wide-field optical image is better given as a radial projection (as proposed for a FITS world coordinate system standard) although a general polynomial distortion residual will still be required. Currently a tangent plane projection is used in combination with a separate text file defining a polynomial distortion function. Another property of wide-field images, such as with the NOAO Mosaic at the 4-meter telescope, which is not obvious at first is that the pixel area (square arc seconds per pixel) may vary significantly. This means the sky and object counts vary with position. Flat-fielding attempts to make the sky counts constant which leads to flux errors. A task based on the WCS is required to restore the correct flux per pixel to flat-fielded data prior to doing any photometry. This only affects the MEF files because the resampling operation (**mscimage**) naturally accounts for the varying pixel areas.

2.1. Pixel Masks

Pixel masks assign integer codes to each pixel. IRAF provides a pixel mask format which is very compact for masks containing regions of constant value. In **mscred** pixel masks are used to identify bad pixels with codes values for cosmetic defects, saturated pixels, and cosmic rays. The masks are assigned to data exposures and the software uses these assignments to determine bad pixel information for the data pixels. The current software supports the first category of predetermined cosmetic defects for replacement by interpolation, avoiding bad data in automatic display scaling, excluding bad data from statistical sampling of scaling factors for combining, and exclusion during the stacking of dithered exposures.

The issues that still have to be addressed are updating other pieces of the the software to add to the mask, such as the flagging of saturated pixels, additional uses of the bad pixel information, such as during resampling, and storage of the bad pixel information in multiextension FITS files. The last topic requires mapping the compact IRAF pixel mask format to a FITS format; most likely as a binary table extension.

2.2. Pixel Uncertainty Information

The propagation of pixel value uncertainty information naturally starts with the raw data. The **mscred** tasks need to be expanded to propagate the pixel uncertainties from the raw data during each step that modifies or transforms the pixel values. There are two development stages that need to be completed. The first is to define the data format representing the pixel uncertainties and the second is to understand how the uncertainties propagate in operations such as flat-fielding, resampling, and combining with pixel rejection.

Research in representing the pixel uncertainties is needed to, hopefully, find a compact description requiring much less than one uncertainty value for each data pixel. Preliminary research suggests a combination of a scaling relative to the pixel data, header keywords, and mapping to a finite set of discrete values that give uncertainties to a useful precision. A key feature of this is the use of pixel masks which can be stored in a compact format as described previously.

2.3. Astrometry

The **mscred** package supports a coordinate system that is quite accurate. The software maintains and propagates this coordinate system. Much of the coordinate system description is fairly static and only terms relating to zero points and rotations need to be calibrated on an individual exposure or run basis. Currently the instrument support personnel provide the static part of the coordinate system description and the **mscred** package provides tools to modify the zero point to yield absolute coordinates and to register overlapping exposures. The problem is that if users want to modify anything but the zero point they have to do a complete astrometric solution which requires a good astrometric field with many stars.

The desired enhancements are to let users to have more control of the coordinate system calibration and to integrate catalog servers to ease the determination of a zero point for absolute coordinates. The first part relates to allowing adjustments of the coordinate system representation short of requiring a compete new astrometric solution. For instance with just a few good astrometric objects users should be able to adjust the scale and rotation in addition to the zero point.

2.4. Data Reduction Agent

The Data Reduction Agent (DRA) is an ambitious part of the NOAO Data Handling System which was described in the original design. It is not directly a part of the **mscred** package. However, this pipeline tool is intended to be portable with the **mscred** package and be closely tied to the **mscred** functionality. As work progresses on the DRA there may be enhancements of the **mscred** package to support the automatic reduction of mosaic data in a data handling system environment.

References

Davis, L. 1998, this volume

Tody, D. 1997, in ASP Conf. Ser., Vol. 125, Astronomical Data Analysis Software and Systems VI, ed. Gareth Hunt & H. E. Payne (San Francisco: ASP), 451

Tody, D. and Valdes, F. 1998, this volume

Valdes, F. 1997a, in ASP Conf. Ser., Vol. 125, Astronomical Data Analysis Software and Systems VI, ed. Gareth Hunt & H. E. Payne (San Francisco: ASP), 455

Valdes, F. 1997b, in ASP Conf. Ser., Vol. 125, Astronomical Data Analysis Software and Systems VI, ed. Gareth Hunt & H. E. Payne (San Francisco: ASP), 459

Astronomical Data Analysis Software and Systems VII
ASP Conference Series, Vol. 145, 1998
R. Albrecht, R. N. Hook and H. A. Bushouse, eds.

Determination of the Permissible Solutions Area by Image Reconstruction from a few Projections: Method 2-CLEAN DSA

Michail I. Agafonov

Radiophysical Research Institute (NIRFI), 25 B.Pecherskaya st., Nizhny Novgorod, 603600, Russia, E-mail: agfn@nirfi.nnov.su

Abstract. We have proposed the 2-CLEAN DSA (Determination of Solution Area) method for the estimation of the area of possible images (from the "obtuse" (smooth) to the "sharp" variants) in complicated cases with constraints and poor *a priori* information. The area of permissible solutions can be determined with the help of two CLEAN algorithms: standard CLEAN and Trim Contour CLEAN (TC-CLEAN). The procedure has high efficiency and simple criteria by errors minimization of initial and control 1-D profiles. We present here a description of some valuable features of the reconstruction technique.

1. Introduction

Iterative algorithms with non-linear constraints are very attractive in image reconstruction with only a few strip-integrated projections (Vasilenko & Taratorin 1986). The process of convergence to solutions for the various realizations of iterative schemes using the different versions of the CLEAN algorithm have already been investigated for this problem (Agafonov & Podvojskaya 1989). Two dimensional image reconstruction from 1-D projections is often hampered by the small number of available projections, by an irregular distribution of position angles, and by positions angles that span a range smaller than about 80 degrees. These limitation are typical of both lunar occultations of celestial sources and observations with the fan beam of radio telescope, and also apply to greatly foreshortened reconstructive tomography. Our previous paper (Agafonov 1997) also contains the basic description of the problem and the features of 2-CLEAN DSA method. This paper contains some examples and useful diagrams and also a valuable addition for the development of the reconstruction technique for this problem.

2. Application

The problem requires the solution of the equation

$$G = H * F \ (+noise) \ , \qquad (1)$$

where $F(x, y)$ is the object brightness distribution, $H(x, y)$ is the fan (dirty) beam, and $G(x, y)$ is the dirty (summary) image. An example of the fan beam

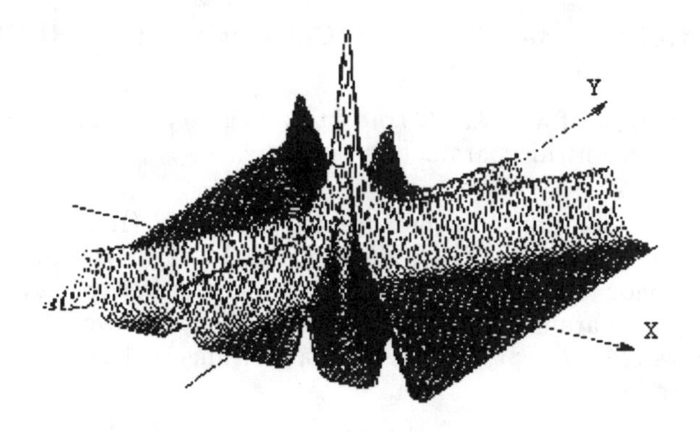

Figure 1. Fan beam $H(x, y)$ for four projections (typical example).

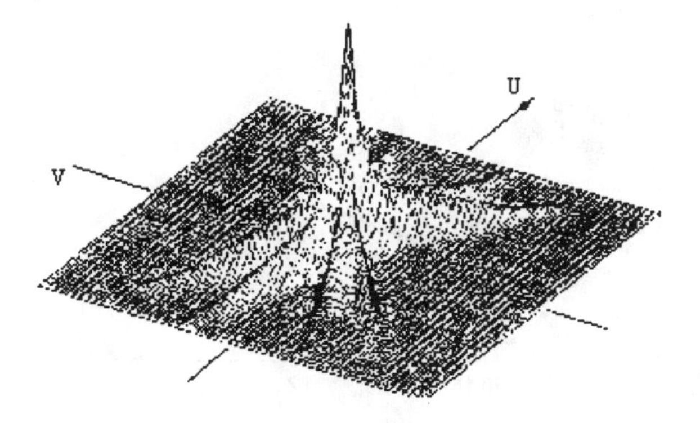

Figure 2. The same fan beam $H(u, v)$ on the UV plane (sampling of about 0.1 of the UV plane in the limit ω_l to the fixed radius of the frequency).

for four projections is shown in Figure 1. The classical case (Bracewell & Riddle 1967) needs a number of projections $N \geq \pi D/\varphi$, where φ is the desired angular resolution, and D is the diameter of the object. The incomplete sampling of $H(u, v)$ (see Figure 2) requires the extrapolation of the solution using non-linear processing methods.

A two dimensional image reconstruction of a complicated object using a poor fan beam function $H(u, v)$ needs to carry out the following procedure:

1. **Test experiment**: evaluation of the possibility and quality of restoration on the similar images versions; modeling for available projections (number N, position angles, signal/noise ratio) and desirable frequency limit ω_l. This point can include following steps:

   - **2-D object model → 1-D profiles → Dirty image**
   - **CLEAN** (λ) or **TC-CLEAN** (λ, TC) using of **Fan beam**

- **Control test** from clean maps: **Calculation of** σ (ERROR of control and initial 1-D profiles)
- **Correction of** λ **or** λ, TC **to** $min\ \sigma$**.** This step can help to determine the best algorithm parameters range.

The process of solutions convergence using CLEAN (Hogbom 1974) and Trim Contour CLEAN (TC-CLEAN) (Steer et al. 1984) has been analyzed (Agafonov & Podvojskaya 1989; Agafonov & Podvojskaya 1990) by optimizing the parameter λ or λ, TC (Trim Contour level) to $min\ \sigma$. A real example of such a process for both algorithms is shown graphically in Figures 3 and 4.

2. **Reconstruction from real observational 1-D profiles.** This process can also include the correction of λ or λ, TC (Trim Contour level) to $min\ \sigma$.

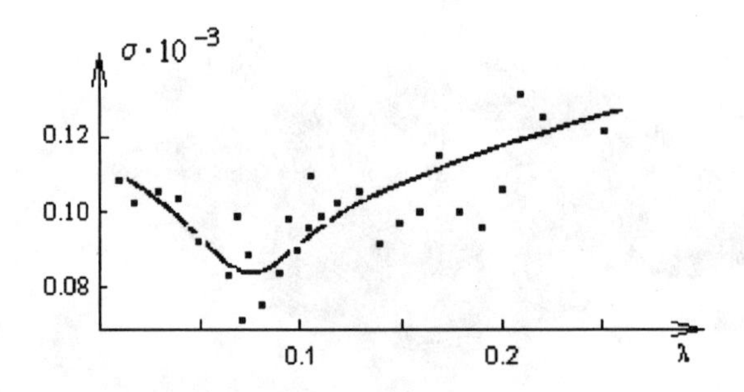

Figure 3. The error of original and control profiles as a function of the loop gain (using standard CLEAN).

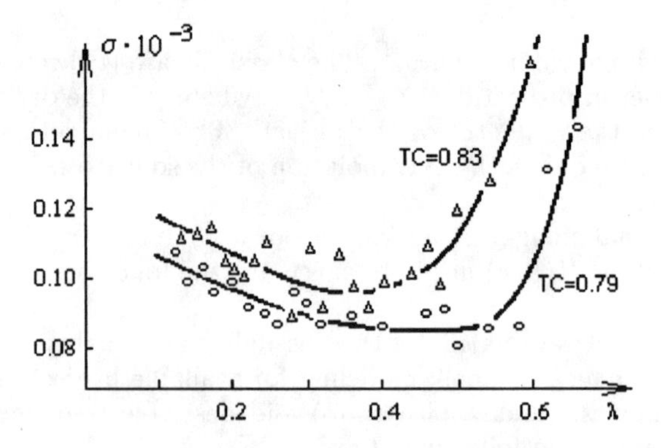

Figure 4. The error of original and control profiles as function of the loop gain loop using two different Trim Contour levels of TC-CLEAN.

3. Conclusions

CLEAN forms the solution from the sum of peaks and the result is the sharpest variant permissible within the established constraints. On the other hand TC-CLEAN accumulates its result from the most extended components that satisfy the constraints, producing the smoothest solution (Agafonov & Podvojskaya 1990).

A simple object (consisting of the peaks) may be successfully restored by the standard CLEAN. The results obtained by both methods are practically identical for a simple object consisting of individual components, but TC-CLEAN is more computationally efficient. For smoothed 1-D profiles with small "hillocks", the solution can be obtained from the isolated individual components (CLEAN), and also from the more smoothed components (TC-CLEAN). CLEAN increases the contrast of small components, but the extended background decreases because of "grooves". **If** $min\ \sigma$ **(CLEAN)** $\cong$ $min\ \sigma$ **(TC-CLEAN)** (see the example shown in Figures 3 and 4), **the solutions will be formally equivalent** for both algorithms, and so we have two choices: (*i*) to prefer the result corresponding to the physical peculiarities of the object in accordance with *a priori* information; or (*ii*) to assume the existence a probable class of solutions between the "smooth" one from TC-CLEAN and the "sharp" one from CLEAN.

The area of permissible solutions of complicated objects can be determined with a help of both algorithms. The 2-CLEAN DSA procedure can show a range of possible images from "smooth" to "sharp" variants satisfying imposed constraints and poor *a priori* information.

The application of TC-CLEAN and CLEAN was presented as a reconstruction of the Crab Nebula map at 750 MHz (Agafonov et al. 1990). The method of 2-CLEAN DSA allowed us to determine that the area of the permissible solutions lies formally between the "sharp" (CLEAN) and "smooth" (TC-CLEAN) variants. Two maps were generally similar. The standard CLEAN increased the contrast of small components while the TC-CLEAN map gave a better agreement with known *a priori* information.

Acknowledgments. I am grateful to the Space Telescope European Coordinating Facility and European Southern Observatory for the support which made possible this presentation and my special gratitude to Rudi Albrecht, Richard Hook and Britt Sjoeberg for their attention and endurance.

References

Agafonov, M. I. 1997, in ASP Conf. Ser., Vol. 125, Astronomical Data Analysis Software and Systems VI, ed. Gareth Hunt & H. E. Payne (San Francisco: ASP), 202

Agafonov, M. I., & Podvojskaya, O. A. 1989, Izvestiya VUZ. Radiofizika, 32, 742

Agafonov, M. I., & Podvojskaya, O. A. 1990, Izvestiya VUZ. Radiofizika, 33, 1185

Agafonov, M. I., Ivanov, V. P., & Podvojskaya, O. A. 1990, AZh, 67, 549

Bracewell, R. N., & Riddle, A. C. 1967, ApJ, 150, 427

Hogbom, J. A. 1974, A&AS, 15, 417

Steer, D. G., Dewdney, P. E., & Ito, M. R. 1984, A&A, 137, 159

Vasilenko, G. I., & Taratorin A. M. 1986, Image Restoration (in Russian), Radio i svyaz', Moscow.

Astronomical Data Analysis Software and Systems VII
ASP Conference Series, Vol. 145, 1998
R. Albrecht, R. N. Hook and H. A. Bushouse, eds.

Image Processing of Digitized Spectral Data

A. V. Boulatov and L. K. Kashapova

Institute of Solar-Terrestrial Physics, Siberian Division of Russian Academy of Sciences, P.O.Box 4026, Irkutsk, 664033, RUSSIA

Abstract.
Every observatory has a library of old pictures taken by photographic cameras. These measurements are not as such high quality as those from modern CCD systems, but could contain important data about rare phenomena or source data for long-period investigations (e.g., concerning the solar cycle).

This paper presents methods for converting photographic images, collected at the Large Solar Vacuum Telescope (Baikal Astrophysical Observatory) to digital images, similar to CCD pictures. Special algorithms were used for this task and have been implemented in the IDL environment. The results of processing real spectral data and a comparison of processed photographic images and images taken with a TEK CCD are presented.

1. Introduction

The Large Solar Vacuum Telescope (LSVT) is one of the exceptional astronomical instruments belonging to the Institute of Solar-Terrestrial Physics, the Siberian Division of the Russian Academy of Sciences (Skomorovsky & Firstova 1996). Because of the excellent spatial (0.3″) and spectral (0.2–0.5Å/mm) resolution of the instrument and also local seeing conditions (the telescope is situated on the shore of Lake Baikal) we have the ability to investigate fine structure of the Sun. These observations are used to investigate short-time phenomena, e.g., solar flares, and also for the study of long-period features, e.g., the solar cycle. For this work series of observations are usually taken.

During the long period of solar maximum we used a photographic camera with 35-mm film for imaging with our telescope. A considerable amount of valuable data (photographic images on films and plates) was taken and then collected for future processing and analysis. During recent years, instead of the photographic camera we have begun to make observations with modern devices such as CCD systems. These have many advantages in comparison with photographic methods: linearity, higher sensitivity, and so on.

To analyze long-period solar features we have to use both old and new observations, i.e., photographic images and frames from the CCD detector. Both photography and CCD techniques have their own characteristics which influence the image. In order to perform reliable analysis we need to convert the different detector data to a homogeneous type.

We have created a set of programs for such corrections based on the IDL language. One of the research fields of our group is the mechanism of energy transport and energy release in solar flares. It is studied using polarization observations. We present the work of these IDL programs using the example of our spectral polarization observations and data analysis.

2. Observations and Preliminary Processing

All observations were obtained at the LSVT as described above. On the one hand the exposure time is determined by the instability of the earth's atmosphere and on the other it is determined by detector sensitivity. So exposure times for the photographic observations were 0.45 sec and the CCD detector allows exposures of about 0.1 sec. We used a Wollaston prism and two $\lambda/2$-plates to separate the ordinary and extraordinary rays.

2.1. CCD Systems

The first CCD system, which we mounted inside the telescope spectrograph, was a CCD system from a St. Petersburg firm with exchangeable CCD heads. There were heads with 800x400 and 370x290 pixels arrays. Using this system we can take frames with very short exposures even with a dispersion of 0.0051 Å/pixel. This CCD system is controlled by microprogramming code and external Pascal programs.

In addition to that system we also use one amateur class CCD system (SBIG ST-6 with the supplied control program) at our telescope and, during the last few months, also another professional CCD system (Spectroscopic Instruments TEK CCD 512x512 with ST-130 controller and WinView). The basic differences between these systems are their array sizes and the rate of obtaining sequential frames (the minimum interval between successive frames). The main calibration procedures (removing the dark current, accounting for the flatfield frames and so on) are similar for all CCD systems (McLean 1989).

2.2. Photographic Methods

As mentioned above, photographic images were taken on the 35-mm film. In Figure 1 a typical view of our polarization data is presented. The two spectral bands correspond to orthogonally polarized spectral bands. You can see the so-called "moustaches" (Ellerman bombs).

Digitizing of the photographic images was performed in 3 alternative ways:

- point scanning with a standard micro-photometer
- scanning by the CCD line on a micro-photometer coordinate device, supplying precise micro-movements
- digitizing with the help of an optical bench and using the CCD array for registration.

After such procedures we can write a digitized frame of the picture taken from the film. To obtain intensity values from the photographic image we have

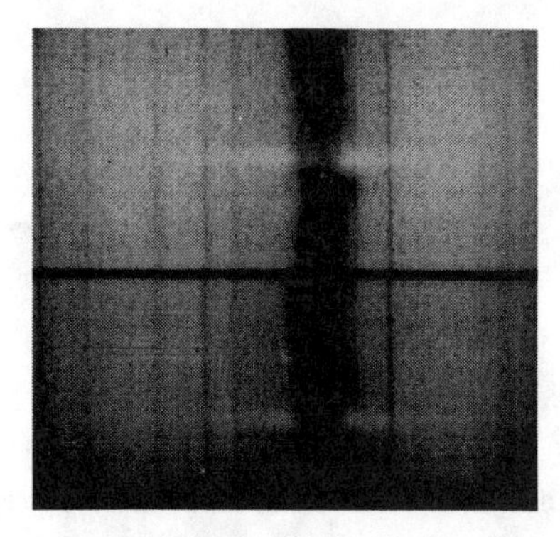

Figure 1. The image of the moustaches in the H_α line. This frame was acquired using the photographic camera with 35-mm film.

to use the characteristic curve for the film and at the same time we should take account of the transfer function of the CCD (used as the digitizer) and the uniformity of light in the microphotometer. This is very important for future accurate analysis of real solar data. Also at this stage we carry out procedures which are specific to every frame, removing errors (e.g., dust or cuts on film), trends and disturbances from the photographic image. After that we can use the digitized image from photographic plate or film as input for astrophysical investigation.

3. Analysis of Digitized Photographs and CCD images

After the calibration the real processing begins, in our example, with mapping of the H_α contour and obtaining the spectral profile. We take sections across or along the dispersion direction over the region being investigated to make effective use of computer memory. We also normally load a colour look-up-table to make the processing of the selected profiles more convenient and clear.

Using spectral atlas data we then define some basic lines close to H_α and, with the help of the computer mouse, mark the position of these lines on the frame and obtain the calculated position of the H_α line centre on the frame and the average dispersion for a set of frames. Once this is done we can operate using wavelength instead of pixels on the frame. For the following polarization calculation two spectral bands are shifted to each other in order to combine the H_α centre from different bands. Results of such processing, showing the corrected spectral profiles of the moustache, are shown in Figure 2. The thin line corresponds to the ordinary ray and the thick line relates to the extraordinary one. The next step is to obtain polarization vectors and additional checks for artificial influences on the spectral image (e.g., high-order trends, instrumental effects). Then we prepare the best visualization of the results.

Recently we have continued these observations with the help of new equipment. In Figure 3 the image frame with the moustache acquired using a CCD - detector is presented. We can use the programs created for processing digitized photographic images in order to process these data as well.

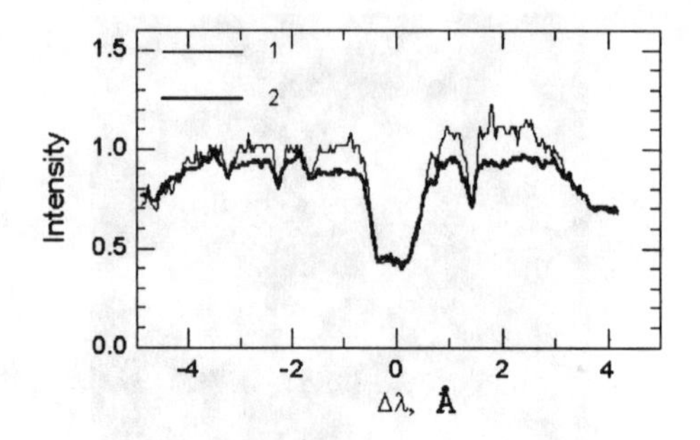

Figure 2. H_α line profiles in the moustaches. 1– the ordinary ray, 2– the extraordinary ray.

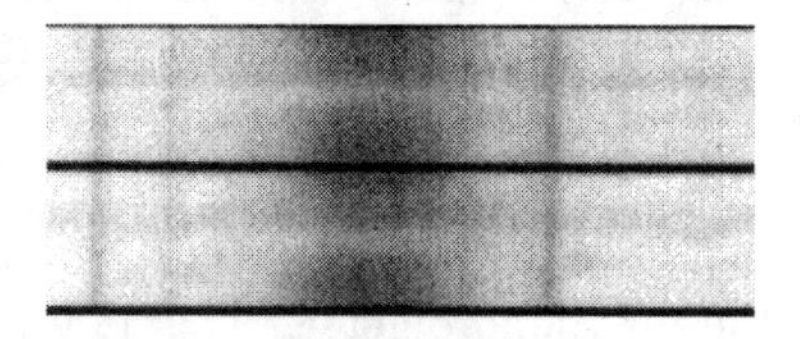

Figure 3. The image of the moustaches, H_α line. The frame was acquired with a CCD detector.

4. Conclusions

The result of this work is a developed set of programs for spectral data processing. We can say with certainly, that on the base of our set of programs, processed photographic images become comparable to real CCD images. After this rather difficult processing is complete we can use the two sorts of data together as a common series as is required for the investigation of long period solar phenomena (in particular) and many other problems of solar physics.

Acknowledgments. We are grateful to Dr. N.M.Firstova, as main observer at LSVT, for her invaluable co-operation during observations.

References

Skomorovsky, V. I., & Firstova, N. M. 1996, Sol.Phys., 163, 209

McLean, I. S. 1989, Electronic and computer-aided astronomy: from eyes to electronic sensors, (Ellis Horwood Limited), 193

Astronomical Data Analysis Software and Systems VII
ASP Conference Series, Vol. 145, 1998
R. Albrecht, R. N. Hook and H. A. Bushouse, eds.

A Parallel Procedure for the Analysis of Long-term Sequences of Light Curves

A. F. Lanza, M. Rodonò[1], U. Becciani and V. Antonuccio Delogu

Osservatorio Astrofisico di Catania, Viale A. Doria, 6 - I 95125 Catania, Italy

Abstract. We present a parallel procedure which allows us to speed up the modelling of photometric and spectroscopic observations of active binary stars with brightness inhomogeneities on their surfaces. The procedure has been implemented using PVM and is suitable to run on a cluster of non-homogeneous, non-dedicated computers. It is optimized to recognize and assign the data sets requiring the largest computational effort to the most powerful CPUs of the cluster, taking into account the evolution of their performances during the calculation in a fully dynamical way. We report on several tests made with the workstation cluster of Catania Astrophysical Observatory and discuss the advantages of this kind of procedure for data analysis in Astrophysics.

1. Introduction

Binary systems belonging to the RS CVn and BY Dra classes show huge cool spots on their photospheres which are regarded as manifestations of intense magnetic fields by analogy with sunspots (e.g., Tuominen et al. 1991, Strassmeier & Linsky 1996).

The analysis of sequences of light curves spanning a long time interval (at least one or two decades) allows us to detect activity cycles, analogous to the solar 11-year cycle, inferring the overall properties of the stellar dynamos. Moreover, long-term data can be used to estimate the effects of starspots on stellar parameters (Rodonò et al. 1995, Lanza et al. 1997).

We address the problem of modelling a long-term sequence of light curves adopting recently developed tools for parallel computing and using an inhomogeneous cluster of CPUs.

2. Light curve modelling

The reconstruction of the surface map of an active star by using photometric data alone is an ill-posed problem. It is possible to find a unique and stable solution

---

[1] Istituto di Astronomia dell'Università degli Studi di Catania, Viale A. Doria, 6 - I 95125 Catania, Italy

if a priori assumptions on the properties of the picture elements (pixels) of the map are adopted such as the Maximum Entropy (hereinafter ME, e.g., Vogt et al. 1987) and the Tikhonov criteria (hereinafter T, e.g., Piskunov et al. 1990).

The optimized map is found by minimizing the objective function Q consisting of a linear combination of the χ^2 and the regularizing function S: $Q = \chi^2 + \lambda S$. The expressions for χ^2 (which gives the deviation between the fluxes computed from the map and those observed) and the regularizing ME or T functions can be found in, e.g., Cameron (1992); the Lagrange multiplier λ measures the relative weights of the a priori assumption and χ^2 in constraining the solution. In our approach the best value of λ is determined by the distribution of the residuals between the observed and the computed fluxes (Lanza et al. 1997; see also Cameron 1992). In any case several solutions are computed for different values of λ in order to find the optimal value through a suitable statistical test.

In the modelling of a light curve sequence it is also of interest to determine the physical parameters of the system components, which may be affected by the presence of spots. In such a case the overall computational problem may become much more complex and time consuming.

3. The parallel procedure

Our parallel procedure was written in standard FORTRAN using the PVM library version 3 and exploits a networked system to perform the analysis of a sequence of N_L light curves. The procedure starts on a main host – the *master-host* – and the single jobs are automatically spawned on all the systems on which PVM is available. This procedure actually generates a *virtual machine* (hereinafter VM). We assign the modelling of each light curve with given values of the system parameters and λ to a single system (host) which, at the end of its task, sends the output back to the master-host.

The minimization of the function Q for all the light curves of a given sequence, for fixed values of the system parameters, represents a cycle of the procedure. In general the analysis is not limited to one cycle because we can be interested in the simultaneous determination of one or more system parameters, such as the luminosity ratio of the system (see Rodonò et al. 1995, Lanza et al. 1997). Thus, considering a typical sequence consisting of a few tens of light curves, a few hundred modelling steps (i.e., Q minimizations) are required if λ is held fixed, and up to a few thousand if λ also is varied (see also Wilson 1993).

The procedure is implemented using a master/slave paradigm. The master is the program controlling the execution of the overall analysis and it is running on the master-host. The master performs the following operations in sequence: a) it identifies all available hosts and adds them to the VM; b) it assigns the values to the system parameters and the λ's for the given cycle; c) it spawns each analysis of the cycle on one of the hosts by choosing it according to the number of normal points in the light curve and the current weights assigned to the hosts; d) it periodically checks whether all the hosts of the VM are running and, in case of a fault, restarts the lost jobs; e) it receives the output files from the hosts which have completed their jobs and updates their current weights; f) at the end of the cycle, according to the task assigned, it ends the execution or starts a new cycle from step b).

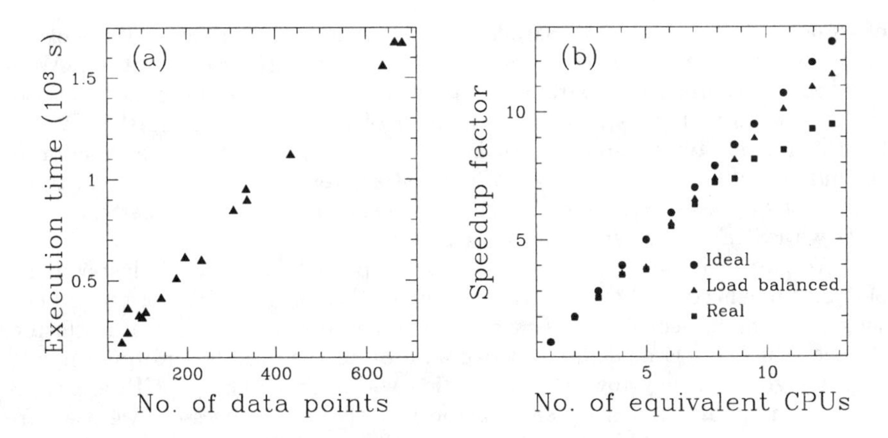

Figure 1. (a) The time of execution on the reference CPU *vs.* the number of data points in the analysed light curve for the linear test. (b) The speedup *vs.* the number of equivalent CPUs on which the parallel procedure is running for an ideal, a perfectly load balanced and our real cluster of workstations, respectively.

The estimate of the weight of each host is initially based on the time it takes to analyse a reference light curve and is updated in the course of the calculation using the elapsed times of the previous analyses.

If a task is anomalously ended, the master detects the fault and tries to restart it on the same host. If the fault occurs again or the host is not available, the master deletes the host from the VM, and the analysis is re-scheduled on the first available free host. After a user-defined period, a deleted host can be checked again and, if it is found available, it can be added again to the VM. Therefore the VM configuration changes dynamically and automatically during the run with hosts being added or deleted by the master according to their current availability status.

4. Tests and results

We have performed several tests on the workstation cluster of Catania Astrophysical Observatory analysing a sequence of 18 light curves of the eclipsing binary AR Lacertae (see Lanza et al. 1997).

The purpose of a first test (linear test) has been to study the variation of the computational complexity, as a function of the number of data points in the light curve. Only one processor has been used to run the procedure. It has been a SuperSparc+ processor (clockspeed 50 MHz) and all the results are reported assuming it as the reference processor. As workstation operating system we used Solaris 2.5 and 64 MB RAM were allocated during the entire execution time. The results are shown in Figure 1a. There is evidence that the execution time, and thus the computational complexity, increases linearly with the number of data points in the analysed light curve. In a second test, we analysed a sequence of 18 light curves running in parallel 4 cycles each with a different value of the luminosity ratio and a fixed $\lambda = 0.5$, adopting the ME regularization. With this test we sampled the parameter space to optimize the luminosity ratio of the components of the system. The number of processors forming the VM has been increased step by step to find how the speedup increases as a function of the number of CPUs. The results of the test are reported in Figure 1b in terms of *equivalent processors*, the reference processor being the CPU SuperSparc+ 50 MHz with SPECint92=76.9 and SPECfp92=80.1.

We plot in Figure 1b also the speedup expected for an ideal cluster and a perfectly balanced application, which assumes that all the CPUs begin and end their jobs simultaneously. This case can be realized only on a dedicated cluster. We see that there is no significant degradation of the overall performance with only a slight tendency toward a saturation when the number of CPUs exceeds ~ 10. This is also a consequence of the fact that the processors we used are rather similar (the maximum difference in their scaled powers is less than 20-25%).

5. Conclusion

We have presented a very general and simple procedure which exploits a cluster of inhomogeneous computers to significantly speed up the modelling of a sequence of light curves. The same procedure, with only minor modifications, can be applied also to the modelling of spectroscopic or polarimetric data (see, e.g., Milone 1993, Vincent et al. 1993).

Our procedure can use a simple Workstation cluster, not necessarily dedicated, or medium-large parallel systems having a message passing software as PVM. In the near future we plan to produce also an MPI version with an X-based user interface to monitor processing and make it available as a public domain software for light curve or Doppler Imaging analysis.

References

Cameron A. C., 1992, in P. B. Byrne, D. J. Mullan (Eds.) LNP 397, (Berlin: Springer-Verlag), 33

Lanza A. F., et al., 1997, A&A, submitted

Milone E. F. (Ed.), 1993, Light curve modelling of eclipsing binary stars, (Berlin: Springer-Verlag)

Piskunov N. E., Tuominen I., Vilhu O., 1990, A&A, 230, 363

Rodonò M., Lanza A. F., Catalano S., 1995, A&A, 301, 75

Strassmeier K. G., Linsky J. L. (Eds.), 1996, IAU Symp. 176, (Dordrecht: Kluwer)

Tuominen I., et al. (Eds.), 1991, IAU Coll. 130, (Berlin: Springer-Verlag)

Vincent A., Piskunov N. E., Tuominen I., 1993, A&A, 278, 523

Vogt S. S., Penrod N. E., Hatzes A. P., 1987, ApJ, 321, 496

Wilson R. E., 1993, PASP, 106, 921

Astronomical Data Analysis Software and Systems VII
ASP Conference Series, Vol. 145, 1998
R. Albrecht, R. N. Hook and H. A. Bushouse, eds.

A Posteriori Guidance for Astronomical Images

P. Melon, M. Guillaume and Ph. Refregier

Laboratoire Signal et Image, Ecole Nationale de Physique de Marseille, Domaine Universitaire de Saint Jerome, 13397 Marseille Cedex 20, France

A. Llebaria, R. Cautain and L. Leporati

Laboratoire d'Astronomie Spatiale du CNRS. Marseille. BP 8, Traverse du Siphon, 133376 Marseille Cedex 12, France

Abstract. Astronomical ultraviolet images were obtained from a balloon-borne telescope equipped with a photon-counting detector. The final images are built from the list of temporal photoevent addresses produced by the detector. We take advantage of the high temporal resolution available, to significantly reduce, during the image reconstruction process, the blur induced by the residual movements of the guidance system.

1. Context of image acquisition

For many years the FOCA experiment (Milliard et al. 1991) has been the main instrument for a balloon-borne middle-UV imaging programme in astronomy. The present version of the experiment consists of an UV camera of 40 cm diameter equipped with a 2D photon-counting device. The image field of view is about 1° of diameter with an angular resolution of 12 arcsec.

The camera is flown in a stratospheric gondola actively stabilized. The guidance error signal is taken from an independent star tracker centered on a guide star (using the visible part of its spectrum). An active system compensates for most of the astronomical field rotation around the central guide star. The residual movements are in the range $2-10$ arcsec. They consist mainly of a slow X-Y drift and of small, random and faster oscillations in X-Y and in angular rotation. Significant power is still present at frequencies above 1 Hz.

In the focal plane the photo-counting device gives the 2D position of each detected photo-event. Position coordinates are digitized and they define a 1024×1024 array of pixels each 3.4 arcsec wide.

Telemetry sends to the ground equipment a sequence of position coordinates in the order of time arrival for each event. They are grouped in sets of photoevents detected in the same short interval of time; each group is called a "frame". An interval of 20 ms per frame is then standard. The full sequence of frames consists of the ordered (in time) set of groups of events. When processed, successive frames are grouped in blocks of the same size (between 1 and 10 frames per block). The block size must be chosen to be small enough to sample fast movements and big enough to include enough events, between 25 and 100 typically. This point is discussed later in this paper. The typical flow of data is about of 2000 events per sec.

The image is built from individual events in a 1024×1024 array by increasing by one the pixel value whose address is defined in the event. If the final image is built without any correction of residual movements of the guidance system a significative blur will be added.

2. Algorithm

Thereafter we consider that an elementary, i.e., a high temporal resolution image, is built from all events included in a block of frames by a simple pile-up of photoevents at their original X-Y address.

We consider the sequence of such elementary images at high temporal resolution. We make the following assumptions about these images:
1) The noise present in the images is photon shot noise and is described by Poisson Law.
2) The scene does not change during the acquisition.
3) The movement is negligible during the block duration.
4) The images are periodic.
5) Rotations are negligible, so we consider only the translations between the different images.
6) We do not have any *a priori* knowledge either about the translations or the imaged astronomical field.
Let $s_p(i)$ denote the intensity of the pth observed image s_p at pixel i where $i \in [1, N]$, and N is the number of pixels of the image (we use one-dimensional notations for simplicity). Let $r(i)$ denote the intensity of the imaged astronomical field at the same pixel i. It is the perfect, non-noisy and non-blurred image of the observed portion of the sky. The observed image s_p is translated by j_p pixels from the reference r. Without any *a priori* knowledge of the movement, we determine the likelihood of the hypothesis that the translation between the images s_p and r is j_p pixels. In a previous article (Guillaume et al. 1997), the value of $r(i)$ is considered known, and it has been proved that the optimal estimation of j_p is obtained by maximizing the intercorrelation between the observed image and the logarithm of the reference image $r(i)$:

$$j_p^{ML} = \arg\max_{j_p} \left[\sum_{i=1}^{N} s_p(i + j_p)\ ln[r(i)] \right] \quad (1)$$

In the present paper, the true value of the reference image $r(i)$ is considered to be not available. In this case, we consider the maximum likelihood estimation of the reference $r_{ML}(i) = \sum_{p=1}^{P} s_p(i + j_p)$, and then the set of relative translations $\bar{J} = (j_1, j_2, ...j_P)$ of the P observed images s_p. Assuming the statistical independence between the images, it can be shown that the maximum likelihood estimate $\bar{J}^{ML}$ of $\bar{J}$ is found as:

$$\bar{J}^{ML} = \arg\min_{\bar{J}} \left[-\sum_{j} r_{ML}(j) \ln r_{ML}(j) \right] = \arg\min_{\bar{J}} S(\bar{J}) \quad (2)$$

The determination of $\bar{J}^{ML}$ is performed by an iterative algorithm which is very close to a steepest descent procedure:

- Choose the size m of the search window and the number of iterations
- For each iteration k:
  - For each image p of the sequence:
    - Calculate the variation $\Delta S(\bar{J}^k)$ of $S(\bar{J}^k)$ for all the $m \times m$ neighbours of j_p^k
    - Choose the value j_p^{k+1} for which $\Delta S(\bar{J}^k)$ is negative and minimum

We take advantage of the low photon level (only few pixels have non-zero value) and we develop a fast algorithm inspired by the Nieto-Llebaria algorithm (Nieto, Llebaria & di Serego, 1987) by calculating $\Delta S(\bar{J}^k)$ on tables of photo-events addresses rather than on images. For example, for 2000 images with a hundred photons per image, the computation time can be reduced from 40 hours for the direct calculation to 5 minutes for the fast algorithm on a Sun Sparc station 10.

We note that:
1) In practice, for all the performed simulations, the convergence for $S(\bar{J})$ has always been attained with less than 10 iterations (i.e., 10 presentations of all the images).
2) The size m of the search window can be adapted to the amplitude of the translations in order to avoid local minima of $S(\bar{J})$.

3. Results

This algorithm has been tried on simulated fields as well as real images from balloon flights. We present here the results produced in an image centered on the M3 globular cluster. The experimental event series includes 24000 frames 1024×1024 pixels size and a mean of 25 photons/frame.

In the algorithm, the only free parameter is the number of frames per block. Previous simulations (Melon 1997) showed (see Table 1) that for each block the probability of *exact* recentering drops as the number of photons per block decreases. As rule of thumb: the threshold is between 50 and 100 photons per block depending on the spatial arrangement and flux distribution of stars, and background intensity.

Nb of blocks	Photons/block	Frames/block	Percentage
5000	110	4	75
10000	45	2	45
20000	22	1	20

Table 1. Probability of exact correction of a block

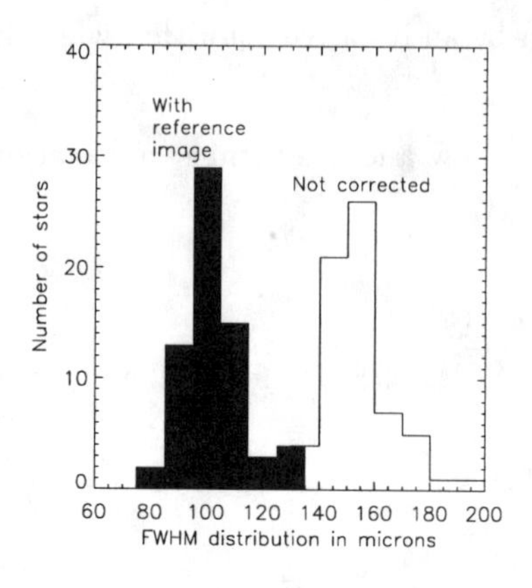

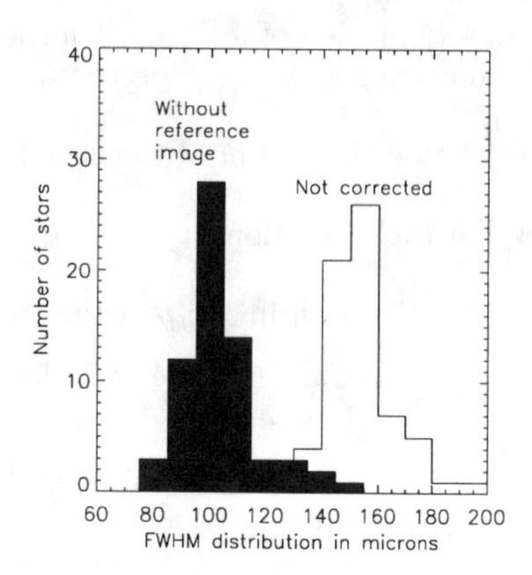

Figure 1. Left: Histograms of FWHM for case 1) & 2). Right: Histograms of FWHM for case 1) & 3)

In our case a block of 2 frames, that is, a 40 ms interval was enough to assure a good bandwidth to sample the movement and to get a correct reconstruction. To characterize the effects of the algorithm we measured for a limited set of stars ($N \simeq 50$) their FWHM (full wide to half maximum). We build the FWHM distribution for: 1) Stars from an uncorrected image, 2) Stars from a corrected image using a reference (see algorithm 1), 3) Stars from a corrected image without reference (see algorithm 2).

Figure 1 shows the histograms of these distributions. On the left side we compare 1) with 2) and in the right side we compare 1) with 3). As can be seen, a decrease of 20 μm over 120 μm is clearly visible for both corrected images. From that and from other work, not shown here, we can conclude that the new algorithm improves resolution as efficiently as the algorithm with reference image. The new algorithm does not show any ringing side-effect and has the major advantage of not needing a reference image

The lack of ringing side effects and the "self-sufficient" use of event series with any loss in accuracy are encouraging results for a large and deep study of this new method.

Acknowledgments. We are grateful to M. Laget and B. Milliard for their valuable help and fruitful discussion of this paper.

References

Milliard, B., Donas, J., & Laget, M., 1991, Adv. Space Res.

J.L. Nieto, A. Llebaria, & S. di Serego Aligheri, 1987, Astron. and Astroph. 178, 301

M. Guillaume, Th. Amouroux, Ph. Refregier, B. Milliard & A. Llebaria, 1997, Opt. Lett. 22, 322

P.Melon, 1997, Rapport de DEA, Ecole Nationale de Physique, Marseille

Astronomical Data Analysis Software and Systems VII
ASP Conference Series, Vol. 145, 1998
R. Albrecht, R. N. Hook and H. A. Bushouse, eds.

Recognition of Anomalous Events

S. Monai[1,2] and F.Pasian[2]

[1]*Osservatorio Astronomico di Brera Merate, Italy*

[2]*Osservatorio Astronomico di Trieste, Italy*

Abstract. We present an original algorithm for the recognition of anomalous events in a CCD frame (i.e., cosmic hits, bad pixels etc.). The algorithm is able to distinguish between these events and real features by comparing the standard deviation and the kurtosis of a synthetic PSF, whose width is given by the user, and the corresponding values in a running window, whose size is chosen by the user, in the actual frame. The anomalous pixels are therefore filtered with a neighbourhood average.

Raw images, as read from a CCD, are often affected by some events which are revealed by anomalous pixel values, i.e., single saturated pixels or pixels at zero intensity (also named hot or bad pixels). Often cosmic rays cross the CCD during the exposure and are revealed as strange features normally affecting a few contiguous pixels. It is not always simple to disentangle automatically these features. Among the approaches to this recognition problem we have: - IRAF tasks cosmicrays; - MIDAS task filter/cosmic; - a locally adaptive modified Haar transform approach due to Richter (1991) and Richter et al. (1991); - a median filter approach due to Meurs et al. (1991) and van Moorsel (1991); - a method using filters based on ordered statistics (Pasian 1991); - an image restoration, followed by flagging badly restored "objects" as bad pixels, and re-restoring (Weir 1991), this last allows the addition of a point spread function (PSF) fitting, coupled with the flagging of poorly restored "objects"; - a three step procedure is used by Yee et al. (1996).
In this work we make only one reasonable assumption about this kind of events: **The anomalous events are smaller than the actual PSF of the instrument.** This is not always true, but if this is the case we think there are no classic criteria, affordable by a classical algorithm, to distinguish these events (different approaches are those of Priebe et al. (1993) and Murtagh (1994), both use object recognition techniques).
When the previous assumption is true (by far the great majority of cases) we have the possibility to establish some precise criteria to identify, and filter, the anomalous events, without affecting the real features (which are normally degraded by a normal filtering technique). First of all we must know the PSF width in pixel units, and this is not very difficult both for imaging and spectroscopical observations. A Gaussian fit across a star or an emission line is sufficient to determine the actual PSF width (the seeing normally prevents us to see the true diffraction pattern). With this datum we are able to compute two important parameters: the standard deviation and the kurtosis in a window centered on a synthetic PSF (the window width is chosen by the user). The first parame-

ter allows recognition of events having an anomalous intensity variation across the specified window, while the second recognizes the event's shape which must have a profile narrower than the PSF (i.e., a greater kurtosis). The standard deviation of the real image derives from the signal variations across the feature plus the image noise; the deviation factor from the corresponding synthetic PSF must be chosen appropriately. This parameter must therefore be tuned: after several experiments on different images and spectra taken with different instruments, we suggest a deviation factor of at least 5. With these criteria we are confident not to identify as anomalous a real feature. When in a window the standard deviation is more than 5 times greater and the kurtosis is larger than the corresponding values of the PSF, the central pixel is identified as anomalous, its coordinates are stored and we can substitute taking twice the average value of its neighbours.
We want to stress that only the contemporaneous satisfaction of both requirements allows us to identify an anomalous event; in fact a large kurtosis alone doesn't mean a narrower profile since the condition is necessary but not sufficient. The double averaging step is necessary because some events could be spread over more than one pixel and a simple filtering could not remove the anomalies, with a second iteration we are averaging the already filtered pixels and we are quite sure to eliminate completely the anomaly. If the anomalous event is spread along more than the window's width (but only in one direction) the algorithm still recognizes it but the neighbour average could not remove it completely. In this case the entire procedure should be repeated once more.

Acknowledgments. We are grateful to Dr. M. Nonino for having offered the image of the Seyfert galaxy. This work has been carried out thanks to a grant of the "Oss. Astron. di Brera Merate" for the realisation of the Data Reduction Software for the DOLORES instrument, in the framework of the TNG project.

References

Meurs E.J.A., Bonifacio V.H., & Lima N.M. 1991, in 3rd ESO/ST-ECF Data Analysis Workshop, 45

Pasian F. 1991, in 3rd ESO/ST-ECF Data Analysis Workshop, 57

Priebe A., Liebscher E., Lorenz H., & Richter G.M. 1993, in ASP Conf. Ser., Vol. 52, Astronomical Data Analysis Software and Systems II, ed. R. J. Hanisch, R. J. V. Brissenden & Jeannette Barnes (San Francisco: ASP), 442

Richter G.M. 1991, in 3rd ESO/ST-ECF Data Analysis Workshop, 37

Richter G.M., Boehm P., Lorenz A., & Capaccioli M. 1991, Astron. Nach. 312,345

van Moorsel G. 1991, ST-ECF Tech. Rep

Weir N. 1991, in 3rd ESO/ST-ECF Data Analysis Workshop, 115

Yee H.K.C., Ellingson E., & Carlberg R.G. 1996, Ap. J. Supp. Ser. 102,269

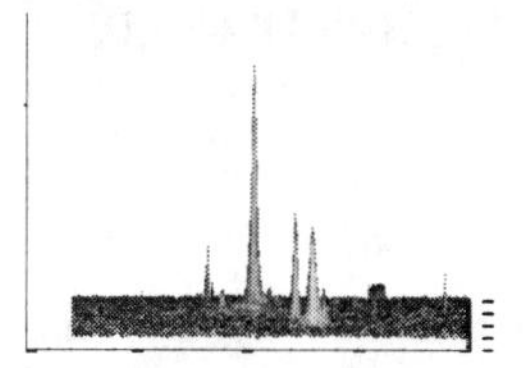

Figure 1. Assonometric projection of a raw image

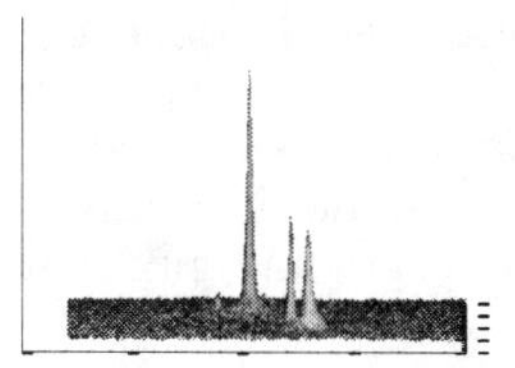

Figure 2. Assonometric projection of the cleaned image

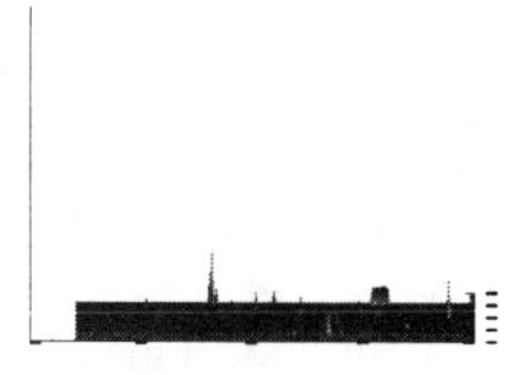

Figure 3. Assonometric projection of the residuals

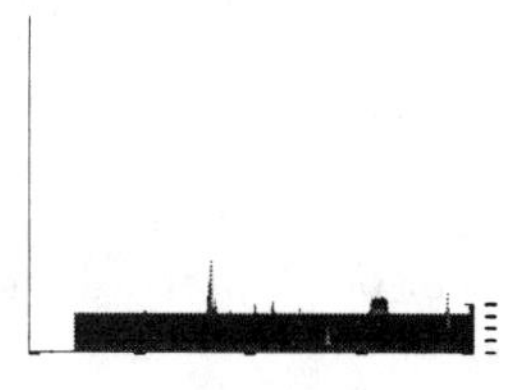

Figure 4. The cross-cut of a cosmic hit filtering steps

Astronomical Data Analysis Software and Systems VII
ASP Conference Series, Vol. 145, 1998
R. Albrecht, R. N. Hook and H. A. Bushouse, eds.

Slitless Multiobject Spectroscopy with FOSC Type Instruments

V. F. Polcaro and R. Viotti

Istituto di Astrofisica Spaziale, CNR, Frascati, Italy

Abstract. We present a simple, effective technique that allows the contemporary spectroscopy of all the objects, up to a limiting magnitude, contained in a $\sim$ 100 arcmin^2 field. This technique makes possible the immediate and unambiguous identification of peculiar objects included in the field. A data analysis procedure that produces the final results in a very short time has also been developed. As a test of this technique, we present the results* obtained for the peculiar young open cluster Berkeley 87. We show that this technique should also be quite useful in emission line and WR star surveys, and illustrate the software procedures developed to make the wavelength and flux calibration of these spectra.

(*) Based on data collected at the Loiano Observatory

Modern hard X-ray satellites allow the real time positioning of transient high energy sources with a precision of $\sim$10′–30′. In some cases, as for the gamma-ray bursts, the fast identification of the optical counterpart is a crucial item to unveil the physical nature of the emitting object. On the other hand, no other way to make their identification in the optical range has been so far found, but making the spectroscopy of all the objects included in the error box, and looking at the objects which show a peculiar (e.g emission line) optical spectrum. Of course, within a sky area of order of some 10^2 arcmin^2, there is an enormous number of objects and it is unlikely that we can obtain all their spectra in a reasonable time period. This fact usually overcomplicates the identification of the optical counterpart of the high energy cosmic sources, at the risk of substantially diminishing the scientific potential of the imaging hard X-ray and gamma-ray telescopes.

For this reason, we have developed a technique, that allows the simultaneous spectroscopy of all the objects included in a square field of view with $\sim$10′ side, using the widely used FOSC-type optical spectrometer and camera focal-plane instruments. The technique is based on the use of a slitless low dispersion grism coupled with a broad-band filter, which perform an "objective prism like" spectrogram of all the objects present in the field, without serious overlap of adjacent spectra. For instance, the use of the very common grisms with $\sim$400-800 nm bandpass and of the Johnson R filter will make an imaging spectroscopy of $\sim$ 200nm centered on Hα, and emission line and peculiar spectrum objects are easily identified.

We tuned this technique on the Bologna Faint Object Spectrometer and Camera -BFOSC- instrument (Merighi et al. 1994), mounted at the Cassegrain focus of the Bologna Astronomical Observatory "G.B. Cassini" 1.52 m telescope, sited near Loiano (Italy) at about 800 m altitude on the Appennine Mountains.

In order to test this technique, we used the peculiar open cluster Berkeley 87, an intriguing object, that we have studied for more than 10 years (e.g., Norci et al. 1988; Polcaro et al. 1989; Polcaro et al. 1991; Manchanda et al. 1996). Actually, most members are young, heavily reddened OB stars, but a few are much more evolved objects, such as the WO star Sand 5 and the M3.5I variable BC Cyg making the evolutionary status of the cluster extremely uncertain. The cluster member n.15 in the Turner and Forbes (1982; hereafter TF82) list is an emission line star (V=11.8) also known as V 439 Cyg and MWC 1015 (Merril and Burwell, 1949). This star dramatically changed its spectrum from late to early type in a few decades; furthermore, some absorption lines that were still present in 1986 and 1987 completely disappeared in 1988. The star is characterized by a strong IR excess and a peculiar position in the HR diagram, suggesting that it should be considered as an intermediate RGS/LBV star (e.g., Polcaro and Norci, 1997 and references therein). Three other cluster members (no. 3, no. 9 and no. 38 in the TF82 list) are emission line stars.

On June 24, 1997 (23:11:55 UT), we took a 30 s R filter exposure of the central (9 x 9 $arcmin^2$) region of Berkeley 87, shortly followed (23:18:34 UT) by a slitless 300 s image through the R filter and the grism no. 4 (Fig 1). The Hα emission of stars no. 3, 9, 15 and 38 and the WR spectrum of Sand 5 where immediately visible after a short tuning of the contrast. The direct overlap of these two images using the IRAF task *imarith* on the recorded FITS files allowed a prompt and unambiguous correlation between the objects and the spectra.

The whole procedure, including the overlap, took less than half an hour.

Notice also that our short exposure allowed the immediate identification of the Hα emission of the V=14 star no. 38 and of the WR spectrum Sand 5 (V=13.8) without saturating the Hα emission of the V=7.81 star no. 3. In absence of such a bright object (or if we do not care if we saturate its spectrum) a longer exposure should allow the identification of emissions also in much fainter objects.

A careful analysis of the slitless spectral image demonstrated that this technique has a field of application much wider than the fast identification of peculiar objects described above.

Actually, after debiassing and flat-fielding (using standard IRAF procedures) by means of a 10 s flat-field taken through the R filter, we obtained a pretty clean image from which 25 good quality, not overlapped spectra of stars up to V=16 were extracted. The comparison of this multiobject spectroscopy with long slit spectra of a few objects in the same field taken on the same night through the same grism, which were wavelength calibrated by means of a He-Ar comparison lamp and flux calibrated using the standard star Kopf 27, shows that the pixel-wavelength relationship was not affected by the lack of the slit and the presence of the photometric filter, but just shifted in pixels, due to the different declination of the stars in the field, and cut between 550 nm and 780 nm due to the combined effect of the filter and grism transmission curves. Thus, the wavelength calibration is easily obtained, with a precision of $\pm$0.5 nm from the position of the stellar image, the 550 nm cut-off and the red telluric absorptions.

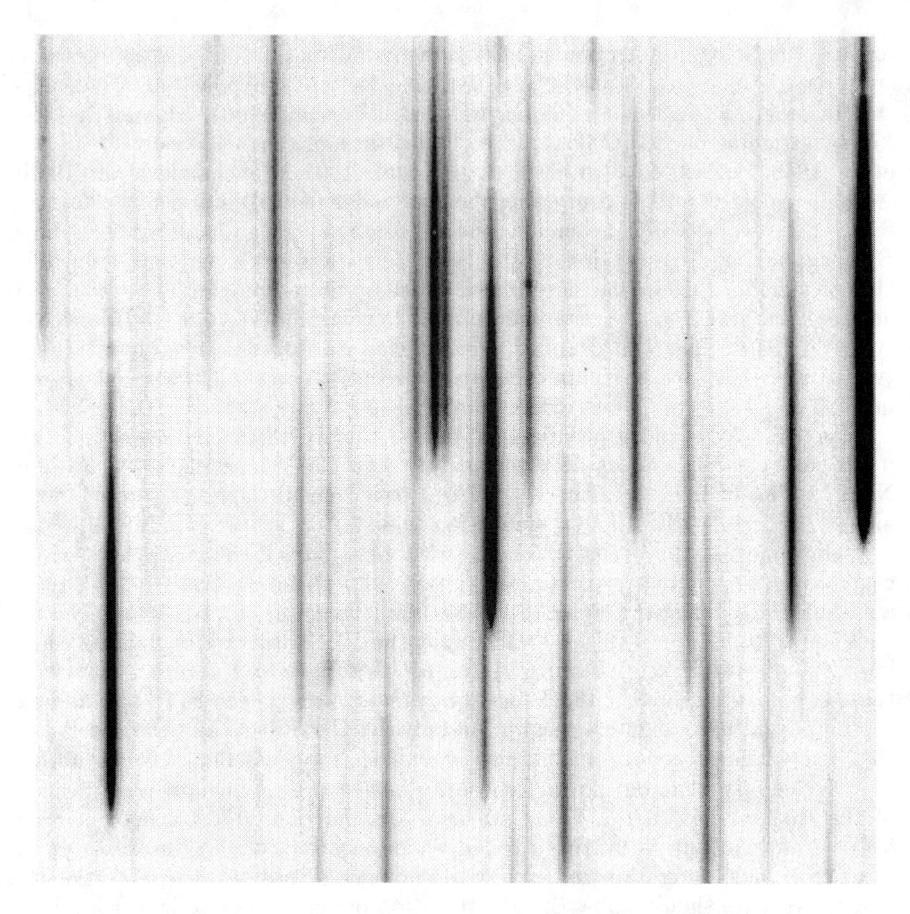

Figure 1. Figure 1: Slitless red spectral image of the open cluster Berkeley 87. The field is 9 x 9 arcmin^2. North is at the top and East to the left. The Hα emission line in star no.38 (top left) and n. 15 (top centre) and the WR spectrum of star n.29 (Sand 5, bottom centre) is clearly visible.

A further benefit was obtained from the slitless technique: the night sky lines (and mainly the antropic lines that are quite strong at the telescope site, due to the close proximity of the city of Bologna) as well as the numerous nebular lines (mainly, but not only, Hα) due to the intracluster matter (see e.g., Polcaro et al. 1991) result in a limited increase of the background level, so that spectra was much cleaner than those obtained in previous runs by means of slit spectroscopy. Furthermore, the absence of the slit eliminates the light loss due to the poor seeing (3.5 arcsec in the night of our experiment) so that the signal-to-noise ratio was much higher than that of "classical" spectra. In this way we were able to unveil narrow Hα emission cores in the cluster members no. 26, no. 31 and no. 32. Furthermore, once the wavelength calibration of each individual spectrum was achieved, we recognized (from the comparison of the debiassed and flat-fielded slitless spectra in reduced CCD counts and of the long slit flux calibrated spectra of the same objects) that, due to the linearity of the CCD detector used in the BFOSC instrument, the flux calibration curve was the same for the whole image. Thus the flux calibration of the slitless spectra was also straightforward.

Our test demonstrates that the use of the slitless spectrometry technique with the modern FOSC type instruments does not only allow the immediate identification of spectroscopically peculiar objects in a field of the same order of magnitude of that of a "typical" hard X-ray instrument. It also shows how this technique can be used to obtain, without the need of either new hardware or of large telescopes or high quality skies, much valuable scientific information: for instance, we identified in a single short test-shot three previously unknown emission line stars.

We now plan to apply the method to a search for emission line and WR stars in young clusters.

We thank the night assistants of the Loiano observatory, Mr. G. Tessicini and Mr. E. Delogu, for their help during our experiment

References

Manchanda R. K., Polcaro V.F., Norci L. et al., 1996, A&A, 305, 457

Merighi R., Mignoli M., Ciattaglia C., et al. 1994, *"BFOSC User's Manual"*, RT 09-1994-05, Bologna Astronomical Observatory.

Merrill, P.M. & Burwell, C.G., 1949 , Astrophys. J., 110, 387

Norci, L., Giovannelli, F., Polcaro, V. F., & Rossi, C. 1988, in "Frontier Objects in Astrophysics and Particle Physics", F. Giovannelli and P. Mannocchi (eds.), Italian Physical Society, Bologna, Italy, 7

Polcaro, V.F., Giovannelli, F., Norci, L., & Rossi, C., 1989, Acta Astronomica, Vol 389, No. 4.

Polcaro V.F., Rossi C., Persi, P. et al. 1991, Mem.S.A.It. 62, 933

Polcaro V.F. & Norci L., 1997, Proc. of the UMIST/CCP7 Workshop on *Dust and Molecules in Evolved Stars*, Manchester, March 24-27

Turner, D.G. & Forbes, D., 1982, Publ. Astr. Soc. Pac., 94, 789 (TF82).

Astronomical Data Analysis Software and Systems VII
ASP Conference Series, Vol. 145, 1998
R. Albrecht, R. N. Hook and H. A. Bushouse, eds.

Substepping and its Application to HST Imaging

Nailong Wu and John Caldwell

Dept. of Physics and Astronomy, York University, 4700 Keele St., North York, Ontario, M3J 1P3, Canada

Abstract. The substepping technique is used for the Hubble Space Telescope (HST) imaging to cope with the problem of undersampling and to improve resolution. In this paper, this technique is first introduced in the language of signal/image processing. Then its application to HST: FOS ACQ imaging and WFPC2 subpixel dithering, is described. Its possible use for NICMOS is also discussed.

1. Introduction

The substepping technique is used to ameliorate the problem of undersampling in data acquisition for an imaging system. For the HST, undersampling means that the size of the pixel in a camera is larger than the critical value determined by the optics of the telescope, and aliasing takes place.

Obviously, the problem of undersampling could be resolved if the pixel size could be reduced. However, in many circumstances, the pixel size is fixed. Then, substepping is the only solution to the problem. In essence, substepping means that data are acquired in steps smaller than the pixel size, and then processed to achieve resolution comparable to the step size.

2. Substepping in Data Acquisition and Reconstruction

In this section one-dimensional (1-D) notation and diagrams are used for clarity. Results can be easily extended to the two-dimensional (2-D) case by changing dimensional subscripts and operations to 2-D.

2.1. Image Formation and Sampling

The brightness distribution, $a_0(x)$, in the field of view is convolved with the point spread function (PSF) of the optics of the telescope to form a continuous image, $a(x)$, which has lower resolution than $a_0(x)$ due to the convolution. Resolution can by improved by deconvolution of $a(x)$ with respect to the PSF to restore $a_0(x)$.

Sampling is an operation to integrate $a(x)$ within a pixel, and to assign the result to this pixel (Figure 1A). Each pixel has its nominal position at the center of the integral interval.

2.2. Substepping

To avoid aliasing, the sampling interval Δx must be equal to (critical sampling) or smaller than (oversampling) its critical value Δx_c, which is determined by the highest frequency component in $a(x)$. If $\Delta x > \Delta x_c$ (undersampling), aliasing will take place.

In order to overcome the problem of undersampling (Figure 1A), we may reduce the pixel size by a factor of N, such that $1/N\Delta x \leq \Delta x_c$ (proper sampling, including critical sampling and oversampling; Figure 1D for $N = 2$). We call this *normal sampling* because both the integral and sampling intervals are equal to the pixel size.

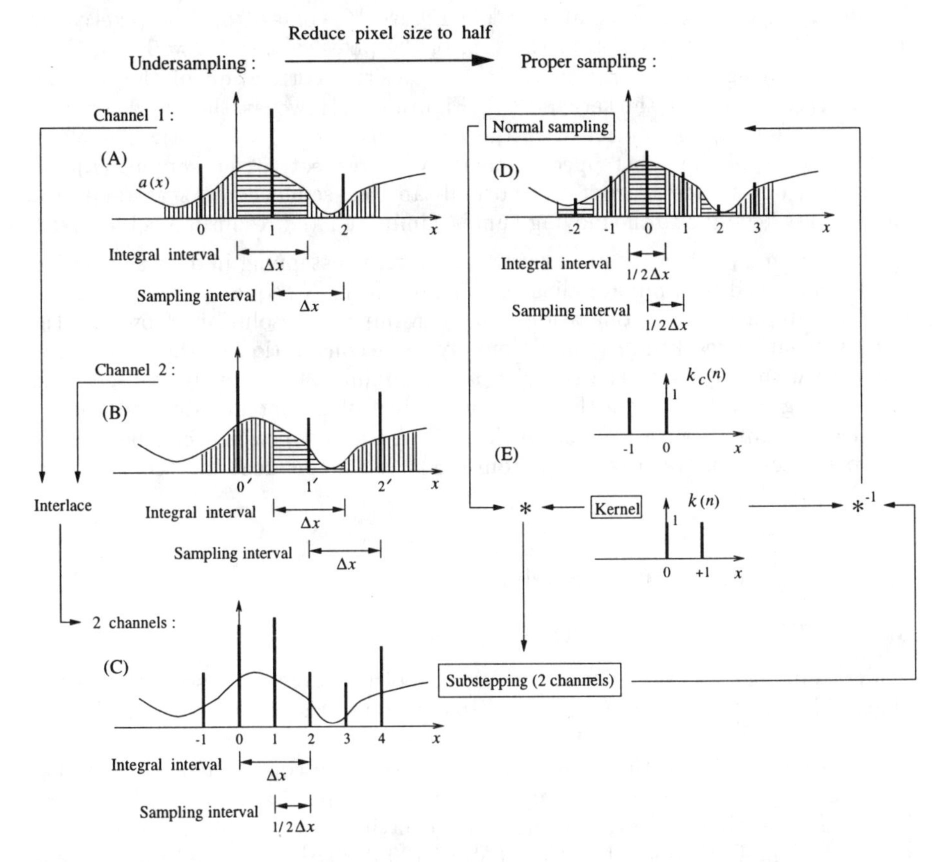

Figure 1. Substepping (N=2) and normal sampling. The vertical bar at a pixel's center represents the integral value. $*$ and $*^{-1}$ denote convolution and deconvolution, respectively.

In the case where reducing the pixel size is impossible, we can use N integral-sampling devices (channels) to acquire data. The resulting N sequences shifted successively by $1/N\Delta x$ (Figure 1A,B) are combined in a manner of interlacing (C), so that the sampling interval becomes $1/N\Delta x$. We call this *subsampling* or *substepping* because the sampling interval is equal to a subpixel size while the integral interval is still equal to the pixel size. If N is sufficiently large such that $1/N\Delta x \leq \Delta x_c$, aliasing will be eliminated, and consequently resolution will be improved.

2.3. Reconstruction Using the Substepped Sequence

Let us compare substepping (Figure 1C) with normal sampling (D), both with the sampling interval $1/N\Delta x$. For the former, the integral interval is larger, and the smoothing effect of integration is stronger. Therefore, its resolution is lower than that of normal sampling.

The substepped sequence is a moving-sum of the normal-sampled sequence. For instance, the pixel value at $x = 1$ in Figure 1C comes from the pixel value at $x = 1$ in A, which is equal to the sum of the pixel values at $x = 0, 1$ in D.

The moving-sum operation is, in fact, a cross-correlation of the normal-sampled sequence with the kernel $k_c(n)$ (Figure 1E). However, this cross-correlation is equivalent to a convolution with the kernel $k(n)$, $k(n) = k_c(-n)$. Therefore, deconvolution of the substepped sequence with respect to the kernel $k(n)$ can be carried out to reconstruct the normal-sampled sequence. As a result, the smoothing effect due to the moving-sum is eliminated and resolution is improved.

In summary, when the pixel size is too large, substepping in data acquisition can be employed to eliminate aliasing. The resulting N sequences from N channels are interlaced. This operation alone can improve resolution. However, the improvement in resolution is made mostly by deconvolution of the substepped sequence with respect to the kernel $k(n)$ to eliminate the moving-sum effect in substepping, i.e., to weaken the smoothing effect of the integration operation.

Deconvolution with respect to the PSF (Sect. 2.1.), which can be used to improve resolution (restore $a_0(x)$ from $a(x)$), is independent of reconstruction of substepped data.

3. Application to HST Imaging

3.1. WFPC2 Subpixel Dithering

Undersampling with WFPC2 occurs because of the large pixel size of the CCD chips. The substepping technique used to overcome this problem is named "subpixel dithering" or simply "dithering" (Figure 2a).

During an observation, the pointing of the telescope is changed so that successive images are shifted along each axis by subpixel amounts. Then, these images are combined to obtain a single image having a smaller pixel size on a finer grid, using the POCS-based method (Adorf 1995), "Drizzling" method (Hook & Fruchter 1997), and deconvolution-based methods like **acoadd**, **crcoad**, **mem**, and **lucy** in IRAF/STSDAS.

3.2. FOS ACQ Imaging

The undersampling problem with FOS arises from the large size of the diodes used for data acquisition. In substepping (Figure 2b), the step of the diode's motion is one quarter of its width in the X-direction, and one sixteenth of its height in the Y-direction.

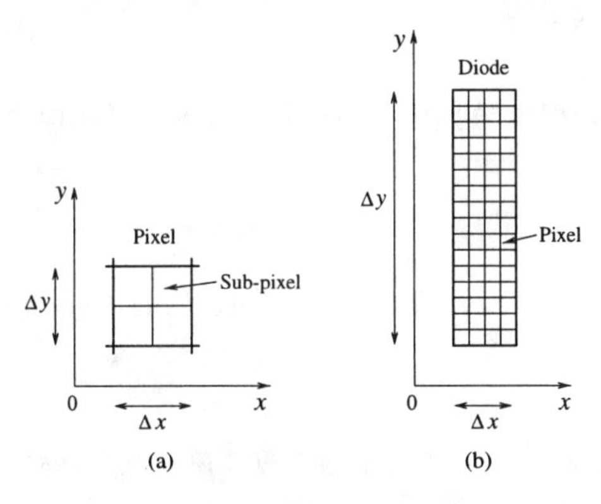

Figure 2. Substepping for HST imaging. (a) WFPC2 dithering ($N_x = N_y = 2$). (b) FOS ACQ imaging ($N_x = 4, N_y = 16$).

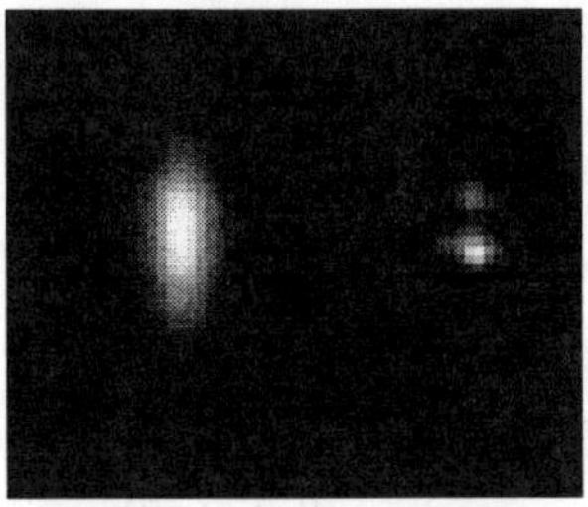

Figure 3. An FOS ACQ image (left) and its reconstruction by **mem** (right).

The diode and the 64 (= 4 × 16) small areas on it would normally be called the pixel and subpixels, respectively, according to standard substepping terminology. However, in this particular case the small areas have commonly, but inaccurately, become known as pixels.

The deconvolution with respect to $k(n_x, n_y)$ is called *reconstruction*. The deconvolution task **mem** or **lucy**, or the direct inversion task **tarestore**, in IRAF/STSDAS can be used for this purpose. **tarestore** results in high level sidelobes (rings) and noise in reconstructed images; **mem** and **lucy** give much better results (Wu & Caldwell 1997).

The smoothing effect due to the large diode size badly reduces resolution in FOS ACQ images. Reconstruction can remove this effect and dramatically improve resolution. Figure 3 shows an image, before and after reconstruction, of a star behind the bar in an FOS barred aperture.

3.3. NICMOS Imaging

In NICMOS imaging, PSFs are critically sampled at $\lambda_c = 1.0$ and 1.75 μm for Cameras 1 and 2, respectively. When a working wavelength is shorter than λ_c, the problem of undersampling occurs. The situation here is similar to WFPC2 in principle. Therefore, the substepping technique can be used.

References

Adorf, H.-M. 1995, in Astronomical Data Analysis Software and Systems IV, A.S.P. Conf. Ser., Vol. 77, eds. R. A. Shaw, H. E. Payne & J. J. E. Hayes (San Francisco: ASP), 456

Hook, R. N. & Fruchter, A. S. 1997, in Astronomical Data Analysis Software and Systems VI, A.S.P. Conf. Ser., Vol. 125, eds. G. Hunt & H. E. Payne (San Francisco: ASP), 147

Wu, N. & Caldwell, J. 1997, in Proceedings of 1997 HST Calibration Workshop, Space Telescope Science Institute, ed. S. Casertano, et al. (in press)

Astronomical Data Analysis Software and Systems VII
ASP Conference Series, Vol. 145, 1998
R. Albrecht, R. N. Hook and H. A. Bushouse, eds.

The Co-Addition Technique Applied to Images of Galaxy Cores

W. W. Zeilinger

Institut für Astronomie, University of Vienna, Türkenschanzstraße 17, A-1180 Wien, Austria

P. Crane, P. Grosbøl and A. Renzini

European Southern Observatory, Karl-Schwarzschild Straße 2, D-85748 Garching, Germany

Abstract. This is a preliminary report on an application of the co-addition technique based upon the Richardson-Lucy algorithm. The combination of high signal-to-noise ground-based images obtained with the ESO NTT using SUSI with high angular resolution HST FOC f/96 images gives new insights on the properties of galaxy cores and their potential time variability.

1. Introduction

Cores of galaxies are one of the main thrusts of present extragalactic research. Answers to questions like the possible presence of massive central dark objects (black holes), central gas/dust disks, nuclear subcomponents and so on are essential to our understanding of the nature and evolution of galactic nuclei and their interrelation with the host galaxy. Many ellipticals and bulges of spiral galaxies show central, unresolved sources which are particularly bright in the UV and significantly less conspicuous or even completely absent in the corresponding visual images. The double nucleus of M31 is such a case: King, Stanford & Crane (1995) found that the brightest peak in the UV is not coincident with the brightest peak in the visual. Attesting to the value of UV observations, we note that *all* the galaxies observed by Crane et al. (1993) which have blue nuclei are also radio sources.

An important discovery was reported by Renzini et al. (1995) The core of the elliptical NGC 4552 contains a *variable* UV source. The galaxy was imaged in the UV using the HST FOC in 1991 and 1993. Over this period the central spike, which is only visible in the UV, brightened by a factor of at least 7. Subsequent observations have revealed broad emission features associated with the flaring region. The UV–bright flare is attributed to tidal stripping of a star during a close encounter with a central black hole. The frequency of such an event is estimated to be of the order of once per 10^4 years in giant ellipticals (Rees 1990). Assuming the complete disruption of the star, the flare is expected to be visible for several years and is predicted to be even brighter than the event observed in NGC 4552. Alternative explanations include the ideas of a collision between two stars, gravitational lensing and the hypothesis of a recent accretion of a gas rich dwarf galaxy whose material settled in the nuclear region. There is some circumstantial evidence to support the latter scenario because extended Hα emission is detected in the central 2$''$ region and because there is a nuclear dust lane. However, based upon only one observed case, all the models are unfortunately not well constrained.

2. Observations and Data Analysis

The most complete and homogeneous survey of galaxy cores in the UV has been carried out by Maoz et al. (1995) who obtained a data sample using the pre-COSTAR FOC in the f/96 mode. The analysis of the de-archived images revealed that the majority of the bulge dominated galaxies contain a central UV bright compact source. In those cases where the signal-to-noise ratio was adequate, the central region was further analyzed and the point source nature established. Nuclear subcomponents such as central dust lanes or ring-like structures were identified in some cases.

If the interpretation of Rees (1990) holds true, the brightness variations caused by the flare event in the galaxy nucleus are expected to be observable in a time scale of less then 2 months. A sample of suitable candidate galaxies is now being monitored in the U band using SUSI at the ESO NTT in order to search for UV variability of the central sources similar to that detected in NGC 4552.

The archival FOC f/96 F220W images from the Maoz et al. (1995) survey serve as reference for deconvolving the core regions of the SUSI images using the Richardson-Lucy technique (Hook & Lucy 1994) as also described by Zeilinger (1994). It has been demonstrated that the co-adding technique not only consolidates the signal of images with different resolutions but also conserves the resolution of the sharpest one. One may therefore take full advantage of the high resolution FOC images by co-adding them with (high signal-to-noise) ground-based data.

Nevertheless the ground-based observing programme carried out at the NTT is clearly seeing dependent: In order to derive meaningful core fluxes in combination with the high-resolution FOC images, the seeing conditions should not be much worse then 0.7$''$ FWHM and the variation of the seeing for each galaxy data set should not exceed 0.15$''$.

The ground-based image is usually aligned with the FOC image in order to leave the image with the weaker signal-to-noise ratio as unprocessed as possible. Because of missing field stars as reference points in the FOC image, the alignment is carried out using the galaxy nucleus itself. An accuracy of typically one FOC pixel (0.044$''$ pixel^{-1}) can be achieved with this method. The selection of an appropriate PSF for the co-addition is a crucial part. For the ground-based images well exposed stellar images are almost always available in the galaxy field to determine the shape of the PSF. In the case of the FOC images one is constrained to know the shape of the PSF in the position of the galaxy nucleus on the detector. Therefore stellar images repeating the exact instrument configuration are almost never available. The software tool Tiny Tim (Krist & Hook 1997) is a solution to this problem. An "artificial" PSF for a given observing date and instrument configuration can be calculated. However, one has to be aware that Tiny Tim is only a tool based upon current best fits of aberration values for the various mirror positions and current best estimates of the obscuration positions and sizes which may still not describe the HST PSF *exactly* (see also Krist 1995).

Typically 100 to 300 iterations of the Richardson-Lucy algorithm are then applied, depending on the signal-to-noise ratios of the input images and the resolution (seeing) of the ground-based image. The latter turns out to be the most serious constraint.

3. Results

We have obtained images of 6 galaxies so far of which one has been observed at two epochs. The images were obtained in service mode under less than ideal conditions. We are proceeding with the analysis.

There are other obvious applications of this technique such as the monitoring variable stars in galaxies.

Acknowledgments. WWZ acknowledges the support by the *Jubiläumsfonds der Oesterreichischen Nationalbank* (grant 6323).

References

Crane, P. et al. 1993, AJ, 106, 1371

Hook, R. & Lucy, L. B. 1994, ST-ECF Newsletter, 17, 10

King, I. R. Stanford, S. A. & Crane, P. 1995, AJ, 109, 164

Krist, J. 1995, in ASP Conf. Ser., Vol. 77, Astronomical Data Analysis Software and Systems IV, ed. R. A. Shaw, H. E. Payne & J. J. E. Hayes (San Francisco: ASP), 349

Krist, J. & Hook, R. 1997, The Tiny Tim User's Manual

Maoz, D. et al. 1995, ApJ, 440, 91

Rees, M. 1990, Science, 247, 817

Renzini, A. et al. 1995, Nature, 378, 39

Zeilinger, W. W. 1994, ST-ECF Newsletter, 21, 29

Astronomical Data Analysis Software and Systems VII
ASP Conference Series, Vol. 145, 1998
R. Albrecht, R. N. Hook and H. A. Bushouse, eds.

Robust, Realtime Bandpass Removal for the HI Parkes All Sky Survey Project using AIPS++

D. G. Barnes

School of Physics, The University of Melbourne, Parkville, VIC 3052, Australia; Email: dbarnes@physics.unimelb.edu.au

L. Staveley-Smith, T. Ye and T. Oosterloo

Australia Telescope National Facility, Marsfield, NSW 2122, Australia

Abstract. We present the algorithm and implementation details for the robust, realtime bandpass removal and calibration routine developed for the HI Parkes All Sky Survey project. The software is based on the AIPS++ toolkit.

1. Introduction

In Barnes (1998) we give an overview of the Parkes Multibeam Software, and the motivation for its development, that is, the desire to have robust, realtime processing software for the neutral hydrogen HI Parkes All Sky Survey (HIPASS). In this paper, we provide more details on the bandpass correction algorithm, and its implementation, based on the AIPS++ toolkit.

2. Robust Bandpass Removal

Traditionally, the bandpass that is present in single-dish 21 cm spectra is removed by observing in a "signal-reference" mode. In this mode, an extended integration (typically 3 min) is acquired while the telescope is pointed at the target position, or *signal.* A second extended integration is then acquired while the telescope is pointed towards a nearby position free of line or continuum sources—the *reference.* The bandpass is removed by dividing the signal spectrum by the reference spectrum. If longer on-source integration times are required, this process can be repeated many times, and the quotient spectra averaged.

For the HIPASS Project, where the telescope is actively driven across the sky at a rate of $\sim 1^\circ \text{ min}^{-1}$, the traditional bandpass removal technique is not suitable. Instead, a statistical estimate of the bandpass at the position and time that a particular spectrum—the *target* spectrum—was acquired is made from a set of earlier and later spectra observed by the same beam—these are the *reference* spectra. The bandpass is then removed from the target spectrum by dividing it by the statistical bandpass estimate.

This process can be formalised as follows: a particular integration will be denoted as $I_{b,p,c}$, where the subscripts b and p represent a beam-polarisation combination[1], and c is the cycle number of the integration. $I_{b,p,c}$ is a 1024-channel flux density spectrum.

For a given beam and polarization pair, the target spectrum is $I_{b,p,c_{\mathrm{target}}}$, and the reference spectra, $\mathcal{R}$, are a set of spectra which can be used to estimate the bandpass for the target spectrum. The suitability of a particular spectrum depends on a number of factors. Most importantly, a reference spectrum must be measured on a part of the sky which is independent (with respect to the telescope beam) of the part of the sky measured by the target spectrum. However, the bandpass can be expected to vary with time, so spectra taken nearby in time to the target spectrum are more useful reference spectra than those taken at greater time separations. Furthermore, any movements of the receiver with respect to the telescope dish, including rotations and axial movements, were found to introduce large phase changes in the bandpass. This led to extreme ripple in the baseline of spectra corrected with reference spectra having varying parallactic angles or axial offsets. Thus, valid reference spectra must be from the same right ascension (or Galactic latitude) scan as the target spectrum.

Having selected the reference spectra $\mathcal{R}$, the estimate of the bandpass ($B_{b,p,c_{\mathrm{target}}}$) is given by

$$B_{b,p,c_{\mathrm{target}}} = \mathrm{median}\,(\mathcal{R}) \tag{1}$$

where the median statistic is taken channel by channel. Other, more sophisticated estimators, such as robust linear interpolation, were tested, but proved too slow for realtime reduction since in general they require minimization of non-linear functions in multi-dimensional parameter spaces.

With $B_{b,p,c_{\mathrm{target}}}$ determined, calibration of the spectral data can be done concurrently with bandpass removal. The system temperature for the target spectrum, T_{target}, is provided in the correlator file, whilst that for the bandpass estimate, T_{bandpass}, is obtained in a fashion analogous to the calculation of a single channel in the bandpass estimate. Thus the bandpass-corrected, calibrated target spectrum is

$$S_{b,p,c_{\mathrm{target}}} = \frac{I_{b,p,c_{\mathrm{target}}}}{B_{b,p,c_{\mathrm{target}}}} \times T_{\mathrm{bandpass}} - T_{\mathrm{target}} \tag{2}$$

2.1. Implementation

The BANDPASS CALCULATOR, which is our AIPS++ implementation of Equation 2, stores incoming spectra in a four dimensional buffer, of the form `Matrix<Matrix<Float>>`. The row and column in the parent matrix select a beam-polarization pair, and the row and column in the sub-matrices select cycle and channel numbers. The spectra are arranged this way to optimise access speeds, yet still provide `row()` and `column()` methods to efficiently extract spectra or time series of individual channels. Both these operations are required for bandpass correction. Since shuffling the data through the matrices would be inefficient, an indexing system is utilised to keep track of the newest spectrum added to the buffer. Spectra are discarded (overwritten) when they are no longer required.

---

[1]Normally, $b \in \{1, 2, \ldots, 13\}$ and $p \in \{1, 2\}$.

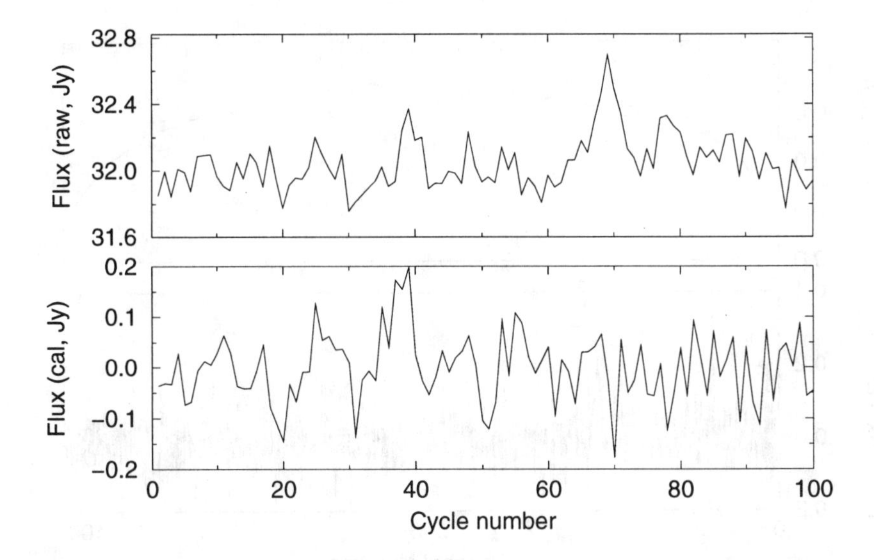

Figure 1. Raw (top) and calibrated flux densities for a selected channel of a HIPASS scan through the interacting galaxy ESO 269-IG 056 (cycles 34–41) and the continuum source PKS 1307-403 (cycles 66–73)—*see text.*

2.2. Examples

Figure 1 shows the uncalibrated and calibrated flux densities for a selected channel of a HIPASS scan through a galaxy with the central beam of the Multibeam. The galaxy appears as a rise and fall in flux density over cycles 35 to 42. This feature is preserved in the calibrated flux densities; the uncalibrated and calibrated spectra corresponding to cycle 36 are shown in Figure 2.

The feature observed near cycle 69 in the raw bandpass (Figure 1) is the continuum source PKS 1307-403 which generates an increased flux density in *all* line channels. To first order, continuum sources are removed by the bandpass removal algorithm, since the entire spectrum is rescaled according to Equation 2, and then the median value of all channels is subtracted from each channel as a simple means of baseline removal. For strong continuum sources though, there are still serious baseline ripple and curvature effects which remain in the calibrated spectra.

3. Discussion and Conclusion

The robust bandpass removal algorithm implemented for the HIPASS Project successfully produces calibrated H I emission line spectra for most of the survey data. The technique is robust to the presence of contamination by radio frequency interference and strong on-axis and off-axis continuum sources in the reference spectra. Unfortunately though, sources which emit in a particular channel over more than $\sim 2^\circ$ of declination, such as the Galaxy or associated

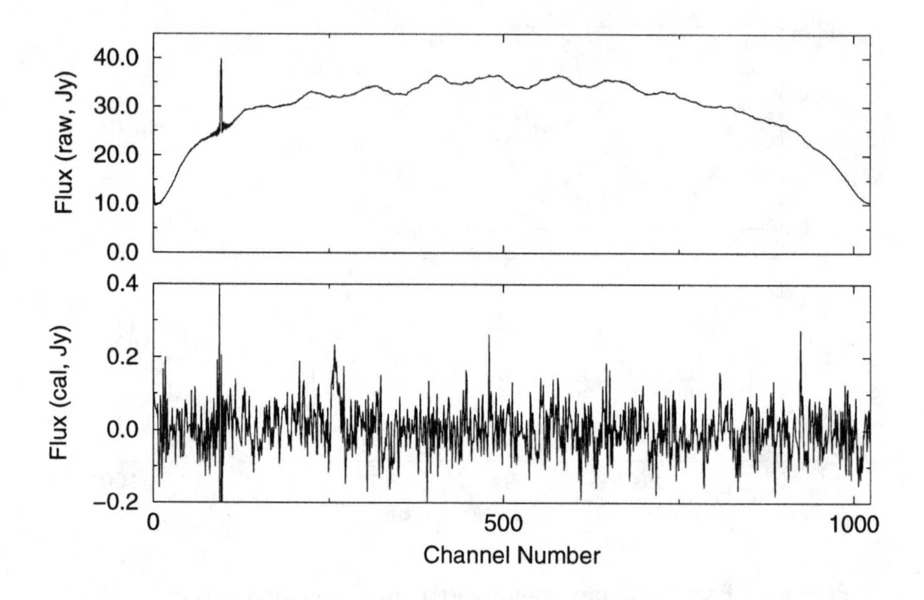

Figure 2. Raw (top) and calibrated flux density spectrum for a single integration (cycle 36) of the same HIPASS scan as shown in Figure 1—*see text.*

high-velocity clouds, are not well calibrated. This is because a trade-off between a larger set of reference spectra (yielding a better estimate of the emission-free bandpass in such cases) and a smaller set of reference spectra (yielding robustness to time-variability of the bandpass) was necessary. Conversely, the calibrated spectra of unresolved sources, which will form the bulk of the sources detected by the HIPASS, are invariably flat and of excellent quality.

Acknowledgments. We were delighted with the frequent and high-quality help provided by the many AIPS++ programmers during the planning and programming of the Parkes Multibeam Software. We are grateful to the ATNF and collaborating institutions for supporting the development of the Parkes Multibeam Receiver. We acknowledge the large investment of time and personnel towards the design and construction of the receiver by the CSIRO Division of Telecommunications and Industrial Physics. DGB acknowledges the support of an APA, and is grateful for the financial assistance to attend ADASS '97 provided by the ADASS POC and a Melbourne Abroad Scholarship.

References

Barnes, D. G. 1998, this volume

Astronomical Data Analysis Software and Systems VII
ASP Conference Series, Vol. 145, 1998
R. Albrecht, R. N. Hook and H. A. Bushouse, eds.

RVSAO 2.0 - A Radial Velocity Package for IRAF

Douglas J. Mink and Michael J. Kurtz

Smithsonian Astrophysical Observatory, Cambridge, MA 02138, Email: dmink@cfa.harvard.edu

Abstract. RVSAO 2.0 is the latest release of a package for calculating apparent radial velocities of celestial objects from observed spectral shifts. There are two main tasks in the package, XCSAO and EMSAO. XCSAO cross-correlates the Fourier transform of an object's spectrum against the transforms of a set of template spectra with known spectral shifts to obtain a velocity and error. EMSAO finds emission lines in a spectrum and computes the observed centers, getting individual shifts and errors for each line as well as a single velocity combining all of the lines. Three tasks which are new in this release are SUMSPEC, which combines spectra after shifting them all to a specified redshift, LINESPEC, which creates a spectrum at a specified redshift from a list of rest wavelengths, and BCVCORR, which computes the correction needed to translate the observed radial velocity to one relative to the solar system barycenter. Full documentation of this software, including numerous examples of its use, is on-line at http://tdc-www.harvard.edu/iraf/rvsao/

1. Introduction

The RVSAO IRAF external package was developed at the Smithsonian Astrophysical Observatory to compute redshifts from spectra in as automatic a way as possible. It has been used by several large redshift surveys and is also used for stellar radial velocity work. An earlier version of the XCSAO task, which computes radial velocities by cross-correlating spectra against templates of known redshift, has been described by Kurtz et al. (1992). The EMSAO task, which automatically identifies emission lines in a spectrum and computes their redshift has been described by Mink and Wyatt (1995). Mink and Wyatt (1992) described how these IRAF tasks could be combined to reduce large amounts of data in a pipeline. Both XCSAO and EMSAO have been improved over the years, and new tasks have been added to prepare template spectra for cross-correlation and to compute velocity corrections for data with different header keywords than are used by the Smithsonian's telescopes.

2. Changes in XCSAO

The cross-correlation algorithms in XCSAO have been changed very little over the years, although the optional elimination of high-frequency filtering has been added to enable the use of templates with narrow emission lines. Additional log format options have been added at the request of users.

To totally eliminate the effects of bad night sky subtraction, or to remove other features appearing at known positions in the observed wavelength space of a spectrum, a new feature has been added to both XCSAO and EMSAO. If the parameter *fixbad* is set to yes, XCSAO replaces sections of the spectrum described in the file designated by the *badlines* parameter with values interpolated from the ends of the sections. A line list is provided to remove the regions around night sky emission lines.

To conform with IRAF conventions for multispec files, that is spectrum files with multiple spectra from multiple apertures, new parameters *specband* and *tempband* have been added to specify the band to be read from each aperture. For example, in standard multispec files, band 1 is the object spectrum and band 3 is the sky spectrum. The *specnum* and *tempnum* parameters, which could specify either the band or the aperture, now specify only the aperture to be used from the file.

3. EMSAO Line Fitting Improves

The major change in the EMSAO task has been to replace the old minimization routine with one which has been used in a different astronomical context to fit multiple Gaussians. It is both faster and more robust than the old subroutine, making it practical to routinely run EMSAO on every spectrum, whether it has emission lines or not. There is always a danger of getting false emission lines, so it is safest to run EMSAO on spectra against which an emission line template has correlated well in XCSAO.

The sky spectrum, which is used to get the observed noise for better error calculations, may come from a different band *skyband* of a spectrum, such as the subtracted sky of an apextracted multispec spectrum, as well as from a separate aperture.

Many parameters which were built-in constants have been turned into task parameters. *mincont* sets a minimum continuum level at which an equivalent width is computed. There are many criteria for whether lines should be kept once they have been fitted. The *lwmin* and *lwmax* parameters set the minimum and maximum variation from the mean line width allowed for a line to be accepted. *lsmin* is the minimum ratio of the equivalent width (or area) to its error for a line to be accepted. A number is now appended to the line rejection flag in tables of results to indicate why the line was dropped.

4. Creating Templates from Line Lists with LINESPEC

By cross-correlating both emission and absorption line objects with XCSAO, a single output line can give both a reasonable redshift and a characterization of the object. Since every emission line object is different, a pure emission line template, with idealized line profiles seemed optimum. LINESPEC was written to use the line profile information provided by a reporting format added to EMSAO to create a spectrum from mean profiles of various identified emission lines.

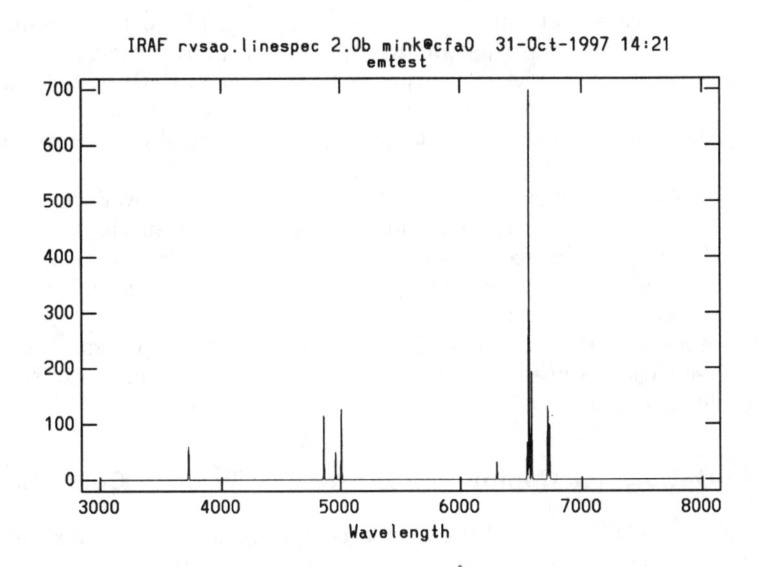

Figure 1. This is an artificial emission line spectrum created by LINESPEC for SAO's FAST Spectrograph

For each line, the center wavelength in Angstroms, the half-width in Angstroms if positive, in km/sec if negative, the height of line in arbitrary units, and the name of line for labeling, are read from a file. For each line in the table, the center is redshifted accordingly by a z (delta lambda / lambda) or apparent Doppler shifting velocity. The linewidth, if it is tabulated in kilometers per second, is converted to Angstroms at the shifted line center. The line width is also broadened appropriately if the line is redshifted. For each line, a Gaussian at the shifted center wavelength, half-width, and tabulated height is added to the spectrum. After all of the lines are computed, a constant continuum level may be added to the spectrum.

The computed spectrum is displayed, as shown in Figure 1, and may be edited before it is written to a disk file. The header of the output spectrum image includes one parameter per emission line with a vector of line characteristics in the format used by EMSAO.

5. Adding Spectra with SUMSPEC

For a long time, SAO has been using composite absorption line spectra as templates for galaxy cross-correlation. To formalize the process of creating such template spectra, SUMSPEC was written. It combines spectra, shifting them to a common redshift. The VELOCITY header parameter of each of these spectra is assumed to be a solar-system-barycenter-corrected velocity, and a barycentric correction (computed by sumtemp or extracted from the BCV or HCV header parameter) is subtracted to get the actual redshift of the spectrum. Each spectrum is shifted and rebinned to the desired wavelength range and bin size, which may be linear in wavelength or in log-wavelength, then added to the summed template. Input may be multispec or twodspec format, but output is always a one-dimensional file. If the desired output velocity is set to INDEF, spectra are redshifted to the solar system barycentric frame so spectra of the same object observed at different times throughout the year may be added to improve signal to noise.

SUMSPEC can automatically find the wavelength range over which all of the spectra to be added overlap. The output binsize may be specified explicitly or computed from the desired number of pixels and wavelength range. The continuum may be subtracted or divided from each spectrum before it is added into the final composite spectrum.

The composite spectrum may include a list of all of the input spectra in its header, so re-creation is possible. This can be turned off if hundreds of spectra are being added together.

6. More Options for Computing Heliocentric Velocity Correction

The XCSAO, EMSAO, and SUMSPEC tasks compute the velocity change needed to correct the observed redshift to the redshift relative to the sun, or more accurately, the solar system barycenter. They read the time of observation, object position, and observatory position from the spectrum image header. Although these tasks check several commonly-used alternative keywords for most of the needed parameters, it is possible that it won't find all of them. A separate task, BCVCORR, has been added to RVSAO to allow several alternate ways of specifying these three major pieces of information. BCVCORR can write its result to the header of the image which it is processing; the other RVSAO tasks will use this value when their *svel_corr* parameter is set to "file".

7. RVSAO's Future

RVSAO will continue to change to meet the needs of astronomers for fast extraction of radial velocities from large numbers of spectra. While RVSAO is ready for current multiaperture and multiple-order spectrographs, new instrumentation will surely require modifications to this software in the future.

References

Kurtz, M.J., Mink, D.J., Wyatt, W.F., Fabricant, D.G., Torres, G., Kriss, G, and Tonry, J.L. 1992, in Astronomical Data Analysis Software and Systems I, ASP Conf. Ser., Vol. 25, eds. D.M. Worral, C. Biemesderfer, and J. Barnes, 432.

Mink, D.J. and Wyatt, W.F. 1992, in Astronomical Data Analysis Software and Systems I, ASP Conf. Ser., Vol. 25, eds. D.M. Worral, C. Biemesderfer, and J. Barnes, 439.

Mink, D.J. and Wyatt, W.F. 1995, in Astronomical Data Analysis Software and Systems IV, ASP Conf. Ser., Vol. 77, eds. R.A. Shaw, H.E. Payne, and J.J.E. Hayes, 496.

Astronomical Data Analysis Software and Systems VII
ASP Conference Series, Vol. 145, 1998
R. Albrecht, R. N. Hook and H. A. Bushouse, eds.

New Developments in the FITSIO and Fv Software Packages

W. Pence

NASA/GSFC, Code 662, Greenbelt, MD 20771, Email: pence@tetra.gsfc.nasa.gov

Abstract. This paper describes recent improvements to the FITSIO subroutine library and the fv FITS file viewer and editor, both of which were initially described at previous ADASS conferences.

1. Introduction

The FITSIO subroutine library for reading and writing FITS format files was last described in the ADASS IV conference proceedings. Since then there have been many significant improvements that are outlined in this paper. Similarly, recent improvements to the fv FITS file viewer and editor are described to update the previous paper in the ADASS VI proceedings. Both packages are available from the HEASARC Web page at http://heasarc.gsfc.nasa.gov/docs/HHP_sw.html.

2. FITSIO/CFITSIO

Since the time of the last FITSIO update (Pence 1995) a new CFITSIO library has been developed, written entirely in C. CFITSIO is easier for C programmers to build and call than the previously available C wrapper routines built on top of the FORTRAN FITSIO library. Both the FITSIO and CFITSIO libraries have now been optimized for maximum data I/O performance and can achieve throughputs of order 5 – 10 MB/s or more when reading or writing FITS files on current generation workstations or PCs.

As originally developed, the FITSIO and CFITSIO libraries provided exactly the same functionality to FORTRAN and C applications programmers, respectively. Recently, however, several new features have been added to CFITSIO which are not available in the FITSIO library. It is quite time consuming to maintain the independent FITSIO and CFITSIO libraries with exactly the same functionality, so in the future it is likely that the FITSIO library will be frozen in its current state and that new features will only be added to the CFITSIO library. To make these new CFITSIO features easily available to FORTRAN programmers, a new set of FORTRAN wrapper routines for the CFITSIO library is currently being tested. These wrappers will provide the exact same subroutine calling sequence and functionality as currently provided by FITSIO, but will actually call the corresponding routine in the CFITSIO library.

One of the important new features that is only available in the CFITSIO library is the ability to read or write FITS files directly in computer memory as well as on disk. As a result of this, programs can now read FITS files piped in via the 'stdin' stream, or write them out to the 'stdout' stream. This enables a sequence of tasks to pipe the output FITS file from one task on to the input of the next task entirely in memory. This eliminates the need to write then read back temporary FITS files on magnetic disk and can greatly reduce the amount of time spent doing data I/O in pipeline processing.

Another important new feature in CFITSIO is the ability to directly read compressed FITS files that have been compressed with the gzip, PKZIP, or Unix 'compress' algorithms. This enables programs to directly read the compressed FITS files that often exist in on-line data archives or are distributed on CDROMs, without first having to uncompress the FITS file.

3. Fv FITS File Editor

The fv program for viewing and editing the contents of any FITS file has undergone many improvements since the original announcement (Pence, Xu & Brown 1997). Fv is written in Tcl/Tk and provides a graphical display of the header keywords, tables, and images in the FITS file. The current release, V2.1, has many spread-sheet type functions for sorting tables, inserting or deleting rows or columns in a table, and recalculating the values in any column. Fv can display FITS images and make line plots of the values in table column(s) using an integrated Tcl/Tk tool called POW. It supports interactive image analysis including panning and zooming, and brightness and contrast manipulation. POW also supports readout of image coordinates using the standard FITS World Coordinate System keywords.

Fv currently runs on most Unix workstations, but by the time this article is published, version 3.0 of fv should be available which will also run on IBM and Macintosh personal computers. This will make fv more accessible to the educational community and to the general public as a tool for easily viewing and analyzing the growing database of recent astronomical discoveries that are stored in FITS format.

Acknowledgments. The fv program has been principally developed by Jian-jun Xu, based on preliminary work by Dr. Jim Ingham. The POW image display and plotting package has been developed by Dr. Laurence Brown. The continuing development of fv and POW is supported by the High Energy Astrophysics Science Archive Research Center (HEASARC) at the NASA Goddard Space Flight Center and has been partially funded by a grant from the NASA Astrophysics Data Program.

References

Pence, W. 1995, in ASP Conf. Ser., Vol. 77, Astronomical Data Analysis Software and Systems IV, ed. R. A. Shaw, H. E. Payne & J. J. E. Hayes (San Francisco: ASP), 245

Pence, W., Xu, J. & Brown, L. 1997, in ASP Conf. Ser., Vol. 125, Astronomical Data Analysis Software and Systems VI, ed. Gareth Hunt & H. E. Payne (San Francisco: ASP), 261

Astronomical Data Analysis Software and Systems VII
ASP Conference Series, Vol. 145, 1998
R. Albrecht, R. N. Hook and H. A. Bushouse, eds.

The NCSA Horizon Image Data Browser: a Java Package Supporting Scientific Images

R. L. Plante,[1] W. Xie,

J. Plutchak,[2] R. E. M. McGrath and X. Lu

National Center for Supercomputing Applications (NCSA), University of Illinois, Urbana, IL 61820, Email: rplante@ncsa.uiuc.edu

Abstract. The NCSA Horizon Image Data Browser is a Java package which includes ready-to-use applets and applications as well as reusable classes for building new applications for browsing scientific images locally or over the network. We present the current status of the development of this package focusing on three general features: 1) flexible support for metadata through a schema-independent design, 2) the implementation of our metadata model to support world coordinate systems, and 3) support for use of Horizon applications within the NCSA Habanero Collaborative environment.

1. Introduction

The NCSA Horizon Image Data Browser[3] is a Java package for browsing scientific images. It is a collection of ready-to-use applets and applications as well as a toolkit of reusable classes that can be mixed and matched to create new applications. The basic goal of the package is to provide tools that can give a "first-cut" look at image datasets read from the network or local disk and act as a smooth pipeline from data repositories to specialized native software. For many users, particularly non-scientists, it can serve as a cheap, platform-independent tool for visualizing real scientific data. The design of the package is independent of data format allowing for the support of multiple formats. Initially, the package will come with support for FITS, HDF, GIF, and JPEG. The package provides such features as zooming, animation, pixel value and coordinate position display, spreadsheet display, and color fiddling. (For a full discussion of the goals of the package see Plante et al. 1997.)

In this paper, we will focus on three specific design features the Horizon package: the support for metadata, world coordinate systems, and collaboration.

---

[1]Astronomy Department, UIUC

[2]Atmospheric Science Department, UIUC

[3]http://imagelib.ncsa.uiuc.edu/Horizon

2. Metadata

Metadata—data about data—are an important part of scientific data. They are the ancillary information that accompanies some primary data set, allowing it to be interpreted intelligently by scientists and the software they use. It plays a critical role when data must be transported between different software systems or formats. In the Horizon package, metadata are a set of name-value pairs where the name is of type `java.lang.String` and the value, `java.lang.Object`.

An important assumption that always exists when handling metadata is the agreement between the creators of metadata and its users on the mapping of a metadatum's name to its value type (e.g., `Double`, `String`, etc.) and its conceptual meaning. A set of agreed mappings of names to types and meanings is often referred to as a *schema*. Different formats and different scientific fields often use different schemata to represent metadata.

At the center of Horizon's support for metadata is the `Metadata` class (Plante 1997). Its design is schema-independent and takes into account that there may be multiple schema in use within a single application. Furthermore, metadata can be hierarchical, so the `Metadata` class allows direct access to any individual metadatum in the hierarchy, or whole sections of the hierarchy. Horizon also supports array data as metadata. The following code illustrates both the array and hierarchical features:

```
Metadata mdata;
...
// retrieve all metadata for coordinate axis 0
Metadata axmdata = (Metadata) mdata.getMetadatum("CoordinateSystem.Axes[0]");

// retrieve a specific coordinate axis sub-metadatum directly
String name = (String) mdata.getMetadatum("CoordinateSystem.Axes[0].name");
```

Other features of the `Metadata` class include support for default values that can be overridden but not erased; this feature is used a way of giving read-only access to metadata that are to be shared with many objects. The class also supports on-demand loading of values; if the process of loading metadata is expensive (e.g., they are read from the network), one would prefer that they only be read if the user explicitly requests them.

The Horizon package defines a "slim" metadata schema which it uses to do basic visualization. Part of the job of the format-dependent reader is to convert selected native metadata into the "horizon" schema. Higher-level Horizon classes use this schema to interact with the image data in a format-independent way, including extracting chunks of data and displaying positions in the data's world coordinate system.

3. World Coordinate Systems

A scientist usually wants to know where a piece of data exists in physical or conceptual space–its *World Coordinate System*–rather than its location in some data array. In the Horizon design, data can exist in a data space of arbitrary number of dimensions which maps to a world coordinate space of the same or

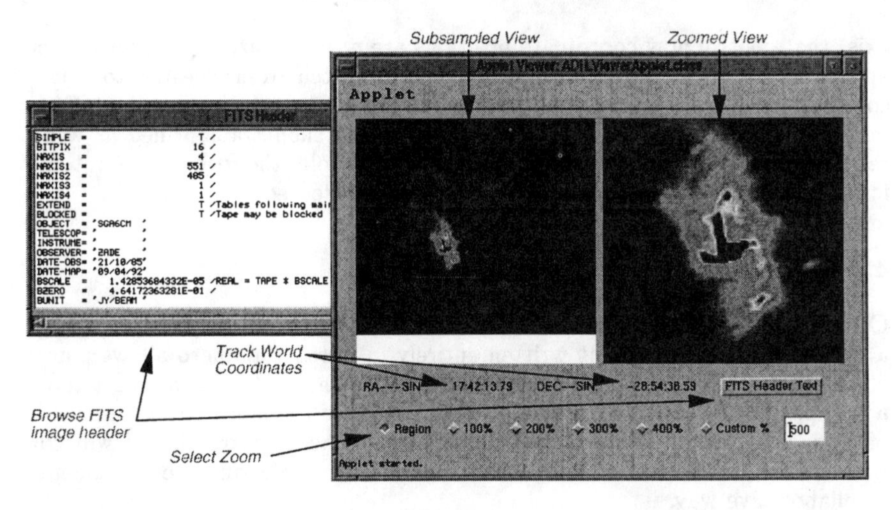

Figure 1. The ADILBrowser Applet: a viewer for browsing images in the NCSA Astronomy Digital Image Library (ADIL).

different number of dimensions. The transformation between data voxels and coordinate positions can be non-linear and even non-reversible.

The two Horizon world coordinate classes used most by the application programmer are `CoordinateSystem` and `CoordPos` (Plante 1997). A `CoordinateSystem` object can be obtained for a particular dataset through its format-independent interface. This object has full knowledge of the dataset's coordinate system and its mapping to the data voxels. Thus, the programmer can give the `CoordinateSystem` a data voxel and get back a `CoordPos` object, an encapsulation of the corresponding coordinate position. The reverse transformation is also supported. One special feature of the `CoordPos` object is that it knows how to print itself; e.g., it knows to print right ascension as *HH:MM:SS.S*, galactic longitude in decimal degrees, and velocity with the appropriate metric units. These default formats can be easily overridden.

The transformation engine within the `CoordinateSystem` class is the `CoordTransform` class, used for converting between two coordinate systems. Horizon comes with a number of specialized `CoordTransforms` that do things like spherical projections, switching between celestial and galactic coordinates, etc. (Some of these have been implemented using FITSWCS, a Java port of the WCS library by Calabretta, see Greisen & Calabretta 1995.) Multiple `CoordTransform` objects can be strung together to produce complex transformations. Thus, every `CoordinateSystem` maintains an internal stack of `CoordTransform` objects through which it can pass data voxels or coordinate positions. Users of the `CoordinateSystem` can attach additional `CoordTransform` objects to the system at any time for on-the-fly switching of coordinate systems.

The combination of `Metadata` and `CoordTransforms` is Horizon's recipe for a smart `CoordinateSystem`. When a `CoordinateSystem` is constructed, it uses `Metadata` gotten from the dataset to configure the needed `CoordTransform` objects. Included in the `Metadata` are `AxisPosFormatter` objects (one for each axis) that control how coordinate positions are printed out. If the user attaches an additional `CoordTransform` object, say to switch from celestial to galactic coordinates, the `Metadata` are consulted to determine which axes the transformation should be applied to. The `Metadata` are then transformed as well to reflect the new coordinates system; thus, for example, the format can automatically be switched from *DD:MM:SS.S* to decimal degrees.

4. Collaboration

Collaboration is supported within Horizon via the NCSA Habanero package,[4] a collaborative environment written entirely in Java. Habanero allows a group of participants located on different machines to interact with a single tool as a group. (See Crutcher et al. 1998 in this volume for more details on the use of Habanero with astronomical applications.) The Horizon classes will come with special hooks that allow them to be run within Habanero environment in a collaborative way.

Figure 1 from Crutcher et al. (1998, this volume) shows an example of running a Horizon applet, the ADILBrowser (upper right corner; see also Figure 1 of this article) within a Habanero session. Participants use the applet to browse images located in the NCSA Astronomy Digital Image Library.[5] The applet tracks world coordinates and allows zooming into subregions of the image. The remaining windows are part of the Habanero environment which include a session manager, a chat window, and a white board.

Acknowledgments. The Horizon development team thanks Jef Poskanzer, Mark Calabretta, and Tom McGlynn for contributed code. The Horizon Image Data Browser package is supported in part by Project Horizon, a cooperative agreement with NASA, and by the NASA Applied Information Systems Research Program (96-OSS-10).

References

Crutcher, R. M., Plante, R., & Rajlich, P. 1998, this volume.

Greisen and Calabretta 1995, Representations of Celestial Coordinates in FITS[6].

Plante, R. 1997, Supporting Metadata and Coordinate Systems for Scientific Data in the Horizon Java Package[7], white paper.

Plante, R., Goscha, G., Crutcher, R., Plutchak, J., McGrath, R., Lu, X., & Folk, M. 1997, in ASP Conf. Ser., Vol. 125, Astronomical Data Analysis Software and Systems VI, ed. Gareth Hunt & H. E. Payne (San Francisco: ASP), 341.

---

[4]http://www.ncsa.uiuc.edu/SDG/Software/Habanero

[5]http://imagelib.ncsa.uiuc.edu/imagelib

[6]ftp://fits.cv.nrao.edu/fits/documents/wcs/wcs.all.ps.Z

[7]http://imagelib.ncsa.uiuc.edu/Horizon/docs/articles/CoordsAndMetadata

Astronomical Data Analysis Software and Systems VII
ASP Conference Series, Vol. 145, 1998
R. Albrecht, R. N. Hook and H. A. Bushouse, eds.

An IRAF Port of the New IUE Calibration Pipeline

Richard A. Shaw and Howard A. Bushouse

Space Telescope Science Institute, Baltimore, MD 21218, Email: shaw@stsci.edu

Abstract. The calibration pipeline that was used for the *IUE* final archive, NEWSIPS, is being ported to IRAF. This port will enable archival researchers to optimize the spectral processing for their scientific needs.

1. Introduction

A new spectral image processing system (NEWSIPS) was constructed for a final, archival reprocessing of the *IUE* data archive. NEWSIPS incorporates many improved processing algorithms and calibrations which greatly enhance the quality, and hence the scientific utility, of *IUE* data for archival research. However, the perceived need to produce a homogeneous archive precluded custom processing options and special calibrations that are essential for many science programs. In addition, limited resources prevented the completion of certain useful calibrations, and limited the scope of efforts to characterize the complex echelle background and ripple correction. Unfortunately, it is not practical for researchers to make use of the production NEWSIPS system, mostly because it depends on vendor-specific software and hardware, and it would have been impossible for users to maintain or enhance the pipeline over the long term.

We are porting the NEWSIPS software to the IRAF environment. The ported pipeline will enable recalibration from the raw data or from certain intermediate stages, will allow for different choices of reference files, and will accommodate the use of more appropriate processing techniques at intermediate stages. We are removing the dependencies of NEWSIPS on licensed vendor software, and replacing them where necessary with existing or new IRAF libraries. Our goal is to retain a high degree of compatibility with the archival NEWSIPS pipeline, given the same input parameters. Placing the NEWSIPS software in the public domain will allow *IUE* archival researchers much greater flexibility for meeting their particular scientific needs, while also promoting greater understanding of the NEWSIPS system by current and future users of *IUE* data. The IRAF system is a good choice for ensuring the longevity, portability, and wide accessibility to the user community of the software, while also providing a very rich environment for analysis of *IUE* data by future archival researchers.

2. Why Recalibrate?

There are many reasons for re-running all or part of the NEWSIPS calibration pipeline, even after all the improvements that have been realized by the original NEWSIPS processing. For example, the spectral extraction processing for low dispersion uses an "optimal" or signal-weighted extraction technique (SWET) to optimize the S/N ratio and exclude outliers. However, this technique is inappropriate if the ionization structure is spatially resolved—e.g., for many planetary nebulae or H II regions. Figure 1 shows how outlier rejection in SWET can cause unexpected results; the solution is to re-extract using a different method.

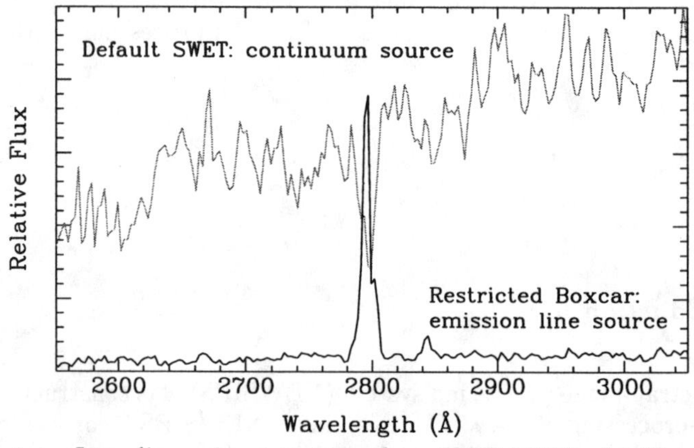

Figure 1. Low dispersion spectrum from LWP 17531 of α Her showing the SWET extraction (*upper*), which excludes a spatially resolved emission feature that can be recovered with a boxcar extraction (*lower*).

The high-dispersion extraction uses a simple boxcar weighting, and as such does not exclude outliers. Figure 2 illustrates the effect of a prominent, grazing cosmic ray that affected several adjacent orders in the SWP high-dispersion spectrum of BD+75°325. Cosmic rays like this could easily be excluded using either a more sophisticated extraction or, when two or more images are available in the archive, by rejecting cosmic rays from the LI image during calibration.

For images obtained in high-dispersion, an imperfect model of the background determination can under some circumstances result in an under- or over-correction. These problems are illustrated in Figure 3, where the SWP spectrum of HD 149438 shows what should be a saturated absorption feature at Ly α. An under-correction of the inter-order background causes this absorption feature not to fall to zero flux in the line core. The opposite problem occurs for HD 163181, where an over-correction for the inter-order background shortward of 1250 Å results in negative fluxes. Improvements in the background estimation will require either a more sophisticated model or a semi-empirical approach.

More ambitious users may need to improve upon the calibration reference files to extract the best science from their data. For example, the most appropriate flat-field data for late-epoch SWP images may be that obtained during the 1992 ITF campaign. But a lack of resources precluded the construction,

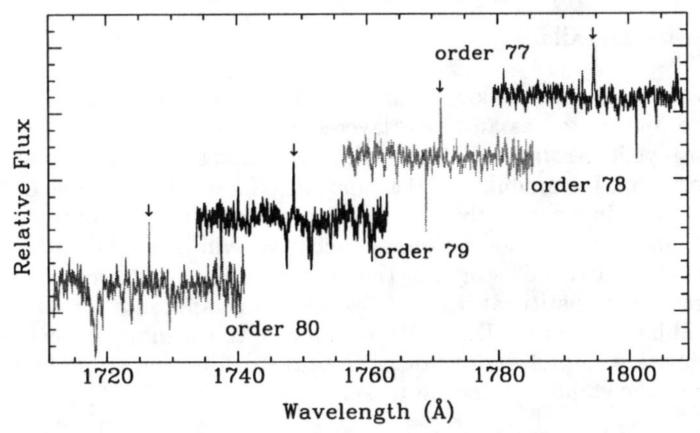

Figure 2. Several adjacent spectral orders are affected by a grazing cosmic ray hit (marked with arrows) on the high dispersion image SWP 35674 of BD+75°325.

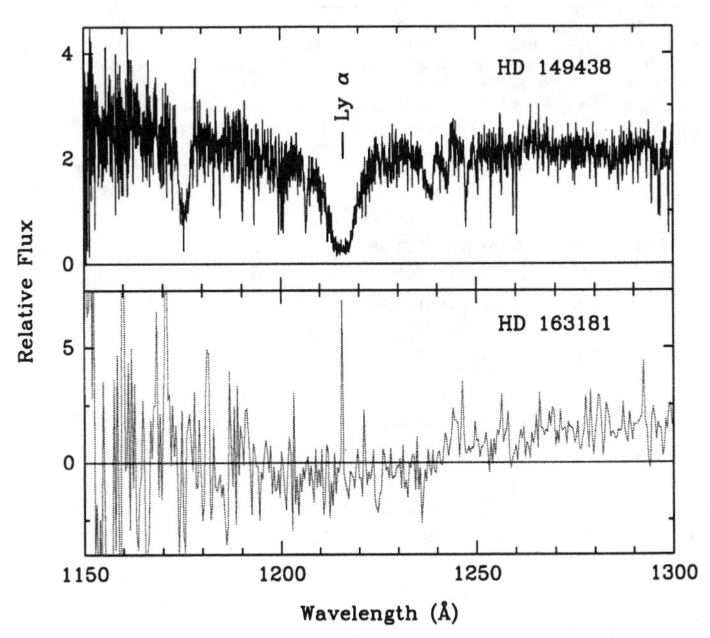

Figure 3. High dispersion SWP spectra of two stars, illustrating the consequences of an imperfect background model in NEWSIPS.

testing, and implementation for final archive processing. For cases like this, the only option is for users to reprocess their images with improved calibrations.

3. Software Details

The NEWSIPS pipeline, various ancillary utilities, and the calibration reference files will be distributed as an IRAF layered package called **iue**. The complete calibration pipeline, *caliue*, will be a high-level script that executes several tasks that comprise major modules of the pipeline. These tasks, listed in Table 1, can also be run independently. The choice of what calibration steps will be performed, and of which calibration reference files will be used, will be managed through a set of keywords stored in the image header, or alternatively through parameters of the constituent tasks. A feature of this approach is that the ported pipeline will be re-entrant—i.e., it will be possible to use different (perhaps user-written) software to perform one or more of the calibration steps, then run the remainder of the standard pipeline tasks.

Table 1. Major Tasks in the **newsips** Pipeline.

Task	Description
ns_init	Populates the raw image header with control keywords
rawscreen	Flags missing data and cosmic rays
ttdc	Computes time- and temperature-dependent dispersion corrections
xcorr	Registers the science image with the ITF
photom	Applies the photometric correction
iuegeom	Rectifies the 2-D spectra and linearizes the dispersion
nsextract	Extracts the 1-D spectrum from the image

One of the goals of this project is to reorganize and better document the calibration reference files used in NEWSIPS, and to provide them to the community in FITS format. The **iue** package will also include tools for constructing the calibration reference files, which will enable archival users to improve upon or extend the calibrations derived for the final archive. The ported pipeline will support FITS format for all input and output files natively, eliminating the dependency on MIDAS-format files. Other enhancements include improved memory management by a factor of $\sim 10^3$ or more, which will enable reprocessing multiple images at once for large archival projects. Finally, an essential element in this port is to eliminate the dependency of NEWSIPS on vendor-proprietary software so that it may be distributed freely to the community.

4. Availability

The first release of the **iue** layered package is planned for early 1998. This release will contain the low-dispersion portion of NEWSIPS pipeline, including some custom processing options. It will also include utilities for creating various calibration reference files. Subsequent releases during 1998 will include the high-dispersion portion of the processing pipeline, additional custom processing options, and new utilities. The **iue** package, including the software, documentation, calibration reference files, and all updates, will be available from the

NEWSIPS home page[1]. Note that the **iue** package will require that IRAF V2.11 and TABLES V2.0 (or later) be installed. The *IUE* archive is being re-hosted to the Space Telescope Science Institute, where support for *IUE* archival data and this NEWSIPS port will continue for the foreseeable future.

Acknowledgments. This software project is funded by the NASA Astrophysics Data Program through grant NAS5–32697 to ST ScI.

---

[1]http://ra.stsci.edu/newsips/

Astronomical Data Analysis Software and Systems VII
ASP Conference Series, Vol. 145, 1998
R. Albrecht, R. N. Hook and H. A. Bushouse, eds.

GIM2D: An IRAF package for the Quantitative Morphology Analysis of Distant Galaxies

Luc Simard

UCO/Lick Observatory, Kerr Hall, University of California, Santa Cruz, CA 95064, Email: simard@ucolick.org

Abstract. This paper describes the capabilities of GIM2D, an IRAF package for the quantitative morphology analysis of distant galaxies. GIM2D automatically decomposes all the objects on an input image as the sum of a Sérsic profile and an exponential profile. Each decomposition is then subtracted from the input image, and the results are a "galaxy-free" image and a catalog of quantitative structural parameters. The heart of GIM2D is the Metropolis Algorithm which is used to find the best parameter values and their confidence intervals through Monte-Carlo sampling of the likelihood function. GIM2D has been successfully used on a wide range of datasets: the Hubble Deep Field, distant galaxy clusters and compact narrow-emission line galaxies.

1. Introduction

GIM2D (Galaxy IMage 2D) is an IRAF package written to perform detailed bulge/disk decompositions of low signal-to-noise images of distant galaxies. GIM2D consists of 14 tasks which can be used to fit 2D galaxy profiles, build residual images, simulate realistic galaxies, and determine multidimensional galaxy selection functions. This paper gives an overview of GIM2D capabilities.

2. Image Modelling

The first component ("bulge") of the 2D surface brightness used by GIM2D to model galaxy images is a Sérsic (1968) profile of the form:

$$\Sigma(r) = \Sigma_e exp\{-b[(r/r_e)^{1/n} - 1]\} \tag{1}$$

where $\Sigma(r)$ is the surface brightness at r. The parameter b is chosen so that r_e remains the projected radius enclosing half of the light in this component. Thus, the classical de Vaucouleurs profile has $n = 4$. The second component ("disk") is an exponential profile of the form:

$$\Sigma(r) = \Sigma_0 exp(-r/r_d) \tag{2}$$

where $\Sigma_0(r)$ is the central surface brightness, and r_d is the disk scale length. The model has a total of 12 parameters: the total luminosity L, the bulge fraction

B/T, the bulge effective radius r_e, the bulge ellipticity e, the bulge position angle ϕ_e, the Sérsic index n, the disk scale length r_d, the disk inclination i, the disk position angle ϕ_d, the centroiding offsets dx and dy, and the residual sky background level db.

Detector undersampling (as in the HST/WFPC2 camera) is taken into account by generating the surface brightness model on an oversampled grid, convolving it with the appropriate Point-Spread-Function (PSF) and rebinning the result to the detector resolution.

3. Parameter Search with the Metropolis Algorithm

The 12-dimensional parameter space can have a very complicated topology with local minima at low S/N ratios. It is therefore important to choose an algorithm which does not easily get fooled by those local minima. The Metropolis Algorithm (Metropolis et al. 1953) was designed to search for parameter values in a complicated topology. Compared to gradient search methods, the Metropolis is not efficient i.e., it is CPU intensive. On the other hand, gradient searches are greedy. They will start from initial parameter values, dive in the first minimum they encounter and claim it is the global one. The Metropolis does its best to avoid this trap.

The Metropolis in GIM2D starts from an initial set of parameters given by the image moments of the object and computes the likelihood $P(w|D, M)$ that the parameter set w is the true one given the data D and the model M. It then generates random perturbations $\Delta\mathbf{x}$ about that initial location with a given "temperature". When the search is "hot", large perturbations are tried. After each trial perturbation, the Metropolis computes the likelihood value P_1 at the new location, and immediately accepts the trial perturbation if P_1 is greater than the old value P_0. However, if $P_1 < P_0$, then the Metropolis will accept the trial perturbation only P_1/P_0 of the time. Therefore, the Metropolis will sometime accept trial perturbations which take it to regions of lower likelihood, and this apparently strange behavior is very valuable. If the Metropolis finds a minimum, it will try to get out of it, but it will only have a finite probability (related to the depth of the minimum) of succeeding.

The step matrix for the trial perturbations $\Delta\mathbf{x}$ is given by the simple equation $\Delta\mathbf{x} = Q \cdot u$ where the vector $\mathbf{u}$ consists of randomly generated numbers between 0 and 1, and the matrix Q is obtained through Choleski inversion of the local covariance matrix. In short, the sampling of parameter space shapes itself to the local topology.

Convergence is achieved when the difference between two likelihood values separated by 100 iterations is less than 3-σ of the likelihood fluctuations. After convergence, the Metropolis Monte-Carlo samples the region where the likelihood is thus maximized and stores the accepted parameter sets as it goes along to build the distribution $P(w|D, M)$. Once the region has been sufficiently sampled, the Metropolis computes the median of $P(w|D, M)$ for each model parameter as well as the 99% confidence limits. The output of the fitting process consists of a PSF-convolved model image O, a residual image R, and a log file containing all Metropolis iterations, the final parameter values and their confidence intervals, and image asymmetry indices.

4. Image Reduction Pipeline

GIM2D has a simple image reduction pipeline for HST (or ground-based) images which makes GIM2D particularly suitable for the analysis of archival data:

- Obtain archival data, combine image stacks and remove cosmic rays. Cosmic rays can be removed using tasks such as STSDAS/CRREJ.
- Run SExtractor (Bertin & Arnouts 1996) on the science image to produce an object catalog and a "segmentation" image. Pixel values in this segmentation image indicate to which object a given pixel belongs. The segmentation image is used by GIM2D to deblend galaxy images.
- For HST images, Data Quality Files obtained with a science image stack can be combined and apply to the SExtractor segmentation image. Bad pixels identified by the DQF images are excluded from the bulge/disk decompositions.
- Create a Point-Spread-Function (PSF) for each object. Tiny Tim can be used for HST images. GIM2D accepts three types of PSFs: Gaussian, Tiny Tim and user-specified.
- Run bulge/disk decompositions. These decompositions are run under an IRAF script which can be sent to multiple computers so that they can simultaneously work on the same science image.
- Create a residual image. The GIM2D task GRESIDUAL creates a galaxy-subtracted version of the input image *à la* DAOPHOT. GRESIDUAL sums up all the output galaxy model images and subtracts this sum from the original image. The resulting residual image is extremely useful to visually inspect the residual images of hundreds of objects at a glance. See Figure 1.
- Determine galaxy selection function. The observed distribution of a parameter (e.g., bulge fraction) derived from a large number of objects in an image is the intrinsic distribution multiplied by the galaxy selection function which gives the probability that an object with a given set of structural parameters will be detected by SExtractor. GIM2D determines the galaxy selection function by taking an empty background section of the science image, adding simulated galaxies to that image section and running SExtractor to determine whether it would have been detected or not. The resulting multidimensional selection function can then be used to "flat-field" the observed structural parameter distributions.

5. Scientific Results

5.1. Hubble Deep Field

Hubble Deep Field F814W images were analyzed with GIM2D (Marleau and Simard 1998). The number of bulge-dominated galaxies is much lower ($\leq 10\%$) than indicated by qualitative morphological studies. Some of these studies quoted bulge-dominated galaxy fraction as high as 30% (van den Bergh et al.

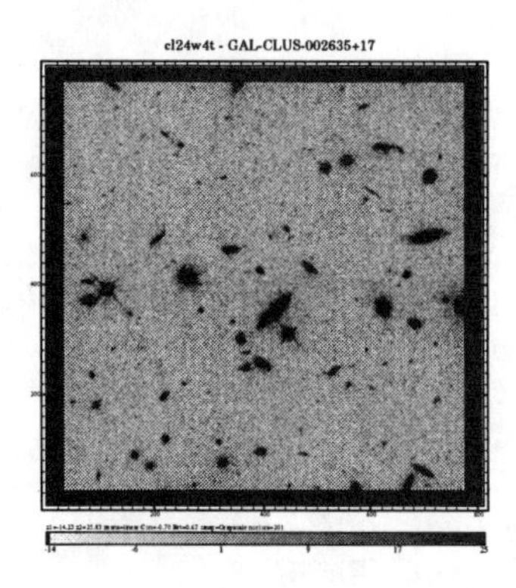

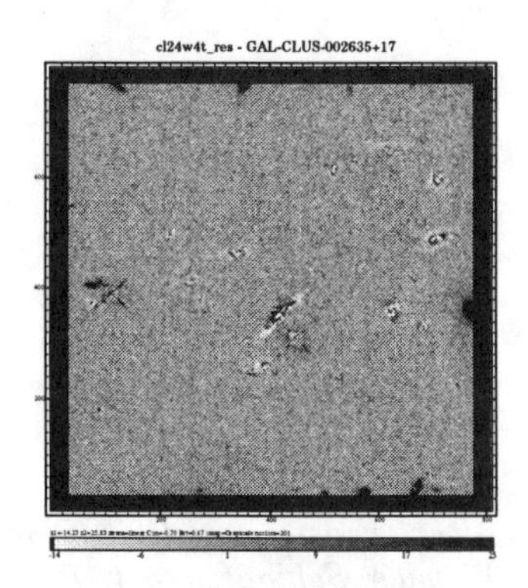

Figure 1. *Left:* HST/WFPC2/WF4 4400 seconds image of a region in the $z = 0.4$ cluster CL0024+16. *Right:* GIM2D residual image. 157 objects were fitted, and galaxies which were too close to the edge of the image were not analyzed.

1996). It was also found that a number of HDF galaxies have surface brightness profiles with Sérsic index $n \leq 1$. Some of them exhibit double exponential profiles which can be easily mistaken for pure bulge systems. However, these systems have spectra resembling those of late-type star-forming galaxies. Other HDF objects are best-fitted with a *single* $n < 1$ component.

5.2. The Distant Cluster CL0024+16

The rich cluster CL0024+16 at $z = 0.4$ was the perfect testing ground for the galaxy image deblending in GIM2D. CL0024+16 had been extensively studied for gravitational lensing effects, but a quantitative morphological analysis had never been performed. A total of 560 objects were analyzed in two colors with GIM2D, and about 260 objects were large and/or bright enough to produce reliable bulge/disk decompositions. Figure 1 shows a WFPC2/F814W image of a section of CL0024+16 and the corresponding GIM2D residual image.

References

Bertin, E. & Arnouts, S. 1996, A&AS, 117, 393

Marleau, F., & Simard, L. 1998, in preparation

Metropolis N., Rosenbluth, N., Rosenbluth, A., Teller, A., & Teller, E. 1953, Journal of Chemical Physics, 21, 1087

Sérsic, J.-L. 1968, Atlas de Galaxias Australes, Observatorio Astronomico, Cordoba

van den Bergh, S., Abraham, R., Ellis, R.S., Tanvir, N.R., Santiago, B.X., & Glazebrook, K.G. 1996, AJ, 112, 359

Astronomical Data Analysis Software and Systems VII
ASP Conference Series, Vol. 145, 1998
R. Albrecht, R. N. Hook and H. A. Bushouse, eds.

Grid OCL : A Graphical Object Connecting Language

I. J. Taylor[1]

Department of Physics and Astronomy, University of Wales, College of Cardiff, PO BOX 913, Cardiff, Wales, UK, Email: Ian.Taylor@astro.cf.ac.uk

B. F. Schutz[2]

Albert Einstein Institute, Max Planck Institute for Gravitational Physics, Schlaatzweg 1, Potsdam, Germany. Email: schutz@aei-potsdam.mpg.de

Abstract. In this paper, we present an overview of the Grid OCL graphical object connecting language. Grid OCL is an extension of Grid, introduced last year, that allows users to interactively build complex data processing systems by selecting a set of desired tools and connecting them together graphically. Algorithms written in this way can now also be run outside the graphical environment.

1. Introduction

Signal-processing systems are becoming an essential tool within the scientific community. This is primarily due to the need for constructing large complex algorithms which would take many hours of work to code using conventional programming languages. Grid OCL (Object Connecting Language) is a graphical interactive multi-threaded environment allowing users to construct complex algorithms by creating an object-oriented block diagram of the analysis required.

2. An Overview

When Grid OCL is run three windows are displayed. A *ToolBox* window, a *MainGrid* window and a *Dustbin window* (to discard unwanted units). Figure 1 shows the *ToolBox* window which is divided into two sections. The top section shows the available toolboxes (found by scanning the toolbox paths specified in the Setup menu) and the bottom shows the selected toolbox's contents. Toolboxes (and associated tools) can be stored on a local server or distributed throughout several network servers. Simply adding the local or *http* address in the toolbox and tool path setup allows on-the-fly access to other people's tools.

---

[1] A post doctoral programmer at Cardiff who has been developing Grid OCL since January 1996.

[2] Professor Schutz is a Director of the Albert Einstein Institute, Potsdam

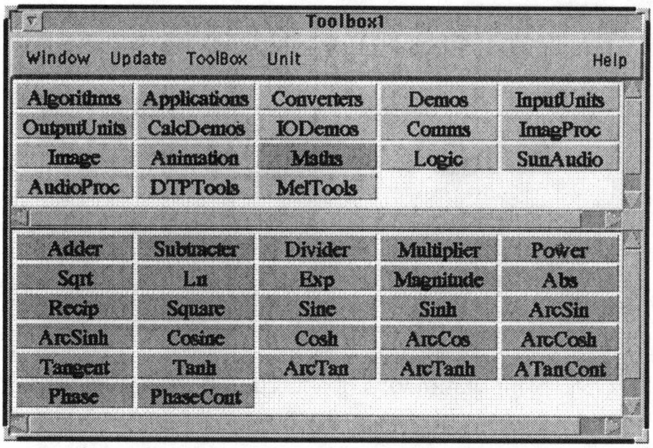

Figure 1. Grid OCL's ToolBox window. Toolboxes can be organised in a similar way to files in a standard file manager.

Units are created by *dragging* them from the ToolBox window to the desired position in the MainGrid window and then connected together by dragging from an output socket on a sending unit to an input socket of the receiving unit. The algorithm is run by clicking on the *start* button (see Figure 2), in a *single step* fashion (i.e., one step at a time) or continuously.

3. New Features and Extensions

Groups of units can now be saved along with their respective parameters. Such groups can also contain groups, which can contain other groups and so on. This is a very powerful feature which allows the programmer to hide the complexity of programs and use groups as if they were simply units themselves. Many improvements have been made to the graphical interface, including compacting the look and style of the toolbox, adding a snap-to cable layout and many more informative windows. The major change however, is that now Grid consists of an object connecting language (OCL) and a separate user interface. This means that the units can be run from within the user interface or as a stand-alone program.

Collaborators are working on tools for various signal and image problems, multimedia teaching aids and even to construct a musical composition system. Currently, in our new release we have toolboxes for various signal processing procedures, animation and a number of image processing/manipulation routines, text processing tools e.g., find and replace, grep and line counting recursively through subdirectories, mathematical and statistical units, a general purpose mathematical calculator (see next section) and a number of flexible importing and exporting units.

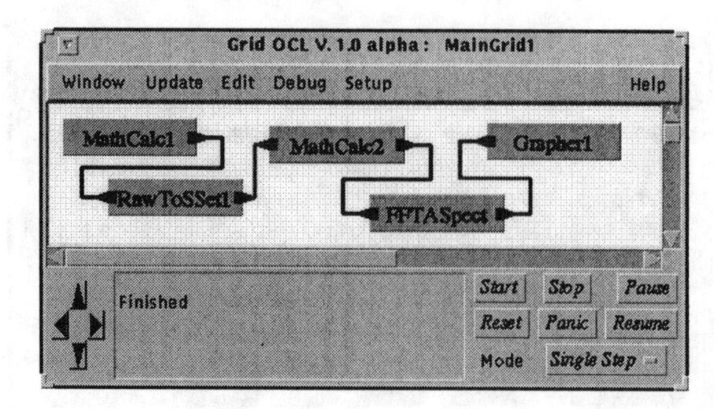

Figure 2. A snapshot of Grid OCL's programming window.

4. MathCalc

The MathCalc unit interprets, optimises and evaluates arithmetic expressions using stream-oriented arithmetic. It recognises a large number of functions and constants. It can be used to evaluate scalar expressions, to process input data sets, or to generate output data sets. All calculations are performed in double-precision real arithmetic.

Stream-oriented arithmetic can be defined as the application of an arithmetic expression to each element of a stream independently. Thus, if B is the sequence $b1, b2, .., bn$, then the function $sin(B)$ evaluates to the sequence $sin(b1), sin(b2), ..., sin(bn)$. MathCalc distinguishes between *constants* and *sequences* or sets. Sets (data sets) are sequences of numbers and constants are single numbers, essentially sequences of length 1. In a MathCalc expression the two can be mixed very freely, with the restriction that all sequences must have the same length. Sequences or constants can be obtained from the input nodes of the MathCalc unit. The example given in the MainGrid window (see Figure 2) demonstrates the flexibility of the MathCalc unit.

The first *MathCalc* unit creates a 125 Hz sine wave by using the equation $sin(((sequence(512) * 2) * PI) * 0.125)$ where the sample rate is 1kHz (MathCalc will optimise this to (2*PI*0.125) * sequence(512)). This is then transformed into a SampleSet type by adding its sampling frequency (i.e., 1 kHz). The second MathCalc unit adds Gaussian noise to its input (i.e., by typing $gaussian(512)+\#0s$). The #0 means node 0 and the s means that it is a sequence as opposed to a c which would a constant. The resultant amplitude spectrum (FFTASpect) is shown from Grapher1 (see Figure 3).

Once the signal is displayed, it can be investigated further by using one of the Grapher's various zooming facilities. Zooming can be controlled via a zoom window which allows specific ranges to be set or by simply using the mouse to *drag* throughout the image. For example, by holding the control key down and dragging down the image is zoomed in vertically and by dragging across from left to right zoomed in horizontally. The reverse operations allow zooming out. Also once zoomed in, by holding the shift key and the control key down the mouse can be used to move around the particular area you are interested in. We also have another powerful zooming function which literally allows the user to drag to the position of interest and the image will zoom in accordingly.

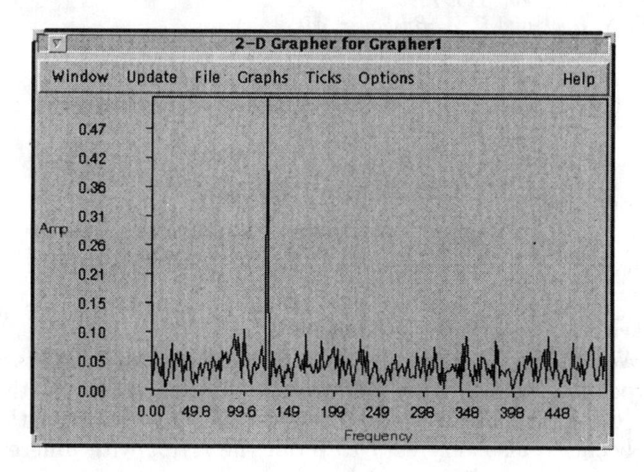

Figure 3. Grapher1's output: any grapher can simultaneously display many signals each with its own colour and line style.

5. Grid OCL : Past and Present

Grid originated from an implementation of the system using C++ and InterViews (Taylor & Schutz 1995) but was abandoned in early 1996. Version two (Taylor & Schutz 1996) was written using the Java Development Kit, JDK 1.0.2 but this was updated in order to be compatible with the new JDK 1.1.x kit. We also re-implemented the base classes and to create OCL. The most recent version of Grid OCL (in November 1997) is a late *alpha* and goes by a different name (*Triana OCL*[4]). Triana OCL will be entering its beta testing stage early next year followed by a final version shortly after. We are in the process of being able to provide a commercial version of the software for which support can be given. None-the-less we will always provide it in a free downloadable from the WWW with a certain time limit (3 or 4 months). Our main goal is to create a very wide user base.

References

Taylor, I. J. & Schutz, B. F. 1995, The Grid Musical-Signal Processing System, International Computer Music Conference, 371

Taylor, I. J. & Schutz, B. F. 1996, The Grid Signal Processing System. in ASP Conf. Ser., Vol. 125, Astronomical Data Analysis Software and Systems VI, ed. Gareth Hunt & H. E. Payne (San Francisco: ASP), 18

[4]http://www.astro.cf.ac.uk/Triana/

Astronomical Data Analysis Software and Systems VII
ASP Conference Series, Vol. 145, 1998
R. Albrecht, R. N. Hook and H. A. Bushouse, eds.

GUI-fying and Documenting your Shell Script

Peter. J. Teuben[1]

Astronomy Department, University of Maryland, College Park, MD 20742, Email: teuben@astro.umd.edu

Abstract.
We describe a simple method to annotate shell scripts and have a preprocessor extract a set of variables, present them to the user in a GUI (using Tcl/Tk) with context sensitive help, and run the script. It then becomes also very easy to rerun the script with different values of the parameters and accumulate output of different runs in a set of user defined areas on the screen, thereby generating a very powerful survey and analysis tool.

1. Introduction

Scripting languages have often been considered the glue between individual applications, and are meant to achieve a higher level of programming.

When individual applications are (tightly) integrated into the scripting language, this offers very powerful scripts, fully graphical user interfaces and a result sometimes indistinguishable from applications. A recent example of this is the glish shell in AIPS++ (Shannon 1996). But of course the drawback of this tight integration is that applications are not always easily accessible to scripts that do not (or cannot) make use of the environment the scripting language was meant for.

Apart from the black art of handcoding, one of the traditional methods to add a GUI to an application is using automatic GUI builders. This has the advantage that application code and user interaction code are more cleanly separated, but this sometimes also limits the flexibility with which the code can be written.

This paper presents a simple implementation where the style of the underlying application is batch oriented, and in fact can be written in any language. The user interface must be cleanly defined in a set of parameters with optional values (e.g., named "*keyword=value*" pairs). Once the input values have been set, the application can be launched, results can be captured and presented in any way the script or application decides.

2. Tcl/Tk: TkRun

The GUI that is created will provide a simple interface to a program that is spawned by the GUI. This program must have a well defined Command Line

Interface (CLI), in the current implementation a "*keyword=value*" interface. Equally well, a Unix-style "*-option value*" could have been used (cf. Appleton's **parseargs** package). The GUI builder, a small 600 line C program called **tkrun**, scans the script for special tags (easily added as comments, which automatically make the script self-documenting), and creates a Tcl/Tk script from which the shell script itself (or any application of choice with a specified CLI) can be launched (see Figure 2 for a schematic).

The added power one gets with this interface builder is the simplified re-execution of the script, which gives the user a powerful tool to quickly examine a complex parameter space of a particular problem.

The input script must define a set of parameters, each with a keyword, an optional initial value, a widget style, usage requirements and one line help. The keyword widget style can be selected from a small set of input styles that standard Tcl/Tk provides (such as generic text entry, file browsers, radio buttons, sliders etc.)

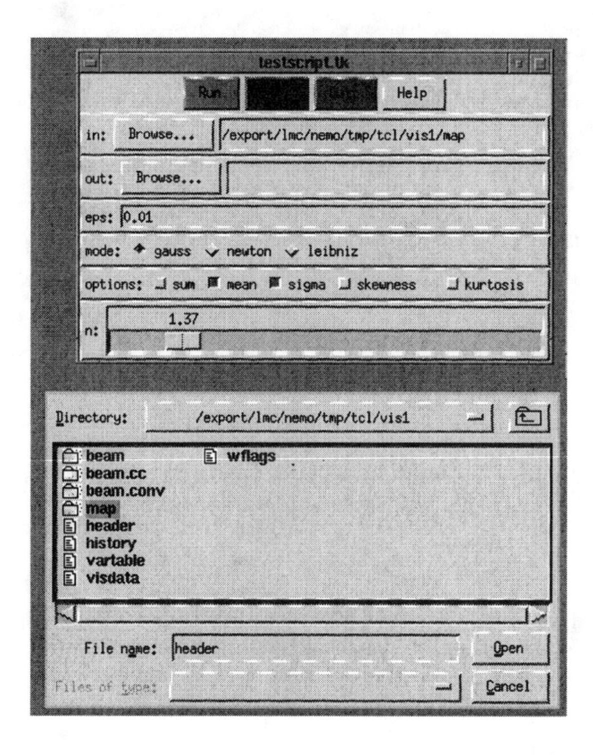

Figure 1. With the command "**tkrun testscript**" the upper panel is created, providing a simple interface to the "*key=val*" command line interface of the script **testscript** (see below). The lower panel is a standard Tcl/Tk filebrowser that can be connected to keywords that are meant to be files. See Figure 2 for a schematic diagram explaining the interaction between the different programs and scripts.

The current implementation has been tested under Tcl/Tk 7.6 as well as 8.0, but is expected to move along as Tcl/Tk is developed further. For example a more modern widget layout technique (`grid` instead of `pack`) should be used. Also keywords cannot have dependencies on each other, for example it would be nice to "grey out" certain options under certain circumstances, or allow ranges of some keywords to depend on the settings of others.

3. Sample Script: testscript

Here is an example header from a C-shell script with which Figure 1 was made. Note that the script must supply a proper "keyword=value" parsing interface, as was done with a simple foreach construct here. The latest version of `tkrun` is available through the NEMO[1] package.

```
#! /bin/csh -f
#                               :: define basic GUI elements for tkrun to extract
#>  IFILE   in=
#>  OFILE   out=
#>  ENTRY   eps=0.01
#>  RADIO   mode=gauss              gauss,newton,leibniz
#>  CHECK   options=mean,sigma      sum,mean,sigma,skewness,kurtosis
#>  SCALE   n=1                     0:10:0.01
#                               :: some one liners
#>  HELP    in      Input filename
#>  HELP    out     Output filename (should not exist yet)
#>  HELP    eps     Initial (small) step
#>  HELP    mode    Integration Method
#>  HELP    options Statistics of residuals to show
#>  HELP    n       Order of polynomial

#                               :: parse named arguments
foreach a ($*)
   set $a
end

#                               :: actual start of code
echo TESTSCRIPT in=$in out=$out eps=$eps mode=$mode options=$options n=$n
#                               :: legacy script can be inserted here or keyword
#                               :: values can be passed on to another program
```

Acknowledgments. I would like to thank Frank Valdes and Mark Pound for discussing some ideas surrounding this paper, and Jerry Hudson for his Graphics Command Manager FLO.

References

Appleton, Brad (parseargs, based on Eric Allman's version)
Judson, Jerry (FLO: a Graphical Command Manager)

---

[1] http://www.astro.umd.edu/nemo/

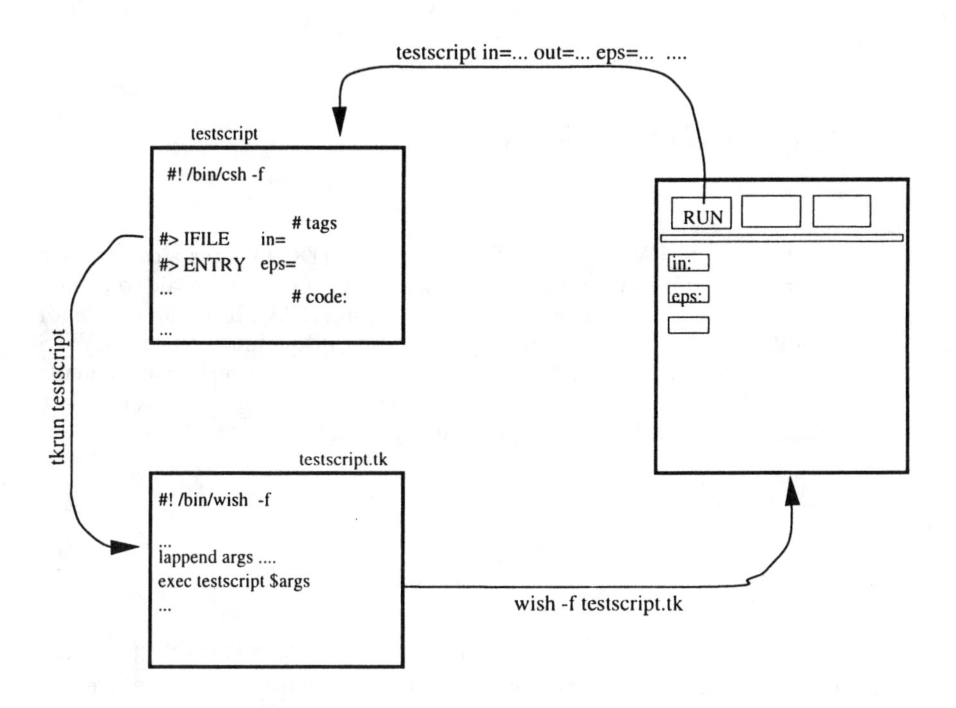

Figure 2. Flow diagram: The command **tkrun** scans the C-shell script **testscript** (top left) for keywords and the Tcl/Tk script **testscript.tk** (bottom left) is automatically written and run. It presents the keywords to the user in a GUI (on the right, see Figure 1 for a detailed view)), of which the "Run" button will execute the C-shell code in the script **testscript**.

Ousterhout, John, 1994, *Tcl and the Tk Toolkit*, Addison-Wesley

Shannon, P., 1996, in ASP Conf. Ser., Vol. 101, Astronomical Data Analysis Software and Systems V, ed. George H. Jacoby & Jeannette Barnes (San Francisco: ASP), 319

Teuben, P.J., 1995, in ASP Conf. Ser., Vol. 77, Astronomical Data Analysis Software and Systems IV, ed. R. A. Shaw, H. E. Payne & J. J. E. Hayes (San Francisco: ASP), 398

Astronomical Data Analysis Software and Systems VII
ASP Conference Series, Vol. 145, 1998
R. Albrecht, R. N. Hook and H. A. Bushouse, eds.

The Mosaic Data Capture Agent

Doug Tody and Francisco G. Valdes

IRAF Group, NOAO[1], *PO Box 26732, Tucson, AZ 85726*

Abstract. The Mosaic Data Capture Agent (DCA) plays a central role in connecting the raw data stream from the NOAO CCD Mosaic controller system to the rest of the data handling system (DHS). It is responsible for assembling CCD data from multiple readouts into a multiextension FITS disk file, for supplying a distributed shared image to real-time elements of the DHS such as the Real-Time Display (RTD), and for feeding data to the IRAF-based data reduction agent (DRA).

1. Overview

A brief design of the data handling system (DHS) for the NOAO CCD Mosaic Camera[2] was presented earlier (Tody, 1997). It is based on a message bus architecture (Tody, 1998) that connects the various components of the DHS. One such component is the *data capture agent* (DCA). The DCA receives streams of readout data from the acquisition system and builds a data object which is shared by other components of the DHS. The image data is stored on disk as a multiextension FITS file (MEF) (see Valdes 1997b).

The DCA is a general purpose network-based data service, using the message bus as the network interface. It receives request messages of various types and, through an event loop, dispatches each request to the appropriate event handler. The request message types currently implemented in the DCA are for control, setting or querying server parameters, readout status, data format configuration, and writing header and pixel data. In addition to servicing client requests, the DCA can broadcast messages to inform clients of the status of the DCA during readout.

The major subsystems of the DCA at present are the message bus interface, an event loop and event handlers for processing message requests, a distributed shared object implementation, a keyword database system, the TCL interpreter, and a TCL-based keyword translation module.

---

[1]National Optical Astronomy Observatories, operated by the Association of Universities for Research in Astronomy, Inc. (AURA) under cooperative agreement with the National Science Foundation.

[2]http://www.noao.edu/kpno/mosaic/

2. DCA User Interface

The DCA may be started by issuing a command at the host level, or invoked as a service via the message bus. This can occur during login to an observing account, using a window manager menu, or by interactively issuing a user command. The command invocation can include DCA server parameters which can also be set or reset after invocation via messages.

The DCA automatically connects to the message bus when it is started. The runtime operation of the DCA is monitored and controlled by clients via the message bus. In the Mosaic system there are two clients; the data feed client (DFC) and the DCA console client (DCAGUI). In general any number of clients can access the DCA. Multiple data feed clients or GUIs can be active simultaneously.

The DCA console client has a graphical user interface based on the IRAF Widget Server (Tody, 1996). It can send control and parameter messages to the DCA and act on messages broadcast by the DCA, e.g., to display the readout status. Currently the DCAGUI executes an autodisplay command for real time display during readout (see Valdes, 1998) and a post-processing command for logging, archiving or other operations once the readout ends.

3. The Readout Sequence

An observation readout is driven by a sequence of DCA message requests. A readout sequence is initiated by the DFC with a message that includes a unique sequence number. Once a readout has started, each subsequent message associated with that readout must be tagged with the sequence number for that readout. Multiple readouts can be simultaneously in progress. Each readout sequence has a separate context identified by its sequence number.

When a readout sequence is initiated the DCA creates a new image-class *distributed shared object* (DSO). The DFC passes information about the size and structure of the object (such as the number of image extensions), allowing the DCA to configure the DSO. At this point the DFC can begin sending pixel and header messages. At image creation time the DCA broadcasts a message so that clients like the DCAGUI can access it if desired. Currently the DCAGUI uses this to initiate an interim real-time display.

When the readout is completed the DCA executes a *keyword translation module* (KTM), an externally supplied, interpreted TCL script which converts detector specific information into standard header keywords. After the KTM finishes the DCA and DSO format the output keywords, write them to the image headers, and generation of the new image DSO is complete. The DCA broadcasts a message when the DSO is complete which can be used to trigger post-processing of the data.

3.1. Pixel Messages

The pixel data messages contain blocks of raw, unprocessed detector pixel data organized into one or more *streams*, one for each CCD amplifier. Each stream directs pixels to a region of an output image extension. This structure allows the data block to simultaneously contain data for several different regions and the data can be arbitrarily interleaved, encoded, flipped, or aliased. Each block of data is processed asynchronously but the client can send a synchronization request periodically to check the output status.

3.2. Keyword Messages

The DCA receives header information via the message bus. This information consists of blocks of keywords organized into named, detector specific *keyword groups*. The keywords are stored in keyword databases (an internal random access data structure), one per keyword group. The set of keyword group names is arbitrary. Some examples for the NOAO Mosaic are ICS for instrument control system, TELESCOPE for telescope, and ACEBn for Arcon controller information for controller n.

The Arcon controller system for the NOAO Mosaic consists of a set of Arcon controllers each of which can readout one or more amplifiers. The current system has four controllers each reading out two CCDs using one amplifier per CCD. Thus there can be controller information for the whole system, for each controller, and for each amplifier readout.

A keyword database library handles creation and maintenance of the keyword database. Note that keyword information does not necessarily have to come only from the DFC, the current mode of operation for the NOAO Mosaic. Other schemes are possible.

4. Keyword Translation Module

The keyword translation module (KTM) is a TCL script called by the DCA at the end of a readout, once all raw header information has been received. The purpose of the KTM is to create the keywords for the global and extension headers. The KTM is passed a list of input keyword database descriptors and it returns a list of output keyword database descriptors, one for each header. The DCA TCL interpreter provides special TCL commands for manipulating (creating, accessing, searching, etc.) the keyword databases. When the KTM finishes the DCA, via the DSO, writes the keywords in the returned output keyword databases to the output MEF FITS file.

The KTM performs a variety of transformations on the input keywords. A keyword can be copied verbatim if no change is desired. The keyword name or comment may be changed without changing the value. New keywords can be added with static information and default values may be supplied for missing keywords. The KTM can compute new keywords and values from input keywords. Identical keywords in each extension may be merged into a single keyword in the global header. The KTM can detect incorrect or missing keywords and print warnings or errors.

Two examples from the keyword translation module for the NOAO CCD Mosaic follow.

1. The data acquisition system provides the keywords DATE-OBS and UT-SHUT giving the UT observation date in the old FITS date format (dd/mm/yy) and the UT of the shutter opening. The KTM converts these

to TIME-OBS, MJD-OBS, OBSID, and the new Y2K-compliant FITS date format.

2. The KTM determines on which telescope the Mosaic Camera is being used and writes the WCS scale, orientation, and distortion information previously derived from astrometry calibrations for those telescopes. The coordinate reference point keywords, CRVAL1 and CRVAL2, are computed from the telescope right ascension and declination keywords.

5. Distributed Shared Objects and Distributed Shared Images

An important class of message bus component is the distributed shared object (DSO). DSOs allow data objects to be concurrently accessed by multiple clients. The DSO provides methods for accessing and manipulating the data object and locking facilities to ensure data consistency. DSOs are distributed, meaning that clients can be on any host or processor connected to the message bus. In the case of the Mosaic DHS, the principal DSO is the distributed shared image which is used for data capture, to drive the real-time display, and for quick look interaction from within IRAF. The distributed shared image uses shared memory for efficient concurrent access to the pixel data, and messaging to inform clients of changes to the image.

The current version of the DCA does not implement the full distributed shared image object. Instead it implements a mapped image file. The image being created appears as a valid FITS multiextension file immediately after the configuration information is received from the DFC. This allows applications to examine the file while the readout is in progress. An example of this is the interim display routine that loads a display server frame buffer as the pixel data is recorded, giving a simple real-time display capability.

6. Current Status and Future Developments

A version of the Data Capture Agent is in production use at two telescopes on Kitt Peak with the NOAO CCD Mosaic Camera. The flexibility of the message bus architecture is used to provide two modes of operation. The standard mode uses two machines connected by fast Ethernet. One machine supports the observing user interface and the interface to the controller system. A DFC running on this host writes to the message bus and the DCA runs on another machine (with a different OS) where the exposures are written to disk and the quick look interaction and any post-processing are done. The second mode is a fallback in case the second computer fails. The DCA can simply be run on the same machine as the UI and controller interface.

Future developments will complete the distributed shared object and console client and add other clients such as the real-time display (Tody, 1997) and data reduction agent (Valdes 1997a).

The design of the DCA (and the whole DHS based on the message bus architecture) is open, flexible, and efficient. It can be used with many data acquisition systems at a variety of observatories. All that is needed is to write a data feed client to connect to the message bus and a keyword translation module appropriate for the data. Currently the lower level messaging system is based on the Parallel Virtual Machine (PVM) library but this can be replaced in the future with other messaging systems such as CORBA.

References

Tody, D. 1996, in ASP Conf. Ser., Vol. 101, Astronomical Data Analysis Software and Systems V, ed. George H. Jacoby & Jeannette Barnes (San Francisco: ASP), 89

Tody, D. 1997, in ASP Conf. Ser., Vol. 125, Astronomical Data Analysis Software and Systems VI, ed. Gareth Hunt & H. E. Payne (San Francisco: ASP), 451

Tody, D. 1998, this volume

Valdes, F. 1997a, in ASP Conf. Ser., Vol. 125, Astronomical Data Analysis Software and Systems VI, ed. Gareth Hunt & H. E. Payne (San Francisco: ASP), 455

Valdes, F. 1997b, in ASP Conf. Ser., Vol. 125, Astronomical Data Analysis Software and Systems VI, ed. Gareth Hunt & H. E. Payne (San Francisco: ASP), 459

Valdes, F. 1998, this volume

Astronomical Data Analysis Software and Systems VII
ASP Conference Series, Vol. 145, 1998
R. Albrecht, R. N. Hook and H. A. Bushouse, eds.

Packaging Radio/Sub-millimeter Spectral Data in FITS

Zhong Wang

Smithsonian Astrophysical Observatory, Cambridge, MA 02138 U.S.A.
Email: zwang@cfa.harvard.edu

Abstract. We report on an experiment to incorporate the multiple extension FITS file format in building data processing software for spectroscopic observations in radio and millimeter-submillimeter wavelengths. By packaging spectral header information into table extensions and the actual spectral data into either table or image extensions, this approach provides a convenient and effective way to organize large amount of spectroscopic data, especially for pipeline-like processing tasks. Tests conducted based on a software package developed for the SWAS mission have demonstrated this to be a viable alternative to the conventional ways of organizing spectral data at these wavelengths.

1. Introduction: Characteristics of Submillimeter Spectral Data

Historically, spectral data were taken roughly one spectrum at a time, and most existing data formats reflect such a methodology. The traditional ways in which FITS can be used for spectral data are less than adequate in meeting the needs of a new generation of instruments that produce large amount of relatively homogeneous data, which nevertheless need to be reduced individually. A set of more efficient, better streamlined reduction procedures are necessary, which in turn require more careful considerations in data packaging. This is particularly important for projects that adopt non-interactive, pipeline-style data processing as their primary mode of operation.

The SWAS[1] (Submillimeter Wave Astronomy Satellite) mission is essentially a small orbiting radio telescope working at relatively high (about 500 GHz) frequencies (Wang 1995). It is designed to make simultaneous spectroscopic measurements at several sub-mm line wavelengths that are inaccessible from the ground. The spectral data have some special characteristics:

- Integration time for each individual "scan" is set to be short (on the order of a few seconds). High signal-to-noise ratios are achieved by co-adding a large number of such scans at the data processing stage;

[1]Details of the SWAS mission can be found on the internet in SAO's SWAS homepage at http://cfa-www.harvard.edu/cfa/oir/Research/swas.html

- Because of the noise properties and frequency shifts with time (due to the motion of the spacecraft as well as instrumental drifts), individual scans need to be calibrated separately before co-adding;

- More than one spectral band is recorded at any given time, thus a complete dataset consists of spectra of several different wavelength ranges from independent detectors. Yet they are of the same source and are taken under very similar observing conditions;

- Observations are broken up into individual "segments" each lasting typically 30–40 minutes. During each segment, many spectroscopic scans (aside from calibration data) are simple repeats under nearly identical conditions, and therefore share values of many header parameters with only slight variations (drifts).

The data reduction plan for the mission calls for a highly efficient pipeline processing scheme in which a large amount of spectroscopic data can be sorted, selected, calibrated and co-added based on some user-changeable criteria. All processing operations have to be recorded and traceable, and are preferably done with minimal human intervention.

2. Software Approach: Packaging Data in FITS

We have selected IRAF as our primary programming platform for the SWAS pipeline software, and use FITS as the main format for our data product. However, to store individual scan data in the conventional one- (single spectrum) or two-dimensional ("multispec" or echelle) format would be very inefficient and even confusing.

Our approach is to make use of the BINTABLE and IMAGE extensions of FITS and store spectral header and pixel data in these two types of extensions in a single FITS data file. The basic rules we adopted are:

- Each FITS file corresponds to a specific time interval of data taking (an observing segment). It contains a single BINTABLE extension and (optionally) several IMAGE extensions. It has a conventional header section containing some none-time-variable, shared parameters of the spectra;

- The table extension has one row per spectral scan taken, and stores all of the *time-variable* spectral header information corresponding to that scan (i.e., time, coordinates, frequency, noise statistics, etc);

- The pixel data of the spectrograph output (the spectrum itself) can be stored in one of two ways: they can either be individual "array elements" of a table row, or an image row similar to the two-dimensional arrays used to save Echelle and multi-fiber spectra in the optical.

  In case of using variable-length, vector (array) columns to store pixel data, a table row can have more than one array element, each representing data of a separate spectral band. The length of an array element is the number of pixels in a spectrum of that band.

In case of images, each image row corresponds to a row in the header parameter table. The number of columns of the image is the same as the number of pixels in a spectrum. There can be multiple image extensions in each data file, with different extensions representing data of different spectral bands.

The choice between using the table extension of FITS alone versus using table plus image extensions depends on the actual application. In principle, saving everything in a single table is structurally more appealing and efficient, but not many existing software tools can accept variable-length vector column FITS tables — which means more development work for basic tools. On the other hand, making use of 2-D images can facilitate data inspection procedures with various tools that examine FITS images. Its shortcoming is that information about a single spectrum is saved in two or more extensions, requiring care in managing them.

3. Advantages and Efficiency Gains

Given the lack of a widely-accepted standard in formatting a largely homogeneous spectroscopic dataset, we find the packaging using FITS extensions to be a viable alternative to the conventional ways in which spectral data are saved:

- It provides a convenient way to organize, sort, edit, monitor and visualize the large spectral dataset (much of it is done with existing tools within IRAF or those directly working with FITS), while preserving all the essential information of each individual spectrum;
- The programming of calibration tasks to process a relatively large, nearly-homogeneous dataset is very much simplified. In particular, the spectral header parameters can be processed either independently or as table columns. It also means that, for example, a subset of the header parameters of a large number of spectra can be easily manipulated without even accessing the actual pixel data;
- The efficiency of data processing is clearly enhanced. Since spectral pixel data are always processed as arrays, the fractional time taken by file I/Os and data buffering is substantially reduced when, for example, those data are accessed as part of an image.

Many of the existing software tools that deal with FITS table and image data are readily available for use with these new data files, sometimes with minor modifications. This can significantly save the programming effort and shorten the software development cycle.

4. Comments and Future Work

Despite the considerable merits of this approach, some problems remain to be addressed, and we are working on future enhancement of our software.

One of the main problems is the interface with existing software tools (such as visualization tools from certain existing tools packages). In practice, we were able to partly circumvent this shortcoming by writing customized interfaces, but the solutions are not always satisfactory. It appears that if a tabular style file format is deemed sufficiently important, a more fundamental (kernel-level) interface for spectral data in FITS should be developed.

The approach described in this paper is an explorative step in enhancing the use of FITS, especially in taking advantage of the multiple extension development and the use of vector columns in FITS tables. With the rapid advancement in database and data warehousing technology, one would ideally like to have a more rigorous database-like framework to deal with the management of the observational data. This, however, needs to be balanced with the requirement of cost-effective and quick prototyping of many projects' practical applications. The latter often means maximizing the use of existing software tools and adopting conventional approaches. The multiple extension FITS adopted by the community and being incorporated by NOAO and STScI's IRAF teams (e.g., Zarate & Greenfield 1996, Zarate 1998) and other software groups is an important and innovative development, which preserves the usefulness of many old tools while allowing new methods to be tested for astronomical data applications. We believe that more efforts are needed in exploring ways to take full advantage of this very useful data format.

References

Wang, Z. 1995, in ASP Conf. Ser., Vol. 77, Astronomical Data Analysis Software and Systems IV, ed. R. A. Shaw, H. E. Payne & J. J. E. Hayes (San Francisco: ASP), 402

Zarate, N. & Greenfield, P. 1996, in ASP Conf. Ser., Vol. 101, Astronomical Data Analysis Software and Systems V, ed. George H. Jacoby & Jeannette Barnes (San Francisco: ASP), 331

Zarate, N. 1998, this volume

Acknowledgments. We are grateful to Phil Hodge and Nelson Zarate for their help in the software development discussed in this paper. Dr. Matthew Ashby has participated in implementation and testing of the SWAS software.

Astronomical Data Analysis Software and Systems VII
ASP Conference Series, Vol. 145, 1998
R. Albrecht, R. N. Hook and H. A. Bushouse, eds.

The IDL Wavelet Workbench

M. Werger

Astrophysics Division, Space Science Department of ESA, ESTEC, 2200 AG Noordwijk, The Netherlands, EMail: mwerger@astro.estec.esa.nl

A. Graps

Stanford University, Center for Space Science and Astrophysics, HEPL Annex A210, Stanford, California, 94305-4085 EMail: amara@quake.stanford.edu

Abstract. Progress in the development of the 1996 release of the IDL Wavelet Workbench (WWB) is shown. The WWB is now improved in several ways, among them are: (1) a smarter GUI which easily directs the user to the possibilities of the WWB, (2) the inclusion of more wavelets, (3) the enhancement of the input and output modules to provide a better interface to the input and output data and (4) the addition of more analysis methods based on the wavelet transform.

1. Introduction

One of the most advanced packages for wavelet analysis is probably *Wavelab*[1] written for MATLAB. New insights have been gained in many other fields by applying wavelet data analysis, thus it was a reasonable task for us in astronomical research to translate most of the code from the *Wavelab* package into IDL (Interactive Data Language, by Research Systems, Inc.). IDL was chosen because of its wide-spread availability in the astronomical community and because of its development environment. The last official version of the so-called *IDL Wavelet Workbench* (WWB) was in the Spring of 1996. It has been made publicly available at the ftp site of Research Systems, Inc.[2].

2. The 1996 version of IDL

The 1996 version of the WWB consists of 111 different modules with approximately 10,000 lines of code in total. Approximately all modules have been written or translated from MATLAB code into IDL by AG. The 1996 version can be run either from the IDL command line or from a graphical user interface (GUI).

---

[1] http://stat.stanford.edu/~wavelab/

[2] ftp://ftp.rsinc.com/

The WWB is written in a highly modularized way to be easily maintained and improved. In the 1996 version, COMMON blocks are used to store important variables for the different routines. These COMMON blocks can be set also from the command line. Therefore, it is possible to use the WWB as a stand-alone package and also as a library to supplement ones own IDL routines.

The 1996 WWB provides simple input and output routines. Its analysis and plotting libraries are sophisticated and employ most of the typical methods used in wavelet analysis like the *Discrete Wavelet Transform*, *Multiresolution Analysis*, *Wavelet Packet Analysis*, *Scalegram*, and *Scalogram*. In addition, the 1996 WWB offers typical routines for de-noising and compression of one- and two-dimensional data. The available set of wavelets is restricted up to four important families: the Haar-wavelet and the families of the Daubechies-wavelets, Coiflets, and Symmlets.

3. Current Developments

The 1996 release the IDL WWB has been widely used for different tasks such as pattern detection, time-series analysis and de-noising of data. A lot of useful routines have been added to the WWB since 1996, or they are foreseen to be included.

- The current version makes use of the most recent changes to IDL (version 5.0.2); now WWB uses pointers to handle arbitrary data arrays. Also, the WWB command line interface and the GUI may be used at the same time.

- The GUI has been simplified; now it includes more possibilities, but with an easier interface and a less complicated dialog structure.

- All necessary variables are now kept in two IDL data structures, those variables also may be set from the command line.

- The data input portion of the WWB has been upgraded to handle FITS-files; the output portion of WWB has been upgraded so that one can use the GUI to set PostScript output.

- More analysis routines are now available. In additional to the forward DWT, now the backward DWT (*IWT*) has been included to show possible differences between the original and transformed data. A continuous wavelet transform using the Gauss, Sombrero, and Morlet wavelets has been added also.

- The capabilities for time-series analysis has been greatly enhanced by adding wavelets and routines which improve period detection. For example, a routine has been added for detecting periods in unevenly-sampled time-series, and eleven new wavelet filters are provided.

- The computations can now allow datasets more than 32767 points long.

- Plotting capabilities of the *Scalogram* have been improved.

- For a better understanding of the wavelet transform, a GUI for manipulating specific wavelet coefficients has been included. This greatly improves the learning and analyzing process.

4. Future Plans

There are some future plans for integrating capabilities to analyze multidimensional data and adding additional routines. Suggestions and contributions from the user community are greatly welcome.

Acknowledgments. The 1996 WWB has been partly funded by RSI, Inc.

Astronomical Data Analysis Software and Systems VII
ASP Conference Series, Vol. 145, 1998
R. Albrecht, R. N. Hook and H. A. Bushouse, eds.

IRAF Multiple Extensions FITS (MEF) Files Interface

Nelson Zarate

National Optical Astronomy Observatories

Tucson, AZ 85719 (zarate@noao.edu)

Abstract. The Multiple Extension FITS (MEF) file interface is an IRAF library providing facilities for general file operations upon FITS multi-extension files. The MEF library has been used as the basis for a set of new IRAF tasks providing file level operations for multi-extension files. These operations include functions like listing extensions, extracting, inserting, or appending extensions, deleting extensions, and manipulating extension headers. MEF supports extensions of any type since it is a file level interface and does not attempt to interpret the contents of a particular extension type. Other IRAF interfaces such as IMIO (the FITS image kernel) and STSDAS TABLES are available for dealing with specific types of extensions such as the IMAGE extension or binary tables.

1. Introduction

The Multiple Extensions FITS (MEF) interface consists of a number of routines to mainly read a FITS Primary Data Unit or an Extension Unit and manipulate the data at a file level. It is up to the application to take care of any details regarding data structuring and manipulation. For example, the MEF interface will read a BINTABLE extension and give to the calling program a set of parameters like dimensionality, datatype, header buffer pointer and data portion offset from the beginning of the file.

Currently the routines available to an SPP program are:

- `mef = mef_open (fitsfile, acmode, oldp)`
- `mef_rdhdr (mef, group, extname, extver)`
- `mef_rdhdr_exnv (mef, extname, extver)`
- `mef_wrhdr (mefi, mefo, in_phdu)`
- `[irdb]val = mefget[irdb] (mef, keyword)`
- `mefgstr (mef, keyword, outstr, maxch)`
- `mef_app_file (mefi, mefo)`
- `mef_copy_extn (mefi, mefo, group)`

- `mef_dummyhdr (fd, hdrfname)`

[irdb]: int, real, double, boolean.

2. Initializing Routine

2.1. mef = mef_open (fitsfile, acmode, oldp)

Initializes the MEF interface. Should be the first routine to be called when performing operations on FITS files using this set of routines. Returns a pointer to the MEF structure.

fitsfile Pathname to the FITS file to be open. The general syntax is:

dir$root.extn[group]

- dir: Directory name where the file resides
- root: Rootname
- extn: (optional) Extension name — can be any extension string
- group: Extension number to be opened

The '[group]' string is optional and is not part of the disk filename. It is used to specified which extension number to open. The extension number is zero based — zero for the primary extension, 1 for the first extension, and so on.

acmode The access mode of the file. The possible values are:

READ_ONLY, READ_WRITE, APPEND, NEW_FILE

oldp Not used. Reserve for future use.

3. Header Routines

3.1. mef_rdhdr (mef, group, extname, extver)

Read the FITS header of a MEF file that matches the EXTNAME or EXTVER keyword values or if not specified, read the extension number 'group'. If no extension is found an error is posted. After reading the header the file pointer is positioned at the end of the last data FITS block (2880 bytes).

mef The MEF pointer returned by mef_open. When the routine returns, all of the elements of the MEF structure will have values belonging to the header just read.

group The extension number to be read — zero for the Primary Data Unit, 1 for the first extension, and so on. If you want to find out an extension by the value of extname and/or extver then 'group' should be -1.

extname The string that will match the EXTNAME value of any extension. The first match is the extension header returned.

extver The integer value that will match the EXTVER value of any extension. If 'extname' is not null then both values need to match before the routine returns. If there are no values to match then 'extver' should be INDEFL.

3.2. mef_rdhdr_gn (mef,group)

Read extension number 'group'. If the extension number does not exist, an error is posted.

mef The MEF pointer returned by mef_open. When the routine returns, all of the elements of the MEF structure will have values belonging to the header just read.

group The extension number to be read — zero for the Primary Data Unit, 1 for the first extension, and so on.

3.3. mef_rdhdr_exnv (mef,extname, extver)

Read group based on the Extname and Extver values. If the group is not encountered, an error is posted.

mef The MEF pointer returned by mef_open. When the routine returns, all of the elements of the MEF structure will have values belonging to the header just read.

extname The string that will match the EXTNAME value of any extension. The first match is the extension header returned.

extver The integer value that will match the EXTVER value of any extension. If 'extname' is not null then both values need to match before the routine returns. If there are no value to match then 'extver' should be INDEFL.

3.4. mef_wrhdr (mefi, mefo, in_phdu)

Append the header from an input PHU or EHU to output file.

mefi The input file MEF pointer returned by mef_open. The header should have been read by now.

mefo The output file MEF pointer returned by mef_open.

in_phdu Boolean value (true, false) stating whether the input header is the primary header or not.

3.5. [irdb]val = mefget[irdb] (mef, keyword)

[irdb]: integer, real, double or boolean.

Get a FITS header keyword value of the specified datatype; for example 'imgeti (mef, "NCOMB")' will return an integer value from the keyword 'NCOMB'.

mef The input file MEF pointer returned by mef_open. The header should have been read by now.

keyword The input string (case insensitive) keyword from which to return its value.

3.6. mefgstr (mef, keyword, outstr, maxch)

Get the string value of a FITS encoded card. Strip leading and trailing whitespace and any quotes.

mef The input file MEF pointer returned by mef_open. The header should have been read by now.

keyword The input string (case insensitive) keyword from which to return its value.

outstr The output string with the value of input keyword.

maxch Length in chars of **outstr**.

4. File Operations

4.1. mef_app_file (mefi, mefo)

Appends a FITS file to an output file. If the file does not exist, a dummy Primary Header Unit is first created.

mefi The input file MEF pointer returned by mef_open. The header should have been read by now.

mefo The output file MEF pointer returned by mef_open.

4.2. mef_copy_extn (mefi, mefo, group)

Copy a FITS extension given by its number 'group' into an output file. If the file does not exists, this extension becomes a Primary Header Data Unit of the output FITS file. If the output file already exists, the input extension gets appended.

mefi The input file MEF pointer returned by mef_open. The header should have been read by now.

mefo The output file MEF pointer returned by mef_open.

group The input extension number to be appended to the output file.

4.3. mef_dummyhdr (fd, hdr_fname)

Write a dummy Primary Header Unit with no data to a new file. Optionally a header file with user keywords can be used.

fd The output file descriptor.

hdr_fname The header filename. This is text file with a FITS header syntax that will be appended to the file. Each FITS card does not have to be 80 characters long. The routine takes care of the correct padding.

Part 3. Computational Infrastructure & Future Technologies

Astronomical Data Analysis Software and Systems VII
ASP Conference Series, Vol. 145, 1998
R. Albrecht, R. N. Hook and H. A. Bushouse, eds.

Cost-Effective System Management

Skip Schaller

Steward Observatory, University of Arizona, Tucson, AZ 85721

Abstract. Quality system management of computer workstations for astronomy can be achieved with relatively small manpower requirements, if the right cost-effective administrative design decisions are made. Recent dramatic changes in the price/performance ratio of computer hardware have modified the model used to distribute computer resources and especially the usage of the network.

1. Introduction

As part of my duties, I manage the Steward Observatory Computer Support Group. Our small group provides programming services to several observatory projects, as well as system administration services for a large number of hosts on our network. We provide full support for approximately 100 Sun workstations, 70 X terminals, 40 network printers, 40 VxWorks real-time computers, and 20 networking boxes, such as routers, switches, and terminal servers. In addition, we provide limited support for some 400 other hosts, which are mostly PCs. Most of these 670 hosts are located in our main building on campus, which has over 200 offices, but others are located in a lab on the other side of campus, and at four different mountain-top observatory sites. This is all done with essentially one system manager.

2. What is Quality System Management?

Despite such a thin staff for such a large responsibility, I am told that the quality of our system management is second to none, by those who have had the opportunity to compare with other places. How does this quality manifest itself? Reliability is one of the best indicators. Our users enjoy 99.9 percent up time on our systems. Interdependence is minimized. The system is designed so that one doesn't depend on four or more hosts, (such as a software server, cpu server, data file server, font server, etc.) to get work done. Another good indicator is that the systems work as advertised. The mail always goes through. Software is not installed on the system in a non-functional or semi-functional form. Users are provided with a default environment that hides gory computer system and network details, so that they can get right to their job or science application. This environment appears the same everywhere on the network. Finally, users get a prompt response to their questions and problems.

3. A Few Good Men

So, how do we accomplish all this? The first thing to do is to hire the right people. In our university environment, we are often pressured to hire students. However, our experience has generally shown that even though one can hire several students for the same money as one highly experienced computer professional, the professional will produce more useful work than all the students combined. A person that is less than 25 years old just doesn't compare with a person that has more than 25 years of experience. Quality outweighs quantity.

4. Homogeneity

Certainly, another way to maximize the useful work produced by a system manager is to make everything, hardware and software, as homogeneous as possible. Several more instances of identical workstations require much less extra work to maintain, than several, even similar workstations that each need to be individually custom-tailored. Reducing the number of degrees of freedom, or the number of variables to deal with, reduces the overall work load. The secret is in choosing exactly which degrees of freedom to eliminate so that the load on the system manager is lightened, without seriously inconveniencing the users. The choice of which variables to eliminate is made by determining which ones are the most burdensome to administer and least useful to the user. This determination is based on experience.

5. Just Say No

How do we achieve this homogeneity? A large part of my job is to just say no. When a user asks me to support a certain hardware or software item, I must determine its money or manpower cost to our group. If the item is just another instance of something we already have or do, then the incremental cost is usually very low, and I can say yes. If the item is new or different, the cost is usually much greater, and if it doesn't fit into my money or manpower budget, I must say no. With a fixed amount of money or manpower resources, supporting a more widely varied list of hardware or software items will reduce the quality of support for those items already on the list. Why have more alternatives to choose from, if that means that fewer or none of them will work very well? Why go to a restaurant that has a very long list of meals on its menu, if none of them taste very good. In the long term, most users will understand that quality outweighs quantity.

6. Trends in Technology

Finally, keeping up with trends in technology is another way to reduce the cost of system management. One important trend currently happening is the rapid decrease in the price of ether switches. Steward Observatory has just recently changed over from a routed net to a switched net. This has given us more effective bandwidth, reduced the cost of spares since switches are cheaper than routers, and has reduced the administrative burden, because switches are easier to manage than routers and because there is now only one subnet to deal with instead of many.

The other important trend is the rapid decrease in the price/performance ratio of CPUs, memory, disks, and tape drives, and the fact that networking, while improving, is not keeping up with the improvements in the other areas. This is leading us to a server-less model of distribution of computer resources. A server that exports software to many clients is no longer necessary when the client can hold all the software it needs on 200 dollars worth of disk space. A tape server is no longer needed when tape drives are so inexpensive and every client can afford one. Not only the purchase and administrative costs of the server are saved, but also the cost of making the underlying network carry that load. The office desk-top workstation soon adopts the model of a home computer, an autonomous machine that has all the resources it needs attached locally.

Astronomical Data Analysis Software and Systems VII
ASP Conference Series, Vol. 145, 1998
R. Albrecht, R. N. Hook and H. A. Bushouse, eds.

Other People's Software

E. Mandel and S. S. Murray

Smithsonian Astrophysical Observatory, Cambridge, MA 02138, Email: eric@cfa.harvard.edu

Abstract. Why do we continually re-invent the astronomical software wheel? Why is it so difficult to use "other people's software"? Leaving aside issues such as money, power, and control, we need to investigate practically how we can remove barriers to software sharing. This paper will offer a starting point for software cooperation, centered on the concept of "minimal software buy-in".

1. Introduction

What shall we do about "other people's software"? There certainly is a lot of it out there. As software developers, we are well aware of the benefit of sharing software between projects. So why do we resist using other people's software? How is it that we find reasons to ignore existing software by repeating the mantra, "of course we should share, but in this particular case ...".

The factors that cause us to ignore the software of others are many and varied. A cynic might argue that it all comes down to money, power, and control. There is truth to such a position, but dealing with economic, social, and psychological factors is well beyond the scope of this short paper – or maybe any paper! Moreover, it may be that these issues always will hinder efforts to share software, unless and until our culture changes or our genes improve. But we still are left with the question of whether we can do anything today to improve the situation regarding software sharing – even while we acknowledge that our efforts may be only chipping away at the edges of a much larger problem.

We can, for example, try to expose technical impediments to sharing software. And surely the first barrier to using other people's software is "buy-in", that is, the effort needed to adopt and use such software. At first glance, it would seem that buy-in encompasses technical issues such as:

- the relative ease of installation of software
- the time and effort required to learn enough about the software to be able to use and evaluate it
- the design changes required to utilize the software – that is, the architectural assumptions made by the software itself

But these factors are not the whole story. Our basic attitude toward software buy-in, our very willingness to consider using the software of others, has changed with changing times.

2. Three Eras of Software Buy-in

From the Dark Ages until the mid-1980's, we lived in a Proprietary Era that demanded total buy-in. This was the age of isolated mini-computers running closed operating systems that were written in assembly language. Given the enormous cost of these machines and the different interfaces they presented to users and developers alike, there was little possibility of utilizing equipment from more than one vendor.

The astronomical software we wrote was influenced heavily by this proprietary culture: it was tailored to individual platforms with little or no thought to portability. A lot of our code was written in assembly language to save memory in these 64 Kb environments and to increase performance on slow CPUs. We also used FORTRAN compilers that incorporated vendor-specific language extensions, program overlays, and other non-portable techniques.

All of these efforts served the aim of optimizing software meant to be used only within individual projects. The central idea behind our efforts was to build the best possible software for our own systems. Indeed, our community was "data-center centered": astronomers visited data centers and telescopes to utilize project hardware and software on their data. Under such circumstances, buy-in required duplication of hardware and thus was a decision made by high-level management.

The Proprietary Era came to an end in the mid-1980's with the rise of workstations based on the Unix operating system. The central strength of Unix was its portability, and by adopting this little known operating system, workstation vendors changed the climate of software development overnight. We witnessed the rise of the Consortium Era, in which portable application programming interfaces (APIs) were developed by consortia of (often competing) organizations. The best example of this sort of effort was the X Consortium, an alliance of 75 large and small companies who agreed to standardize graphics and imaging for workstations on the X Window System. In doing so, they made possible unparalleled opportunities for developing portable software on a wide spectrum of machines.

In the astronomical community, the new push to portability led to the development of "virtual" analysis environments such as IRAF, AIPS, MIDAS, and XANADU. These systems offered sophisticated analysis functionality for almost all of the popular machines used in astronomy. Designed to be complete environments for user analysis, they offered buy-in at the architectural/API level. Once an analysis environment was chosen for a given project, buy-in was accomplished by mastering that environment and then tailoring the software design to exploit its strengths and evade its weaknesses. API-based buy-in established a new set of expectations for software development and use.

We believe that the Consortium Era is over, and that we have entered into a new Free-For-All Era. The most visible evidence of this change is the recent demise of the X Consortium. But this trend away from consortia-based software also is seen in the astronomical community, where there has been a weakening of long-term alliances between development groups. This weakening is an outgrowth of the maturity of our current analysis systems: with an overwhelming amount of software increasing the overlap between systems, it has become harder to choose between them. Furthermore, new development tools such as Java and Tcl/Tk have increased the pace of software creation, while shortening individual program lifetimes. The current watch-word seems to be "let's run it up the flag pole and see who salutes". Software is created, offered, and abandoned with startling rapidity. It is little wonder that consortia cannot keep pace with this explosion of software. Their decline has given way to temporary alliances that exploit the latest technology offering.

3. Minimal Buy-in Software

In such a fast-paced world, everyone is hedging their bets. It is becoming increasingly difficult to choose between software offerings whose longevity is questionable. It is even harder to invest time and effort in rapidly changing software. With so much software and so much uncertainty, we try to use everything and commit to nothing. Having little time or patience to investigate new software, we demand that software be immediately usable, with no buy-in at all, before we are willing to try it out.

For example, distributed objects are being hailed as the key to using other people's software. And indeed, the concept of hundreds of black-box services being available on a wide area "message bus" is very appealing. It promises a new world in which individual missions and telescopes can offer services specific to their data, while in turn making use of services provided by other groups.

But the reality of distributed objects does not match the advertising. Current schemes (CORBA, ToolTalk, OLE) require substantial architectural or even hardware buy-in. These systems have complex APIs, and some of them even have new language requirements. Such a high buy-in cost presents a dilemma: who will commit first to a complex distributed object architecture? Who is willing to build software that only will run, for example, on a ToolTalk-enabled platform, thereby shutting out users who do not run ToolTalk (for example, nearly all Linux users)? Such a design decision simply is not viable in a distributed community such as ours, where portability is taken for granted. We are lead to the conclusion that community buy-in is necessary in order for current distributed object schemes to succeed – and this brings us back to the original problem!

Perhaps we need a new way of looking at the problem of other people's software. Perhaps we need "minimal buy-in" software, that is, software that is too easy *not* to try out.

The key to minimal buy-in software is that it seeks to hide from its users the complexity of complex software. This means striking a balance between the extremes of full functionality (in which you can do everything, but it is hard to do anything in particular) and naive simplicity (in which it is easy to do the obvious things, but you can't do anything interesting). Minimal buy-in acknowledges that design decisions must be made up-front in order to achieve this balance.

Another way of expressing this is to say that minimal buy-in software caters to developers as if they were users. It achieves ease of use for developers by emphasizing simplifications such as:

- easy installation with auto configuration: untar, make, go ...
- no special system set-up, and especially no need for root privileges
- immediate functionality on the desktop, so that it can be tried and evaluated easily
- integration with already-familiar user and developer tools
- use of familiar files (ASCII, FITS)

These concepts often will lead to design and implementation features that are different from the usual conclusions of software engineering and computer science. For example, applying minimal buy-in concepts to message buses and distributed objects might lead to "non-classical" requirements such as:

- no need for a special intermediate message-passing process; use "whatever is around"
- configure using ASCII files
- send and receive messages/data at the command line
- utilize a familiar pattern-matching syntax for broadcasting

We need more research in the area of minimal buy-in software. We do not know, for example, how an effort to develop minimal buy-in software relates to more sophisticated implementations. Would such software be seen as precursors to a full system? Or would it be accepted as a full replacement? The balance between functionality and ease of use needs to be explored further to gain experience with minimal buy-in techniques.

At SAO, we are working on minimal buy-in messaging, using the X Public Access (XPA) mechanism as a base-line. We are extending XPA's point-to-point functionality to support broadcast messaging using well known data formats and pattern-matching syntax. We will maintain XPA's popular command-line support (xpaset and xpaget), which also provides a simple programming interface. Our evolving efforts are available at http://hea-www.harvard.edu/RD/.

It should be emphasized once again that the concept of minimal buy-in is only one step toward the software cooperation that is so elusive to our community. Issues of money, power, and control still loom large in the background of any discussion of software sharing. But it remains true that we need to explore such partial solutions while working on the larger issues.

Acknowledgments. This work was performed in large part under a grant from NASA's Applied Information System Research Program (NAG5-3996), with support from the AXAF Science Center (NAS8-39073).

Astronomical Data Analysis Software and Systems VII
ASP Conference Series, Vol. 145, 1998
R. Albrecht, R. N. Hook and H. A. Bushouse, eds.

Message Bus and Distributed Object Technology

Doug Tody

IRAF Group, NOAO[1], PO Box 26732, Tucson, AZ 85726

Abstract.
In recent years our applications have become increasingly large and monolithic, despite successful efforts to structure software internally at the class library level. A new software architecture is needed to break these monolithic applications into reusable components which can easily be assembled to create new applications. Facilities are needed to allow components from different data systems, which may be very different internally, to be combined to create heterogeneous applications. Recent research in computer science and in the commercial arena has shown us how to solve this problem. The core technologies needed to achieve this flexibility are the *message bus*, *distributed objects*, and *applications frameworks*. We introduce the concepts of the message bus and distributed objects and discuss the work being done at NOAO as part of the Open IRAF initiative to apply this new technology to astronomical software.

1. Overview

In this decade we have seen our software become increasingly large and complex. Although our programs may be well structured internally using hierarchically structured class libraries, the programs have grown large and monolithic, with a high degree of interdependence of the internal modules. The size of the programs and their relatively high level, user oriented interface makes them inflexible and awkward to use to construct new applications. As a result new applications usually have to be constructed at a relatively low level, as compiled programs, an expensive and inflexible approach. The high degree of integration characteristic of programming at the class library level makes it difficult to construct heterogeneous applications that use modules from different systems.

The key technology needed to address this problem, being developed now by academia and commercial consortiums, is known as *distributed objects.* Distributed objects allow major software modules to be represented as objects which can be used either stand-alone or as components of distributed applications. Tying it all together is the *message bus*, which provides flexible services and methods for distributed objects to communicate with one another and share

[1]National Optical Astronomy Observatories, operated by the Association of Universities for Research in Astronomy, Inc. (AURA) under cooperative agreement with the National Science Foundation.

data. Applications are built by linking precompiled components and services together at runtime via the message bus. This paper presents the message bus and distributed object framework being developed by NOAO and collaborators as part of the Open IRAF initiative. This project is funded in part by the NASA ADP and AISR programs.

2. Message Bus Concepts

The message bus is a facility used to bind, at runtime, distributed objects (program components) to construct applications. Components, which are things like data services, display or graphics services, computational services, or program interpreters, execute concurrently and communicate at runtime via the message bus. Components can dynamically connect to or disconnect from the message bus at runtime. The usage of the term "bus" is analogous to the hardware bus in a computer: components are like cards that plug into the bus, and the application program executes in an interpreter component analogous to the CPU. In effect the message bus framework is a virtual machine, built from highly modular, interchangeable components based on an open architecture, with applications being the software available for this virtual machine.

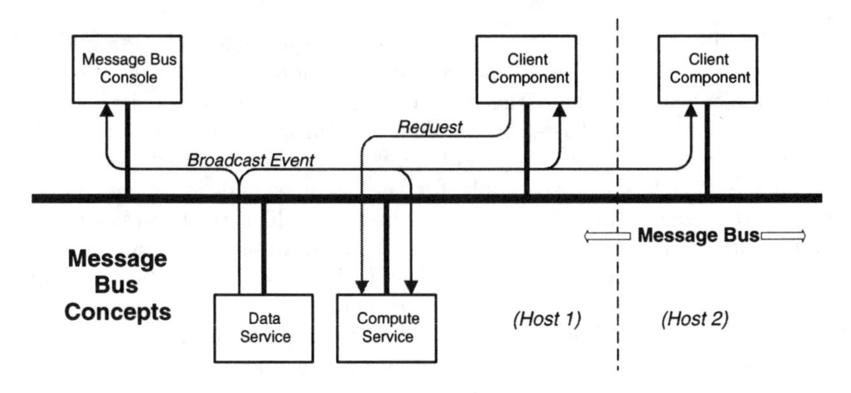

Figure 1. Message Bus Concepts

The message bus is also a means of structuring software, in that it defines a system-level software architecture. Classical structuring techniques are used to construct the individual components, which can be very different from one another internally, for example written in different languages. Applications are composed at a high level, relying upon components for most of their functionality. Components can be large, substantial programs in their own right, capable of independent execution, although from the point of view of the message bus architecture they are merely interchangeable software components which share a standard interface to the message bus. Other components will be custom built compute modules (e.g., implementing science algorithms) developed for the application. *Applications frameworks* are integrated suites of components providing a complete environment for building some class of applications.

3. Message Bus Capabilities

In addition to providing a range of messaging capabilities, the message bus provides facilities for service registration and lookup, autostart of services, and distributed execution. Message bus clients (e.g., components or services) can execute on any host computer connected to the message bus, allowing distributed applications to be easily constructed. The message bus can be either started when an application is run, or "booted" stand-alone in which case a subsequent client will automatically connect to and use the existing message bus. A console client is used to examine and control the state of the message bus and monitor system activity.

As illustrated in Figure 1, most messages fall into one of two primary classes. *Requests* are messages sent to a particular component or object to request that it perform some action (i.e., to invoke a method). Requests may be either synchronous or asynchronous. A synchronous request is similar to a remote procedure call. *Events* are messages which are broadcast by a "producer" client, and received by zero or more "consumer" clients. Clients subscribe to the classes of events they which to receive. The message bus keeps track of the event subscription list for each client and uses it to construct a distribution list when broadcasting an event. In general a producer does not know what other clients, if any, may be consuming the events it generates. Events are inherently asynchronous.

At the simplest level the message bus is responsible only for classifying messages by type and handling message distribution and delivery. The actual content of a message, that is the data contained in a message, is determined by the *messaging protocol* used by the clients. Multiple messaging protocols are possible on the same bus although not necessarily recommended.

The message bus is responsible for the reliable delivery of messages and ensures that messages are received in the order in which they were sent. There is no fixed limit on the size of a message, although other data transport mechanisms such as shared memory may be useful if bulk data is to be transmitted, particularly if it is to be shared by multiple clients. The bus will queue messages for delivery if needed. Point-to-point links are possible in cases where a high data rate is required and only two clients are involved.

4. Existing Messaging Implementations

The first thing we did when we started this project was research on all the existing implementations we could find of anything resembling a message bus. Approximately two dozen were examined, including CORBA, OLE/COM, Tooltalk, some preliminary Java-based implementations, PVM, MPI, ACE, KoalaTalk, EPICS, IMP, Glish, XPA, and others. We were surprised to find that despite all the work in this area, no adequate solution existed. CORBA probably comes closest: it is comprehensive and is based on some very good computer science, but it is expensive, proprietary, bloated, implementations differ, and it is only available on a subset of the platforms we use. CORBA might be a good choice for a local system but it is problematic for use in a product which will be widely distributed and freely available. The chief competition is OLE/COM from Microsoft, which is only available for Windows. Of the freeware implementations we found PVM (Parallel Virtual Machine) to be the most interesting. Although it was developed by the physics community for parallel computation and it is not really a message bus, it is compact, portable, and efficient; the basic facilities provided are good and provide much of the low level functionality needed for a message bus.

5. Open IRAF Message Bus Research

The NASA ADP-funded Open IRAF initiative (http://iraf.noao.edu/projects/) will eventually replace IRAF by a more modern, open system. This will be composed of products which will be usable stand-alone or as modules in other systems. The existing IRAF applications will be migrated to this new framework. The message bus architecture discussed here is being used to develop Open IRAF. In keeping with the Open IRAF approach, the message bus software will be a separate product usable independently of the future IRAF system.

Since no suitable message bus implementation is currently available and the technology is likely to continue to evolve rapidly in any case, our approach has been to develop a message bus API which can be layered upon existing standard low level messaging products. A prototype based on PVM has been developed and is in use now within the Mosaic Data Handling System at NOAO. We are also developing a *distributed shared object* (DSO) facility which will allow data objects such as images to be simultaneously accessed by multiple clients via the message bus. Distributed shared memory techniques are used to provide efficient access to bulk data. Messaging is used to provide the access interface used by clients, to inform clients of any changes to the data object, and to ensure data integrity.

6. Conclusions

The message bus, distributed shared object technology, and applications frameworks based on the message bus provide a powerful way to structure large applications and data systems to control complexity. Applications can be developed at a high level, relying upon sophisticated, well tested components for much of their functionality.

Systems based on the message bus architecture are inherently modular and open, allowing very different sorts of components to be intermixed. Since components can be large and complex products, with few restrictions on how they are implemented internally, it becomes easier for a large number of people to make significant contributions to a data system, reducing the dependence on key personnel.

Frameworks and applications suites layered on a common message bus have the potential to allow different data analysis packages to share the same low level infrastructure. Systems based on the message bus and distributed objects have the potential to combine the resources of disparate data systems groups as well as individual developers in the astronomical community.

Astronomical Data Analysis Software and Systems VII
ASP Conference Series, Vol. 145, 1998
R. Albrecht, R. N. Hook and H. A. Bushouse, eds.

Building Software from Heterogeneous Environments

M. Conroy, E. Mandel and J. Roll

Smithsonian Astrophysical Observatory, Cambridge MA 01801

Abstract. The past decade has witnessed a movement within the astronomical software community towards *Open Systems.* This trend has allowed projects and users to build customized processing systems from existing components. We present examples of user-customizable systems that can be built from existing tools, based on commonly-used infrastructure: a parameter interface library, FITS file format, Unix, and the X Windows environment. With these common tools, it is possible to produce customized analysis systems and automated reduction pipelines.

1. Introduction

Users and developers are confronted everyday with the challenge of cobbling together existing software to solve problems. However, much of the available software has features, limitations, and architectural assumptions that make it useless for new applications. It always is worth reminding ourselves that the primary purpose of software is to solve the users' problem: *Subject Oriented Software* (Coggins 1996) and not just to use trendy technology.

Currently, there is no set of uniform interfaces for the large body of existing astronomical software. Re-using a piece of software is complicated by the architectural buy-in of large systems which require mutually exclusive environments. (Mandel & Murray 1998).

Controlling complexity is the major problem facing software projects. The best watchword for developers is: *keep it simple.* Developers must keep tasks, their interfaces and execution environment as simple as possible because the lifetime of user software may be *one* execution. Therefore the most important software design features are ease of modification and adaptability.

2. System Components and Architecture

Software developers must address issues such as portability, platform independence, and freedom from licensing restrictions if they wish to free the users from these concerns so that the latter can solve analysis problems on their desktop. Users automatically gain the benefit of easily exchanging both software and data with collaborators independent of local environments. Toward this aim, the MMT Instrumentation project[1] surveyed the existing options and selected

[1]http://cfa-www.harvard.edu/mmti/

open-systems components to prototype several of our important applications. The critical components we have identified are:

- **Environment** A command line interface with a scripting language is essential for rapid prototyping and automated systems: a GUI is nice but not sufficient. POSIX components provide the architecture and platform independence for scripting languages, tools and software libraries. e.g., Korn shell, ANSI C and C++ libraries, FORTRAN90. POSIX also provides communication mechanisms such as pipes, shared memory and mapped files which can be used to efficiently pass data between processes.

- **Analysis Tools and Libraries** A tool-box of parameter driven tasks is needed to supply the necessary science algorithms. We have developed a system that we call UnixIRAF[2], that allows non-interactive IRAF tasks to be wrapped with generic Korn shell wrappers to emulate Open-IRAF tools. Starbase[3] provides another toolkit consisting of a full-featured relational data base (RDB) system. This POSIX-compliant toolkit can be used to construct complex data selection, extraction and re-formatting operations by piping together sequences of tools. These tasks are linked with the SAO Parameter Interface.

  Open-IRAF will provide C and C++ bindings for the IRAF libraries. SLALIB[4] will provide the world coordinate system libraries.

- **Visualization Tools/GUI** GUI-driven imaging and graphing applications are essential to aid users in understanding both the raw data and the analysis results. Tcl/Tk provides an architecture and machine independent widget set and a layered GUI written in a scripting language. We use SAOtng[5] for imaging and plan to re-use the Starcat catalog widget to add catalog position overlays.

- **Data Files** The tool box needs to run directly on standard machine independent data files, to free the users from the additional concerns of data conversion, archival formats and transportability. FITS *bintable* and *image* extension formats and ASCII tables are a necessary (but not sufficient) condition for providing machine independent formats. Additional conventions need to be added to FITS for metadata, world coordinate systems and other related information. This issue is complex and is covered in more detail in the Data Model work by several groups.

  FITS files are supported as native format both in IRAF for images and the TABLES layered package for (bin)tables. Starbase is being extended to support FITS bintable.

---

[2] http://cfa-www.harvard.edu/mmti/mmti/

[3] http://cfa-www.harvard.edu/~john/starbase

[4] http://star-www.rl.ac.uk/libs/slalib/mainindex.html

[5] http://hea-www.harvard.edu/RD/

- **Glue** There needs to be a mechanism by which these components can communicate with one another. The *SAO Parameter Interface* is such a mechanism, providing an API, interactive parameters, dynamic parameters and automatic parameter set configuration. When combined with XPA, it can connect analysis tools to visualization tools to build GUI-driven analysis applications (Mandel & Tody, 1995).

3. SAO Parameter Interface

Most of the items cited above exist in a variety of freely available implementations. However, the currently available parameter file systems have serious limitations when inserted into a heterogeneous environment. We therefore developed backward-compatible extensions to the traditional IRAF interface to create an *SAO Parameter Interface*[6] that allows multi-layered options for configuring applications and automating test scripts and pipelines:

- **Default Parameter File Override** We have added the ability to override the default parameter file specification. This allows multiple default parameter files to exist and be selected at runtime. E.g. radial_profile @@hst_prf or radial_profile @@rosat_prf
- **Common Data Sets** We have added a *Common Data Set* database to define dynamically configurable sets of parameter files. This provides the capability of automatically switching parameter files between pre-defined configurations based on the current environment: e.g., different time-dependent calibrations, several filters for the same instrument, or dozens of observations (and filenames).
- **Dynamic Parameter Values** There are many situations where the best parameter value is a function of other parameters or data. The parameter interface provides a mechanism to invoke an external tool to dynamically calculate a parameter value, returning the result to a program when it accesses this parameter at run-time.

3.1. Pipeline Applications

These parameter enhancements allow the developer to write generic pipelines that can be re-configured for different instruments and configurations. Meanwhile the user sees only a simple, explicit, reproducible batch script with no configuration dependencies because a complete, explicit record of the *as-run* parameters is saved.

The default parameter specification permits all the as-run parameters to be preserved even when the same program is run more than once. The common data set specification allows the pipeline to to be reconfigured when settings change, such as: filter in use, CCD-binning or instrument calibration. The dynamic parameters allow quantities such as bias and gain to be calculated from the current dataset and used as parameters by the calibration tools. These features

[6]http://cfa-www.harvard.edu/mmti/mmti

also allow pipelines to be reconfigured for different instruments so they can be re-used for new projects.

3.2. Interactive Analysis Applications

Dynamic parameters are very useful for coupling interactive tasks, allowing analysis to be driven easily the from image display. A simple scenario might be: image the data with SAOtng, draw regions of interest with the mouse, invoke analysis tools on the selected file and region. In this case *Dynamic Parameters* are used by the analysis tool at runtime to determine both the current file and the selected region to analyze. The dynamic values are determined by small scripts that invoke XPA to query the image display for the current file and the current region.

3.3. User Applications

Users often need to sequence several tools. The difficulties in making these user-scripts generic and re-usable stem from the fact that the filename changes at each step of the script and often the same parameter quantity has different names and/or units in each of the independent tools. *Common Data Sets* allow users to define an ASCII table to alias different tool-name:parameter-name pairs. Dynamic parameters can be used to automatically perform unit conversions.

4. Conclusions

The *MMT Instrumentation* group has used these components in all phases of the project, from instrument control and data acquisition to automated reduction pipelines and visualization. The toolbox consists primarily of existing IRAF analysis tools, special purpose instrument control tools and ICE tools. UnixIRAF enables the ICE data acquisition software to be controlled by simple POSIX-compliant Unix shell scripts in the same way as the instrument control software and the pipelines. Pipelines have been developed for CCD data reductions, spectral extractions and wavelength calibrations. Multi-chip CCD data are reduced efficiently by running multiple parallel pipelines for each chip. SAOtng and XPA are used to visualize mosaiced CCD data.

This approach has been highly successful. But it presents some challenges to the astronomical community: Who will contribute tools and components? Are developers rewarded for producing adaptable software?

References

Mandel, E. & Murray S. S. 1998, this volume

Mandel, E. & Tody, D. 1995, in ASP Conf. Ser., Vol. 77, Astronomical Data Analysis Software and Systems IV, ed. R. A. Shaw, H. E. Payne & J. J. E. Hayes (San Francisco: ASP), 125

Coggins, J. M. 1996, in ASP Conf. Ser., Vol. 101, Astronomical Data Analysis Software and Systems V, ed. George H. Jacoby & Jeannette Barnes (San Francisco: ASP), 261

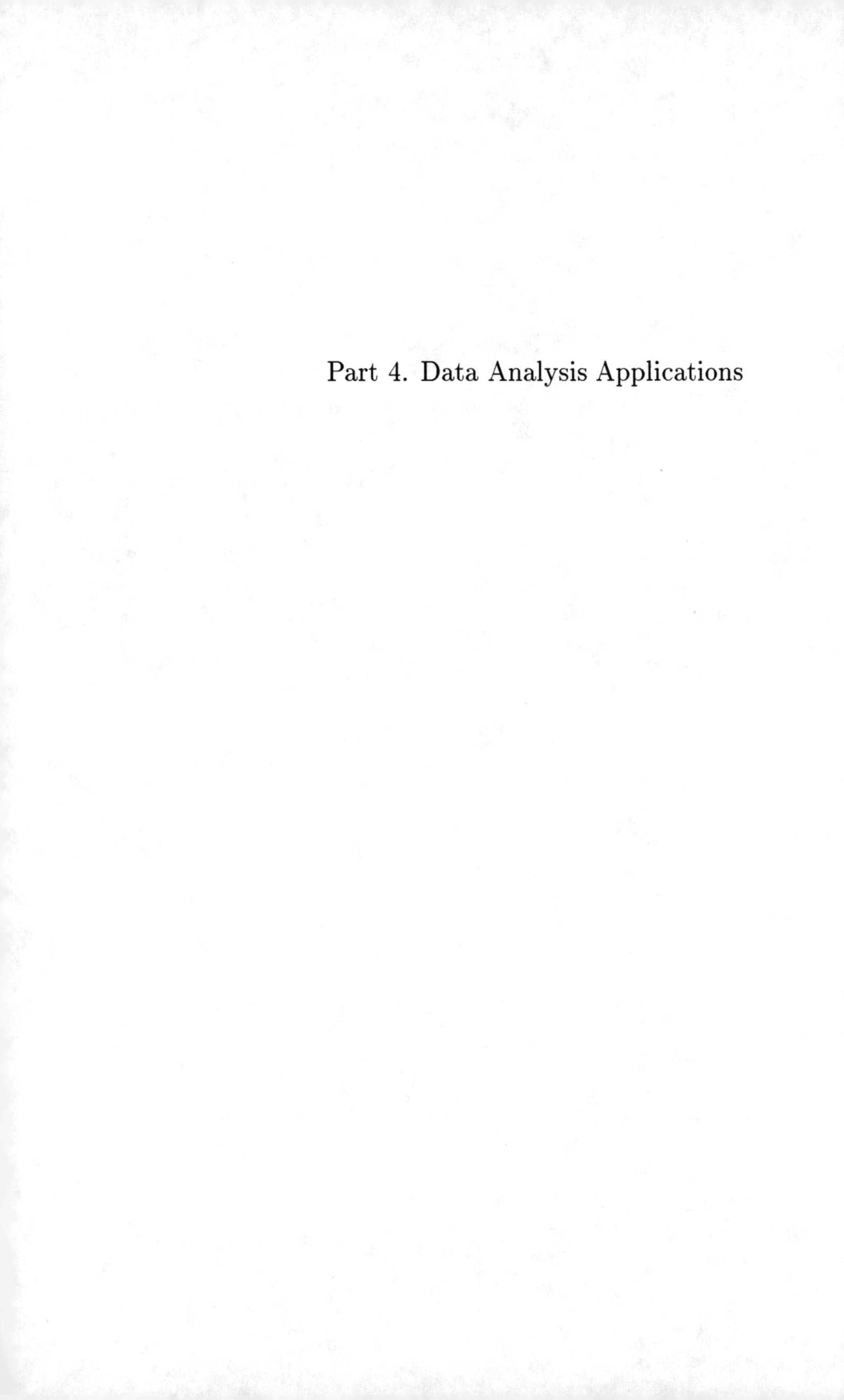

Part 4. Data Analysis Applications

Astronomical Data Analysis Software and Systems VII
ASP Conference Series, Vol. 145, 1998
R. Albrecht, R. N. Hook and H. A. Bushouse, eds.

Fitting and Modeling of AXAF Data with the ASC Fitting Application

S. Doe, M. Ljungberg, A. Siemiginowska and W. Joye

AXAF Science Center, MS 81, Smithsonian Astrophysical Observatory, 60 Garden Street, Cambridge, MA 02138 USA

Abstract. The AXAF mission will provide X-ray data with unprecedented spatial and spectral resolution. Because of the high quality of these data, the AXAF Science Center will provide a new data analysis system–including a new fitting application. Our intent is to enable users to do fitting that is too awkward with, or beyond, the scope of existing astronomical fitting software. Our main goals are: 1) to take advantage of the full capabilities of the AXAF, we intend to provide a more sophisticated modeling capability (i.e., models that are $f(x, y, E, t)$, models to simulate the response of AXAF instruments, and models that enable "joint-mode" fitting, i.e., combined spatial-spectral or spectral-temporal fitting); and 2) to provide users with a wide variety of models, optimization methods, and fit statistics. In this paper, we discuss the use of an object-oriented approach in our implementation, the current features of the fitting application, and the features scheduled to be added in the coming year of development. Current features include: an interactive, command-line interface; a modeling language, which allows users to build models from arithmetic combinations of base functions; a suite of optimization and fit statistics; the ability to perform fits to multiple data sets simultaneously; and, an interface with SM and SAOtng to plot or image data, models, and/or residuals from a fit. We currently provide a modeling capability in one or two dimensions, and have recently made an effort to perform spectral fitting in a manner similar to XSPEC. We also allow users to dynamically link the fitting application to their own algorithms. Our goals for the coming year include incorporating the XSPEC model library as a subset of models available in the application, enabling "joint-mode" analysis and adding support for new algorithms.

1. Introduction

The AXAF is NASA's "Advanced X-ray Astrophysics Facility," scheduled for launch on 27 August 1998. The AXAF Science Center Data Systems will provide the astronomical community with software to reduce and analyze AXAF data. In particular, our team is working on a fitting and modeling application suitable for analysis of AXAF data. We have two main design goals: to provide modeling in up to 4 dimensions, for functions truly $f(E, x, y, t)$; and, to package together

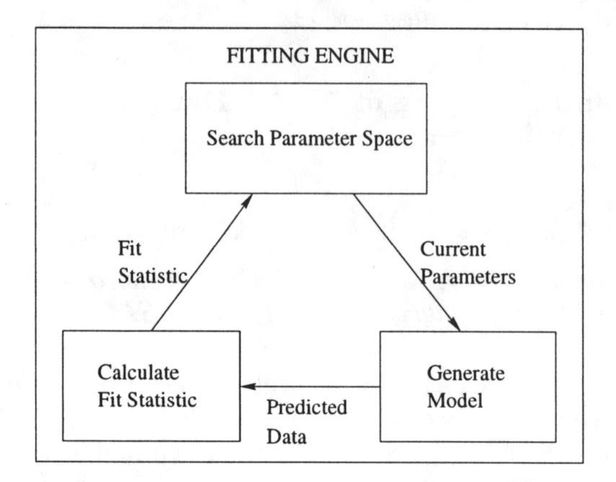

Figure 1. The main fitting process.

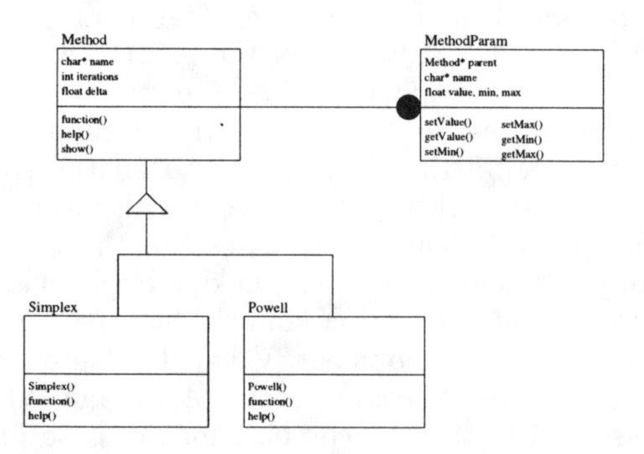

Figure 2. The Method Class.

a wider variety of optimization algorithms and fit statistics. Meeting these goals will enable users to take advantage of the quality of AXAF data.

2. Design of the Fitting Application

Figure 1 shows the main fitting process that occurs within the application. The data are modeled by some function $F(\vec{X}, \vec{P})$, where $\vec{X}$ is the data space and $\vec{P}$ is the vector of model parameters. A fit statistic is then calculated by comparing the data to the predicted data calculated with the model. The fit statistic is used by the optimization algorithm to determine the next $\vec{P}$ to try; the application iterates through this process until the convergence criteria are reached, at which point $\vec{P}$ contains the best fit parameters.

We have taken an object–oriented approach to the design of the fitting application. Figure 2 shows an example; these are the classes associated with the "Search Parameter Space" box shown in the previous figure. Since the design for the classes associated with the other two boxes is substantially the same, we show here only the "Method" class.

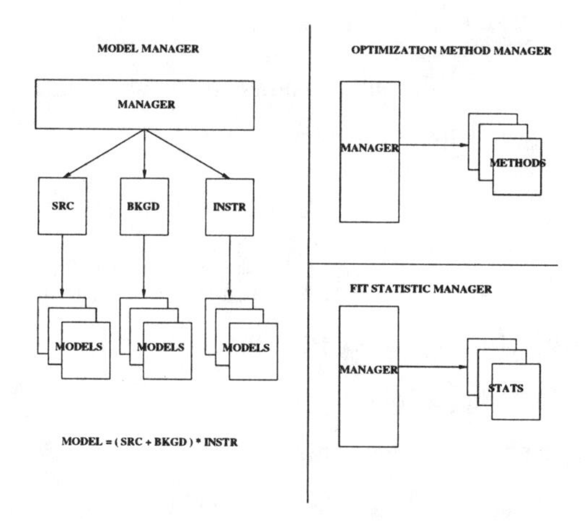

Figure 3. The Model, Method, and Statistic Managers.

We provide a number of different optimization algorithms (e.g., the simplex method, Powell minimization, grid search, etc.); each of these are contained in their own separate derived class as shown in Figure 2. All of these classes inherit from the base "Method" class, which contains information all the derived classes use. This structure has made it particularly easy for us to "plug–in" new algorithms as they are requested by our scientists. There is also a class associated with the Method class called the "MethodParam" class; this class contains the convergence criteria used by the various optimization algorithms.

Thus, we have three types of base classes to manage—the Method, Model and Statistic classes. (A "model" is composed of an arithmetic combination of functions $f(\vec{X}, \vec{P})$ derived from the Model classes; the fit statistics of the Statistic classes are used to compare the data with the "predicted data" generated by models.) Figure 3 shows the managers we use to keep track of the algorithms and models currently in use. The Method and Statistic managers need merely to point at the algorithm currently in use; the Model Manager, on the other hand, needs to construct the model by evaluating three model stacks, the source, background, and instrument model stacks, and then combining them appropriately (i.e., the background model is added to the source model, and the result is convolved with an instrument model). The source model alone may be used, if no background or instrument models have been defined.

3. Current Implementation

We have implemented a version of the fitting application which is currently undergoing alpha testing at the ASC. This version has capabilities which include the following:

- Command-line interface

  —Command completion, history, vi, emacs keymaps, escape to shell.

  —Execute scripts, create logs.

- Multiple data sets
  —E.g., fit data from different missions to the same model.
- Modeling in 1 and 2 dimensions
- Filtering of data in memory
- Modeling "mini-language"
  —Build model out of arithmetic combination of functions.
- Interaction with SAOtng, GNUPLOT, SM
- Selected XSPEC compatibility
- Dynamic linking to user algorithms

This version will serve as the foundation of the flight version.

4. Continuing Development

The first release of ASC software is scheduled for June 1998. For the June release, we must add the following capabilities to the fitting application:

- Implementation of "joint-mode" analysis
- Modeling in up to 4 dimensions for functions truly $f(E, x, y, t)$
- Additional algorithms and statistical approaches
- Interface with ASC Data Model
- Support for X-ray gratings on AXAF
- Support for external AXAF simulators
- GUI

Adding this functionality to the fitting application will provide AXAF users with a powerful tool for the analysis of AXAF data.

Acknowledgments. This project is supported by NASA contract NAS8-39073 (ASC). We would like to thank Mark Birkinshaw for making his OPTIM library available at the ASC.

References

Birkinshaw, M., 1995, CfA internal memo

Doe, S., Conroy, M., & McDowell, J., 1996, in ASP Conf. Ser., Vol. 101, Astronomical Data Analysis Software and Systems V, ed. George H. Jacoby & Jeannette Barnes (San Francisco: ASP), 155

Doe, S., Siemiginowska, A., Joye, W., & McDowell, J., 1997, in ASP Conf. Ser., Vol. 125, Astronomical Data Analysis Software and Systems VI, ed. Gareth Hunt & H. E. Payne (San Francisco: ASP), 492

Astronomical Data Analysis Software and Systems VII
ASP Conference Series, Vol. 145, 1998
R. Albrecht, R. N. Hook and H. A. Bushouse, eds.

The ISO Spectral Analysis Package ISAP

E. Sturm[1], O.H. Bauer, D. Lutz, E. Wieprecht and E. Wiezorrek

Max-Planck-Institut fuer extraterrestrische Physik, Postfach 1603, 85740 Garching, Germany

J. Brauer, G. Helou, I. Khan, J. Li, S. Lord, J. Mazzarella, B. Narron and S.J. Unger

IPAC, California Institute of Technology, MS 100-22, Pasadena, CA 91125, USA

M. Buckley, A. Harwood, S. Sidher and B. Swinyard

Rutherford Appleton Laboratory, Chilton, Didcot, Oxon OX11 0QX, UK

F. Vivares

CESR, BP 4346, 9 av du col. Roche, 31028 Toulouse Cedex, France

L. Verstraete

Institut d'Astrophysique Spatiale, Universite Paris-Sud - Bat.121, 91405 Orsay Cedex, France

P.W. Morris

ISO Science Operations Center, P.O. Box 50727, 28080 Villafranca/Madrid, Spain

Abstract. We briefly describe the ISO Spectral Analysis Package ISAP. This package has been and is being developed to process and analyse data from the two spectrometers on board ISO, the Infrared Space Observatory of the European Space Agency (ESA). ISAP is written in pure IDL. Its command line mode as well as the widget based graphical user interface (GUI) are designed to provide ISO observers with a convenient and powerful tool to cope with data of a very complex character and structure. ISAP is available via anonymous ftp and is already in use by a world wide community.

[1]Email: sturm@mpe-garching.mpg.de

1. Introduction

ISO, the Infrared Space Observatory of the European Space Agency, was launched in November 1995. It carries 4 different instruments including the short and long wavelength spectrometers SWS and LWS (see Kessler et al. 1996). The ISO Spectral Analysis Package, ISAP, plays an important role in ISO data analysis. It is a software package, written in pure IDL[2], for the reduction and scientific analysis of the ISO SWS and LWS Auto Analysis Results (AARs). AARs are the end product of the official automatic pipeline processing, which processes the raw data, as they are received from the satellite, via a number of intermediate products to this AAR stage, which is then sent to the observer. Being the end product of the pipeline of the spectrometers an AAR should be - at least in principle - a "valid" spectrum which is appropriate for immediate scientific analysis. However, the spectra are heavily affected by glitches, detector transients, memory effects and other phenomena, caused, e.g., by cosmic ray hits. The best way to cope with these effects is to keep as much redundancy as possible (i.e., all elementary measurements) in the pipeline products, to enable an appropriate, interactive, post-processing. For this reason elaborate software has been developed (see e.g., Wieprecht et al. 1998, and Lahuis et al. 1998, for a description of the SWS pipeline and Interactive Analysis System), and a lot of instrument and expert knowledge is needed to treat the AARs correctly.

Hence, there was a clear need for a software package that could be given to the observers in addition to the AARs, to impart the expertise, and to enable the observers to process the data further and eventually analyse them. Since both SWS and LWS are spectrometers with AARs that are at least very similar, the LWS and SWS consortia decided in the summer of 1995 to start a collaboration to develop such a package as a common tool for both spectrometers.

2. ISAP

A specific emphasis was put on the ability of the computing environment to easily and immediately plot and visualize data. A widget toolkit was also mandatory. Since IDL fulfills all the requirements, and since a large part of ISO's data processing software, like the Interactive Analysis Systems, had already been written in IDL, it was decided to use it as computing environment for ISAP as well.

ISAP operates on the final pipeline products, and one of its primary functions is to improve these products (AARs). It therefore offers and simplifies all the necessary steps for post-processing the data. Several functions for data input and output are available, as well as elaborate plotting tools for visualization. Bad data such as outliers can be masked out, erased or corrected and replaced. Spectra can, for instance, be averaged, flatfield corrected, normalized, rebinned and smoothed. Removal of fringes and many more operations can also be performed.

---

[2]IDL is a trademark of Research Systems Inc. (RSI)

The result of the first part of a typical ISAP session is expected to be a "simple spectrum" (single-valued and resampled to uniform wavelength scale if desired) that can further be analyzed and measured either with other ISAP functions, native IDL functions, or exported to a different analysis package (e.g., IRAF, MIDAS) if desired. ISAP itself provides many tools for detailed analysis, like line and continuum fitting and flux measurements, unit conversions, conversions from wavelength space to frequency space, dereddening or synthetic photometry and models including a zodiacal light model to predict and subtract the dominant foreground at SWS wavelengths.

All these tools and routines are accessible via two different modes: a command line mode and a Graphical User Interface (GUI). The command line mode is embedded in the basic IDL environment. Full access to all the ISAP routines is provided, and, in addition, all standard IDL routines are available in the usual manner. Suitable command sequences can be combined in scripts and automatically performed. It is, hence, the most powerful mode of ISAP, but it needs some expert knowledge about ISAP and IDL. A more user friendly - and in particular more beginner friendly - mode is the GUI mode, a graphical user interface that is built around the ISAP commands. Figure 1 shows ISAP's main GUI. The first row provides some general information, such as object name or type of observation. The FITS header can be viewed, help pages accessed, etc. A number of buttons in the following rows give some instrument specific informations about, e.g., the detectors used in the observation, or the number of scans, and can be used to extract subsets of the data.

In another column direct access is given to buttons which allow some basic operations, like data input and output, plot control, and special functions such as unit conversion, which operate on the AAR as a whole.

The heart of the GUI is the plot window, offering a completely mouse oriented way of data visualization and selection: zooming in, getting information about single data points and selecting parts (or all) of the data for subsequent processing can be performed by clicking (and dragging) with one of the mouse buttons. A large number of plotting styles can be chosen from a menu. After selecting some data (with the right mouse button) a toolbox pops up, presenting a variety of ISAP tools which can be applied to the selected data. The more complex applications are simplified through separate, dedicated graphical user interfaces, which become available on request. The "average" GUI, e.g., guides the user through the necessary decisions for the type of averaging (across scans for each detector individually, or across detectors), the bin size, or the averaging technique (e.g., mean or median, with and without clipping., etc.). The result can be immediately examined in two separate plot windows, which display the data before and after application of the operation. Again, zooming in with the mouse and individual adjustment of the plot style is possible, just like in the main GUI.

ISAP is currently being developed by less then 10 programmers and scientists (part time) at different sites in Europe and the United States, on different machines under different operating systems (Unix and VMS), demonstrating the high efficiency of programming in IDL and the high portability of IDL programmes. It also makes use of the astronomical users' library of IDL. A few programmer's guidelines form the basis of very simple configuration control.

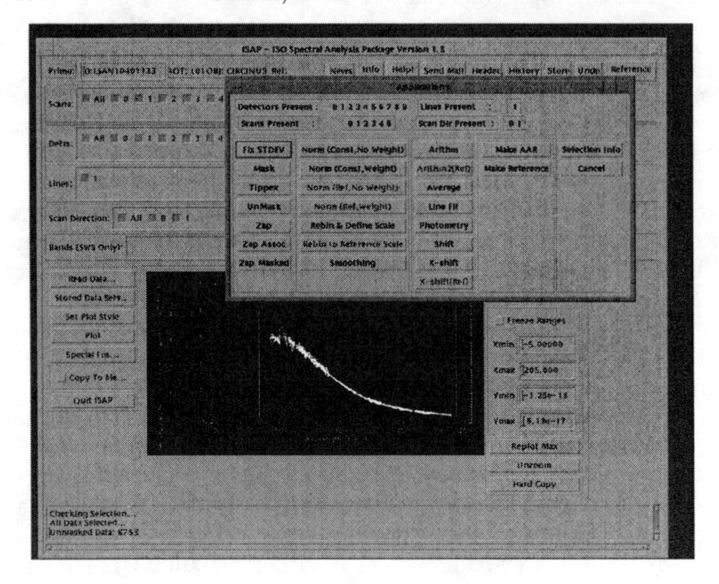

Figure 1. The main ISAP GUI

One main advantage of this approach is, that every user can write his or her own application. Provided the users stick to these guidelines, they can easily add their own routines to their ISAP installation - or ask the ISAP team to officially include them in the next release.

ISAP is publicly available via anonymous ftp and it is already in use worldwide. Currently well over 100 astronomers have downloaded it from the official ftp sites. For many of them it has already become the main processing and analysis tool for ISO spectra. After the proprietary period for ISO data, i.e., when all ISO data will be public, an even larger interest is to be expected.

ISAP has certainly some potential for future astronomical (space) missions. Inevitably a large part of it is very instrument specific, the basic principles and algorithms, however, as well as the whole interface, could easily be adapted for other spectrometers.

More information about ISAP (and how to get it) can be found on the World Wide Web: see the ISAP homepages at
http://www.ipac.caltech.edu/iso/isap/isap.html or
http://www.mpe-garching.mpg.de/iso/observer/isap/

References

Kessler, M.F. et al. 1996, A&A 315, L27

Lahuis, F., et al. 1998, this volume

Wieprecht, E., et al. 1998, this volume

Astronomical Data Analysis Software and Systems VII
ASP Conference Series, Vol. 145, 1998
R. Albrecht, R. N. Hook and H. A. Bushouse, eds.

News on the ISOPHOT Interactive Analysis PIA

C. Gabriel

ISO Science Operations Centre, Astrophysics Division, Space Science Department of ESA, Villafranca, P.O. Box 50727, 28080 Madrid, Spain, Email: cgabriel@iso.vilspa.esa.es

J. Acosta-Pulido[1]

Max-Planck-Institut für Astronomie, Heidelberg, Germany

I. Heinrichsen[1]

Max-Planck-Institut für Kernphysik, Heidelberg, Germany

Abstract. The ISOPHOT Interactive Analysis system, a calibration and scientific analysis tool for the ISOPHOT instrument on board ESA's Infrared Space Observatory (ISO), has been further developed while ISO is in operation. In this paper we describe some of the improvements achieved in the last few months in the areas of scientific analysis capabilities, documentation and in related services to the astronomical community.

1. Introduction

The ISOPHOT Interactive Analysis (PIA) is an IDL based software tool developed for calibration and scientific analysis of data from ISOPHOT (Lemke et al. 1996), one of the instruments on board ISO (Kessler et al. 1996). The general features of PIA, its use as a calibration tool and the built-in mapping capabilities were presented at the ADASS VI conference (Gabriel et al. 1997a, 1997b and 1997c).

The development of PIA, up to the initial phase of the mission, was described by Gabriel et al. (1996). After 2 years of ISO operations considerable experience has been gained in the use of PIA, which has led to several new features in the package. This experience has been achieved by the ISOPHOT Instrument Dedicated Team in its tasks which include calibration, instrument performance checking and the refinement of analysis techniques, and also by a large number of ISOPHOT observers in more than 100 astronomical institutes all over the world.

---

[1]at ISO Science Operations Centre, Astrophysics Division, ESA-SSD, Villafranca, P.O. Box 50727, 28080 Madrid, Spain

PIA has been widely distributed since July 1996 to all astronomers wishing to use it for ISOPHOT data reduction and analysis. The feedback from the different users is reflected not only in the extension of the analysis capabilities but also in a more friendly graphical interface, better documentation, and easier installation.

2. PIA Improvements

In the last year of the ISO mission the PIA capabilities have been continuously enhanced. All the areas of data processing routines, automatic analysis sequences, graphical user interface, calibration menus, configuration, distribution and information have been affected by changes. In this paper we describe the developments in three sections:

- transients modeling implementation, showing changes in the areas of data processing routines and GUI,
- documentation improvements, for the areas of information and maintenance, and
- system testing procedures implementation, for the areas of reliability and support.

2.1. Transients Modeling

ISOPHOT detectors have long stabilization times, which is a known feature affecting IR detectors (Si:Ga, Si:B and Ge:Ga) operating at very low backgrounds. The measurement times are usually shorter than the required stabilization, and this represents a major impediment to the calibration of the instrument and to obtaining a good signal to noise ratio. The transient behaviour depends on several parameters including the flux level to be measured (varying over several orders of magnitude), the previous flux history (flux down or flux up with all levels of difference) and the detector type of the sub-system used.

The approach followed is to use physical and empirical models to fit the data and predict the final flux which would have been obtained for a fixed sky point if measured for sufficiently long. Depending on the sub-system detector different functions and model parameters are proposed:

- physical models for silicon based detectors (Schubert et al. 1995; Fouks et al. 1996)
- empirical models for Ga:Ge detectors (Fukiwara et al. 1995; Church et al. 1996)

The dependencies on flux level and history are reflected in the starting parameters of the different models (which can be fixed for the fitting procedure).

Fitting of a signal distribution corresponding to the measurement of a sky position begins with the determination of the starting parameters corresponding to the type of data and function to be used. The minimization of the Chi-squared or least squares fitting is done first with a downhill simplex method[2] for a first approximation varying the parameters. The parameters with the values thus obtained are then used as starting parameters for a second fit using a Gradient-expansion algorithm for a least squares fit to a non-linear function[3].

The graphical user interface includes the possibility of: a) choosing among the different functions with the corresponding parameter initialisations, b) general parameters for the analysis can be set, such as selection criteria, tolerance level for the fit, weighting method to be used, c) specific function parameters can be reset after initialisation, varied or fixed, tolerance limits can be set.

All the start parameters can be tested and adjusted before analysing a full measurement using portions of the data. After full analysis PIA also provides the opportunity of reassessing the results with partial refitting, using for example, the Chi-squared or the adjusted parameter distributions of all the fitted data for evaluation. This makes the analysis of large data sets very efficient, while permitting fast and deep data manipulation.

2.2. Documentation

We have created the PIA homepage[4] for several reasons:

- to give users faster access to PIA,
- to create an efficient feedback route from the astronomical community to the PIA developers,
- to provide the community with better maintenance and service.

The PIA homepage contains the latest version of PIA together with the release notes and the PIA user's manual both in its HTML and PostScript versions. Publications on PIA can also be seen or retrieved and frequently asked questions and answers are listed. A mailbox is attached to receive bug reports, comments, etc. Especially useful for advanced PIA users are the listings of routine headers, which can be used by external calls and allow re-use of the PIA routines.

2.3. System Testing Procedures

Procedures for testing PIA are applied prior to every new release (main versions are released annually, while sub-versions are every two to three months). The automatic sequences built within PIA are used for running over a huge representative dataset containing all major ISOPHOT observing modes with good reference data. The sequences run through all the data reduction steps performing all the default corrections and saving data in the different formats at all reduction levels. These procedures can not only check the reliability of the new version but also are used for testing the level of calibration accuracy for a new sub-version and/or new calibration files. These tests are also used for checking the reliability of the software under different IDL versions and different machine architectures.

---

[2] as given by the IDL function AMOEBA.

[3] we use a slightly changed version of the IDL routine CURFIT

[4] http://isowww.estec.esa.nl/manuals/PHT/pia/

3. Outlook

The end of the ISO mission, due to Helium boil-off, is foreseen for April 1998. PIA will be further developed during the "Post-operational phase", starting then and lasting at least 3.5 years. The huge amount of ISOPHOT data collected, the excellence of their scientific content, the quality of the instrument and the further consolidation of its calibration result in scientific work for a wide community for years to come. One of the reasons for this is that the ISO archive will be open to everyone for archive research once the proprietary rights expire. PIA plays a major rôle as the tool for data reduction and scientific analysis of ISOPHOT data, and new requirements are continuously arising from the analysis experience of its users. Concrete planning for further PIA development includes the provision of imaging enhancement techniques as well as coherent maps co-addition, polarimetry analysis, time and orbit dependent calibration, etc..

References

Church, S. et al. 1996, Applied Optics, Vol. 35, No. 10, 1597

Fouks, B., & Schubert, J. 1995, Proc. SPIE 2475, 487

Fukiwara, M., Hiromoto, N., & Araki, K. 1995, Proc. SPIE 2552, 421

Gabriel, C., Acosta-Pulido, J., Heinrichsen, I., Morris, H., Skaley, D., & Tai, W. M. 1996, "Development and capabilities of the ISOPHOT Interactive Analysis (PIA), a package for calibration and astronomical analysis", Proc. of the 5th International Workshop on "Data Analysis in Astronomy", Erice, in press

Gabriel, C., Acosta-Pulido, J., Heinrichsen, I., Morris, H., Skaley, D., & Tai, W. M. 1997a, "The ISOPHOT Interactive Analysis PIA, a Calibration and Scientific Analysis Tool", in ASP Conf. Ser., Vol. 125, Astronomical Data Analysis Software and Systems VI, ed. Gareth Hunt & H. E. Payne (San Francisco: ASP), 108

Gabriel, C., Acosta-Pulido, J., Kinkel, U., Klaas, U., & Schulz, B. 1997b, "Calibration with the ISOPHOT Interactive Analysis (PIA)", in ASP Conf. Ser., Vol. 125, Astronomical Data Analysis Software and Systems VI, ed. Gareth Hunt & H. E. Payne (San Francisco: ASP), 112

Gabriel, C., Heinrichsen, I., Skaley, D., & Tai, W. M. 1997c, "Mapping Using the ISOPHOT Interactive Analysis (PIA)", in ASP Conf. Ser., Vol. 125, Astronomical Data Analysis Software and Systems VI, ed. Gareth Hunt & H. E. Payne (San Francisco: ASP), 116

Kessler, M. et al. 1996, A&A, 315, L27-L31

Lemke, D. et al. 1996, A&A, 315, L64-L70

Schubert, J., et al. 1995, Proc. SPIE 2253, 461

Astronomical Data Analysis Software and Systems VII
ASP Conference Series, Vol. 145, 1998
R. Albrecht, R. N. Hook and H. A. Bushouse, eds.

How to Piece Together Diffracted Grating Arms for AXAF Flight Data

A. Alexov, W. McLaughlin and D. Huenemoerder

AXAF Science Center; Smithsonian Astrophysical Observatories, TRW, and MIT, Cambridge, MA 02138

Abstract. The Advanced X-ray Astrophysics Facility's (AXAF) High and Low energy transmission gratings (HETG, LETG) data require new tools and data structures to support x-ray dispersive spectroscopy. AXAF grating data files may be a hundred megabytes (MB) in size, however, they will typically only be a few MB. We are writing data analysis software which can efficiently process the data quickly and accurately into wavelengths, orders and diffraction angles for each event. Here we describe the analysis procedure as well as some of the technical constraints we had to overcome in order to process the tasks efficiently.

1. Data Processing

1.1. Standard ACIS/HRC Event Processing

Initial data processing applies transformations from detector to sky coordinates for each photon in the data set. However, with grating data, additional event processing must be performed before data analysis can commence.

1.2. Select Zero Order Sources/Find Observed Target

In order to find where the dispersed photons lie, the center of the zero order source position on the sky must be determined, since this is the origin of the spectral coordinates. Source detection must not be fooled by emission lines located away from the center of the field; several methods exist to discriminate bright emission lines from zero order sources. Primarily, emission lines have a small variance in the PHA (energy) spectrum, while zero order has a large variance since it encompasses all of the source energies. Alternatively, the PSF (Point Spread Function) can be used instead of PHA, to weed out emission lines. Once zero order source positions are found, the target source is identified by matching the sky positions with an observation target list.

1.3. Identify Spectrum Parts Geometrically

Grating data events may be categorized by their part of the spectrum. This is done by creating mask regions for each grating part and by checking every photon for inclusion within the mask regions. For the HETG grating data, the relevant parts for each source are: zero order, MEG photons, HEG photons, and background photons. Regions are defined as rectangles in diffraction coordinates, in which the zero order is the origin, and one axis is parallel to the spectrum (dispersion direction), and the other perpendicular to it (cross-dispersion direction). The width of the rectangle is calculated using the effective PSF (mirror psf, Rowland geometry astigmatism, instrument de-focus, and aspect). These regions are translated into sky coordinates using each zero order source position, and by rotating the region by the grating angle (known) plus the mean observed telescope roll angle. Any event inside the rectangle is then assigned to that part of the spectrum (HEG or MEG, or LEG). Zero order photons are assigned by being within some radius of the zero order centroid. Each photon is tagged with one or more source ID's as well as grating part(s), depending on the number of regions into which it falls (overlaps).

1.4. Compute Linear Diffraction Coordinates

To calculate the diffraction angle (r) of a photon, work needs to be done in the geometric system in which the photon was diffracted. The sky coordinates, which are referenced to the telescope mirror node and therefore independent of the grating reference node, are only useful for the imaging of zero order and the filtering of grating arm photons. Grating diffraction coordinates are referenced to the grating assembly node. Reverting back to the original chip coordinates (chip number, chip position, grating node and zero order position) allows grating diffraction coordinates to be calculated for each time interval.

1.5. Compute "first order" Wavelengths

Now that diffraction angles have been determined parallel to the dispersion angle, the basic grating equation, $m\lambda = Pr$, can be applied to determine $m\lambda$ for each photon. Here, m is the integral diffraction order, P is the known mean period of the set of gratings (MEG, HEG, LEG), and r is the diffraction angle.

1.6. Resolve Orders (ACIS only)

The ACIS detector provides moderate spectral resolution via a pulse-height (PHA, or PI) for every photon. This resolution is enough to determine, with high confidence, what the spectral order is at any r or $m\lambda$, since only integral multiples are allowed at any given position (grating equation, diagram below). Order sorting is also useful for photons in overlapping mask regions. These events can be resolved as belonging to one arm/source versus another through order sorting. Since the HRC detector has no energy resolution, a method has yet to be determined for resolving orders or any overlapping regions in data taken by this instrument. The grating equation is used to calculate an estimate of the grating order (m_est), using physical constants and previously calculated values: $m_{est} = Pr(PI)/hc$. Here, hc is a physical constant; PI is the "Pulse Invariant" energy from the CCD; r is the calculated diffraction angle; and, P is the known grating period.

The ACIS instrument response determines the distribution of PI. This information permits the calculation of an allowed range of m for each photon. If m_est is within this range, then the grating order is equal to the rounded value of m_est. Otherwise, the order is unresolved. For ACIS overlap regions, order is calculated for each region. If the order can be uniquely identified as belonging to a single region then it is resolved. Otherwise, the order is left as unresolved.

1.7. Bin into 1D spectra

Now that the data are complete with identifiers for the part of the spectrum, diffraction angles, orders, wavelengths, and counts spectra vs λ can be created. These spectra are used for further data analysis (i.e., emission line identification and flux analysis).

2. Software Solutions

Processing grating data comprised of multiple sources is a non-trivial task. Many design tradeoffs were considered in attempting to create efficient and effective analysis software. This section identifies some of the considerations taken into account when designing the gratings software.

The possibility that parts of different sources may overlap imposes a great challenge to the data analysis software. All overlapping regions need to be identified and the software must be able to decompose these regions into the component source parts. While standard IRAF region software can easily support overlapping regions of a few sources via region algebra, a mechanism is still necessary for keeping track of the source parts in a given overlap region. As the total number of regions (including overlaps) has an exponential growth (5 to the n for HETG), this method is impractical when dealing with several sources. For instance, the worst case scenario for 10 HETG sources is over 9.5 million source parts and overlap combinations to track. Realistic scenarios may contain approximately 10 sources, but these sources will typically be more spread out (i.e., Orion Trapezium).

To circumvent this problem, the software has been designed to maintain tables for all of the parts of each source (resulting in 5n tables for HETG). Photons are checked for inclusion in each table and source bits are set to indicate the sources and parts to which the photon may belong. Since the geometry of the regions being utilized is limited to circles and rectangles, the tables simply contain sky coordinate range boundaries. To save memory, the axis spanning the minimum distance for any given source part is offset to zero and utilized as the index. For cases where the instrument roll angle equals 45 degrees, the index axis is arbitrarily chosen as neither of the axes provides an advantage.

3. Future Challenges

AXAF data fields may contain multiple sources. In order to be able to detect, mask out, coordinate transform, and order sort all these sources correctly is quite a challenge. We have made the software flexible for the users to be able to specify source specific characteristics/mask widths and response matrices. We hope that these extra features will allow severely crowded fields to be scientifically analyzed in the same context as more common single source fields.

For more details on the AXAF grating flight processing software, see:
http://space.mit.edu/ASC/analysis/L1.5_overview/L15.html

Acknowledgments. We are grateful to Maureen Conroy for initial design work on the XRCF grating software, which has lead to the flight version. This project is supported by NASA contract NAS8-39073 (ASC).

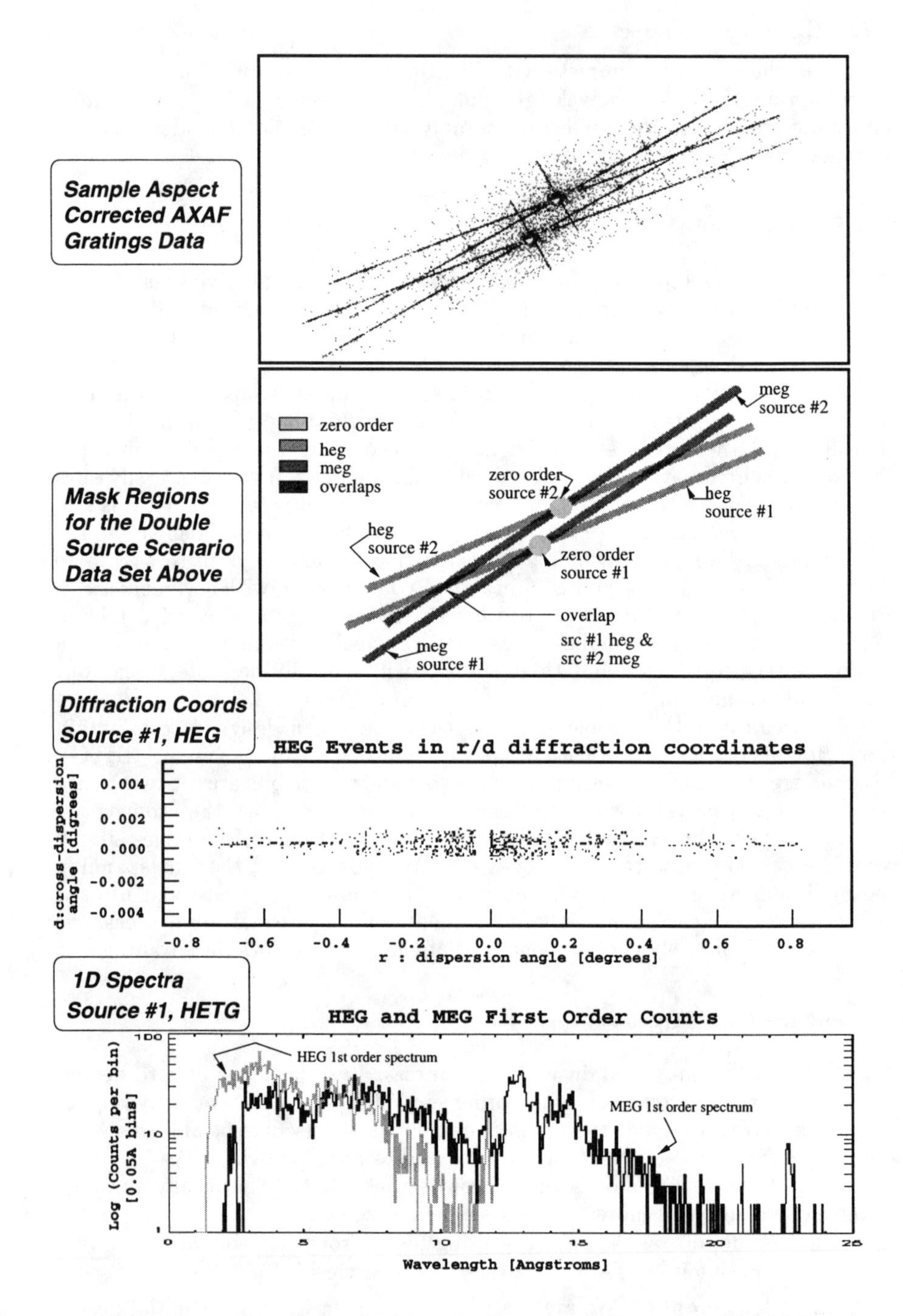

Figure 1. Cumulative look at the Processing Steps using Mock Data

Astronomical Data Analysis Software and Systems VII
ASP Conference Series, Vol. 145, 1998
R. Albrecht, R. N. Hook and H. A. Bushouse, eds.

Data Analysis Concepts for the Next Generation Space Telescope

M.R. Rosa[1], R. Albrecht[1], W. Freudling and R.N. Hook

Space Telescope European Coordinating Facility, European Southern Observatory, D-85748 Garching, Germany

Abstract. Several data analysis concepts, developed in response to the nature of HST data, are applicable to the optimum exploitation of data from NGST as well. Three examples are discussed in this paper: simulation of PSFs in combination with image restoration techniques, automatic extraction of slitless or multi–object spectra, and predictive calibration based on instrument software models. These tools already greatly enhance HST data analysis and the examples serve to demonstrate that S/W solutions to many of the challenging data-analysis requirements of NGST can be derived straight forwardly from the HST software environment.

1. Introduction

The volume of data flowing from the Next Generation Space Telescope (NGST) will in all likelihood exceed by huge factors that currently being received from HST. The anticipated operational scenarios will at the same time enforce even more the reliance upon calibration pipelines, automatic data analysis pipelines and software supported observation planning at the users level.

The NGST core science program serves as an example of the type of science exposures regularly to be obtained with such an instrument. This program implements the recommendations of the HST-and-Beyond (Dressler) Report with emphasis on targets at high redshifts. The main observing modes will be very deep imaging and low resolution multi (several hundreds) object spectroscopy. Targets are typically galaxies, barely resolved stellar clusters and luminous point like sources such as supernovae in high redshift galaxies. Embedded sources and faint cool objects in the immediate neighborhood of bright stars, i.e., planets and very low mass stars, will certainly also have a share in the schedule. The study assumed a 10 % mission time overhead for calibrations, implying maximum retrieval of information from raw data. This low overhead and stringent demands can only be met if loss-less data combination and analysis is combined with robust, noise-free calibration strategies.

The budget to be allocated for NGST operations including observation planning, data calibration and data analysis is ultimately linked to the complexity

[1]Affiliated to the Astrophysics Division of the Space Science Department of the European Space Agency

of the observational procedures and to the requirements imposed by operational aspects. To name a few: field rotation and plate scale change between successive re-observations of the same field, field position dependent PSFs, non-availability of calibration reference sources during certain periods, low ceilings on overheads available for calibration observing time.

We present below three areas of software development for the current HST data exploitation which lend themselves directly to the support of optimum NGST science output without charging NGSTs budget .

2. PSF Modeling and Image Restoration

The combination of the rather broad PSF in the IR, the faint limit of the NGST, and the fact that the deep images will be crowded with fore- and background galaxies will make photometry with NGST image data difficult.

We have developed a preliminary NGST PSF generator based on Tiny Tim, the widely used PSF generator for the HST (Krist & Hook 1997). At this time the PSF is generated for the NGST "reference design". Under operational conditions such PSFs will be predicted using future developments of the code relying on information about the wavefront errors. Such highly accurately simulated PSFs will allow the application of techniques which have been developed and are successfully in use for HST data. These are in particular:

Photometric restoration, a software package producing the independent restoration of point sources as well as a non-uniform background. It also handles the problem of crowding, which will be much more significant for NGST deep images than it is for HST, mainly because of the broader IR PSFs and because of the limiting magnitudes (Lucy 1994; Hook & Lucy 1994)

Improving the resolution of the NGST at wavelengths where a deployable primary mirror is not diffraction limited, i.e., the classical image restoration domain.

Homogenization and combination of images taken with different PSFs. Given a low mass, deployable mirror, NGSTs PSF might vary with time. Data fusion techniques such as those developed for HST will allow the lossless combination of these frames (Hook & Lucy 1993).

3. Software for NGST Multiple Slit Spectroscopy Planning

One of the most important scientific motivations of NGST is the spectroscopy of very faint galaxies detected in a previous deep imaging exposure. The current instrument plans include a multi–object spectrograph fed by a micro-mirror array. This device will provide flexible, software controlled apertures of arbitrary shapes and locations within the imaged field.

The operational concept will routinely make use of the deep survey images of the field for subsequent multi-object spectroscopy of objects selected according to special criteria (eg. blue color drop-outs). Because of the huge number of

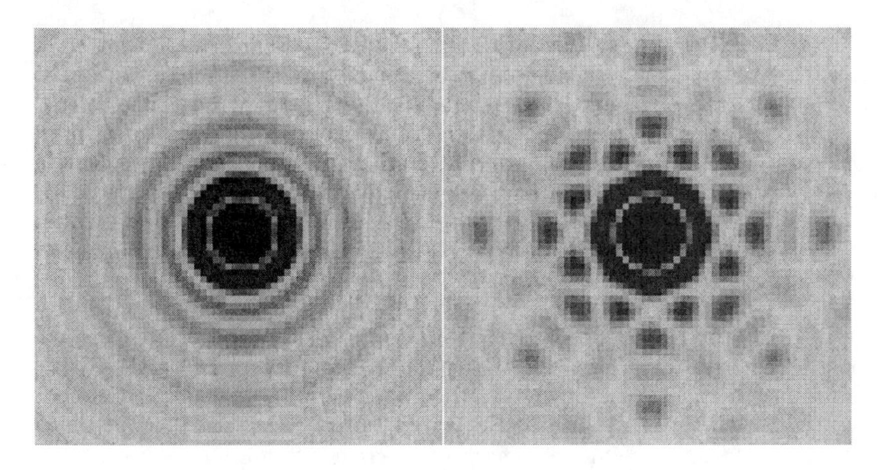

Figure 1. NGST PSFS at 2 μm - generated with a modified version of the HST PSF tool Tiny Tim. Left for an 8 m filled pupil, right for the notched GSFC design.

sources to be selected and identified, and because the field will be accessible in general only for a few weeks, object selection and the configuring data for the micro-mirror array have to be performed within a very short time (days at most). This can only be achieved by completely automatic image analysis and subsequent object classification.

The software requirements are very similar to those for batch processing of slitless spectroscopy, which uses direct imaging for object detection, wavelength calibration and weighting of spectrum extraction by size and orientation of the objects (Freudling 1997). A completely automatic "pipeline" to process such data for the HST NICMOS camera has been developed (Freudling & Thomas 1997), and this program (Calnic-C) could be very easily adapted to become both the micro-mirror configuration program and the spectral extraction code.

4. Calibration of NGST Multiple Slit Spectroscopy Data

The multi-object spectroscopy concept mentioned above serves the scientific requirements very well. However, such highly flexible instrument configurations are very demanding on calibration. It is obvious that classical calibration concepts will not be able to cope with an almost unlimited variety of on/off target apertures that can be constructed across the field. Clearly, one can not possibly hope to obtain useful sensitivity curves by observing standard stars even in only a limited subset of these slit-lets. The aperture-mirror configuration for the field to be studied will destroy the current "instrument setup" that was unique to a particular position in the sky. Along the same lines, dispersion relations and wavelength zero points will become obsolete as soon as the micro-mirrors are reorganized.

The predictive calibration methods currently developed for HST (Rosa 1994) and ESO VLT (Rosa 1995) instrumentation are, however, ideally suited to this situation. The kernels are instrument software models based on first principles (e.g., grating equations). Configuration data, usually measurable engineering quantities, are regularly verified on dedicated calibration exposures for a few selected instrument configurations. Such models demonstrably permit very accurate predictions of dispersion relations and sensitivity curves for modes not actually covered by calibration data (Rosa 1997; Ballester & Rosa 1997).

Once "calibrated" on empirical calibration data, the software models can be made integral parts part of the data calibration and data analysis pipelines in two ways:

Predictive calibration: Calibration reference data (eg. dispersion relations, sensitivity curves) are generated on-the-fly for the observational data set under processing, permitting high-performance pipelines. Only configuration data need to be accessed, rather then multiple reference files for each mode. In addition, the model-generated reference data are noise free.

Forward analysis: An even more advanced application is the use of these software models to match astrophysical ideas and observations in the raw data domain. For example, an optimization package would browse an archive of theoretical galaxy spectra at various evolutionary stages, very accurately simulate raw observational data with eg. luminosities and redshifts as free parameters, and seek for statistically significant matches.

Predictive calibration and forward analysis are currently being explored for HSTs STIS. Utilizing these methods NGST operations in multi-object spectroscopy mode will need to allocate only very small amounts of time for specific calibrations. Only wavelength zero-point shifts need to be verified observationally for a given instrumental setup. Aperture specific sensitivity curves require only infrequent checks on configuration data that have been obtained for a suitable subset of the micro-mirrors from standard stars.

References

Ballester, P., & Rosa, M.R., 1997, A&AS, in press (ESO prepr No 1220)

Freudling, W., 1997, ST-ECF Newsletter, 24, 7

Freudling, W., & Thomas, R., 1997, "http://ecf.hq.eso.org/nicmos/calnicc/calnicc.html"

Hook, R.N., & Lucy, L.B., 1993, ST-ECF Newsletter, 19, 6

Hook, R.N., & Lucy, L.B., 1994, in "The Restoration of HST Images and Spectra", R.J. Hanisch & R.L. White, Baltimore: STScI, 86

Krist, J.E. & Hook, R.N., 1997, "The Tiny Tim Users Manual", STScI

Lucy. L.B., 1994, in "The Restoration of HST Images and Spectra", R.J. Hanisch & R.L. White, Baltimore: STScI, 79

Rosa, M.R., 1994, CAL/FOS-127, Baltimore: STScI

Rosa, M.R., 1995, in "Calibrating and Understanding HST and ESO Instruments", P. Benvenuti, Garching: ESO, 43

Rosa. M.R., 1997, ST-ECF Newsletter, 24, 14

Astronomical Data Analysis Software and Systems VII
ASP Conference Series, Vol. 145, 1998
R. Albrecht, R. N. Hook and H. A. Bushouse, eds.

Mixing IRAF and Starlink Applications – FIGARO under IRAF

Martin J. Bly and Alan J. Chipperfield

Starlink Project[1,2], Rutherford Appleton Laboratory, Chilton, DIDCOT, Oxfordshire OX11 0QX, United Kingdom. Email: bly@star.rl.ac.uk, ajc@star.rl.ac.uk

Abstract. The Starlink Project has introduced a scheme which allows programs written for the Starlink Software Environment to be run in the IRAF environment. Starlink applications can be controlled from the IRAF CL, reading and writing IRAF `.imh` data files. Conversion to and from the Starlink Extensible N-Dimensional Data Format (NDF) is done on-the-fly, without the need for a separate data-conversion step. The intention has been to make the Starlink applications appear as much like native IRAF applications as possible, so that users can intermix tasks from the two environments easily. The general-purpose data reduction package FIGARO is now available to be run from the IRAF CL, and more Starlink applications will follow in due course.

1. Introduction

Modern astronomical instruments demand flexible data-analysis facilities which allow complex processing chains. Single packages, or sets of packages under one environment, may not provide all the necessary steps or may not do specific tasks as well as a package from another environment. However, the increased sophistication of software has isolated users of one package from the facilities of other packages.

Two ways to break down the barriers are:

- To enable applications from one system to be run from the user-interface of another.
- To perform 'on-the-fly' conversion of data formats.

A system has been developed to enable packages written for the Starlink Software Environment to be run from the IRAF CL so that an IRAF user can mix and match application tasks from the two environments as required to process data, all under the control of the IRAF CL.

---

[1]Managed by the Council for the Central Laboratory of the Research Councils on behalf of the Particle Physics and Astronomy Research Council

[2]http://star-www.rl.ac.uk/

Figure 1 shows how the components of the Starlink/IRAF inter-operability system fit together.

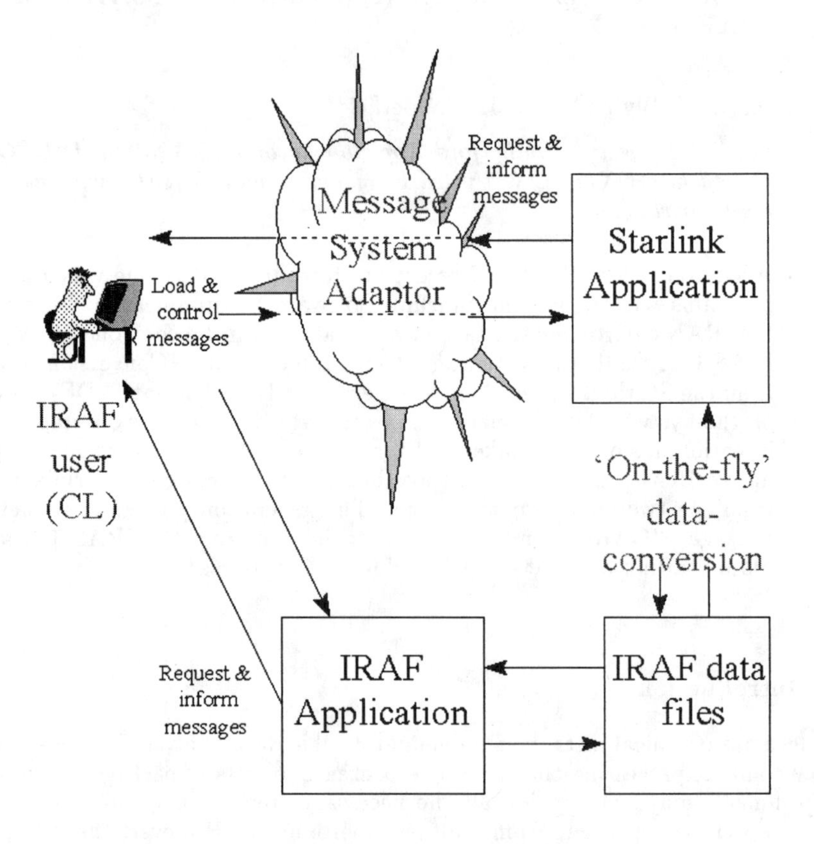

Figure 1. The Starlink/IRAF Inter-operability System.

2. The Message System Adaptor

At the heart of the system is the Message System Adaptor (Terrett 1996). IRAF and the Starlink Software Environment have similar architectures whereby the user-interface communicates with the task by means of messages. The user-interface sends messages to control the task and the task replies with information or requests for parameters *etc.* The Message System Adaptor is a Tcl script which sits between the IRAF CL and the Starlink task, intercepting the messages flowing between them and converting between IRAF CLIO protocol and Starlink ADAM protocol. In this way the adaptor looks like an IRAF task to CL and like a Starlink user-interface to the Starlink task.

3. Data Conversion

Starlink packages use Starlink's Extensible N-Dimensional Data Format (NDF) for data storage. The data access layer for programs is provided by the NDF subroutine library, which has the built-in ability to use externally-provided data conversion tasks to convert foreign data formats to and from the NDF format 'on the fly'. This facility is used, together with tasks from Starlink's CONVERT package, to convert between NDF format and IRAF `.imh` or FITS format when running from IRAF. There is no need for a separate data-conversion step in the processing and the potential is there to handle many other data formats.

One pitfall with this system is that not all components of all NDFs have equivalents in the other formats, so data may be lost in the conversion process. This is particularly true of error estimate data, which are an intrinsic part of the Starlink NDF design but are absent from the other formats.

4. The User's View

Tasks may be invoked and parameters set in the normal IRAF ways. The main differences a user will see are:

- A Starlink-style graphics display is used by Starlink applications. Each graphics application has a parameter defining the device to be used and, if the device is 'xwindows', a special Starlink window will be used.
- Starlink applications tend to have richer sets of parameters.
- More diagnostic messages appear in the event of errors.

5. Generic Package Definition

The Starlink Software Environment requires that application packages have an associated Interface File, which defines the parameters of its applications, and Package Definition Files which define the commands available and the source of help *etc.* Other environments, such as IRAF, require similar information presented in a different way.

So that all the files required by various environments can be generated from a single source, a generic package definition format, known as the Interface Definition Format (IFD), has been developed. The format allows environment-specific inclusion or exclusion of elements of the description, for example where an application required in one environment is not relevant in another. Tcl scripts have been written to process a package IFD file and produce the files required by the Starlink and IRAF environments. The scheme could be extended for other environments.

6. Availability

At present, the general purpose data reduction package FIGARO is the only Starlink package in production use under the IRAF CL. More, including the kernel applications package KAPPA and the CCD data reduction package CCD-PACK, will be available soon.

The software is freely available for non-profit research use on the Starlink CD-ROM or from the Starlink Software Store at:

`http://star-www.rl.ac.uk/cgi-store/storetop`

Further information may be found in the Starlink documents:

- SUN/217 – Running Starlink Applications from IRAF CL,
- SUN/220 – IRAFFIG – Figaro for IRAF,
- SSN/35 – IRAFSTAR – The IRAF/Starlink Inter-operability Infrastructure, and
- SSN/68 – IFD – Interface Definition Files.

7. Conclusions

This method of running packages designed for one environment from the user interface of another works well and increases the range of procedures available to users for data processing without the cost of writing new packages. Comments from users have been favourable. The system could be expanded to cover other environments, particularly where architectures and parameter systems are similar.

Tcl is a powerful tool from which to build a protocol translation system and language converters.

Following the release of the next batch of Starlink application packages for use with IRAF we intend to monitor user reaction to the products before embarking on further work in this area.

Acknowledgments. This work was done in the UK by the Starlink Project at The Central Laboratory of the Research Councils and funded by the Particle Physics and Astronomy Research Council. We are grateful for the contribution of the IRAF Programming Group at NOAO in bringing this work to fruition.

References

Terrett, D. L. 1996, in ASP Conf. Ser., Vol. 101, Astronomical Data Analysis Software and Systems V, ed. George H. Jacoby & Jeannette Barnes (San Francisco: ASP), 255

Astronomical Data Analysis Software and Systems VII
ASP Conference Series, Vol. 145, 1998
R. Albrecht, R. N. Hook and H. A. Bushouse, eds.

An Infrared Camera Reduction/Analysis Package

S. J. Chan
Institute of Astronomy, University of Cambridge, Madingley Road, Cambridge, CB3 0HA, United Kingdom

Abstract. An infrared-array reduction and analysis package which is written in the IRAF environment has been developed to handle the huge amount of data obtained at the Teide Observatory using the 1.5 m Carlos Sánchez Telescope with the new infrared Camera/Instituto de Astrofísica de Canarias(IRCAM/IAC). This is one of the near-infrared observational projects of the author. Several tasks are being developed which are written in CL or SPP in the IRAF environment for efficiently reducing and analyzing near-infrared data.

1. Introduction

Several barred spiral galaxies and galactic H II regions associated with small molecular clouds were observed with the IR-Camera on the 1.5 m Carlos Sánchez Telescope (CST) on the Teide Observatory during October 1996. This project is one of the near-infrared observation projects of the author.

2. IRCAM Package

2.1. IRCAM/IAC

The IRCAM/IAC is based on a 256×256 NICMOS 3 array and has a plate scale of 0.4″/pixel. Currently, the camera contains 7 filters covering the wavelength range from 1.2 μm to 2.32 μm. The 3 σ 60 seconds limiting magnitudes for point sources were 18.8 mag at J, 18.5 mag at H, and 16 mag at K.

2.2. General Description of the IRCAM package

The IRCAM/analysis package (Chan & Mampaso 1997) has been expanded into a large package for reducing and analyzing infrared data. Currently, there are 3 packages which are for reduction (IRCREDUCT), analysis (IRCPHOT) and for general usage and file handling (IRCUTIL). One of the advantages of this package is that it can easily handle the large amount of data. Furthermore, after the data were processed by tasks in IRCREDUCT, the side-effects due to problems from the CST (guiding problems for high declination objects) and from the IRCAM/IAC (the "drip noise" produced by the camera; see Figure 1a in Chan & Mampaso 1997)are reduced dramatically. The most important improvements are that the point spread function of stars becomes well-defined and the resolution of the images is improved (Figure 1 and Figure 2) The package

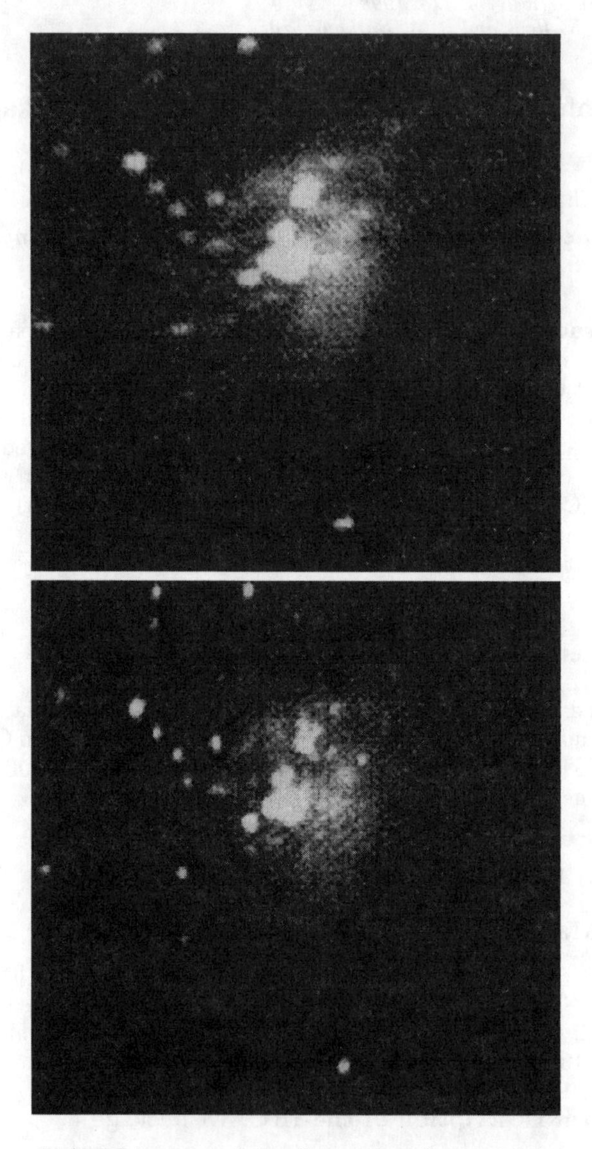

Figure 1. H band images of the H II region S152. In the top figure, the side-effect has not been corrected. In the bottom figure, the side-effect has been corrected and the PSF of stars becomes well-defined.

is now working well. IRCUTIL and IRCREDUCT which have been released to users in the IAC, also have full help utilities.

Acknowledgments. S.J.Chan thanks the Institute of Astronomy, University of Cambridge for offering her financial support to attend the Conference and the hospitality of the Instituto de Astrofísica de Canarias during her stay.

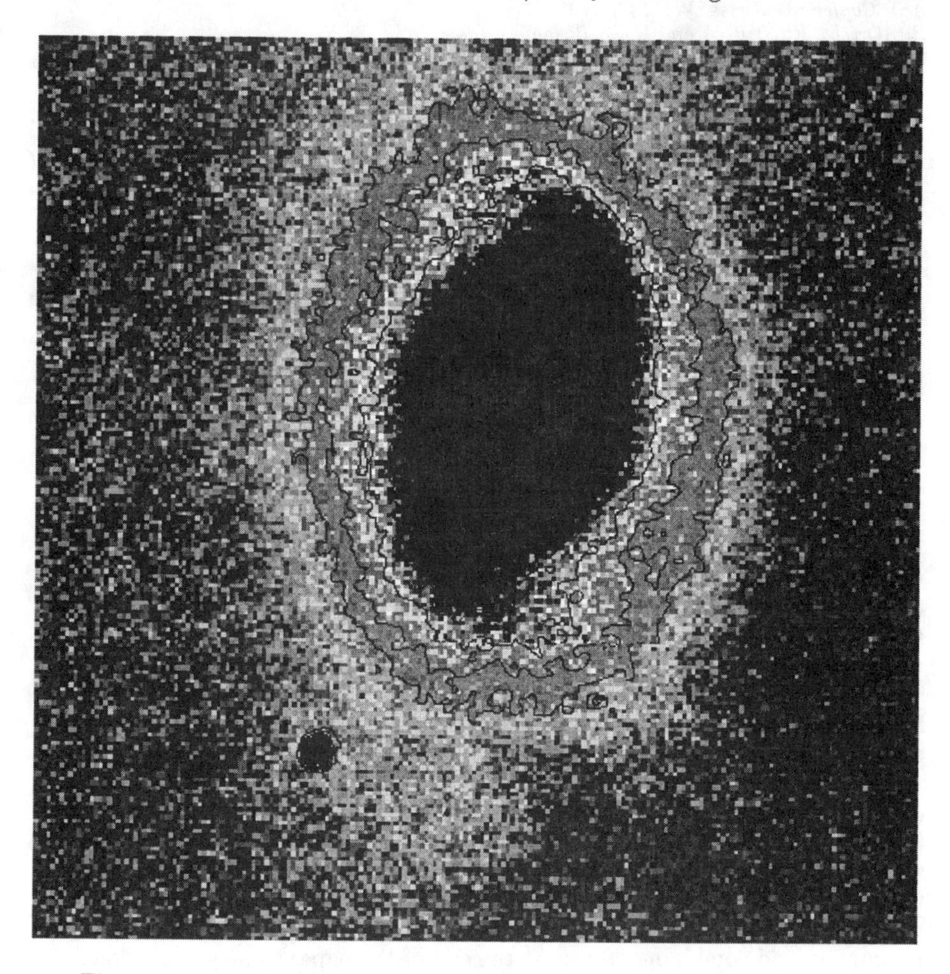

Figure 2. J band image of NGC2273. It is of good enough quality to allow ready detection of the inner bar. The "*effective*" resolution in working mode of the CST can now be better than the 1 ″ limit after image processing using the IRCAM package.

References

Chan, S. J., & Mampaso, A. 1997, in ASP Conf. Ser., Vol. 125, Astronomical Data Analysis Software and Systems VI, ed. Gareth Hunt & H. E. Payne (San Francisco: ASP), 93

Astronomical Data Analysis Software and Systems VII
ASP Conference Series, Vol. 145, 1998
R. Albrecht, R. N. Hook and H. A. Bushouse, eds.

The Astrometric Properties of the NOAO Mosaic Imager

L. E. Davis

National Optical Astronomy Observatories, Tucson, AZ 85719

Abstract. The astrometric properties of the NOAO Mosaic Imager are investigated using the new IRAF MSCRED and IMMATCHX packages and observations taken during recent engineering runs.

1. Introduction

The NOAO Mosaic Imager is an 8K by 8K pixel camera composed of 8 individually mounted 2K by 4K CCDs separated by gaps of up to 72 pixels. Each CCD has its own amplifier and is read into a separate image in a multi-extension FITS file. A set of dithered exposures is required to obtain a complete observation of a given region of the sky. Reassembling an observation into a single image requires selecting the reference coordinate system of the final combined image, resampling each exposure to the selected reference coordinate system, and combining the resampled images.

In this paper the new IRAF MSCRED and IMMATCHX packages and data obtained during recent Mosaic Imager engineering runs on the NOAO 4m (FOV 36 arcminutes, scale 0.26 arcseconds per pixel) and 0.9m (FOV 59 arcminutes, scale 0.43 arcseconds per pixel) telescopes are used to investigate the functional form of the plate solutions, the accuracy of the plate solutions before and after resampling and combining, and the accuracy of the flux conservation in the resampling and combining steps. At the end of the paper some speculations are offered on the feasibility or otherwise of reassembling the Mosaic Imager data in real time at the telescope.

2. The Plate Solutions

The plate solutions were computed using published astrometry for Trumpler 37 (Marschall and van Altena 1987) and data from recent engineering runs. Fits to both gnomonic projection (TAN) plus polynomial in x and y models (Kovalevsky 1995), and zenithal polynomial projection (ZPN) models (Greisen and Calabretta 1996) were performed.

2.1. 4m Plate Solutions

A theoretical 5th order ZPN model for the 4m prime focus plus corrector optical system was available at the time of writing (Vaughnn 1996). This model predicted pincushion distortion with a maximum scale change of 6.4% and maximum pixel area change of 8.5% across the FOV. Fits of ~400 stars in Trumpler

37 to the theoretical model produced good plate solutions for each CCD with residuals for the 8 detectors averaging ~0.10 and ~0.07 arcseconds in α and δ respectively. Marginally better residuals of ~0.09 and ~0.06 arcseconds were obtained with TAN projection plus cubic polynomial models. The residuals from the latter fits showed no evidence for the predicted 5th order term, most probably due to a combination of the limited angular size of the individual CCDs, and the precision of the published astrometry.

2.2. 0.9m Plate Solutions

An accurate theoretical model for the 0.9m f/7.5 Cassegrain focus plus corrector system was not available at the time of writing. However TAN projection plus cubic polynomial model fits to ~800 stars in the field produced good plate solutions, with residuals for the 8 CCDs averaging ~0.07 arcseconds in both α and δ and no systematic trends in the residuals. The polynomial fits revealed the presence of a 1.8% maximum scale change and 2.0% maximum pixel area change over the FOV. These results were used to derive the equivalent "theoretical" ZPN model. Fits of the derived ZPN model to the data produced good plate solutions with residuals of ~0.09 arcseconds in each coordinate.

3. Reassembling a Single Mosaic Observation

The TAN projection plus cubic polynomial plate solutions and bilinear interpolation were used to combine the 8 pieces of the mosaic into a single 8K by 8K image with an undistorted TAN projection coordinate system. Because the individual images were flat fielded before resampling, no additional flux correction during resampling was required. Empirical rather than theoretical plate distortion models were used in order to test the validity of the empirical approach, and because they produced marginally better fits. Bilinear interpolation was chosen for efficiency, The TAN output coordinate system was chosen because it is the standard projection for small field optical astrometry. Other options are available.

3.1. Astrometric Accuracy

New plate solutions were computed for the resampled 4m and 0.9m data. In both cases TAN projection plus first order polynomials in x and y produced good plate solutions with residuals of ~0.08 / ~0.06 and ~0.10 / ~0.09 arcseconds in the α / δ coordinates fits for the 4m and 0.9m images respectively. In all case no discernible distortion remained, and the accuracy of the original plate solutions was almost recovered.

3.2. Flux Conservation

Before and after resampling aperture photometry of the astrometric stars was obtained and compared with flux correction model derived from the ZPN radial distortion models. Agreement for the 4m data was excellent with <correction (observed) - correction (predicted)> = 0.0007 +/- 0.01 magnitudes, and no observed trends with distance from the center of the image. The corresponding numbers for the 0.9m data were, <correction(observed) - correction(predicted)>

= 0.0004 +/- 0.018 magnitudes, and no trends with distance from the center of the image. Therefore as long as the coordinate transformation used to resample the image models the geometry of the optical system accurately, the resampling code used to recombine the images will accurately conserve flux.

4. Combining Dithered Mosaic Observations

The dithered resampled observations must be combined to fill in the gaps in the mosaic and produce a single image with complete sky coverage. Combining the dithered observations requires precomputing the offsets between the frames in a dither set (before resampling), and resampling the dithered images to the selected reference coordinate system (TAN with no distortion) in such a manner that the images are separated by integer pixel offsets, and combining the intensities in the overlap regions. At the time of writing only a single 0.9m set of dithered Trumpler 37 observations was available for investigation.

4.1. Astrometric Accuracy

A new plate solution was computed for the combined images. A TAN projection plus first order polynomials in x and y model produced an excellent fit with residuals of ~0.09 arcseconds in α and δ. Therefore the combining step did not introduce any distortion into the image geometry and the accuracy of the original plate solutions was approximately recovered.

4.2. Flux Conservation

The image combining algorithm employed was median with no rejection. Before and after aperture photometry of ~800 astrometric stars around the combined frame and one of the resampled frames produced a mean value of <mag (resampled) - mag (combined)> = -.008, and no trends with position in the image. The small but real offset appears was caused by changing observing conditions which was not corrected for in the combining step.

5. Automatic Image Combining at the Telescope

5.1. Plate Solutions

Automating the image combining step to run in close to real time requires that either the plate solutions are repeatable from night to night and run to run, or a mechanism is in place to automatically compute new plate solutions at the telescope. Thus far only the first option has been investigated, although the second is also being considered. Preliminary tests suggest that the higher order terms in the plate solutions are very repeatable, but that small adjustments to the linear terms are still required. More rigorous testing with the system in a stable condition is required to confirm this.

5.2. Efficiency

On an UltraSparc running SunOS 5.5.1, a set of nine dithered images currently takes approximately ~2 minutes per image to read out, ~4 minutes per image to reduce, and ~10 minutes per image to resample. A further ~18 minutes is required to combine the 9 resampled dithered images. Cpu times are ~1/2 to ~1/3 of the above estimates, depending on the operation. Although a factor of 2 improvement in efficiency may be possible through system and software tuning in some cases, it is obvious from the time estimates above that only some observing programs can realistically consider combining images at the telescope in real time and under current conditions.

5.3. Software

Much of the image combining process has already been automated, however some more automation in the area of adjusting the plate solutions for zero point and scale changes needs to be done.

6. Conclusions and Future Plans

The individual mosaic pieces can be recombined with high astrometric accuracy using either the empirical TAN projection plus cubic polynomial models or the theoretical ZPN models. The latter models have fewer parameters than the former and are a more physically meaningful representation of the data, but the former still produce lower formal errors.

As long as the computed plate solutions are a good match to the true geometry of the instrument flux conservation during resampling is very accurate. If this is not the case another approach must be taken, such as using a precomputed "flat field" correction image.

Automating image combining at the telescope is currently only feasible for some types of observing programs. The main limitation is the computer time required. The repeatability of the plate solutions from run to run and the issue of computing plate solutions at the telescope are still under investigation.

Acknowledgments. The author is grateful to Frank Valdes for assistance with the use of the MSCRED package and to Taft Armandroff and Jim Rhoades for providing test data.

References

Greisen, E. R. & Calabretta, M. 1996, Representations of Celestial Coordinates in FITS, fits.cv.nrao.edu fits/documents/wcs/ wcs.all.ps

Kovalevsky, J. 1995, in Modern Astrometry, Springer-Verlag, 99

Marschall, L. A. & van Altena, W. F. 1987, AJ, 94, 71

Vaughnn, D., 1996, private communication

Astronomical Data Analysis Software and Systems VII
ASP Conference Series, Vol. 145, 1998
R. Albrecht, R. N. Hook and H. A. Bushouse, eds.

NICMOSlook and Calnic C: Slitless Spectra Extraction Tools

N. Pirzkal, W. Freudling, R. Thomas and M. Dolensky

Space Telescope - European Coordinating Facility,
Karl-Schwarzschild-Str. 2, D-85748 Garching, Germany

Abstract. A unique capability of NICMOS is its grism mode which permits slitless spectrometry at low resolution. Extracting spectra from a large number of NICMOS grism images requires a convenient interactive tool which allows a user to manipulate direct/grism image pairs. NICMOSlook and Calnic-C are IDL programs designed for that purpose at the Space Telescope – European Coordinating Facility[1]. NICMOSlook is a graphical interface driven version of this extraction tool, while Calnic C is a program which performs the same functionality in a "pipeline" approach.

1. NICMOSlook and Calnic C

Typically, the NICMOS grism mode involves taking a pair of images which includes both a direct image and a grism image. The direct image is used to locate objects of interest, to find the position of the appropriate spectra in the grism image, to determine the optimal extraction technique to use, and to calibrate the wavelength range of the extracted spectra. NICMOSlook, which has an IDL user interface based on STISlook from Terry Beck, is designed to allow a user to interactively and efficiently extract spectra from a small batch of data. The user maintains full control over the extraction process. NICMOSlook was designed to be most useful to users who want to quickly examine data or fine tune the extraction process. The stand-alone version, Calnic C, is meant to be used on larger sets of data and requires only a minimum amount of user input. Both programs can be fully configured and extended through the use of a few configuration and calibration files.

2. Features Overview

2.1. Image Display

NICMOSlook allows a direct image and corresponding grism images to be read from FITS files and be displayed. A variety of display options are available to the user including a variety of color tables, selectable zoom factors, and the ability

[1] http://ecf.hq.eso.org

to blink the images successively. Some basic image filtering and processing capabilities are also provided to allow the user to locate and identify objects in the direct image (or spectra in the grism image).

2.2. Object Identification

In NICMOSlook objects can be identified automatically using the DAOFIND algorithm or can be provided manually by the user (through the use of the cursor or of a text file containing the coordinates of the objects). The spatial extent of objects can also be determined either automatically or through additional user input. In Calnic C objects in the direct image are located without user input: the location, orientation, and size of each object are determined using SExtractor (Bertin 1996).

2.3. Determination of the Location of the Spectrum

Once the position, orientation, and shape of an object have been determined, they are used to infer the position, orientation, and extent of object's spectra in the grism image. This process is completely automatic but it can, nevertheless, be fully customized by the user.

2.4. Spectral Extraction

Extracting a spectrum can be done using either weighted or unweighted extractions and using either the Point Source mode (where the cross dispersion spectral profile is taken to be that of a point source), or the Extended Object mode (where the cross dispersion spectral profile is first determined from the shape of the object in the direct image). Once extracted, the spectrum is wavelength calibrated, and a response function is used to produce a calibrated spectrum in mJy or erg/s/$cm^2/\AA$ as a function of μm.

2.5. Deblending

An attempt can be made to automatically remove the spectral contamination caused by the presence of nearby objects. The method used uses only the assumption that the shape of the object is not wavelength dependent.

2.6. Spectral Lines Search

Emission and absorption lines are automatically identified in the resulting spectra. The continuum emission is automatically determined.

2.7. Result Output

In NICMOSlook the extracted spectrum is plotted on the screen. The user can interactively modify the continuum background fit and examine spectral features. The user can also export the spectrum as a binary FITS table and PostScript plots. Calnic C outputs a binary FITS table and PostScript plots of each of the extracted spectra.

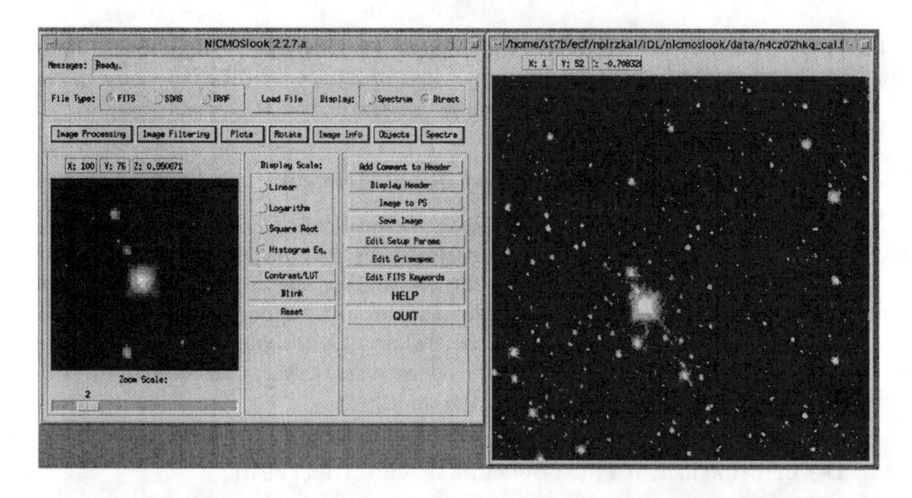

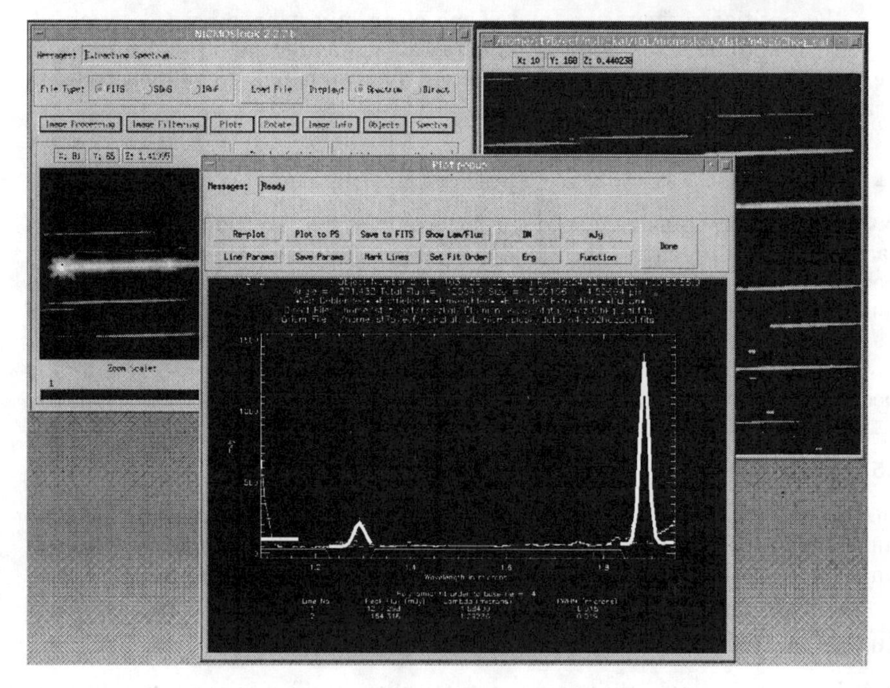

Figure 1. The NICMOSlook User Interface

3. How to Obtain NICMOSlook and Calnic C

NICMOSlook, Calnic C, and their respective User Manuals can be downloaded from *http://ecf.hq.eso.org/nicmos*
NICMOSlook and Calnic C require IDL 4.0 or 5.0.

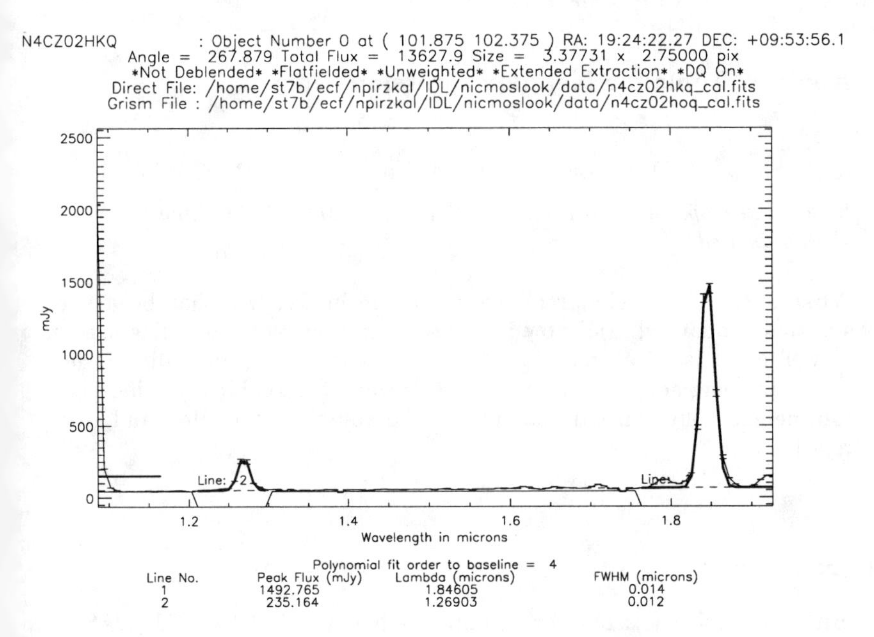

Figure 2. The paper output of NICMOSlook and Calnic C

References

Bertin, E., & Arnouts, S., 1996, A&A, 117, 393

Astronomical Data Analysis Software and Systems VII
ASP Conference Series, Vol. 145, 1998
R. Albrecht, R. N. Hook and H. A. Bushouse, eds.

Analysis Tools for Nebular Emission Lines

R. A. Shaw, M. D. De La Peña, R. M. Katsanis, and R. E. Williams

Space Telescope Science Institute, Baltimore, MD 21218, Email: shaw@stsci.edu

Abstract. The nebular analysis package in STSDAS has been substantially enhanced, and now includes several new ions and diagnostics, as well as updated atomic data for all supported ions. In addition, new tasks are being added to compute, from ratios of recombination lines, certain nebular physical parameters and the abundances of He^+ and He^{+2} relative to H^+.

1. Introduction

The **nebular** analysis package is a set of applications within IRAF/STSDAS for computing physical parameters in emission line nebulae (such as electron density, N_e, and temperature, T_e), as well as ionic abundances of several elements relative to ionized hydrogen. **Nebular** also provides utilities for exploring the range of validity of the diagnostics themselves, and for employing them in the context of a very simple nebular model. Several enhancements to the package have been implemented since the original descriptive paper by Shaw & Dufour (1995) was published, and the major new features are described here.

2. New Ions and Atomic Data

Thirteen new ions of C, N, O, Ne, Na, Mg, Al, Si, S, Cl, K, and Ca have been added to the set of 21 previously supported in **nebular**. As a consequence, 15 new diagnostics for N_e and T_e are available, and they span a much greater range of density, temperature, and ionization than in the previous version. The full set of supported ions and diagnostics can be found from the **nebular** home page[1]. Several new features have also been added, including the capability to compute collisional line emissivities from up to eight (rather than the five previously supported) atomic levels, depending upon the availability of the supporting atomic data. These low-lying levels arise from the same electron configurations as the ground level. The atomic data for the various lines have been updated to the best, most recent available as of mid-1996. These data have been appearing in the literature at a rapid rate, owing to the success of the IRON project (Hummer et al. 1993), a concerted international effort to compute precise atomic data for iron-group ions of astrophysical interest. The collision strengths in particular

[1]http://ra.stsci.edu/nebular/

have been computed for most ions over a much wider range of T_e, and are more accurate by factors of 3 to 10. This improvement in the data quality permits the calculation of reliable diagnostics over a much greater range of T_e and N_e. The references to the atomic data are too numerous to include here (though they are given in the **nebular** help documentation), but in general are at least as recent as those given in the compilation by Pradhan and Peng (1995).

The T_e and N_e diagnostics of collisionally excited lines that are typically used in the literature derive from ratios of specific line intensities (or ratios of sums of intensities) which have a reasonably high emissivity, consistent with being very sensitive to the diagnostic in question. These line intensity ratios are generally the same for ions with a given ground state electron configuration. Table 1 lists the line ratios that are traditional in this sense, where the notation $I_{i \to j}$ refers to the intensity of the transition from level i to j. The correspondence of wavelength and transitions between energy levels for a given ion can be found by running the *ionic* task.

Table 1. Traditional Diagnostic Ratios

Ground Configuration	Diagnostic	Traditional Ratio
s^2	N_e	$I_{4\to1} / I_{3\to1}$
	T_e	$(I_{4\to1} + I_{3\to1}) / I_{5\to1}$
p^1	N_e	$I_{5\to2} / I_{4\to2}$
p^2	T_e	$(I_{4\to2} + I_{4\to3}) / I_{5\to4}$
p^3	N_e	$I_{3\to1} / I_{2\to1}$
	T_e	$(I_{2\to1} + I_{3\to1}) / (I_{4\to1} + I_{5\to1})$
	T_e[a]	$(I_{2\to1} + I_{3\to1}) / (I_{5\to3} + I_{5\to2} + I_{4\to3} + I_{4\to2})$
p^4	T_e	$(I_{4\to1} + I_{4\to2}) / I_{5\to4}$

[a]For N^0 and O^+.

Note that these traditional diagnostic ratios are not the only viable ratios to use: some ratios, though perhaps not as sensitive to the diagnostic in question, are useful if the spectral coverage or resolution is limited; others are simply better for some purposes, such as the T_e-sensitive [O III] ratio $I(1660+1666)/I(4363)$. Some **nebular** tasks, such as *temden* and *ntcontour*, now allow the user to specify a transition other than the traditional (default) one.

3. Recombination Lines

The **nebular** package will soon include new tasks to accommodate the analysis of recombination lines. Atomic data for H^+, He^+, and He^{+2} have been incorporated from Storey & Hummer (1995). Specifically, we adopt their tabulations of emissivities, ϵ, separately for Case A and Case B recombination. (The code interpolates in log-T_e, log-ϵ for intermediate values within their grid.) The *recomb* task will solve for the interstellar reddening and/or T_e, given a list of recombination line intensities for a single ion. The *rec_abund* task will compute the abundance of these ions with respect to H^+ from a list of recombination lines. The *abund* task, which computes abundances from collisional lines in the context of a simple model, is being enhanced to include the ionic abundance calculations from recombination lines as well.

4. Exploration of the Diagnostics

The *ntcontour* task has been substantially enhanced, and is now more useful for exploring new or traditional diagnostics from collisionally excited lines. This task computes and plots curves that show the range of T_e, N_e, and/or intensity ratios that are consistent with a specified diagnostic. A family of secondary curves may optionally be plotted, where each curve may be specified explicitly or as a set of successive, small differences from the reference ratio, giving the appearance of contours. Though for all ions there are default diagnostics for N_e and/or T_e, it is possible to customize the diagnostic to the ratio of any of the supported transitions. In addition, the diagnostics may be plotted as N_e vs. T_e, N_e vs. I_{line}, and T_e vs. I_{line}. This task may be run interactively, so that it is possible to investigate many diagnostics quickly. *Ntcontour* is particularly useful for determining the range of N_e and T_e where a particular diagnostic is sensitive, for investigating non-traditional diagnostics, and for estimating the consequences of a given level of uncertainty in an observed line ratio. Figure 1 shows an example plot of the default density-sensitive ratio for Al II.

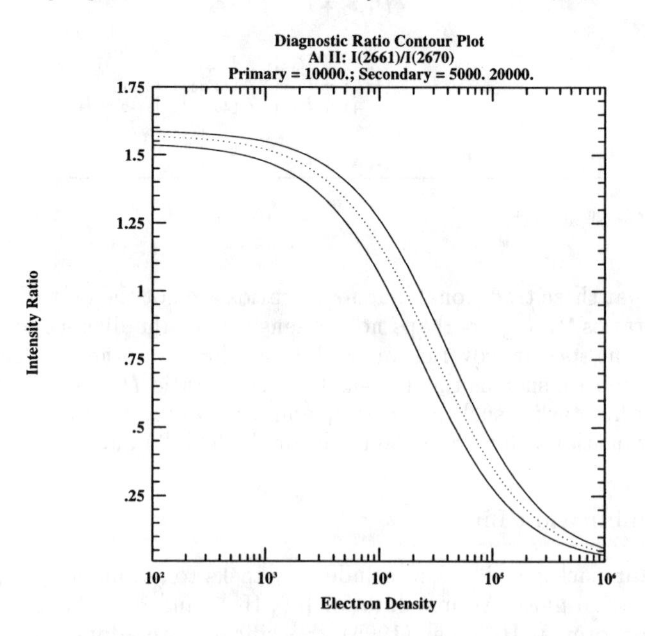

Figure 1. Ratio of [Al II] $I(2661)/I(2670)$ vs. N_e using *ntcontour*.

5. Software Features

The **nebular** package has been substantially re-designed, and is largely table-driven. For example, the atomic reference data are encapsulated in FITS binary tables, and the data pedigree (including literature references) are documented in header keywords. Fits to the collision strengths as a function of T_e are evaluated at run-time, so that the reference tables contain data in a form much as it appears in the literature. These features provide great flexibility for anyone to substitute improved atomic data, and allow for easy extensibility of the package to accommodate high-quality data for more ions as they become available.

The data and associated functional fits for one or more atoms/ions of interest are collected into persistent in-memory objects during the course of program execution. The diagnostics for N_e and T_e are stored internally as strings containing algebraic relations (in a form not unlike those given in Table 1), and are evaluated at run-time. These features allow much greater flexibility in the applications to make use of relations between ions, such as the comparison of physical diagnostics from similar ions, or the computation of relative abundances.

6. Software Availability

In their sum, the applications in the **nebular** package are a valuable set of tools for nebular analysis: the derived quantities can be used directly or as input to full photo-ionization models. Most of the enhancements described here have already been incorporated into V2.0 of the **nebular** package; the new tasks for analysis of recombination lines will be available in the next release. **Nebular** is publically available, and is included in the STSDAS external package; the new version will be found under the **analysis** package in STSDAS V2.0 and later. Users who do not have STSDAS may obtain the new version for personal installation from the **nebular** ftp[2] area. Note that **nebular** V2.0 requires that IRAF V2.10.4 (or later) and TABLES V2.0 (or later) be installed. Some of the tasks may alternatively be run via the Web from an HTML forms interface: view the **nebular** home page for details.

Acknowledgments. Initial support for this software was provided through the NASA Astrophysics Data Program. Support for these enhancements was provided by an internal ST ScI research grant.

References

Hummer, D. G., Berrington, K. A., Eissner, W., Pradhan, A. K., Saraph, H. E. & Tully, J. A. 1993, A&A, 279, 298

Pradhan, A. & Peng, J. 1995, in Proc. of the ST ScI, No. 8, eds. R. E. Williams & M. Livio (Baltimore, Space Telescope Science Institute), 24

Shaw, R. A. & Dufour, R. J. 1995, PASP, 107, 896

Storey, P. J. & Hummer, D. G. 1995, MNRAS, 272, 41

---

[2]ftp://ra.stsci.edu/pub/nebular/

Astronomical Data Analysis Software and Systems VII
ASP Conference Series, Vol. 145, 1998
R. Albrecht, R. N. Hook and H. A. Bushouse, eds.

The Future of Data Reduction at UKIRT

F. Economou

Joint Astronomy Centre, 660 N. A'ohōkū Place, University Park, Hilo, HI 96720, USA

A. Bridger and G. S. Wright

Royal Observatory Edinburgh, Blackford Hill, Edinburgh EH9 3HJ, United Kingdom

N. P. Rees and T. Jenness

Joint Astronomy Centre, 660 N. A'ohōkū Place, University Park, Hilo, HI 96720, USA

Abstract. The Observatory Reduction and Acquisition Control (ORAC) project is a comprehensive re-implementation of all existing instrument user interfaces and data handling software involved at the United Kingdom Infrared Telescope (UKIRT). This paper addresses the design of the data reduction part of the system. Our main aim is to provide data reduction facilities for the new generation of UKIRT instruments of a similar standard to our current software packages, which have enjoyed success because of their science-driven approach. Additionally we wish to use modern software techniques in order to produce a system that is portable, flexible and extensible so as to have modest maintenance requirements, both in the medium and the longer term.

1. Background

UKIRT[1] has been using automated data reduction for one of its main instruments, CGS4, for some years. The benefits of data reduction in near-real time are many, including more efficient use of telescope time and a higher publication rate for data. However, the program CGS4DR (Daly 1995 & 1997) that was used for this purpose, despite its successes proved to also have its drawbacks. As part of the ORAC project[2] (Bridger et al. 1998) and the preparation for the arrival of two new instruments, the future data reduction at UKIRT is being reassessed in the light of our experiences with CGS4DR. In particular, while we wish to continue to provide near publication quality data at the telescope, we also want to:

---

[1]http://www.jach.hawaii.edu/UKIRT/

[2]http://www.roe.ac.uk/abwww/orac/

- Ensure that operational staff are not left with a large body of code to support
- Be able to make changes easily to data reduction sequences to reflect a change in the performance or our understanding of an instrument
- Have a system that can be speedily ported to other platforms
- Be well placed to take advantage of new astronomical software and techniques

The above requirements are not, of course, unique to UKIRT; and it seems that the astronomical software community in recent years has shifted from large instrument-specific programs to data reduction pipelines.

2. The ORAC Data Reduction Pipeline

Our proposed data reduction pipeline consists of five major parts:

- An *algorithm engine*: One or more generally available packages containing suitable algorithms for data reduction
- A *pipeline manager*: A program written in a major scripting language that initiates and oversees control of the pipeline
- A *messaging system*: An interface between the pipeline manager and the algorithm engine
- A *recipe bank*: A catalogue of data reduction recipes that are associated with the data (by the data file header) and the observation sequence used to obtain it (by the observation preparation system)
- A *data reduction dictionary*: A translation of in-house data reduction recipes to actual scripts in the language of the algorithm engine

Of these, it is our intention that only the pipeline manager, the recipe bank and the dictionary would be supported by the local staff, whereas the other two components would already be supported by another organization.

The aim is that in the future any one of these components could be upgraded or changed without affecting all other parts of the system; for example we could chose a different algorithm engine without needing to change the code of the pipeline manager.

2.1. The Algorithm Engine

At least at first, our algorithm engines will be some major packages maintained by the Starlink[3] organization (KAPPA, FIGARO, CCDPACK etc), who support astronomical computing in the UK. These come in the form of monoliths that

[3] http://star-www.rl.ac.uk/

can be loaded in memory once and then triggered to execute commands sent to them via the ADAM messaging system (see §2.3.) without the start-up overheads imposed by their (more usual) Unix shell invocation.

Starlink packages use the hierarchical extensible N-dimensional Data Format (NDF) as their native data format which is already in use at UKIRT. Its main attraction to this project is its HISTORY component, which contains a list of operations that were performed on the dataset and its output. Similar components can be used to record processing instructions, so that data carries its own data reduction recipe with it.

Moreover, the NDF data format has quality and error as well as data arrays which are correctly propagated by the majority of Starlink packages.

2.2. The Pipeline Manager

The pipeline manager's tasks are:

- To detect the arrival of new data
- To understand the nature of the data (object, calibration frame etc) and whether it fits in the expected data reduction sequence
- To parse the recipe(s) appropriate for the data
- To send the necessary messages to the algorithm engine to process the data

The pipeline manager is also expected to be robust, have good error recovery, propagate meaningful error messages to the observer and under no circumstances interfere with the data acquisition.

2.3. The Messaging System

Using a messaging system rather than, for example, generating scripts to be executed by the Unix shell, has the following advantages:

- Fast execution because the data reduction tasks are memory resident
- Keep track and parse the status returned by the individual data reduction tasks
- Control the flow of informational and error messages to the user

The choice of the messaging system is dependent on the algorithm engine (for example to use IRAF as the algorithm engine one would chose the IRAF message bus); however we hope to introduce a messaging layer that would enable alternative messaging systems (and their algorithm engines) to be used with relative ease.

The ADAM messaging system used by Starlink packages has been long used at UKIRT as part of the instrument control system and has proved very reliable. We have interfaces to it from both major scripting languages (perl and tcl).

2.4. The Recipe Bank

It is envisaged that each standard observation sequence will be associated by one or more data reduction recipes. These contain high level commands (eg SUBTRACT_BIAS, DIVIDE_BY_FLAT) that represent conceptual steps in the reduction of the data. The instrument scientist would be expected to specify what these steps should do in the particular context of an observation, and the software engineer would then make appropriate entries in a data reduction dictionary.

2.5. The Data Reduction Dictionary

As an important part of divorcing the specifics of the data reduction implementation (algorithm engines and messaging systems) from the pipeline manager as well as the end user, we are introducing a data reduction vocabulary that maps into specific commands. For example the data reduction command DIVIDE_BY_FLAT could map into one or more actual commands (eg the KAPPA command `div` or the IRAF command `imarith /` .

This method has the further advantage of reducing user documentation load - even if the implementation of DIVIDE_BY_FLAT were to change, the user documentation, that does not delve to this lower level, remains unaffected.

Furthermore, the actual scripts can be shipped with the data for the observer's future use.

3. Delivery

UKIRT is expecting the arrival of two new instruments – the UKIRT Fast Track Imager (UFTI) in early 1998 and the Mid-infrared Echelle (MICHELLE) at the end of the same year. The ORAC project is required to deliver a functional system for UFTI and the full system by the arrival of Michelle.

References

Bridger, A., Economou, F., & Wright, G. S., 1998, this volume

Daly, P. N., 1995, in ASP Conf. Ser., Vol. 77, Astronomical Data Analysis Software and Systems IV, ed. R. A. Shaw, H. E. Payne & J. J. E. Hayes (San Francisco: ASP), 375

Daly, P. N., 1997, in ASP Conf. Ser., Vol. 125, Astronomical Data Analysis Software and Systems VI, ed. Gareth Hunt & H. E. Payne (San Francisco: ASP), 136

Astronomical Data Analysis Software and Systems VII
ASP Conference Series, Vol. 145, 1998
R. Albrecht, R. N. Hook and H. A. Bushouse, eds.

The IRAF Client Display Library (CDL)

Michael Fitzpatrick

IRAF Group[1], *NOAO*[2], *PO Box 26732, Tucson, AZ 85726*

Abstract. The Client Display Library is a new interface that allows host (non-IRAF) C or FORTRAN programs to display images or draw into the graphics overlay using the XImtool image server and image viewers which emulate the XImtool client protocol such as SAOtng, SAOimage, and IPAC Skyview. The CDL provides most of the algorithms and functionality found in standard IRAF tasks such as DISPLAY and TVMARK, but provides these in library form for use in host level user programs, much as the IMFORT library provides access to IRAF images on disk. High-level procedures allow simple display of IRAF and FITS images, automatic image scaling, and a wide variety of graphics markers and scalable text. The IIS protocol used is easily replaced allowing the library to be ported for use with other display servers.

1. Introduction

For more than a decade IRAF has relied on the use of a *display server* as the primary means for image display. IRAF client tasks connect to the server and send or read data using a modified IIS[3] Model 70 protocol, originally via named fifo pipes but more recently using Unix domain or inet sockets. The advantage to this approach was that IRAF client tasks can make use of an image display without duplicating the code needed for actually displaying the image in each task. The longtime disadvantage however was that the IIS protocol used was arcane and undocumented, making the display server largely unavailable to applications or programmers outside of the IRAF project. Earlier attempts to solve this problem, such as the SAO-IIS library (Wright and Conrad, 1994), have shown there was a need in the community for such an interface but never managed to provide all the features and generality necessary to be widely used.

The CDL is meant to provide an easy-to-use, fully featured application interface which can be easily evolved for future display servers, communications schemes, or display functionality. It is independent of IRAF itself (as are the dis-

---

[1]Image Reduction and Analysis Facility, distributed by the National Optical Astronomy Observatories

[2]National Optical Astronomy Observatories, operated by the Association of Universities for Research in Astronomy, Inc. (AURA) under cooperative agreement with the National Science Foundation

[3]International Imaging Systems

play servers) so client tasks can be written for any discipline or application. CDL client programs can connect to the server using named pipes (fifos), Unix domain sockets or inet (Internet TCP/IP) sockets. The default connection scheme is the same as with IRAF (domain sockets then fifos) but client programs can specify any connection desired. With the C interface it is possible to open multiple connections to one or more servers, allowing clients to display both locally and to a remote server which can be anywhere on the Internet.

2. Features

2.1. Image Display

- Connections on (Unix or inet) sockets and fifo pipes
- Hi-level display of IRAF, FITS or arbitrary rasters of any datatype
- Image cursor read/write
- Frame buffer subraster I/O
- Automatic 8-bit scaling of images
- Automatic frame buffer selection
- Hardcopy from the client program

2.2. Overlay Graphics

- Marker Primitives
  - Point (10 basic types providing dozens of combinations)
  - Line (arbitrary width, 3 line styles)
  - Box (fill optional)
  - Circle (fill optional)
  - Circular Annuli
  - Polyline
  - Polygon (fill optional)
  - Ellipse (arbitrary size and rotation)
  - Elliptical Annuli (arbitrary size and rotation)
  - Text (arbitrary size and rotation, 4 fonts including Greek, in-line sub and superscripting capability)
- Erasure of markers
- 14 static colors plus 200 scalable graylevels

Figure 1. Example image demonstrating some available marker types and capabilities. Text in the lower corner was drawn using a filled box marker to erase the image pixels overlayed with text showing the use of in-line font changes and sub/superscript capability.

The CDL runs on all Unix platforms and is compatible with the XImtool, SAOimage, SAOtng and the IPAC Skyview display servers. The distribution includes all sources, full documentation and example tasks in both C and FORTRAN. An experimental SPP interface is available which lets IRAF developers (or occasional programmers) quickly prototype new applications without worrying about the details of image display. A "Virtual XImtool" dummy display server is also available which can be used as a "proxy server", taking input from one client and re-broadcasting it to a number of other servers, effectively creating a 'tee' for display programs.

3. Communications Interface

The CDL currently uses a modified IIS Model 70 protocol to communicate with IRAF display servers such as XImtool. This protocol is isolated from the rest of the library by a generic communications interface (providing I/O functions for the cursor, image rasters, WCS, frame configuration and erasure, etc) which can be rewritten to use a different protocol without affecting the rest of the library. Servers which provide much more functionality through the client interface can still be used with only small modifications needed to extend the CDL.

Figure 2. A demonstration of what can be done with image subraster I/O. The zoomed feature was created entirely by the client program which reads back the image pixels after display and writes out a zoomed subraster to another part of the frame buffer. Line and box markers are used to connect the endpoints of each region.

4. Future Work

Image sizes are growing to the point of straining the modern machines used to do the reductions and have long since past the point where they can be displayed at full resolution on a typical monitor. The CDL, like most software, could benefit from some memory optimization to better handle these large images. New display servers are currently in development meaning the communications protocol will likely need to be extended if not changed entirely; we expect that the CDL will be made compatible with the new IRAF message bus architecture and the next generation display servers.

The graphics markers available provide all the primitives needed to to do graphics of any complexity however it is up to the client program to still create the overlay. For simply marking a few positions this is a straightforward task, but more powerful applications could be developed more easily by providing high-level routines for axis labeling or improving the speed required to draw many markers by doing the overlays in memory prior to display. Users with suggestions or questions are encouraged to contact the author at *iraf@noao.edu.*

5. Acknowledgments

CDL development was supported in part by the Open IRAF initiative, which is funded by a NASA ADP grant.

6. References

Wright, J. R. & Conrad, A. R., 1994, in ASP Conf. Ser., Vol. 61, Astronomical Data Analysis Software and Systems III, ed. Dennis R. Crabtree, R. J. Hanisch & Jeannette Barnes (San Francisco: ASP), 495

Astronomical Data Analysis Software and Systems VII
ASP Conference Series, Vol. 145, 1998
R. Albrecht, R. N. Hook and H. A. Bushouse, eds.

Recent Developments in Experimental AIPS

Eric W. Greisen

National Radio Astronomy Observatory[1]

Abstract. An experimental version of AIPS is being developed by the author with the intent of trying new ideas in several areas. Copies of this version are made available periodically. For interferometric data, there have been significant developments in spectral-line calibration, data editing, wide-field imaging, self-calibration, and analysis of spectral cubes. For single-dish data, there have been improvements in the imaging of "on-the-fly" spectral-line data and new programs to analyze and image continuum, beam-switched data. Image display techniques which retain their state information to any other computer running this version have been developed. Results from several of these areas are presented.

1. Introduction

The Astronomical Image Processing System ("AIPS", Greisen 1990; van Moorsel et al. 1996) was developed at the NRAO beginning in the late seventies. Since then, we frequently encountered the conflict between the desire to develop new ideas and capabilities and the desire to provide to the users a stable, reliable software package. Some new capabilities are simply new programs added with little or no effect on current AIPS usage. However, since AIPS is in many areas a tightly connected system, some new ideas are potentially disruptive since they affect, for example, how all interferometer data are calibrated or how image models are computed. The conflict between stability and the desire to continue to develop new ideas led the author, beginning in April 1996, to develop a separate, openly experimental version of the AIPS package called AIPS CVX. This version contains a variety of new and improved capabilities developed by the author as well as the full functionality of the version that is sent out by NRAO. Periodically, a copy of CVX is frozen in `tar` files and made available for down-loading over the Internet. The current state of CVX is described on the author's home page[3] which includes instructions, comments, and links for installing the most recent copy of CVX. The remainder of this paper is used to describe some of the highlights of the new code.

---

[1]The National Radio Astronomy Observatory is a facility of the National Science Foundation operated under cooperative agreement by Associated Universities, Inc.

[3]http://www.cv.nrao.edu/~egreisen

2. Changes in infrastructure

The ability to allocate memory at run time for use by FORTRAN routines was added to AIPS just before the initiation of CVX. The memory is actually allocated by a C-language procedure with `malloc` which passes the start address of the memory back as a subscript into an array whose address was provided by the FORTRAN calling routine. A top-level "Z" routine acts as an interface between the programmer and the C procedure and manages the allocation/deallocation in an attempt to avoid memory leaks. Dynamic memory is now used where appropriate in CVX in order to scale memory to the size of the problem and to avoid, for example, burdening continuum users with large spectral-line arrays.

The AIPS "TV" image display has always been "stateful." A program could query the system to determine which portions of the display are visible and what the coordinates and other header parameters are for the image(s) displayed. This was implemented through disk files containing device descriptions and other files containing the display catalog. This technique tied the displays to a particular local-area-network, a particular computer byte ordering, and a hand-maintained set of link files for the cooperating workstations. In CVX, for the now ubiquitous X-windows form of TV display, I have eliminated both sets of disk files, placing the device and catalog information inside the software (XAS) which implements the display. I have simplified the machine-dependent routines to remove all knowledge of the meaning of the bytes being passed. Instead, the packing and unpacking of data to/from the TV (including conversion to/from network-standard integers and floats) is done by the routines which must know about the TV, namely XAS itself and the XAS version of the TV virtual device interface (the "Y" routines). Among numerous benefits, this simplifies the addition of new capabilities and allows passing of data words of any length and type. To implement the remote catalog, machine-independent routines convert the AIPS header structure to/from standard FITS binary forms. XAS was enhanced to do initialization and vector and character generation internally, reducing the amount of data which has to be sent from the calling programs.

XAS communicates with programs running on the same and different computers through an assigned Internet socket. It was found that, when the socket is busy for one program, a second program's request for the socket is apparently honored even when XAS is coded not to queue requests. The second program then hangs waiting for the socket to respond to its opening transmission. In the past, this undesirable delay was avoided by doing file locking on the image device file (a shaky proposition over networks) which will respond immediately when the file is already locked. To avoid this hanging in CVX in which there are no disk files for locking, a new TV lock server was written. The Internet socket to it may still hang, but the lock server does so little that the hanging time is negligible. In this way, any computer running CVX may display images on any workstation anywhere in the world running CVX's XAS server. Collisions are noted immediately so that calling programs need not wait for busy displays. Pixel values, coordinates, and other parameters of the displayed images are then available to the compute servers. Thus, a scientist in Sonthofen might reduce data in a large compute server in Socorro while interacting with displays run on the local workstation. The only limitations are security restrictions on Internet sockets (if any) and available data rates.

3. Processing of interferometric data

A major area of experimentation in CVX has been the calibration of interferometric data, especially multi-spectral-channel data. Because the software did not handle time dependence properly, users were required to handle the antenna-based, channel-dependent complex gains (the "bandpass") as if they were constant or only slowly varying in time. I changed the bandpass solutions to use weighting of antennas in solving for and averaging bandpasses and corrected the bandpass application code to apply time-smoothed calibrations properly. These two corrections have allowed overlapped spectra to be "seamlessly" stitched together for the first time (J. M. Uson and D. E. Hogg, private communications). It has been long believed that VLA spectral-line images should, but do not, have noise which is independent of channel. Using a variety of new techniques to record and report the failure of calibrator data to contain only antenna-based effects, I have found that these closure errors are a function of spectral channel that closely matches the noise spectra found in the final image cubes. Since some configurations of the VLA correlator have very much worse closure failure than others we hope to find and correct the hardware cause of the problem.

Better calibration is only useful if bad data are properly excluded. With instruments as large as the VLA and VLBA, it is very time consuming to examine all the data and bad data are not always readily apparent (*i.e.*, unusually small amplitudes tend to be invisible). Using object-oriented programming in FORTRAN (Cotton 1992), I have developed an editor class which allows the user to interact with the visibility data, or various tables associated with those data, to prepare tables of deletion commands. It has been found that editing based on the antenna system temperatures is not particularly time consuming and detects a very large fraction of the data which should be deleted. The editor class uses the AIPS TV display as a multi-color graphics device to display a function menu, the data to be edited, and data from other antennas/baselines for reference. The editor class is implemented in normal AIPS CVX tasks to edit *uv* data using the system temperature and the *uv* data themselves. Since it is a class, I inserted it as a menu choice in the iterative imaging and self-calibration task `SCMAP` by adding only about 100 lines of code some of which pass the new user parameters needed by the editing. An example screen from `SCMAP` may be found in the author's home WWW page previously cited.

AIPS has long supported the imaging and deconvolution of ≤ 16 fields surrounding the direction toward which the telescopes were pointed. Among other things, this allows for removal of sidelobes due to distant sources from the field of primary interest. Previously, the phases of the data were rotated to the center of each field before imaging, but no other geometric corrections were made. This makes each field parallel to the tangent plane at the antenna pointing direction. Such planes separate from the celestial sphere very much more rapidly with angle from their center than would a tangent plane. Therefore, I have changed CVX to offer the option of re-projecting the data sample coordinates for each field so that all fields are tangent planes capable of a much larger undistorted field of view. (In direct comparisons, the re-projection surprisingly adds virtually no cpu time to the job.) This option requires each field to have its own synthesized beam image and to be Cleaned in a somewhat separated manner. In CVX, up to 64 fields may be used; source model components may

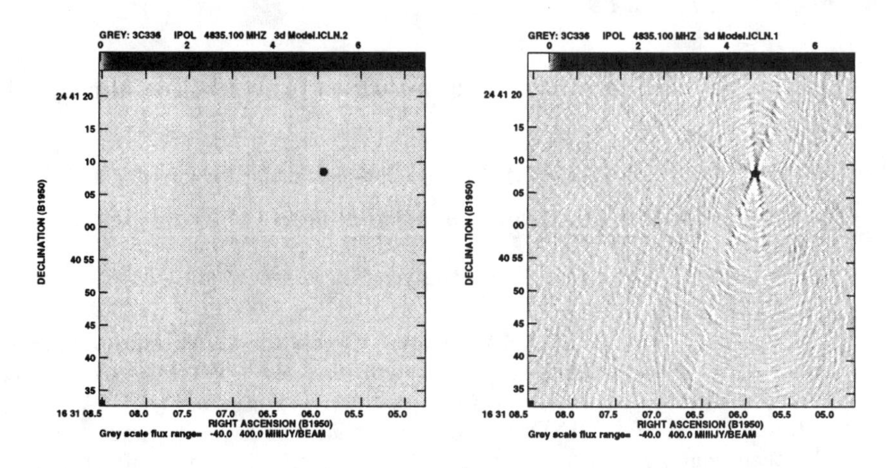

Figure 1. Model image with (left) and without 3D correction.

be restored to all fields in which they occur, rather than just the one in which they were found; and a Clean search method may be used to minimize the instability that arises if the same direction is searched in more than one field. The results of using the "3D" option are illustrated in Fig. 1 which shows the Cleaned image of a model source separated from the pointing direction by 2° (18575 synthesized beams) and separated from the field center by 14.9 arcsec (36 synthesized beams). The non-signal rms of the uncorrected image is 35 times that of the corrected one. The source model found with 3D-Cleaning may be used in all other AIPS tasks in CVX, in particular for self-calibration. Since sidelobes of distant sources are frequency dependent, the ability to make and subtract accurate models of such sources greatly increases the spectral dynamic range and the field of view free of systematic errors in multi-channel images.

References

Cotton, W. D., 1992, "Object-Oriented Programming in AIPS FORTRAN," Aips Memo No. 78,[4] NRAO.

Greisen, E. W., 1990 "The Astronomical Image Processing System," in *Acquisition, Processing and Archiving of Astronomical Images*, eds. G. Longo and G. Sedmak, (Osservatorio Astronomico di Capodimonte and Centro di Formazione e studi per il Mezzogiorno), Naples, Italy. Also appeared as AIPS Memo No. 61, 1988,[3] NRAO. See also Aips Memo No. 87

van Moorsel, G., Kemball, A., & Greisen, E. W. 1996, "AIPS developments in the Nineties," in ASP Conf. Ser., Vol. 101, Astronomical Data Analysis Software and Systems V, ed. George H. Jacoby & Jeannette Barnes (San Francisco: ASP), 37

---

[4] http://www.cv.nrao.edu/aips/aipsmemo.html

Astronomical Data Analysis Software and Systems VII
ASP Conference Series, Vol. 145, 1998
R. Albrecht, R. N. Hook and H. A. Bushouse, eds.

ASC Coordinate Transformation — The Pixlib Library, II

H. He, J. McDowell and M. Conroy

Harvard-Smithsonian Center for Astrophysics
60 Garden Street, MS 81
Cambridge, MA 02138, Email:hhe@cfa.harvard.edu

Abstract. Pixlib, an AXAF Science Center (ASC) coordinate library, has been developed as the continuing effort of (He 1997). Its expansion includes, handling of the High Resolution Mirror Assembly (HRMA) X-ray Detection System (HXDS) stage dither and the five-axis mount (FAM) attachment point movements, correction of misalignments of the mirror mount relative to X-ray calibration facility (XRCF) and to the default FAM axes, as well as solution of sky aspect offsets of flight, etc. In this paper, we will discuss the design and the configuration of the pixlib system, and show, as an example, how to integrate the library into ASC data analysis at XRCF.

1. Introduction

The work of He (1997) established a preliminary framework for the pixlib system, including the parameter-interface data I/O structure, matrix calculation algorithm, and coordinate transformation threading baselines. Since then, the library has undergone thorough re-organization and expansion to meet the AXAF on-going requirements of both ground calibration and flight observation. At the time of writing, the library is about 95% completed with approximate 6000 source lines of codes. It was successfully integrated and built during the XRCF calibration phase.

In this paper, we will highlight the system design and architecture of the library, complementary to the early work, and describe the system configuration in terms of user application. The complexities of coordinate transformation at XRCF and the resolutions will be discussed.

2. Architecture of the ASC Coordinate Library

The building blocks of the Pixlib library are three sub-systems, core, auxiliary, and application interface (API), and the foundation of the library is built with the parameter-interface structure. Figure 1 sketchs the architecture of the library.

As discussed in He (1997), the design of pixlib is modular to allow system expandibility, easy maintenance and simple ways to incorporate new scientific knowledge. The core sub-system, which includes 8 modules (see Figure 2 for

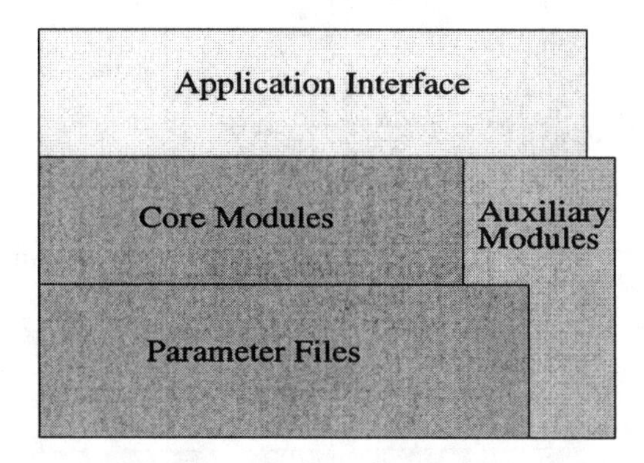

Figure 1. Pixlib library architecture, constructed on three sub-systems which are layered on the parameter-file structure.

details), builds the ASC coordinate frameworks of grating, FAM motion, sky aspect offsets, telemetry (raw) reads, detector-chip tiling, and coordinate transformation among chip pixels and celestial angles. Because of the common needs of generic data sources, handy utilities, module-to-module communication, etc., the library is supported with a 4-module auxiliary sub-system, as shown below.

```
pix_errstatus.c     -- error handling
pix_utils.c         -- utility functions
pix_common.c        -- common data sources to all modules
pixlib_hiden.c      -- internal configuration, bookkeeping
```

The upper-level interface of the library is implemented in the module `pixlib.c`, which distributes functions between the lower-level modules. `pixlib.c`, in large part, provides setup functions for system configuration, and other API functions are implemented locally without the need for cross-module function calls. All the API functions are identified by the "`pix_`" prefix.

The data in-stream of the parameter-interface approach simplifies system configuration and data readability. The number and organization of those data files have remained almost same as described in He (1997) with few updates. `pix_pixel_plane.par`, substituting the original `pix_size_cnter.par`, groups 2-D pixel system parameters of focal plane, tiled detector, grating dispersion together; `pix_grating.par` is added to define dispersion period and angle of grating arms.

3. System Configuration

Prior to application program execution, the library needs to be configured properly. The system configuration is optionally either static or dynamic, as illustrated in Figure 2. A set-parameter-value to a parameter file, `pix_coords.par`, handles the static configuration and the user can set values for the following parameters.

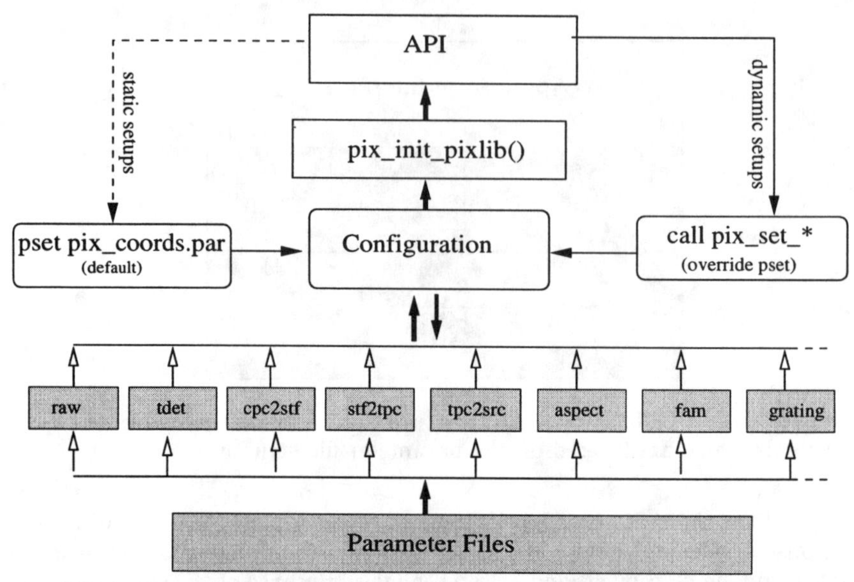

Figure 2. Pixlib data flow and system configuration. Each smaller box above represents a module. For example, the "raw" and "cpc2stf" boxes denote the modules of pix_raw.c and pix_cp2stf.c, respectively.

```
 flength = Telescope focal length in mm
aimpnt   = Name of aim point  of detector
   fpsys = Focal Plane pixel system
 tdetsys = Tile Detector pixel system
  gdpsys = Grating Dispersion pixel system
 grating = Grating arm
   align = FAM    misalignment angle in degrees(pitch, yaw, roll)
  mirror = mirror misalignment angle in degrees(pitch, yaw, roll)
```

In the course of the system initiation, executed through `pix_init_pixlib()`, internal functions lookup the parameter table to parse the information down to relevant modules, which are then configured accordingly.

An alternative way to configure the system is to make function, "`pix_set_*`", calls in application program following the initiation. `pix_set_flength(int)`, for instance, is equivalent to the pset "flength" for `pix_coords.par`, and `pix_set_fpsys(int)` to the pset "fpsys", to name a few. The consequence of those calls is to override the static configuration which is the system defaults.

4. Coordinate Transformation at XRCF

Coordinate transformations at XRCF need to be carefully handled when the FAM feet move and the HXDS stage dithers. In the default, boresight configuration the FAM axes are parallel to the XRCF (and Local Science Instrument, LSI) axes, but they may undergo some movements in addition to the HXDS stage dithering and possible mirror mount movement. Therefore, those effects, as listed below, must be accounted for before coordinate transformations between focal plane and LSI system are made:

- misalignments of the default FAM axes from the mirror due to FAM attachment point motion,
- motion of the HXDS stage from the default FAM reference point,
- possible misalignments of the mirror mount relative to the XRCF prescribed by pitch-yaw angles,
- misalignments of the FAM axes from the default due to FAM feet motion.

The following two functions, in addition to other generic configurations, effectively supply the system configuration for coordinate transformation at XRCF. The routine

```
pix_set_mirror (double hpy[2],        /* in degrees */
                double stage[3],      /* in mm      */
                double stage_ang[3]) /* in degrees */
```

corrects misalignment from the mirror axis by measuring its displacement from the boresight configuration of the default FAM frame (`stage_ang`) for a given mirror position (`hpy`) in mirror nodal coordinate system. The `hpy` is measured in HRMA pitch and yaw axes, and the HXDS stage position (`stage`) monitored relative to the default FAM reference point. The routine

```
pix_set_align(
    double mir_align[3], /* (yaw, pitch, roll), in degrees */
    double stg_align[3]) /* (yaw, pitch, roll), in degrees */
```

serves to assess

- misalignments of mirror mount (`mir_align`) relative to XRCF axes are measured in the given yaw-pitch-roll Euler angles in the mirror nodal coordinate, and
- misalignments of the default FAM (`stg_align`) relative to XRCF axes are corrected in terms of yaw-pitch-roll Euler angles in the default FAM frame.

The system configuration above was successfully applied to and integrated into ASC data analysis during the X-ray calibration.

Acknowledgments. We gratefully acknowledge many fruitful discussions with ASC members. This project is supported from the AXAF Science Center (NAS8-39073).

References

McDowell, J., ASC Coordinates, Revision 4.1, 1997, SAO/ASCDS.

He, H., McDowell, J., & Conroy, M., 1997, in ASP Conf. Ser., Vol. 125, Astronomical Data Analysis Software and Systems VI, ed. Gareth Hunt & H. E. Payne (San Francisco: ASP), 473

Astronomical Data Analysis Software and Systems VII
ASP Conference Series, Vol. 145, 1998
R. Albrecht, R. N. Hook and H. A. Bushouse, eds.

A Software Package for Automatic Reduction of ISOPHOT Calibration Data

S. Huth[1,2] and B. Schulz[2]

(1) Max-Planck-Institut für Astronomie, Königstuhl 17, 69117 Heidelberg, Germany

(2) ISO Science Operations Center, Astrophysics Division of ESA, Villafranca, Spain

Abstract. ISOPHOT is one of the four focal plane instruments of the European Space Agency's (ESA) Infrared Space Observatory (ISO) (Kessler et al. 1996), which was launched on 17 November 1995. The impurity IR-detectors do not behave ideally but show varying responsivity with time. Also the measured signals are affected by detector transients, altering the final results considerably. We present the design and the main features of a software package, developed in order to support the absolute calibration of the instrument. It enables a homogeneous processing and reprocessing of a large amount of calibration observations and provides automatic transient correction and quality flagging. Updates to the data processing method can be immediately applied to the full dataset.

1. Introduction

The Si or Ge based infrared (IR) detectors of ISOPHOT (Lemke, Klaas et al. 1996) show varying relations between incident flux and measured signal (responsivity) over the orbit of the satellite, triggered by influences like high energy radiation impacts or changing IR-flux. Internal fine calibration sources (FCS) provide a reproducible reference flux, that is absolutely calibrated during the mission by comparison with known celestial standards (stars, planets, asteroids). The FCS signal is then used to determine the detector-responsivity at the time of an observation. The software package presented here was developed in order to support the absolute calibration of the FCSs, which is one of the tasks of the Instrument Dedicated Team (IDT) in the ISO-ground-station in Villafranca near Madrid (Spain).

2. Concept

The program was designed to allow for evaluation of a large number of ISOPHOT calibration observations automatically and in a consistent way. Thus new developments in data processing and calibration can be accommodated easily. Immediate reprocessing of the calibration database is performed without user interaction. The intentional omission of a GUI helps to keep a maximum of flexibility, which is essential for this kind of work, where changes are frequent, the users are experts and research and development go hand in hand.

The software is based on the Interactive Data Language (IDL) (Research Systems Inc., 1992) and the PHT-Interactive Analysis (PIA) (Gabriel et al. 1997), using a simple command line driven interface. The data is processed in units of a single measurement, e.g., data that were taken at a fixed instrument configuration. In this context pointings within a raster are also considered to be single measurements. Detector pixels are taken as independent units and are processed individually. While the standard-way of PIA-processing is followed in general, some improvements were introduced, such as non-interactive empirical drift fitting and quality flagging. Under certain conditions, e.g., when long-term transients are likely to influence a sequence of measurements, the full sequence is corrected together (baseline fit).

The derived signals are then combined, guided by formatted ASCII-files, that indicate the sequence of the different targets, such as source, background or FCS. The resulting FCS-fluxes are stored in formatted ASCII-tables, allowing for easy examination in editors and step-by-step checking of the intermediate values and are used as database for further research. In addition a PostScript record of signal-diagrams is saved during processing and used in the subsequent analysis and quality checks.

3. Automatic-drift-fit

Figure 1 shows a typical detector response after a flux-change, with a constant flux over the following 128 seconds. The measured signal is asymptotically drifting towards an upper limit. This non-ideal detector behaviour is observed in intrinsic photoconductors operated at low illumination levels.

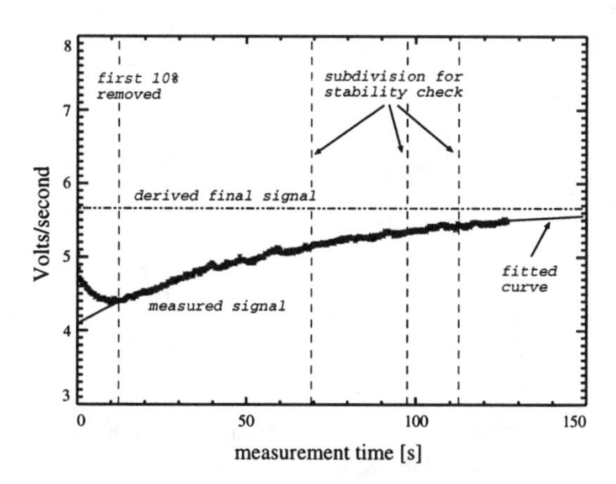

Figure 1. A typical detector response after a flux change.

In order to derive the final signal the first 10% of the data is discarded, thereby avoiding the so called hook-response, which is not described properly by transient models so far (Fouks & Schubert 1995). A statistical trend-check is performed on the remaining portion of the data. In case the dataset is not stable within a confidence limit of 95%, the data is split into two parts and the test is repeated on the latter one, until either the remaining part of the measurement is considered to be stable or one of the following three conditions is met in the next iteration: a) the dataset comprises less than 8 datapoints b) the dataset comprises less than 8 seconds c) the number of iterations exceeds 5. If any of these three conditions is met, the stability test is considered to have failed and an empirical drift fit is tried on the full dataset with the first 10% removed. The measurement is divided into three parts and weighted with the values 0.2, 0.8 and 1 respectively, to improve the fit convergence. Again a quality check using criteria on the curvature is performed on the result. The limits were found empirically and tuned with real data on a trial and error basis. A further check is performed on the final signal deduced from the successful fit. The result is rejected if the average of the last 30% of the measurement is more than a factor of 2 different from the final signal. A final signal is always rejected if negative. In case the fit is not successful, the average of the last 30% of the data is taken as the final signal. The outcome of the various tests, is coded in a two digit quality flag attached to the result.

4. Baseline Fit

In the case of small signals on a large background, whole measurement sequences can be strongly disturbed by long lasting transients. Figure 2 left shows a sequence of measurements of a faint source on a high background. The sequence is bounded by two measurements of the FCS, to fix the responsivity for the

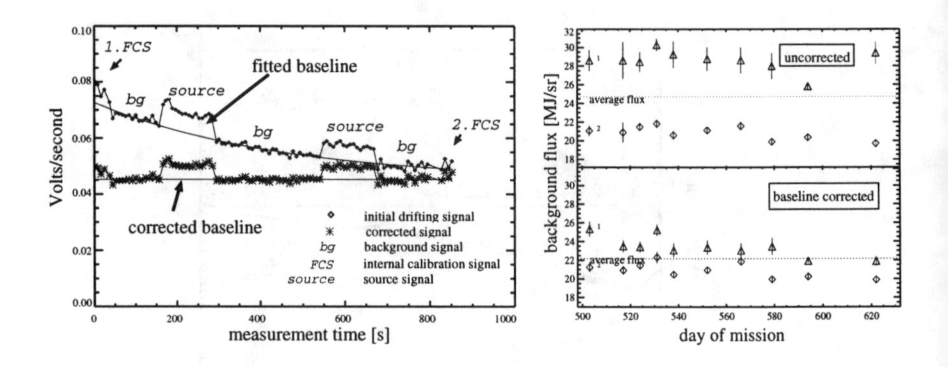

Figure 2. (left) Measurement sequence which is affected by a longterm transient and the applied correction method (baseline correction). (right) Stability monitoring: the upper and lower plots show fluxes derived without and with baseline correction respectively.

time of the observation. The downward drift superposed over all signals causes strong deviations in the two FCS signals and therefore in the two calculated responsivities. In order to eliminate the longterm drift, an empirical drift curve is fitted to the signals of all four background measurements. All signals are divided by this baseline. This correction leads to a much better agreement of individual signals measured at equal flux levels and the consistency of the finally derived fluxes is restored. Figure 2 right shows an example of the improvements achieved by this method. The diagrams show evaluated fluxes from a long-term monitoring program, repeating measurements of the type described above (see Figure 2 left) on the same source every two weeks. The two different symbols represent individual calibrations of the final background signals, using the first or the second FCS-signal. The triangle corresponds to the first FCS measurement. The calibrated fluxes are clearly much more consistent in the lower diagram, where the baseline correction was applied.

5. Discussion and Future developments

The development of this software has considerably improved the data quality in the area of ISOPHOT flux calibration, which forms a cornerstone of the overall calibration of the instrument. As long as theoretical transient-models are only available under certain conditions, empirical models provide a useful means to achieve better results. The ASCII-output tables including the quality flags have proved to be a good basis for further analysis of the data, reducing the necessity for checking other datasets from earlier processing levels, in case questions arise. Further improvements, except the ones coming from improved transient-models, are corrections taking into account the actual pointing of the satellite and the beam-profiles.

Acknowledgments. We thank our colleagues in the PIDT in VILSPA for their valuable suggestions and support.

References

Kessler, M.F., Steinz, J.A., Anderegg, M.E., et al. 1996, A&A, 315, L27

Lemke, D., Klaas, U., et al. 1996, A&A, 315, L64

Gabriel, C., et al. 1997, in ASP Conf. Ser., Vol. 125, Astronomical Data Analysis Software and Systems VI, ed. Gareth Hunt & H. E. Payne (San Francisco: ASP), 108

Fouks, B., & Schubert, J. 1995, Proc. SPIE 2475, 487

IDL 1992, IDL is a registered Trademark of Research Systems, Inc. ©1992

Astronomical Data Analysis Software and Systems VII
ASP Conference Series, Vol. 145, 1998
R. Albrecht, R. N. Hook and H. A. Bushouse, eds.

Reducing SCUBA Data at the James Clerk Maxwell Telescope

T. Jenness and J. F. Lightfoot[1]

Joint Astronomy Centre, 660 N. A'ohōkū Place, University Park, Hilo, HI 96720, USA

Abstract. The Submillimetre Common-User Bolometer Array (SCUBA) is now operational at the James Clerk Maxwell Telescope (JCMT). This paper describes the SCUBA User Reduction Facility (SURF) data reduction software that has been developed for use with SCUBA.

1. Introduction

The Submillimetre Common-User Bolometer Array (SCUBA) is the recently commissioned instrument on the James Clerk Maxwell Telescope (JCMT)[2] on Mauna Kea. SCUBA is a bolometer array which means that for the first time JCMT has imaging capabilities in the submillimetre.

In order to maximize the number of bolometers in the array, they were packed on a hexagonal (rather than rectangular) grid. This means that the software has to reconstruct the data into an image. Moreover, because the bolometers are larger than half the beam width, the image cannot be fully sampled in a single exposure. In order to fully sample the image, the secondary mirror is "jiggled" to several adjacent positions and the software has to keep track of what sky position is associated with each data measurement. Finally, because of how SCUBA is mounted (on the Nasmyth platform), the sky rotates as the telescope tracks; this is also something the software must correct for.

2. SCUBA Observing modes

There are three major SCUBA observing modes, and these need some individual treatment from SURF, the Scuba User Reduction Facility, which is run off-line by the user:

Photometry: This mode is used to measure the flux of a point source. In its simplest guise the observation involves pointing a single bolometer at the source, measuring the signal, chopping and nodding to reduce the effect of sky emission, and integrating to build up the signal-to-noise. SCUBA also

---

[1]Royal Observatory Edinburgh, Blackford Hill, Edinburgh, EH9 3HJ, United Kingdom

[2]http://www.jach.hawaii.edu/

allows for 2 or 3 bolometer photometry (chopping on the array), simultaneous photometry using the long and short wave arrays, and jiggling on source to reduce the effects of seeing.

Jiggle-mapping: This is the main imaging mode for sources which are smaller than the array (i.e. less than about 2 arcmin). As the SCUBA arrays are not fully-sampled and not on a rectangular grid, images can not be taken in the same way as for an optical CCD array. At least 16 individual secondary mirror, or 'jiggle', positions (each of 1 second) are required to make a fully sampled image (64 jiggles are required if both the long and short wave arrays are being used simultaneously). The SURF data reduction package must take these data, combine them and regrid them onto a rectangular grid.

Scan-mapping: Scan mapping, as the name suggests, is performed by the telescope scanning across an area of sky while simultaneously chopping. During the scan, care is taken that enough bolometers measure the same patch of sky so that the image is fully sampled, so it does not require jiggling or additional scans. This mode is suitable for sources that are extended. SURF combines the two beams that result from the chopping into a single-beam map.

3. SURF – The SCUBA User Reduction Facility

The real-time SCUBA observing system first demodulates data in the transputers – it takes data at 128Hz but only supplies data every second. It also provides a 'quick-look' display of the data based on some rudimentary data reduction done in the transputers.

The demodulated data is then "raw data" as far as SURF, the off-line system, is concerned. SURF aims to provide publication quality data reduction and currently requires user input and feedback.

The 'quick-look' display provided by the transputers can also be eavesdropped by a remote observer using a WORF system (Economou et al 1996; Jenness et al 1997).

SURF (like the real-time system) is written in the Starlink[3] software environment and therefore runs on Sun Solaris, Digital Unix and Linux. Computationally intensive routines are written in FORTRAN as ADAM tasks that form part of a monolith. Applications such as data reduction pipelines (calling multiple tasks from the monolith), string handling routines (e.g., generating observing logs) and small data reduction tasks were more suited to modern scripting languages than to FORTRAN and it was decided to use the Perl programming language (Wall et al. 1996) for these.

Since the software is written in the Starlink environment, Starlink's N-Dimensional Data Format (NDF) is used for all file I/O. This format provides for error arrays, bad pixel masks, storage of history information and a hierarchical extension mechanism and SURF makes full use of these features. In order to

---

[3]http://star-www.rl.ac.uk/

get the most out of using Perl it was necessary to write a Perl interface to the FORTRAN NDF and HDS libraries. This module (distributed separately from SURF) provides complete access to all of the routines in these libraries allowing for full control of the NDF and HDS structures from Perl.

To improve maintainability of the package all source code is kept under revision control (RCS) and subroutine libraries are shared with the on-line observing system.

SURF provides the tasks necessary to convert raw demodulated SCUBA data into a regularly-sampled image:

- flatfielding the array

  The different response of each bolometer is removed. Note that the flatfield remains constant with time and does not need to be re-measured every night.

- correcting for atmospheric extinction

  This task simply calculates the extinction due to the atmosphere by using the zenith sky opacity (tau) and calculating the elevation of each bolometer during each 1 second sample.

- removing spikes (manually and in software)

  Occasionally spikes are present in the data (instrumental glitches, cosmic rays etc) and these can be removed manually or by using software.

- removing sky fluctuations from JIGGLE/MAP and PHOTOM data.

  The submillimetre sky is very sensitive to sky noise caused by fluctuations in the emissivity of the atmosphere passing over the telescope. These variations occur in atmospheric cells that are larger than the array and the resultant noise is seen to be correlated across the array. At present this sky noise is removed by analysing bolometers that are known to be looking at sky and removing this offset from the data.

- generating a rectangularly sampled image.

  As described earlier, SCUBA data is under-sampled (for jiggle data), taken on an hexagonal grid and subject to sky rotation. SURF calculates the position of each bolometer during each 1 second sample (in a number of coordinate systems: RA/Dec, Galactic, Nasmyth, Azimuth-Elevation or moving centre (planet)) and regrids these data onto a rectangularly sampled image.

Once the data have been regridded the image can be converted to FITS (a full astrometry header is provided), if necessary, and analysed in the normal way.

The latest version of SURF is freely available from the SURF home page[4]. Additionally the software is distributed by Starlink and available on their CD-ROM.

---

[4]http://www.jach.hawaii.edu/jcmt_sw/scuba/

4. Future

This software is being developed actively at the Joint Astronomy Centre and we hope to implement the following features in the next few months:

- Improve the automated data reduction by using a perl interface to the ADAM messaging library (already in development) and the UKIRT ORAC system (Economou et al. 1998). This will be implemented at the telescope, triggered by the SCUBA observing system, and off-line, triggered by the observer.
- At present sky noise removal has not been implemented for SCAN/MAP (since, unlike jiggle-map data, there is no concept of a 'sky' bolometer). It is hoped to implement a system for handling sky noise in due course.
- Implementing an Maximum Entropy reconstruction algorithm for SCAN/-MAP data (e.g., similar to the DBMEM implementation for the single pixel UKT14 (Richer 1992)) in addition to the standard Emerson, Klein and Haslam algorithm. A maximum entropy implementation would treat the reconstruction of the map as a direct linear inversion problem and would result in an image that is deconvolved from the beam. This technique has the advantages of enforcing positivity in the image, a proper treatment of correlations in the two dimensional data and allowing for the possibility of resolution enhancement for high signal-to-noise data. It is also hoped that this implementation will be able to handle sky-noise removal by treating sky-noise as additional free parameters for the reconstruction.

Acknowledgments. TJ would like to thank Frossie Economou for her helpful discussions during the development of this software.

References

Economou, F., Bridger, A., Daly, P. N., & Wright, G. S., 1996, in ASP Conf. Ser., Vol. 101, Astronomical Data Analysis Software and Systems V, ed. George H. Jacoby & Jeannette Barnes (San Francisco: ASP), 384

Economou, F., Bridger, A., Wright, G. S., Rees, N. P., & Jenness, T., 1998, this volume

Emerson, D. T., Klein, U., & Haslam, C. G. T., 1979, ApJ, 76, 92

Jenness, T., Economou, F., & Tilanus, R. P. J., 1997, in ASP Conf. Ser., Vol. 125, Astronomical Data Analysis Software and Systems VI, ed. Gareth Hunt & H. E. Payne (San Francisco: ASP), 401

Richer, J. S., 1992, MNRAS, 254, 165

Wall, L., Christiansen, T., & Schwartz, R. L., 1996, Programming Perl, 2nd edn. (Sebastopol, CA: O'Reilly[5]) (http://www.perl.org/)

---

[5]http://www.oreilly.com/

Astronomical Data Analysis Software and Systems VII
ASP Conference Series, Vol. 145, 1998
R. Albrecht, R. N. Hook and H. A. Bushouse, eds.

ISDC Data Access Layer

D. Jennings, J. Borkowski, T. Contessi, T. Lock, R. Rohlfs and R. Walter

INTEGRAL Science Data Centre, Chemin d'Ecogia 16, Versoix CH-1290 Switzerland

Abstract. The ISDC Data Access Layer (DAL) is an ANSI C and FORTRAN90 compatible library under development in support of the ESA INTEGRAL mission data analysis software. DALs primary purpose is to isolate the analysis software from the specifics of the data formats while at the same time providing new data abstraction and access capabilities. DAL supports the creation and manipulation of hierarchical data sets which may span multiple files and, in theory, multiple computer systems. A number of Application Programming Interfaces (APIs) are supported by DAL that allow software to view and access data at different levels of complexity. DAL also allows data sets to reside on disk, in conventional memory or in shared memory in a way that is transparent to the user/application.

1. Introduction

INTEGRAL[1] is an ESA Medium Class gamma-ray observatory mission scheduled for launch in 2001. It consists of four co-aligned instruments: two wide field gamma-ray detectors (IBIS, SPI) an x-ray monitor (JEM-X) and an optical monitor (OMC). The IBIS imager, SPI spectrometer and JEM-X x-ray monitor all employ coded mask[2] detection technology which, amongst other complexities, requires the spacecraft to constantly "dither" its pointing position in order to accumulate the required number of coded sky images for analysis.

The dithering strategy creates some unique issues for data organization and analysis. Each celestial *observation* will consist of a collection of *pointings* (5 minute to 2 hour fixed position integrations); conversely, it is possible that a given pointing may belong to several observations at once. A pointing is itself a collection of science, instrument housekeeping and auxiliary data sets, all of which may be grouped into various combinations for reasons of efficiency and conceptual elegance. Thus, the INTEGRAL data analysis system must be capable of supporting distributed data sets composed of many individual files and exhibiting *one to many* and *many to many* associations between the individual data structure elements.

---

[1] http://astro.estec.esa.nl/SA-general/Projects/Integral/integral.html

[2] http://lheawww.gsfc.nasa.gov/docs/cai/coded.html

2. ISDC Data Model

The complicated relationships between observations, pointings, and pointing data set components implies a natural *hierarchy* to the INTEGRAL data. This has led to the current ISDC Data Model which generalizes the concepts of pointings and observations into *Data Objects* and *Data Elements.*

A Data Object is an association, or collection, of one or more Data Elements. A Data Element may itself be Data Object (i.e., a sub-collection of data elements with lower position in the hierarchy) or a terminal *Base Element* containing an atomic data structure. There are three classes of terminal Base Elements that define the atomic data structures: collections of data in tabular (row, column) format known as *TABLE* elements, N dimensional data sets of homogeneous data type known as *ARRAY* elements, and human readable and/or bulk data (e.g., programs, text, GIF images, PostScript output) known as *INFO* elements. In addition to the atomic data structures, there is a fourth class of Base Element used to define compound (i.e., non-atomic) data structures known as *GROUP* elements.

The recursive nature Data Elements allows for the construction of *unbounded hierarchical associations* of atomic data structures. At the opposite extreme, a single atomic data structure may also be considered a Data Object in its own right. Note that it is also possible for a Data Element to belong to many different Data Objects, thus allowing Data Objects to share a given collection of data.

3. The DAL (Data Access Layer)

To implement the ISDC Data Model within the INTEGRAL analysis system, ISDC is currently constructing the DAL, or *Data Access Layer.* The DAL allows applications to create and manipulate data sets at the Data Object and Data Element level of abstraction.

The DAL consists of four logical layers. The physical format I/O modules CFITSIO[3] (for FITS) and SHMEMIO (for shared memory resident data) make up the first layer. These modules handle all the details of the particular storage formats available though DAL. Since the format specific details are isolated in this manner it is possible to add other data format capability with no change to the higher level software layers.

Above the physical format modules are the driver interface modules: FITSdriver, MEMdriver and SHMEMdriver. All these drivers contain an uniform set of interface functions that implement the respective data storage methods (FITS resident data, memory resident data and shared memory resident data). By using the driver level modules it is possible to provide the higher level DAL layers with consistent interface calls regardless of the storage medium in use.

The next DAL layer, and the first used in application programming, is the base element API level. Each of the four base elements supported by DAL (ARRAY, TABLE, INFO and GROUP) has its own Application Programming

---

[3] http://heasarc.gsfc.nasa.gov/docs/software/fitsio/fitsio.html

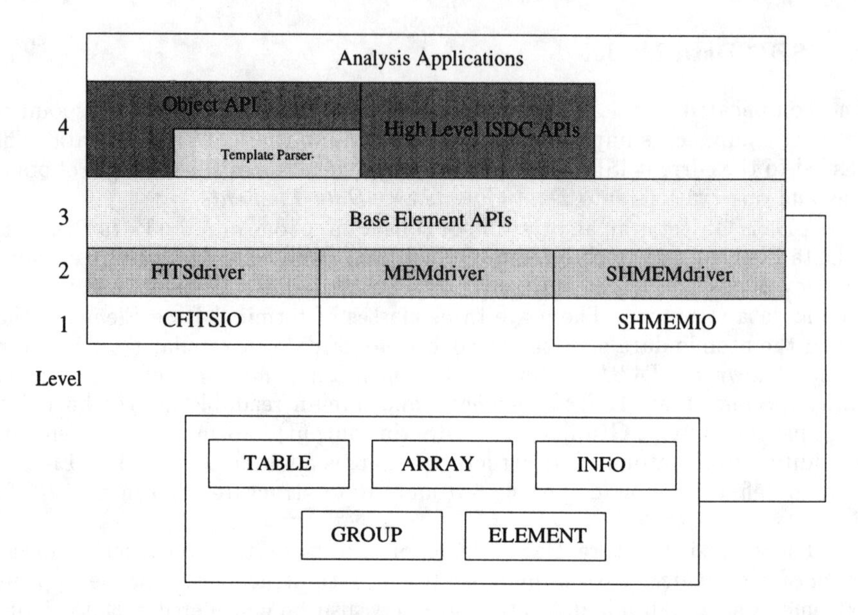

Figure 1. The Data Access Layer four level API structure, with the analysis applications residing above.

Interface that implements the element type. There is also a fifth API at this level, the ELEMENT API, that allows applications to operate upon individual data elements regardless of the base type.

The top level DAL layer is a collection of APIs that allow analysis applications to make efficient use of the base element APIs. The Object API implements hierarchical associations of data elements (i.e., the data model Data Objects), and the high level ISDC APIs implement scientific data-specific collections of data elements (e.g., spacecraft attitude and orbit, spectra, event lists).

4. DAL Data Objects: Data Format vs. Data Model

In order for DAL Data Objects, specifically the associations between the data elements, to be *persistent* with time the physical data formats must support the ISDC Data Model. For Data Objects stored on disk in FITS format the FITS Hierarchical Grouping Convention[4] is utilized to achieve Data Format - Data Model conformity.

Each DAL Data Element is stored in a FITS file as a HDU (Header Data Unit). Associations between Data Elements are stored in *grouping tables* as defined by the Grouping Convention. DAL manages all the details involved in creating and updating the grouping tables when a Data Object is written to disk in FITS format, and it attempts to locate and read grouping tables when opening an existing FITS-stored Data Object.

[4]http://adfwww.gsfc.nasa.gov/other/convert/group.html

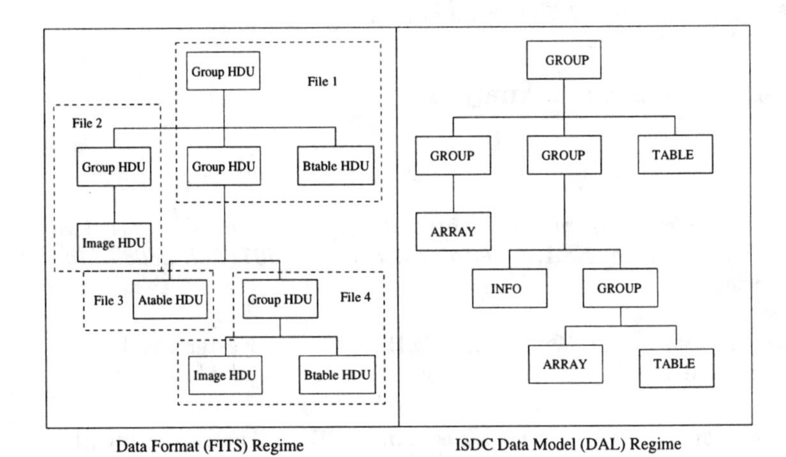

Figure 2. Data format – data model conformity. The lines connecting the FITS HDUs (left) and DAL data elements (right) define the hierarchical relationships between the data structures.

5. Conclusion

The DAL concept as presented here has several important implications that are of general utility to astronomical software.

First of all, the data format specifics are hidden from the software application, thus allowing the same API functions to be used for a variety of data formats and access methods. Data residing in memory, for instance, may have a different storage paradigm than data residing in disk files. Also, many different file-based formats (FITS, HDF, CDF) may be used transparently by the analysis applications.

Secondly, it is possible for Data Objects to span multiple data files in a way that is *inherent to the data structure itself.* The relationship between data structures persists even when the supporting archival infrastructure changes (e.g., port to a new OS) or is eliminated entirely (i.e., as can happen at the end of a space mission), thus providing for a *self-contained data organization.* Data Objects may also in theory span multiple computer file systems in a way that is transparent to the analysis applications. The implications for distributed network data archives are significant.

Lastly, different institutes may construct their own DAL-like data interface packages to achieve their own specific goals, but still cooperate to promote commonality across the data formats/structures. The commonality may be added or modified without changing higher level software.

The latest releases of DAL software and documentation may be obtained on line from the INTEGRAL Science Data Centre[5] Web site. Questions or comments regarding DAL should be directed to Don.Jennings@obs.unige.ch.

[5]http://obswww.unige.ch/isdc/

Astronomical Data Analysis Software and Systems VII
ASP Conference Series, Vol. 145, 1998
R. Albrecht, R. N. Hook and H. A. Bushouse, eds.

ISO–SWS Data Analysis

F. Lahuis[2], E. Wieprecht[3],

ISO Science Operations Centre, Astrophysics Division, Space Science Department of ESA, Villafranca, P.O.Box 50727, E-28080 Madrid, Spain

O.H. Bauer[3], D. Boxhoorn[2], R. Huygen[4], D. Kester[2], K.J. Leech[1], P.R. Roelfsema[2], E. Sturm[3], N.J. Sym[2] and B. Vandenbussche[1,4]

Abstract. We present aspects of the Data Analysis of the Short Wavelength Spectrometer (SWS) on-board ESA's Infrared Space Observatory (ISO). The general processing from the raw telemetry data of the instrument, up to the final calibrated product is described. Intermediate steps, instrument related aspects and data reduction techniques are highlighted.

1. The SWS Instrument

The Short-Wavelength Spectrometer (de Graauw et al. 1996) is one of the four instruments on board ESA's Infrared Space Observatory (Kessler et al. 1996), launched on November 17, 1995 and operational till approximately April 1998. SWS covers the wavelength range from 2.38 to 45.2 μm with two nearly independent grating spectrometers with a spectral resolution ranging from 1000 to 2000. In the wavelength range from 11.4 to 44.5 μm Fabry-Perot filters can be inserted which enhance the resolution by a factor 10 to 20.

2. SWS Data Processing

The processing of the SWS data can roughly be split into three distinct parts. The first part is the processing from the raw telemetry data to Standard Processed Data (SPD). The second part starts with the SPD and results in a spectrum of wavelength versus flux, the Auto Analysis Result (AAR). The last stage

[1]ISO Science Operations Centre, Astrophysics Division, Space Science Department of ESA, Villafranca, P.O.Box 50727,E-28080 Madrid, Spain

[2]Space Research Organization Netherlands, Postbus 800, NL-9700 AV, Groningen, The Netherlands

[3]Max Planck Institut für extraterrestrische Physik, Giessenbachstrasse, 85748 Garching, Germany

[4]Katholieke Universiteit Leuven, Instituut voor Sterrenkunde, Celestijnenlaan 200 B, B-3001 Heverlee, Belgium

involves post-processing and scientific analysis using the AAR. The first two steps form part of the ISO Standard Product Generation 'pipeline' and the products mentioned (e.g., ERD, SPD and AAR) form part of the data sent to the ISO observer. This processing is also part of the SWS Interactive Analysis System IA^3 which is described in detail by Wieprecht et al. (this volume).

2.1. ERD to SPD

The raw data are stored in FITS files containing the Edited Raw Data (ERD) and status and house-keeping parameters. Additional files containing information on e.g., the pointing of the satellite and the status of the instrument and satellite exist and can be used as well if required.

The ERD data consist of digital readouts at a rate of 24 per second for each of the 52 SWS detectors (four blocks of 12 detectors for the grating and two blocks of 2 detectors for the FP). The data is checked for saturation, corrected for reset pulse after-effects, linearized (AC correction), cross-talk corrected and converted to voltages. Next a slope is fitted for each detector and reset interval and wavelengths are assigned for each of these points. The data are extracted per reset interval and stored in a SPD.

2.2. SPD to AAR

For each reset interval the SPD contains a value for the slope in μV/sec and a wavelength in μm plus ancillary information about timing, instrument settings, data quality etc.

As a first step memory effects should be corrected. For the time being this is still in an experimental phase and not included in the standard processing. Next the dark currents are determined and subtracted. The relative spectral response of the instrument (RSRF) is corrected and the absolute flux calibration is applied (i.e., conversion from μV/sec to Jy). The wavelengths are corrected for the ISO velocity towards target and finally the data is sorted by wavelength, extracted and stored in an AAR.

2.3. AAR Processing

The processing using the AAR involves steps like aligning the signals of different detectors which cover the same spectral region, sigma clipping, rebinning, line and continuum fitting etc. All this is possible within IA^3 and the ISO Spectral Analysis Package (ISAP) (Sturm et al. this volume).

3. Edited Raw Data (ERD)

Figure 1 shows ten seconds of data from one detector. The data is taken with a reset time of two seconds. This means that we have 48 data-points per reset per detector. This plot illustrates the correction steps required at this stage of the processing. The two main corrections are the correction of the ramp curvature and the correction of glitches, caused by the impact of charged particles.

One source of curvature is the after-effect of the applied reset pulse (here at 0,2,4,6, and 8 seconds). This effect can be fit with a decaying exponential and can be determined from the data (this is required since the pulse-shapes appear

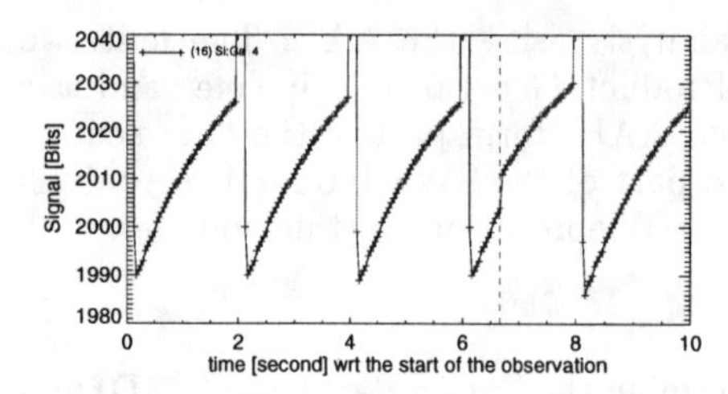

Figure 1. *Sample of Edited Raw Data (ERD) data illustrating the curvature in the uncorrected slopes and an example of a glitch (in the fourth slope) as a result of a charged particle impact.*

to change from observation to observation). The pulse-shape is an additive term to the slope and thus introduces an offset to the calculated signal. Since this affects all slopes in a similar manner it has no direct impact on the signal since it will be corrected by the subtraction of the dark currents. The main reason for correcting for the pulse-shape is to maximally linearize the slopes. This helps to improve the detection and correction of glitches. The other source of curvature can be found in the electronics in the form of a high-pass filter. The time constant for this is known to a fairly high accuracy and experiments have shown that this time constant appears to be fairly constant from observation to observation (for most detectors less than 5-10%).

In the hostile environment of outer space, there is a constant impact of charged particles. These show up in the data as a sudden increase in the raw detector signal (as illustrated in figure 1). These are recognized, corrected and flagged in the data. The flagging gives the user the ability to check the validity of affected slopes in a later stage of the processing.

4. Relative Spectral Response

An important step in the processing from SPD to AAR is the application of the Relative Spectral Response Function (RSRF). Part of the long wavelength section of the SWS (from 12 μm to 29 μm) the spectrum is highly affected by instrumental fringing as a result of Fabry Perot effects within the instrument.

Directly applying the RSRF to the data quite often gives unsatisfactory results. The fringe residuals are significantly large and this limits the detection and analysis of weak spectral features. The main reason for this is that the RSRF has a limited accuracy, the fringes as they appear in the data are sometimes shifted with respect to the fringes in the RSRF when the source is not exactly in the center of the aperture or if the source is extended. Also the width of the features (i.e., the resolution of the data) can change depending on the extent of the source.

A number of methods have been developed within IA^3 to adapt, change or correct the effects mentioned above. These can reduce the residuals down to a level of 1–2%. The following techniques can be applied:

- shift correlation/correction between the RSRF and the data
- adapting the resolution of the RSRF
- fitting cosine functions to the RSRF corrected data
- Fourier filtering of the RSRF corrected data
- modeling the instrumental FP effects

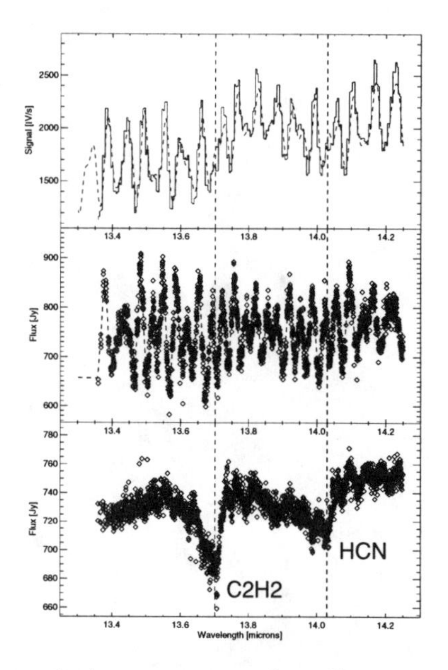

Figure 2. *The impact of fringes in part of the spectrum. In the top plot, the signal from one of the twelve detectors covering this spectral range is shown. Overplotted with the dashed curve is the scaled RSRF signal. The second plot shows the result for all twelve detectors as produced with the standard processing: large fringe residuals remain. In the last plot the result of improved data reduction is shown. The fringe residuals are low and allow detection of weak spectral features. Here the detection of the C_2H_2 ν_5 and the HCN ν_2 vibration-rotation bands towards a deeply embedded massive young star are shown (Lahuis et al. 1998).*

Figure 2 shows an example utilizing two of the techniques mentioned here. In this case first a shift correction and enhancement of the RSRF was applied. After this the residual fringes were removed by fitting cosine functions.

5. Future Developments

The data analysis of SWS has not reached its final stage. The knowledge about the instrument is continuously increasing and consequently the software is evolving. After the satellite has ceased to function a post-operations phase starts in which the improvement of the calibration and data-reduction of SWS will continue.

References

de Graauw, Th. et al. 1996, A&A 315, L49

Huygen, R & Vandenbussche, B., 1997, in ASP Conf. Ser., Vol. 125, Astronomical Data Analysis Software and Systems VI, ed. Gareth Hunt & H. E. Payne (San Francisco: ASP), 345

Kessler, M.F. et al., 1996, A&A 315, L27

Lahuis, F. et al., 1998, in *First ISO Workshop on Analytical Spectroscopy with SWS, LWS, PHT-S and CAM-CVF*, ESA SP-419, European Space Agency

Leech, K. et al., SWS IDUM, SAI/95-221/Dc

Roelfsema, P.R. et al., 1993, in ASP Conf. Ser., Vol. 52, Astronomical Data Analysis Software and Systems II, ed. R. J. Hanisch, R. J. V. Brissenden & Jeannette Barnes (San Francisco: ASP), 254

Sturm, E. et al. this volume

Wieprecht, E. et al. this volume

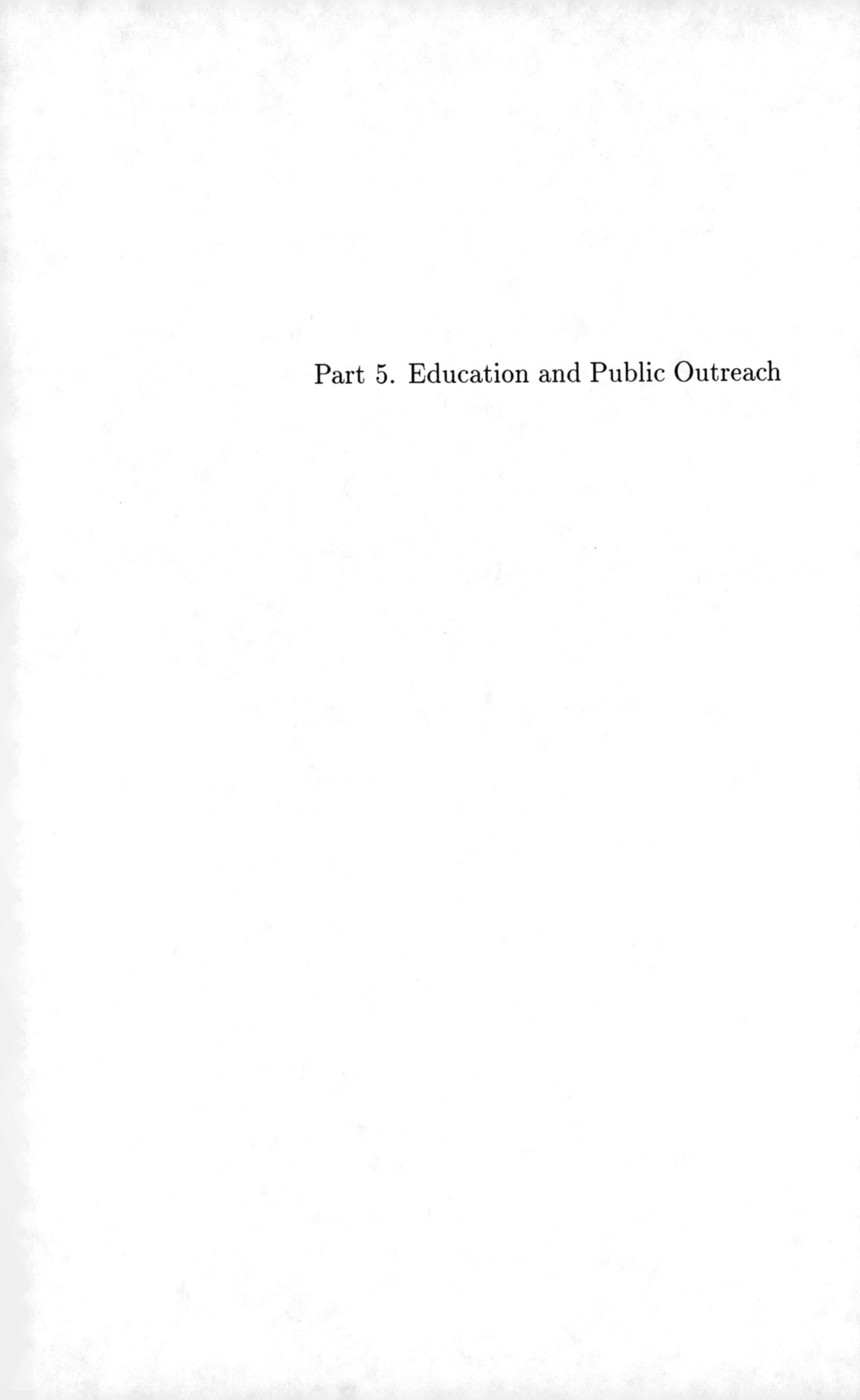

Part 5. Education and Public Outreach

Astronomical Data Analysis Software and Systems VII
ASP Conference Series, Vol. 145, 1998
R. Albrecht, R. N. Hook and H. A. Bushouse, eds.

Cyber Hype or Educational Technology? What is being learned from all those BITS?

C.A. Christian[1]

Office of Public Outreach Space Telescope Science Institute, Baltimore, MD 21209, Email: carolc@stsci.edu

Abstract. This paper discusses various information technology methods being applied to science education and public information. Of interest to the Office of Public Outreach at STScI and our collaborators is to investigate the various techniques through which science data can be mediated to the non-specialist client/user. In addition, attention is drawn to interactive and/or multimedia tools being used in astrophysics that may be useful, with modification, for educational purposes. In some cases, straightforward design decisions early on can improve the wide applicability of the interactive tool.

1. Introduction: What is the Big Picture?

As recipients of federal funding, the interest and pressure has increased on U.S. agencies to provide to taxpayers some direct benefit from their investment in our research and technological developments. Clearly, part of this effort was intended for direction towards the improvement of science, mathematics and technical education. The increase in attention by the public in examining the benefits of research has resulted in some redirection of funding to produce quality educational materials and public repositories of information, but also offers the research community an opportunity not only to share scientific results with the public and to encourage them to continue their investment (funding), but also to give them a vested interest in our scientific enterprises.

Through appropriate conduits, scientific and technical data can be mediated in such a way so as to be useful to a wide audience. One such channel, cited in this paper, is the Office of Public Outreach (OPO) at Space Telescope Science Institute (STScI), which translates scientific results to the public through a variety of mechanisms: 1) news releases, 2) science and technical background information, 3) online curriculum support materials and 4) a variety of ancillary products. This paper addresses instructional technology used in astrophysics-related public programs and will explore issues related to the clientele, resource design and budgetary issues.

[1]Director, Office of Public Outreach, STScI

2. The Audience

It has been said that the largest global information enterprise is Education. This idea is worrisome considering the apparent poor public understanding of science, but it does offer a context for the development of at least educational resources (curriculum support) and therefore it is worth taking some time to understand the relevant user clientele.

2.1. The Classical Approach to Formal Education

Those who have adopted the *Classical Approach* to formal education often adhere to the following principles:

- Student development occurs as a result of the injection of content via the educational system
- The teacher is the sole subject matter expert
- Evaluation of the success of the classical method has been demonstrated through rigorous content testing via recall or rote exercises
- The classical approach is further proven through learning of the "classics" - that is, logic, mathematics, and Latin

Therefore, in the classical approach, "instructional technology" (IT)is seen as a threat because it can only replace teachers – clearly an undesirable and philosophically distasteful result. In this view, computers, obviously inferior to human educators, have no place in the classroom. Sometimes, IT is grudgingly accepted if proponents can convince the classicists that: 1) information access is the only problem needed to be addressed with IT, 2) textbooks fill specific educational needs, so the "books on line" represent the IT sufficient to serve education, 3) IT can provide quick reference material for teacher/expert and 4) the only IT development required is hyptertext linkage, good content and a decent index.

2.2. Other Approaches

Fortunately other views are held also:

1. *Individualistic* - education focus is on realizing the potential of each individual by providing development of skills, abilities and interest
2. *Social* - education is intended to create a literate workforce. This is a reformist approach based on raising student awareness concerning technical, ethical and legal issues.
3. *Process* - education develops the individual's ability to think

Happily, each of these approaches can rationalize or even embrace the use of IT.

3. Specifications for Resources

3.1. Framework

Accepting the above, a framework for creating resources based on science and technical information can be adopted, containing several components.

- Information, data and algorithms are bricks in the structure of all resources
- Interactivity is needed to engage the user and to make available specific tools and environments
- Presentation and design of information about a discipline are important, especially for the non-expert
- NASA missions such as HST must consider the broad audience, eg., national/international distribution and use
- Resources must be modular and easily used separately by users and other developers
- Resources must be adaptable to educational contexts and support rather than replace substantive activities. That is, the resources offered are curriculum supplements, not surrogates.
- Resources must be tested in different situations with a variety of demographics

3.2. Science Applications

Where should the research community start? For example, consider astronomical "information bases" which are characterized by the following:

- On-line living archives
- "Standard" formats - astronomy has a real advantage in the early adoption of standard data exchange formats which can be readily converted (certainly with some loss of information) to popular formats
- Globally distributed infrastructure - the astronomical community adopted, at least philosophically that data is geographically distributed and should be globally accessible
- Search engines, location services, security mechanisms and conversion services - critical items for *robust* data location and access
- Client-server and peer-to-peer technology

These attributes in data systems are laudable. They also are critical but not sufficient characteristics for developing public interfaces based on scientific information repositories. Therefore the further implications for developing access methods for existing and future scientific data systems should be considered.

First, the data must be mediated to a palatable form for the target clientele. For example, simply providing a conversion of "FITS" to "GIF" as a front end to an archive is insufficient. While access to digital libraries is crucial, it is not a sufficient condition for satisfying the requirements of a broad audience because every piece of precious data, each enchanting bit of research and clever algorithm is not inherently educational or of public interest or utility.

Furthermore, developers are wise to make judicious use of multimedia and use new technologies when and if they are appropriate and useful in context. Specifically just because a technology is *cool* does not mean it is appropriate. The balance of data, algorithms, technology and information must be integrated wisely to be useful.

3.3. Interactivity

Many terms in the field of instructional technology (including the term "instructional" or "educational technology") vary widely with the context and particular community which use those terms. The word *interactive* is no exception. The types of interactivity referenced here include:

- Computer mediated communications
- Real time or near real time
- Collaborative learning
- Distributed learning
- Multimedia systems
- Simulations, modeling and games
- Intelligent agents
- Adaptive tutoring systems
- Virtual reality based learning

In this paper, systems are characterized as *interactive* if they provide feedback which depends upon user input. That is, the result of the interaction is not deterministic, or if it is digital or parametric, it involves a significantly large number of possible responses.

4. Interface Design

The design of the interface pertaining to the visual presentation of information is often the component that is taken least seriously in scientific information systems. However on the Web, the use of an engaging presentation is a time honored tactic to grab attention and increase transient clientele. It also is effective for *retaining* users if the services offered are innovative, unique, and implement imaginative processes built on interactive tools. Further, the public interfaces into scientific and technical resources should relate to the user. Part of the presentation should give users a context and a specific connection to not only the content but also to the individuals involved in the research or technology being made available. For example, profiles and interviews (audio, video, real-time) with observatory scientists and engineers are useful: Who are they? Why do these individuals pursue scientific research and what technical challenges related to telescopes, instrumentation and methods are encountered?

Figure 1. The user (eg., students) provide test velocities to the astronaut for hitting a golf ball while standing on the surface of a planet, asteroid or comet in the Amazing Space module on gravity and escape velocity. If the astronaut hits the golf ball with a sufficient stroke, the golf ball, naturally, is launched into orbit.

Interfaces should provide some user control over selection of resources, data, and information rather than appearing to be restrictive and proscribed. The presentation, though engaging and clever, should not interfere with or obscure the usability or relevance of the resources. These principles are not trivial to implement and are far more important for users who are not experts in the subject matter being presented. Such users do not generally "need" access to the resources, and therefore interface design is critical for retaining clientele and reducing their frustration at finding relevant information in a consistent way. "Web hits" on the graphically engaging "top page" are a far cry from exhibiting actual user interest in a suite of resources.

4.1. Further Design Considerations

Once the user is "committed" to delving deeper into the resource structure, the underlying content must continue to be marketed well. The use of each resource encountered must be clear, and structures which are tiered in complexity and interactivity are recommended. Designers must plan for high bandwidth, but create low bandwidth options. Clearly browse products must be included so that users can identify precisely that the data and information is what is expected. Scientific developers should recognize that raw data online is NOT interesting or useful to the non expert user. Science data must be made relevant within the context of the resource and for the target audience.

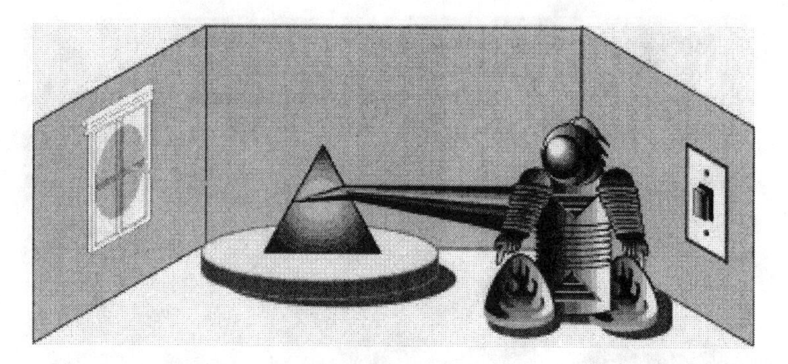

Figure 2. The robot is a mascot who guides interactive modules for understanding the basics of various kinds of waves in the Amazing Space resource: "Light and Color". The robot actively performs various activities (generating waves and measuring temperature, etc.). In this graphic, the robot uses a prism to disperse sunlight. The switch at the robot's left opens the window shade

5. Content Creation Best Practices

As discussed previously, the public audience often does not perceive a "need" for the science and technical content that the research community can provide. The user base is drawn from a population of varied demographics, that is, wildly heterogeneous in expertise, motivation, interest, and sensitivity. Scientific and technical fields are highly competitive, and are drawn from a population which is often introspective. These two communities often have divergent approaches, concepts and motivations (and sometimes actual mistrust) which should be addressed in building public interfaces to scientific content.

At the Office of Public Outreach / STScI we find that direct involvement of the user in the design of services is critical. This philosophy is espoused in other venues, including industry, but it is often actually just "lip service". Our programs in OPO insist that representative users are not just temporary critics of resources, but rather, collaborators and co-authors. In this way, compromise and balance are achieved which suit users from varied backgrounds. The collaborative teams we create include representative users (eg., teachers, science museum personnel, journalists – depending upon context), scientists, engineers, programmers, graphic artists and animators.

Note that often, knowledge of this *team*, face-to-face approach frequently tempts the software engineer or scientist ("who knows better" and who would prefer to lecture on content rather than collaborate on resource creation) to disengage from the process before it starts! Managing collaborative projects aimed at producing useful public access to scientific research is a challenging proposition and must include talented brokers to engage and mesh the expertise of the various participants productively.

It should be noted that an initial period of building of intra- team trust and rapport is essential. Careful consideration of design issues is worth the effort, otherwise expensive multi-media resources may be only of fleeting interest to the intended users no matter how compelling or astonishing the technology to be used appears to be.

5.1. Evaluation

Evaluation of publicly accessible resources is a key issue for determining effectiveness, for reviewing level of effort to be devoted to resource creation, and for planning new development. At STScI/OPO, we test resources in a variety of environments with users both local to our area, but also across the United States. Some resources are piloted internationally. Key topics addressed in the evaluation of specifically curriculum online materials are:

- Product concept
- Overall design, and design of specific modules or objects in the resource
- Usability by target clientele and other users, and relevance to users requirements and unforeseen needs
- Pedagogical approach for curriculum resources
- Customer feedback - acquired through *in situ* testing at workshops, seminars, and classrooms, and further through a network of remote testers
- Ancillary usage: modules used in new ways in varying environments and as integrated into new products

Note that educational hard copy products are evaluated similarly. The specific evaluation procedures and results of resources are the subject of separate papers.

5.2. Copyrights

Copyrights are a growing, serious problem and are a concern at many levels. A full discussion of the issues is well beyond the scope of this paper. Clearly educational resources must obtain copyright permissions for material reproduced from external sources and this procedure is not trivial. Conversely, resource authors must consider carefully what copyright permissions will be allowed. At STScI/OPO the copyright policy [2] is derived from the NASA policy and basically gives reasonably liberal permissions to research and educational users for content re-use.

6. Budget Model - Can you afford it?

The creation of meaningful and useful resources based on scientific research data and results is non-trivial and demands consideration of many issues including budget. The typical costs to be considered include:

---

[2] **http://www.stsci.edu/web/Copyright.html**

– Overall development time including pre-planning
– Longevity of products and resources
– Cost of maintaining currency including ease of upgrade
– Overall maintenance and other overheads
– Marketing costs (even the design of an enticing frontispiece may be attributed to "marketing" costs)

6.1. Cost Effective Strategies

The cost effective strategies that have been tried and true in resource development are to create usable and easily upgradable modules. The modularity also should encourage re-use by both the original authors and by new users who add value to each item. In addition, avoidance of duplication where possible, resisting the temptation to reinvent resources with small modifications that may be transparent or even useless to the user. Well designed resources separate the interface layer from the underlying content, code and objects and include rational handles to encourage re-use through changeable interfaces.

6.2. What is NASA Doing?: The Leverage Power Model

The NASA community has considered seriously attempting to understand how resources are created and disseminated, and has taken a look at the "leverage power" of projects and programs. This model is used widely in commercial ventures particularly for cost effective sales and marketing.

Consider that for a single module or activity A is created. That activity, if created efficiently, optimally and with high quality (an exercise left to the reader) has a particular leverage power depending upon how it is mediated to the audience. For example, if the activity can be replicated and further amplified through one-to-many dissemination methods and then made available to a wider audience through connections and networking, the item can be characterized with a leverage power:

$$Leverage\ Power = L = \rho\alpha\gamma \tag{1}$$

$\rho = replication$ including re-use and duplication
$\gamma = connections/networking$ through cross links
$\alpha = amplification$, that is, one-to-many
(e.g.., content source $\rightarrow$ master $\rightarrow$ teacher) $\rightarrow$ student

For example, a classroom activity presented in a workshop to 30 regional science curriculum supervisors who in turn transmit the information to 30 local science teachers and these teachers in turn share the information with other (2) science teachers, then the leverage power is 1800. If the workshop is replicated in 10 regions, the leverage is 18,000 for the one example module. Clearly the "leverage power per student" is multiplied by the number of students in a class (conservatively, 25).

$$\alpha = 30,\ \gamma = 2,\ \rho = 30 \times 10$$
$$LeveragePower = 18,000$$

However if the same "activity" is presented by an outsider, visiting one classroom of 25 students, then the leverage is 1 x 25 rather than 18,000 x 25. Therefore, the leverage power of "scientists in the classroom" is low, particularly if the teacher in the classroom is unable to replicate the activity presented by the scientist or engineer in the future.

6.3. Brokers

It is clear that a variety of networking, replication, re-use and amplification techniques are possible. Through the use of organized mediators and brokers, the leverage power of educational resources made available from astrophysical sources can be impressively large. Within the NASA Office of Space Science, a system of "Education Forum" and "Brokers" have been created to serve the networking, replication, and dissemination roles, α and γ above. Each Forum provides handshakes between the public and the science community through a variety of means, including electronic multimedia. Brokers provide dissemination networks, spreading the word and demonstrating use of Space Science content (hosted and linked) through the Forum. This system is intended to provide effective methods that allow public access to scientific data, technology and expertise in a useful way which does not disable the research enterprise.

6.4. Launching a Project

STScI/OPO is the site of the NASA OSS "Origins Education Form", and thereby has the mandate to help researchers find ways to craft effective educational programs through pragmatic approaches. The Forum provides models for effective, tested processes and advice on best practices. There are also mechanisms for obtaining funding and links to existing successful programs as examples.

7. Summary

Information technology methods are being applied and evolved to provide exemplary materials for science education, curricula and general public information about research and technology. A variety of successful techniques are emerging and interactive resources are now being offered and tested to allow users with an interest in science to experience scientific principles and research environments. It is clear that the algorithms, archives and data representations in use within astrophysics are necessary building blocks, but must be mediated and culled to address the needs of the user. Design considerations and presentation must be weighed carefully when planning the budget and necessary human resources to be devoted to the public interface.

References

Recker, M., 1997, *Educ. Technol. Rev.*, 7, 9

Reeves, J., 1997, *Educ. Technol. Rev.*, 7, 5

Astronomical Data Analysis Software and Systems VII
ASP Conference Series, Vol. 145, 1998
R. Albrecht, R. N. Hook and H. A. Bushouse, eds.

Using Java for Astronomy: The Virtual Radio Interferometer Example

N.P.F. McKay[1] and D.J. McKay[1]

Nuffield Radio Astronomy Laboratories, University of Manchester, Jodrell Bank, Macclesfield, Cheshire SK11 9DL, UK

Abstract. We present the Virtual Radio Interferometer, a Java applet which allows the demonstration of earth-rotation aperture synthesis by simulating existing and imaginary telescopes. It may be used as an educational tool for teaching interferometry as well as a utility for observers. The use of the Java language is discussed, and the role of the language, within the astronomical software context, reviewed.

1. Introduction

In 1996, the Australia Telescope National Facility (ATNF) was awarded a grant to upgrade its existing interferometer — the Australia Telescope Compact Array (ATCA). The ATCA is a 6 kilometre earth-rotation aperture synthesis telescope, consisting of six movable 22 metre antennas located on an east-west railway track (JEEEA 1992). Each antenna can be moved and placed at various locations along this track, known as stations. Part of the upgrade involves the construction of new stations on the existing east-west track and the addition of a north-south track. These extra antenna positions will improve the imaging capability of the instrument.

To help astronomers and engineers at the various institutions experiment with and visualise the effects of different positions of the new stations, a design and simulation software package was required. This program had to be intuitive to use, widely available and quickly developed. The resulting program was called the Virtual Radio Interferometer[2] (VRI).

2. The Virtual Radio Interferometer

The idea behind the VRI project was to create an interactive, user configurable, simulated interferometer, providing a tool for determining the synthesised aperture of real and hypothetical telescopes. It fosters an intuitive feel for antenna positions, uv-coverage and the resultant point-spread function. VRI also proved

[1]Formerly: Australia Telescope National Facility, CSIRO, Paul Wild Observatory, Locked Bag 194, Narrabri NSW 2390, Australia

[2]http://www.jb.man.ac.uk/vri/

to be a useful teaching tool, in addition to its original design role of investigating new antenna arrays and experimenting with "what-if" scenarios.

As well as exploring the properties of several existing instruments, the user has control of a number of telescope parameters, allowing the creation of hypothetical interferometers. These include the latitude of the site and the number of array elements which form the interferometer, together with their diameter, elevation limit and position on a two dimensional plane. The declination of the source, as well as the frequency and bandwidth of observation can also be specified. Once these parameters are set, a corresponding *uv-plot* is produced by the program, showing the aperture-plane coverage.

Given the plot of the uv-coverage, we can simulate its effect on various objects. An image of the source is loaded onto one of four display panels, a Fourier transform performed on it, and also displayed. This can be then masked with the uv-coverage. When the inverse Fourier transform is applied, the original image of the source is replaced by the image obtained by "observing" it with the simulated interferometer. Finally, uv-plots generated by a number of antenna configurations, or even different observatories, may be combined to generate an image composed of multiple "observations".

3. Implementation

The development path of the project, and hence the choice of programming language, was governed by several constraints. Firstly, the program had to be easy to use and have a short learning curve. It was decided that VRI would be interactive, present a predominantly graphical interface and be mouse-driven. To encourage people to use the it, the application had to be easily accessible, so the World Wide Web was chosen as the transport medium and, as we had no control over the platform that our audience was using, the code had to be portable. Above all, rapid development was required.

Java was chosen as the coding language, because it satisfied these requirements. As a language, it is high-level, strictly object-oriented, and easy to learn. Classes which facilitate the writing of software for Internet distribution form part of the standard Java Development Kit. Hence, setting a program up to run on the Web is not a complex exercise. The source code is partially compiled into a set of *byte codes* before released for use. These are platform independent, hence eliminating the need for multiple versions of the software. The byte code is downloaded onto the client computer and, at run time, is interpreted into native instructions and executed by the host machine. This alleviates congestion on the organisation's HTTP server, and once the program is downloaded onto a local machine, provides a faster program for users.

4. Appraisal of the Project

On the whole, Java proved to be an ideal development language for this particular project. This is not to say that no problems were encountered.

Being a semi-interpreted language, programs run slower than their native code counterparts. For example, a 256×256 point Fourier transform takes 15 seconds to be executed by the Java interpreter, on a Sun Ultra-1 running at

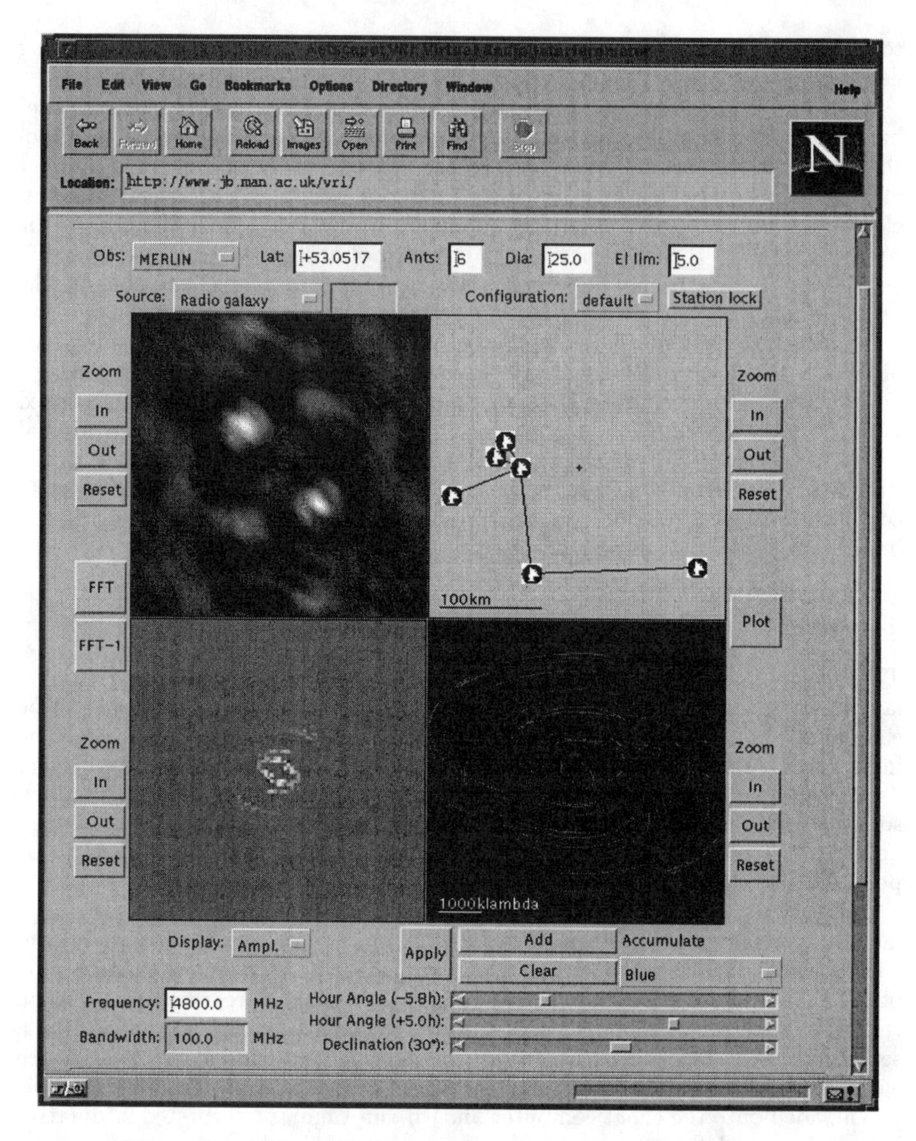

Figure 1. The VRI Java applet, run from a Web browser.

140 MHz, which can be compared with figures of less than one second for native code. Other potential reasons for avoiding the language include the fact that Java is still somewhat immature and evolving, the lack of complete control over registers, memory, etc., provided by other languages such as C and C++ and the problems posed by integrating a new language with legacy code.

Nevertheless, the portability of the language, the rich vocabulary of standard class functions, as well as Java's easily extended object-oriented design, greatly outweigh these disadvantages.

Table 1. Software development metrics for VRI

Java Development Kit Version	1.0.2	
Software classes	26	
Java instructions	1565	
Development time	80	hours
Size of byte code	1.3	Mbyte

5. Future Development

VRI has proved to be a successful software project and users have provided favourable feedback. Improvements have been planned, and subsequent versions of VRI will include deconvolution, three-dimensional arrays and the ability to import real astronomical data. As Java has demonstrated itself to be a model environment for the coding of this application, it will continue to be utilised.

The authors are of the opinion that the Java programming language has a useful role to play in general astronomical software. We would not, however, recommend its use for all applications. Its strength lies in the regime of off-line applications, interactive tools and, of course, user interfaces.

Acknowledgments. The authors would like to thank Mark Wieringa for his contribution to this project, as well as Andrew Lyne, Jim Caswell, John Reynolds and Dave Shone.

The Australia Telescope is funded by the Commonwealth of Australia for operation as a National Facility managed by CSIRO.

References

June 1992, JEEEA (Journal of Electrical and Electronics Engineering, Australia), 12, 2

Astronomical Data Analysis Software and Systems VII
ASP Conference Series, Vol. 145, 1998
R. Albrecht, R. N. Hook and H. A. Bushouse, eds.

Teaching Astronomy via the Internet

L.Benacchio, M. Brolis and I. Saviane

Astronomical Observatory, vicolo Osservatorio 5, I-35122 Padova, Italy

Abstract.
In the framework of a partnership between the Astronomical Observatory of Padova and Telecom Italia, the most important Italian Internet and Intranet provider, we are experimenting with the educational possibilities of the Internet. A Web site *entirely* devoted to the teaching of Astronomy to secondary school students has been designed and realized. Graphics and animations are widely used to introduce the concepts, and special care has been devoted to the language used in the text.

A group of junior and senior high schools (8-12 grades) will be connected to a server located in the Observatory, and the server will in turn be a gateway to the whole Net. Access will be given to a selected number of external astronomical Web sites.

The actual use of this tool in the selected classrooms will be followed up in the next scholastic year (October 97–June 98). A strong and continuous interaction with the teachers is also, obviously, foreseen to ensure a strong positive feed–back. On one hand, this project will give us important suggestions for the further development of a specific educational methods suitable for the Web. On the other, it will give the students basic astronomical knowledge and the possibility to become familiar with the Internet environment.

1. Guidelines for this Project

There are several reasons why astronomy is an appropriate choice for a project like this. Astronomy and astronomical images have immediate appeal to a large number of young people, so they can be used to attract people to science in a general sense. Moreover, images are an obvious way to build an attractive Web site, and a search on the Web does indeed reveal that there is a huge amount of astronomical information available. On the other hand, there are no comprehensive projects which lead a student through astronomy from the very basics, particularly among Italian Web sites. Furthermore, the information that one can find is also sometimes wrong or misleading. The aim of our project is therefore first to provide introductory and *reliable* astronomical material to 8 -to-12 grades students, and, as a second step, to be an informed guide through the existing external Web sites.

With regard to the first point, the new medium allows one to overcome the limitations of traditional textbooks: the students can establish by themselves the pace of their 'lessons', and they can experiment in real time with the concepts they have learned, interacting with the environment that we provide. Furthermore, the Web site also contains a guided first contact to the Internet and explains how to exploit its wealth of astronomical information.

We must also point out that teachers themselves are not always aware of the possibilities of the Net, and textbooks are generally lacking in topics suitable for effective hands-on activities. Another important task of our project is therefore to assemble a complete database of reference hands-on activities to be performed in the classroom. Guidelines for their usage and simulations of the results are also given.

2. Theoretical Approach to the Educational Method

It is now widely accepted that the best strategy to introduce new knowledge to students is the so–called Karplus' cycle (Atkin & Karplus 1962; Karplus et al. 1976). It consists of three steps. First, the students are given some 'environment' where they can explore a given phenomenon, make their own observations and draw some first conclusions. Second, these preliminary conclusions are discussed with the teacher and incorrect interpretations are removed. Finally, once the new and correct explanation is acquired, new concepts are used in the exploration of a new situation, and the *learning cycle* repeats itself.

We also bear in mind that new knowledge should be introduced after the students' misconceptions have been removed. If this task is not performed, it is possible that no actual change in the individual reasoning will happen. Her/his knowledge is not *restructured* (see e.g., Vosniadou & Brewer 1987). The new information will be somehow adapted to the old mental schemes, so that distortions and misunderstandings will result. Moreover, the knowledge is never incorporated into one's vision of the world and soon forgotten (Ausubel et al. 1978).

Once the educational method has been defined, the contents must be selected and organized. The basic approach is to put the subjects in a hierarchical structure. To this aim, several methods can be devised in order to identify which concepts are prior to the others, and conceptual pyramids can be built. We followed the results of the project STAR, which are discussed in by Sadler (1996, see his Figure 6).

After this first learning phase, another approach to the educational material can be given, which is that of unifying concepts. It is useful for the students to see how the same principles and methods are at work throughout all the subjects they have encountered.

Educational content and unifying concepts are now briefly discussed.

2.1. Educational Content

The subjects have been selected from a standard list (see e.g., Schatz 1996) and organized in a multilevel structure. The starting points are the concepts of the Earth's shape and motion, then we move on to gravity and its consequences, periodic phenomena and orbits, the laws of planetary motion and the Solar System. Radiation properties are introduced before studying the Sun and the stars. The masses, luminosities, radii, temperatures and distances of the stars are discussed, and care is always used to supply all data in a comparative way.

At the higher levels, our current vision of the structure of the Universe is given, emphasizing the role of galaxies as the fundamental components of the large scale structure. Finally, the Big Bang theory is introduced and the past, present and future of our Universe are discussed.

During a normal school year only a few themes can be explored in depth, so the teacher will have to select a subset of all the subjects, which should be consistent with the actual level of her/his students.

2.2. Unifying Concepts

After introducing the fundamental idea of science as inquiry, other conceptual structures are highlighted, such as energy balance and conservation, the equilibrium and evolution of a physical system, and the clustering properties of the Universe. It is shown how the Universe is ordered and organized from micro to macro worlds. The relation between distances and time scales is emphasized, and between the mass, radius and luminosity of stars and galaxies. The concept of order of magnitude of physical quantities is also explored.

3. Application to the Internet Environment.

The misconception removal phase and the learning cycle have been reproduced by means of tests. In the first case, the questions reproduce the ones used in many experimental investigations on the children's learning processes and are connected to a list of answers which contains the most common misconceptions. If a wrong answer is selected, then its consequences are shown to the student, and eventually he will get to the point of self-contradiction. From this point he will get back to the initial test, and when the right answer is chosen the correct explanation is reinforced by means of other examples, images, films etc.. This process makes the student interested in the forthcoming material and prepares him/her for the next subject.

When the test is connected to the Karplus' cycle, it is used to perform the three phases. First, the exploration of a phenomenon is proposed by means of introductory text and images, and some observations and/or hands-on activities are suggested. These observations are usually possible with very simple aids, and the teacher is expected to guide the students through the various steps. If for some reason the observations cannot be made, or the user has no time for them, then simulations are provided. The simulations are also useful as a guide to the correct interpretation of the observations. Java scripts are widely used in this phase.

The students are then requested to give their interpretation of the observations and/or simulations. A set of alternative explanations for the phenomenon is listed and the students are required to choose one. When the student's choice is incorrect, they are led to a self-contradiction.

The third phase (application of new concepts) is performed when the correct answer is chosen. The phenomenon is further discussed and its consequences highlighted, and it is then used to introduce the next subject in the hierarchy. CGI scripts have been devised in order to manage the tests and the related actions, and to take care of the user database and access log.

The site is organized in a multiple level structure. The single levels can be accessed either from a general table-of-contents or through an ordered path. A log of all the user's accesses is kept. If the user wishes, his/her performance can be evaluated using the number of correct vs. wrong answers. In that case, access to the single levels is given with a password and is flagged in the log file.

4. Conclusions and Future Developments

This project allowed us to establish a template in the teaching of Astronomy via the Internet. The most effective educational strategy has been identified and it has been adapted to the new medium. The key feature of this medium, i.e., interactivity, has been fully exploited.

The educational contents have been carefully selected and organized in a hierarchical structure. Therefore, we ensured both that the basic astronomical knowledge is provided and that no previous elements are needed to fully exploit the Web site. The basic unifying concepts have also been identified and are stressed throughout the various levels.

Browsing through this material, the students will encounter a new way to learn things. A lively way which overcomes the traditional static textbook limitations, and which gives the students the chance to wander around along their preferred path at a pace they will establish according to individual capabilities. They will learn by doing things themselves, which is the best way to learn.

At the same time, the Web site provides a starting point to become familiar with the Net environment and to establish a minimum set of astronomical concepts. Using this background astronomical knowledge and the basic navigation skills, they will be able to access the available material on the Web with full awareness.

In the near future, we will extend the database of interactive experiments by means of a Java laboratory, and VRML will be implemented. The response of real classes will be tested and we will take advantage of the students' and teachers' suggestions in order to improve the site (a 6 month phase is expected). The Web site can be reached via the Padova Astronomical Observatory[1].

References

Atkin, J.M., & Karplus, R. 1962, The Science Teacher, 29, 45

Ausubel, D.P., Novak, J.D., & Hanesian, H. 1978. In Educational Psychology: A Cognitive View. New York, Holt, Rinehart and Winston

Karplus, R., et al. 1976, AESOP Pub. ID-32, Univ. of California, Berkeley

Mestre, J.P. 1991, Phys.Tod., Sep 91, 56

Sadler, P.M. 1996, ASPC, 89, 46

Schatz, D. 1996, ASPC, 89, 33

Vosniadou, S., & Brewer, W.F. 1987, Review of Educational Research, 57, 51

---

[1]http://www.pd.astro.it/

Astronomical Data Analysis Software and Systems VII
ASP Conference Series, Vol. 145, 1998
R. Albrecht, R. N. Hook and H. A. Bushouse, eds.

Astronomy On-Line - the World's Biggest Astronomy Event on the World-Wide-Web

R. Albrecht[1], R. West, C. Madsen

European Southern Observatory, Garching, Germany

M. Naumann

Serco c/o European Southern Observatory, Garching, Germany

Abstract.
This educational programme was organised in a collaboration between ESO, the European Association for Astronomy Education (EAAE) and the European Union (EU) during the 4th European Week for Scientific and Technological Culture. Astronomy On-Line brought together about 5000 students from all over Europe and other continents. Learning to use the vast resources of tomorrow's communication technology, they also experienced the excitement of real-time scientific adventure and the virtues of international collaboration.

1. Introduction and Background

While the main task of the European Southern Observatory (ESO) is to provide state-of-the-art observing facilities to the European astronomical community and otherwise support European astronomy, there have also been efforts to advance the knowledge about astronomy among the general public, and, in particular, among students.

ESO provides, on a regular basis, press releases to the European media. There are arrangements at all ESO facilities for the general public to visit. The "Open Door Day" at the ESO European Headquarters (Garching, Germany) regularly draws thousands of visitors. Various "ESO Exhibitions" are shown in different European countries and material is provided to science museums and planetaria. This is a formidable task, given the necessity to not only cater to different levels of education, but also to audiences with different languages and cultural backgrounds.

2. Astronomy On-Line

The programme is a collaboration between the European Association for Astronomy Education (EAAE) and the European Southern Observatory. It is

[1]Affiliated to the Astrophysics Division, Space Science Department, European Space Agency

sponsored by the European Union (EU) via the European Commission (EC) through its Directorates XII (Science and Technology) and XXII (Education) and was initiated in conjunction with the 4th European Week for Scientific and Technological Culture in November 1996. Chile was included in the arrangements as host country for the ESO La Silla and Paranal Observatories.

During Astronomy On-Line, a large number of school students and their teachers from all over Europe and other continents as well, together with professional and amateur astronomers and others interested in astronomy, participated in a unique experience that made intensive use of the vast possibilities of the World-Wide-Web (WWW). It is obvious that such a programme cannot be a series of lectures with the goal of bringing the full audience to a well defined level of knowledge. Indeed, the main challenge of the programme was to attract interest on many different levels, yet to allow students to evolve from the level of aroused curiosity to the level of personal contact with professional astronomers.

Through the WWW, the participants 'meet' in a 'marketplace' where a number of different 'shops' are available, each of which tempt them with a number of exciting and educational 'events', carefully prepared to cater for different age groups, from 12 years upwards. The events cover a wide spectrum of activities, some of which have been timed to ensure the progression of this programme through its three main phases. It is all there: from simple, introductory astronomy class projects to the most advanced on-line access to some of the world's best telescopes, from discussions with peer groups to on-line encounters with leading professionals.

Astronomy On-Line is not just about 'trivial' data retrieval or about enhancing the seductive drive into virtual reality. For example, through the possibility of designing and conducting a real observing run on some of the major, professional telescopes, it offered the opportunity for hands-on experience to students even in the most remote areas. In addition, they were able to 'meet' some of the professional astronomers at the participating observatories on the WWW and discuss subjects of mutual interest.

Apart from its astronomical and natural sciences component, a particularly fascinating perspective of the project was that it significantly contributed to an understanding of the usefulness and limitations of the communication technologies that will increasingly govern all our daily lives. Other side benefits, of course, included stimulating schools to go on-line and prompting international cooperation among young people. Another important aspect is that the programme did lead to the natural involvement of business and industrial partners in the local areas of the participating groups. Moreover, its unique character and international implications was very inviting for extensive media coverage, both in human and scientific/technological terms.

3. Steering Committees and Participants

A preparatory meeting of the Executive Council of the EAAE and EAAE National Representatives in 17 countries was held at ESO and an International Steering Committee (ISC) was established. The ISC was responsible for the planning of the main activities in Astronomy; it met in September 1996 to evaluate the progress and to define the further actions and goals. The EAAE

National Representatives set up National Steering Committees (NSC) to coordinate the Programme in their respective countries. More countries, in particular in Eastern Europe, joined in in the meantime.

The NSCs consist of educators, scientists, representatives of leading planetaria, Internet specialists and, in some places, also representatives from sponsors (Internet providers, PC hardware suppliers etc.). Most NSCs established good liaisons with their National Ministries (of Education).

The ISC prepared a detailed programme description together with basic guidelines that served to co-ordinate the work of the NSCs. They in turn provided organisational and technical information (e.g., computer and communication link specifications) to the participating groups, sponsors and supporters of the programme.

The first task of the NSCs was to issue a call for participation to interested schools, astronomy clubs and other astronomy-interest groups in their respective countries. This was done during the summer of 1996 and continued after the beginning of the new school year.

The participating groups consisted of a teacher and his/her students or of one or more astronomy enthusiasts. Groups of young astronomy-enthusiasts without a teacher and amateur astronomers were also welcome and many joined. Each participating group had to register officially via the Astronomy On-Line WWW pages. A summary of technical requirements for access to the WWW was available on the Web. In those cases where access was not yet available at the school, this was sometimes arranged by 'sponsors' in the local area (planetaria, institutes, businesses or private benefactors).

The full information was made available on the two central computer nodes of the Programme which were continuously updated as the elements were specified in increasing detail. The Astronomy On-Line WWW Homepages can still be reached at http://www.eso.org/astronomyonline/

Most communication via the WWW took place in English. However, the appropriate areas of the National Astronomy On-Line Homepages were often in the local language and when communicating with other groups in their language area, some groups did use their own language.

4. Implementation

The World Wide Web provides a mechanism which is both convenient and widely accessible. While maximizing the visibility of the project it also acquainted the target audience with modern computing concepts, which was not only beneficial for the students, it has also been one of the foundations of astronomy as an almost all-digital science.

The Astronomy On-Line Programme was based on the concept of a WWW 'marketplace' with 'shops' that could be consulted by the participants. The 'shops' were 'open' at specified times, some from the beginning of the program on October 1, 1996, and others later. The 'shops' displayed a variety of 'goods' (activities) at different levels of complexity in order to attract participants of different age groups. The following shops were installed:

1. General information: Information about the Programme, and Help facilities. List of participating groups. Links to related Web sites. 2. Collaborative

projects: Projects which required observations by many groups, all over the continent, thereby leading to 'joint' results. 3. Astronomical observations: Preparation of a real observing programme, submitted and executed by telescopes at participating, professional (and in some cases, amateur) observatories. 4. Astronomical software: Use of a variety of general astronomical software (orbits, eclipse predictions, etc), which could also be taken over for future use at the schools. 5. Use of astronomical data on the WWW: Retrieval of data (images, text, archive data), available on the WWW at different sites. This shop also included educative 'Treasure Hunts' on the Web. 6. Prepared exercises (Try your skills): A variety of prepared, astronomical exercises of different level. 7. Talk to the professionals: Talk over the WWW to professional astronomers and educators. 8. Group communication: Connect to other Participating Groups. 9. Newspaper: Publication on the WWW of the results of the various activities, etc. Announcements about the Programme and its progress.

Astronomy On-Line was divided into three phases, lasting from early October to November 22, 1996, and reflecting the gradual development of the associated activities. During this period, a number of special projects took place, for instance in connection with the Partial Solar Eclipse on October 12, and the amount of information on the Astronomy On-Line Web pages grew continuously.

The NSCs established national computer nodes for the Astronomy On-Line Programme. In many cases, this was done in collaboration with a national university/observatory or with a (sponsoring) Internet provider. In several places, it was done in conjunction with the already existing EAAE Nodes.

The National Astronomy On-Line Home Pages had two main features:

1. A national component, dedicated to the activities in that country, and
2. A mirror of the 'ESO Astronomy On-Line Home Page' (which acted as the central 'European Homepage').

In addition to their function as carriers of information, these WWW nodes plus the national home pages also acted as an on-line advertisement for the Programme. ESO produced a colour poster which was distributed by the NSCs. ESO also provided VHS tapes with a specially prepared feature video that was used to promote the Astronomy On-Line Programme.

5. Conclusion and Outlook

The programme met all expectations by stimulating interest in astronomy and related sciences. The participants experienced in a very direct way how science is done and acquainted themselves with important aspects of the scientific working methods.

On the educational side, many participants were introduced to the benefits of the WWW for the first time and they became familiar with the incredible opportunities of communication and information extraction which are available through this new medium.

At the same time it was noted with great satisfaction that the Ministries of Education in several European countries took this opportunity to begin the implementation of systems for continuous Internet and Web access from the schools.

Part 6. Dataflow and Scheduling

Astronomical Data Analysis Software and Systems VII
ASP Conference Series, Vol. 145, 1998
R. Albrecht, R. N. Hook and H. A. Bushouse, eds.

Nightly Scheduling of ESO's Very Large Telescope

A. M. Chavan, G. Giannone[1] and D. Silva

European Southern Observatory, Karl Schwarzschild Str-2, D-85748, Garching, Germany

T. Krueger, G. Miller

Space Telescope Science Institute, Baltimore, MD 21218, USA

Abstract.
A key challenge for ESO's Very Large Telescope (VLT) will be responding to changing observing conditions in order to maximize the scientific productivity of the observatory. For queued observations, the nightly scheduling will be performed by staff astronomers using an Operational Toolkit. This toolkit consists of a Medium and a Short-Term Scheduler (MTS and STS), both integrated and accessible through a common graphical user interface. The MTS, developed by ESO, will be used to create candidate lists of observations (queues) based on different scheduling criteria. There may be different queues based on "seeing", or priority, or any other criteria that are selected by the staff astronomer. An MTS queue is then selected and supplied to the STS for detailed nightly scheduling. The STS uses the Spike scheduling engine, which was originally developed by STScI for use on the Hubble Space Telescope.

1. Introduction

An Operational Toolkit (OT) is being developed at ESO as a front-end component of the Data Flow System (DFS). It is designed to react quickly to evolving weather, taking the best advantage of the current observing conditions. It will enable the observers to take the most valuable science data at any given time, with the goal of maximizing the observatory's scientific return. While it is aimed at supporting service mode operations, the toolkit can also be used by Visiting Astronomers.

The toolkit will make all observation preparation data available to the operator, and will allow the definition of *queues* and *timelines* of Observation Blocks (OBs). The OT will also interface to the VLT Control Software, providing observation instructions and recording OB-related run-time events. Finally, the toolkit will act as the interface to the OB repository for the whole Data Flow System.

---

[1] Serco GmbH, at the European Southern Observatory

2. Operating the front end of the Data Flow System

Observation Blocks are modular objects, joining target information with the description of the technical setup of the observation. OBs are created by investigators using another DFS front-end tool, the Phase II Proposal Preparation System (P2PP; Chavan 1996), then checked into the ESO repository, where they are verified for correctness by ESO.

The operations team then uses the Operational Toolkit to browse the repository and build queues, based on object visibility, expected observing conditions, and user-defined scheduling constraints. A queue is a subset of the available OBs: it usually covers one to a few nights of observation, and several queues can be defined for the same night (e.g., based on different expectations of weather, or equipment availability). An OB can belong to more than one queue at a time.

Later, as the observing night begins and progresses, the staff observer can switch back and forth among the available queues, and build timelines (observing plans) for each individual queue. Timelines are based on the current weather conditions and OB priority, and can be built on several possible scheduling strategies.

Events originating from the VLT Control Software or other DFS subsystems — such as a change in the current seeing conditions, the termination of an OB, an instrument configuration change, or a failure while reducing the science frames — are also received by the OT, and fed back into the scheduling process and the OB repository.

3. Architecture of the Operational Toolkit

Users of the OT will see a single graphical user interface, and access OT functionalities through a unified set of commands and display widgets. However, several independent components will cooperate behind the scenes to provide such features, as shown in Figure 1.

- The OB repository will be implemented on top of a commercial DBMS, running on dedicated servers in Garching (Germany), La Silla, and Paranal (Chile). Database replication will be handled transparently to the users, and will ensure that the same up-to-date information is used for operations throughout the system.
- Repository browsing, queue management, and interaction with the VLT Control Software and other DFS subsystems will be provided by the Medium-Term Scheduler (MTS). The browser will enable the OT user to select OBs from the repository, based on target coordinates, object status, requested instrument and instrumental configuration (such as filters and grisms) and scheduling constraints (see below). OBs can also be selected according to observation type (such as imaging, spectroscopy, or calibration) and specific observing programmes or users. The amount of information to be displayed in the browser for each OB can be customized by the operator.

  OBs extracted from the repository will then be appended to queues (to provide a certain degree of scheduling flexibility, queues will normally over-

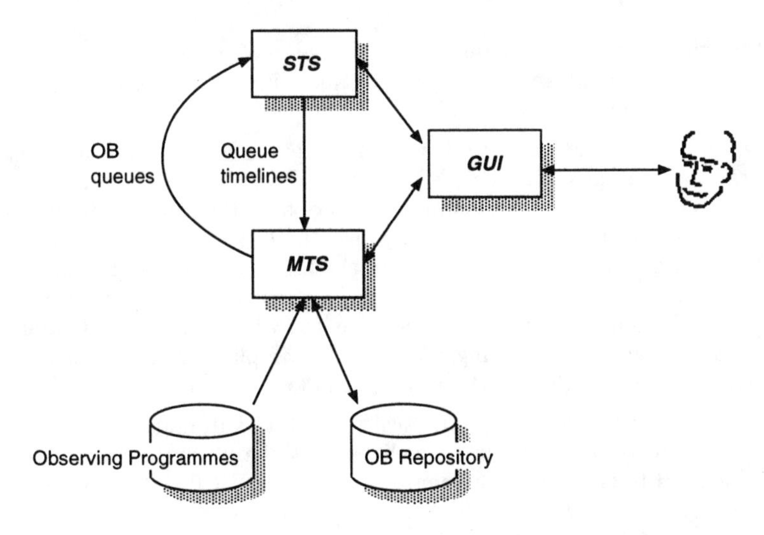

Figure 1. System architecture of the OT

subscribe the available time by a factor of two). The user will then be able to "zoom" in and out of the OBs, selectively displaying all the available information in a hierarchical fashion. Queues can be ordered according to a number of criteria, printed out, and sent to the STS for timeline computation. The currently selected OBs within a queue can be pulled over by the VLT Control Software for execution.

Finally, the MTS will have (read-only) access to the ESO Programme ("Phase I") database as well.

- Timelines will be computed by the Short-Term Scheduler (STS), a derivative of the scheduler used for the Hubble Space Telescope. The STS is based on the Spike core (Johnston & Miller 1994), adapted for use in ground-based observatories. It is being jointly developed by ESO and the Space Telescope Science Institute in Baltimore, MD. Timeline computation will based on scheduling constraints and strategies.

  OB scheduling is implicitly constrained by target visibility and availability of configurable resources: for example an OB cannot be scheduled if it needs a filter that is not currently mounted. OB authors will be able to further constrain execution by specifying exactly when and/or under which conditions the observations can take place: the scheduler will honor timing constraints, sequencing ("chains") of OBs, weather constraints (such as seeing, sky transparency and IR emissivity) and moon-related constraints (such as fractional lunar illumination and moon angular distance).

  Several scheduling strategies will be available, based on a combination of factors including OB priority, weather evolution and a number of preference values: observations should be performed as close as possible to the zenith, away from the moon, etc. Operators will be able to compare schedules generated with different strategies, and chose the one optimizing the current queue. A number of different metrics and GUI tools will be available to build and compare schedules, including a graphic display of the set of scheduling constraints.

OB priority is an important scheduling parameter. ESO observing programmes are ranked by ESO's Observing Programmes Committee (OPC); when creating queues, the computation of an OB's priority starts from the OPC rank of the programme to which the OB belongs. However, priorities may change dynamically during the night, due to weather evolution or programmes nearing completion, and the scheduling engine needs to keep track of them — as a result, the schedule will be highly dynamical, and change several times during a night. This implies important performance requirements for the Short-Term Scheduler.

Note that the STS is a support system, not an automatic scheduler: it can be used to compute several timelines, but the ultimate responsibility for which OB to execute, and when, rests with the VLT Data Flow operations staff.

- The usability of system as complex as the Operational Toolkit depends largely on the friendliness of its user interface. We'll try to make sure that (a) all OB and schedule information is readily available and easy to read, (b) the user is always able to override the system's suggestions, and (c) all of the tool's features are "one click away" (no double operations).

4. Perspective

A version of the MTS has been in operations at ESO's New Technology Telescope (NTT) in La Silla since the beginning of 1997, including a large fraction of the above listed features; feedback from early users proved invaluable in shaping the tool's behavior and looks. The first prototype of the STS was released in November 1997 for in-house testing, with field tests (at the NTT) foreseen for the beginning of 1998. We plan to have a fully operational OT by the time that service observing at the VLT starts.

References

Chavan, A. M. and Albrecht, M. 1997, in ASP Conf. Ser., Vol. 125, Astronomical Data Analysis Software and Systems VI, ed. Gareth Hunt & H. E. Payne (San Francisco: ASP), 367

Johnston, M. and Miller, G. 1994, in Intelligent Scheduling, ed. M. Zweben & M. S. Fox (San Francisco, Morgan Kaufmann), 391

Astronomical Data Analysis Software and Systems VII
ASP Conference Series, Vol. 145, 1998
R. Albrecht, R. N. Hook and H. A. Bushouse, eds.

VLT Data Quality Control

P.Ballester, V.Kalicharan, K.Banse, P.Grosbøl, M.Peron and M.Wiedmer

European Southern Observatory, Karl-Schwarzschild Strasse 2, D-85748 Garching, Germany

Abstract. The ESO Very Large Telescope (VLT) will provide astronomers with the capabilities to acquire large amounts of high signal-to-noise, multi-dimensional observational data. In order to fully exploit these data, the entire chain of the observation process must be optimized by means of operational procedures and control tools. As modern observatories evolve to production units delivering data products and calibrated scientific data to the end users, new challenges arise in assessing the quality of data products.

1. Introduction

Quality control is concerned with the quantitative assessment of calibration and science data. Controlling a quantity involves a measurement procedure and the comparison of the measure to a pre-determined target value. Taking the analogy of speed control on a motorway we can describe the system as a device (radar) collecting measurements of the system performance (passing cars). Measures are compared to a target value (the speed limit). If the system identifies a discrepancy, an operator (the policeman) takes a corrective action.

In an astronomical context, values will be measured by the pipeline on raw and reduced exposures, as well as with the Astronomical Site Monitor tracking ambient conditions parameters. Target value include user requested parameters, initial performance solutions, and modeled performance.

Quality control is essentially a distributed activity. It takes place at different locations and moments during the life cycle of an observation. Target values are produced off-line and must be transfered to the on-line control system. Results are recorded and summarized for evaluation and trend analysis.

2. On-line Quality Control

2.1. Verification of Science Data

Astronomers preparing their observation with the Phase II Proposal Preparation system (P2PP) can request a range of ambient conditions, including airmass, seeing, or moon phase. The observation scheduler takes into account the ambient conditions prevailing before starting the exposure. The on-line quality control system verifies these conditions after the exposure has been realized. The observation is flagged if the ambient conditions do not match the user requested values. Additional verification is performed on the raw frames, for example pixels saturation or read-out noise.

2.2. Ambient Conditions

The Astronomical Site Monitor provides information about weather conditions and other environment parameters (Sarazin & Roddier 1990). The site monitor data are based on cyclical reading of a variety of sensors and measuring equipment and calculation of derived data. The measurements include seeing, scintillation, atmospheric extinction, cloud coverage, meteorological data, and all-sky images. These measurements are compared with the user requested values by the quality control system. Independent seeing measurements are made on raw images during the off-line quality control stage.

The image quality measured at the focal plane of the instrument is usually larger than the ASM value because of the internal seeing resulting from dome, telescope and instrument thermal effects. The QC system will therefore accumulate data and allow the correlation of external ASM seeing measurements with instrumental image quality.

2.3. Verification of calibration data

One of the direct applications of quality control is the verification of instrument performance based on the analysis of calibration exposures. Indeed, in this case we observe a reference source of known characteristics, performance parameters can be measured accurately, and the images are repeatable exposures taken in standard conditions which make them adequate for automatic processing. It will therefore be possible to check that the observation equipment is working normally by analyzing calibration or reference targets exposures.

Observations taken at ground-based observatories are affected by diverse sources of variability. The changing characteristics of optical and electronic systems and atmospheric effects make it necessary to frequently re-calibrate equipment. By performing regular monitoring of the characteristics of the calibration solutions, it will be possible to discriminate between the stable, slowly varying and the unstable components of the solutions, and therefore to learn about the characteristics of the instrument. The stable part of the calibration solution can usually be explained by physical models (Ballester & Rosa 1997).

On the on-line system, calibration data are verified against the reference solutions. A graphical window is regularly updated to display the measurements (Figure 1). The system was tested as a prototype at the New Technology Telescope (NTT).

Performance measurement is an essential step in making progress in the inherent conflict between the need for calibration data and the time available for scientific data taking. Regular monitoring makes it possible to decide which calibration data are actually required and on which timescale they need to be updated.

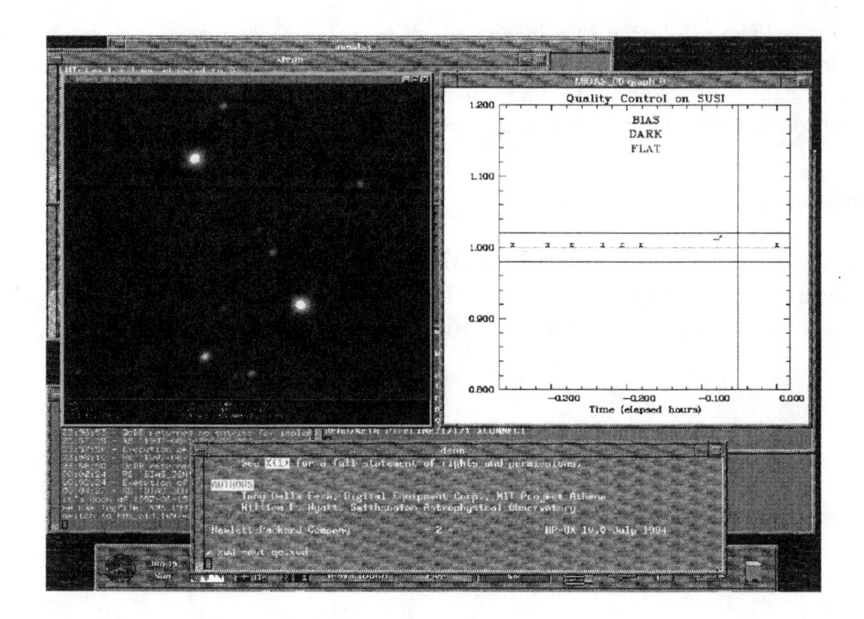

Figure 1. The prototype on-line quality control system at the New Technology Telescope

3. Off-Line Quality Control

Technical programs are scheduled by application of the instrument calibration plan, describing the type and frequency of calibration exposures required to monitor the characteristics of the observation equipment. The calibration data resulting from technical programs are delivered to the users in addition to their scientific data, and used by the Data Flow Instrument Responsibles (DFIR) to prepare master calibration solutions and monitor instrument performance.

The calibration observation blocks are pre-processed by the instrument pipeline in order to generate a preliminary calibration solution as well as quality measurements. The Data Flow Instrument Responsible is notified after the execution of the calibration observation block. The pre-processed data are retrieved from the temporary area and reduced to produce the master calibration data. After certification, the master data is included in the central archive and distributed to all local copies of the calibration database.

Exposure time calculators and other models are used as references for the instrument performance. They are also used as observation preparation tools, and made available on the Internet.

Among the parameters monitored with the off-line system are the instrument variability (e.g., flexure, drifts, throughput) and peculiarities (e.g., non-linearities, scattering and reflexions). The accuracy of the calibration solutions is verified by the DFIR and documented in the quality control report database.

For the validation of science user data, a more complete verification of the user requested parameters is made off-line, and the science data are associated with calibration solutions. The information collected during on-line verification concerning the calibration accuracies is added to the data.

References

Ballester, P., & Rosa, M., 1997, ESO Preprint 1220, in press.

Sarazin, M., & Roddier, F., 1990, "The E.S.O Differential Image Motion Monitor", Astron. Astrophys. 227, 294

Astronomical Data Analysis Software and Systems VII
ASP Conference Series, Vol. 145, 1998
R. Albrecht, R. N. Hook and H. A. Bushouse, eds.

Astro-E's Mission Independent Scheduling Suite

A. Antunes and A. Saunders
Hughes-STX, Goddard Space Flight Center, Greenbelt, MD 20707, Email: antunes@lheamail.gsfc.nasa.gov

P. Hilton
Hughes International/ISAS, 3-1-1 Yoshinodai, Sagamihara, Kanagawa 229, Japan

Abstract. The next generation of Mission Scheduling software will be cheaper, easier to customize for a mission, and faster than current planning systems. TAKO ("Timeline Assembler, Keyword Oriented", or in Japanese, "octopus") is our in-progress suite of software that takes database input and produces mission timelines. Our approach uses openly available hardware, software, and compilers, and applies current scheduling and N-body methods to reduce the scope of the problem. A flexible set of keywords lets the user define mission-wide and individual target constraints, and alter them on-the-fly. Our goal is that TAKO will be easily adapted for many missions, and will be usable with a minimum of training. The especially pertinent deadline of Astro-E's launch motivates us to convert theory into software within 2 years. The design choices, methods for reducing the data and providing flexibility, and steps to get TAKO up and running for any mission are discussed.

1. Scheduling Defined

The basic concept of scheduling is simple. You take a list of targets, and make a calendar. The rest is efficiency and bookkeeping, which is akin to saying "to make a painting, just put paint on a canvas." In reality, it is a bit more complicated. The three main factors to consider are calculated quantities (i.e., unchangeable facts of the situation), derived quantities (soft constraints, ratings of good and bad events, and politics), and capacity constraints (such as time and telemetry).

For mission scheduling, the general calculated quantities are based on the platform. For satellites, this includes: sun angle, orbital position, earth limb angle, roll angle, and others. For ground based, several analogs are day/night, latitude/longitude, and elevation angle. These have no particular weight to them, but are simply physical facts about the observing situation.

From these, we create the derived quantities, which determine whether a given observation is feasible. This includes (for satellites) issues like allowable sun condition, occultation, thermal profile, in-SAA region, bright/dark earth, ground-station contacts, and star tracker acquisition. Some of these are functions of the calculated quantities, and others are qualitative opinions based on the calculated quantities. Some derived quantities can be entirely political, and imposed from the outside (scientific priority, for example).

Capacity constraints include, first and foremost, time, generally defined as "you can only look at one thing at a time." Telemetry is a large concern for satellites, and other user-defined resources vary from mission to mission.

Tracking all of these is the principal task of scheduling, that is, to place targets such that no constraints are violated and all resources are used to maximum efficiency without overbooking. The goal is ultimately to maximize the scientific routine.

So tools to manipulate this output not just "a schedule", but also an evaluation of its overall suitability (and stability) for the problem at hand. The interface must present all the quantities above to the user when hand-editing is required. Finally, the automatic scheduling routines (ranging from simple "good time bad time" limiting, to AI routines that make sequences) must interface with the editing options available to the user.

TAKO ("Timeline Assembler, Keyword Oriented") is our multimission software suite for achieving this goal while simplifying training and operations (and thus saving time and expenses). Our goal is that TAKO will be easily adapted for many missions, and will be usable with a minimum of training. Rather than having a single monolithic program, we are building an integrated suite (similar to the FTOOLs/XANADU approach) that handles all input, parameterization, scheduling, and output. This provides flexibility both for a variety of missions, and for changes during an mission lifetime.

Further, our design is modular to allow customization of relevant sections for different missions. Feedback from the Astro-E science team and from other schedulers has been crucial in determining the necessary capabilities for this suite. The first benchmark goal for the Astro-E first edition is to be able to schedule at least 400 items over a 1-year baseline at 1 minute resolution, and to manipulate this schedule in real time without alienating the user.

2. TAKO Design

The focus for users is the intuitive GUI for interaction and scheduling. Scheduling is an intensely visual task, and much of the development time is for this GUI. The actual scheduling engine deals only with the generic time-ordered quality function and header, without having restrictions on what the header elements are. The intermediary code between this flexible GUI and generic scheduler is what makes TAKO specifically useful for satellite and observational scheduling.

For a given mission, the input file currently specifies how to custom-build single observations in a highly flexible way. It uses ASCII files for storage and output, allowing filters to be written and input/output to be integrated with other packages (including spreadsheets). Thus the entire suite can be dropped as a single unit into the mission's science operations center, with pre- and post-processors written to make the linkage.

TAKO uses standard hardware, operating systems, and compilers (i.e., any workstation with gcc and Tcl/Tk 8.0). And, TAKO is modular by design, to allow for use with different missions. Switching missions or handling updated requirements requires adjustments to a handful of the discrete modules, rather than an entire code overhaul. And much of the mission-specific specifications can be done at the user level, in the input files, rather than requiring a recompile.

To improve communication, all our documentation is produced as the modules are written, in HTML, so that the entire package design can be understood and easily accessed by Web. The design work and user manuals alike will be available from the Web. After integration, training manuals will also be available. In the absence of dense technical details within this paper, we therefore refer you to these Web pages (URL given below).

3. TAKO Implementation

For TAKO, we set several basic design precepts. The GUI is an on-screen editor that ties it all together. Keyword-value pairs are used to define data elements and relations generically. Each data element has a header, and acts like an object. This object-oriented design means that new elements can be defined on the fly, much as with a database. And, parameters can be adjusted during runtime or before each run, with saved definitions allowing mission customization.

Mission-specific code is largely in orbital prediction routines ("calculated quantities"). Derived quantities are specified in terms of these, using the keyword-value pair schema. There are four primary types of structures ("target information", "constraints", "resources", and "event list"). "Target information" is the information from the science proposals. "Constraints" is the interpretation of the science objectives into operations terms, i.e., "avoid SAA" in people-speak becomes an on-screen curve showing what times are allowed. Constraint curves are very similar to SPIKE "suitabilities" (for those familiar with HST's package). "Resources" includes items like the actual schedule (how much time is spent on target) and telemetry. And "event lists" are outputs like the mission timeline itself, lists of calibration observations, and so on.

4. TAKO Buy-In

To work for many missions, a scheduling tool must be flexible, accept post-launch changes, be able to handle many situations, and be better/faster/cheaper[1]. In the past, it took as long to customize an old tool as to write a new one, and most solutions required post-launch customization. Also, each mission required new training for new tools, often at a programmer level, increasing costs and time spent doing everything except actually scheduling. So TAKO is designed to be easily adapted for many missions, and to require a minimum of training at the user level (substituting "common sense" instead)[2].

Buy-in assumes that the software is subject-oriented rather than focusing on the specific algorithms or performance benchmarks. In short, the user comes

---

[1]NASA motto

[2]*It takes an engineer to design a car and a mechanic to fix it, but anyone can drive.* R. A. Heinlein

first. A good design is presumed to already use appropriately chosen algorithms and to achieve requisite benchmarks; what matters most is that the software not merely function, but be genuinely usable. Therefore, the problem it solves is "maximize science return"– to consistently produce good mission schedules over a long period of operations.

Looking at it from a system perspective, again focusing on the user base, we find that it must install easily and require no special software, must provide immediate functionality, and be easily integrated with familiar tools and files. From a long-term perspective, it should be flexible, adaptable, and easily modified. And, it should be evolutionary (taking the best of previous packages) rather than implementing new ideas simply to be different– there should be a familiarity to the overall methodology.

To achieve these goals, we've defined two approaches, "internal" and "external". Internally, the base-level algorithm and programming work is being done with standard professional methodologies, to select the best approach for the given precepts and implement it (nothing terribly novel there). Externally, the GUI is the predominant way the user will interact with TAKO, and is being designed by a programmer/analyst (p/a), receiving constant feedback from three different mission schedulers. Thus the interface is essentially being designed by the users, with the p/a acting as adjudicator and implementer. TAKO was designed in detail before coding began, so that ad hoc coding and integration problems are generally minimized.

Astro-E launches early in 2000. For more information, visit the Astro-E URL (http://lheawww.gsfc.nasa.gov/docs/xray/astroe) and the TAKO subpages (http://lheawww.gsfc.nasa.gov/docs/xray/astroe/tako).

Acknowledgments. We are grateful to Pamela Jenkins for scheduling advice, Larry Brown for programming support, and Glenn Miller for moral support.

Astronomical Data Analysis Software and Systems VII
ASP Conference Series, Vol. 145, 1998
R. Albrecht, R. N. Hook and H. A. Bushouse, eds.

ASCA: An International Mission

P. Hilton

Hughes International/ISAS, 3-1-1 Yoshinodai, Sagamihara, Kanagawa 229, Japan

A. Antunes

Hughes STX/GSFC Greenbelt, MD 20771

Abstract. The ASCA X-ray satellite mission involves scientists from Japan, America, and Europe. Each year more than 400 targets are observed by ASCA. The process starts with the electronic submission of a proposal and ends with the delivery of a data tape. A successful observation depends on organization within the operations team and efficient communication between the operations team and the observers. The methods used for proposing, scheduling, coordinating observations, quick-look plots, and data delivery are presented. Cooperation is the key to success in an international mission

1. Introduction

ASCA is an X-ray satellite which was launched by Japan in 1993. While the platform is Japanese, the hardware was a joint effort by US and Japanese teams. The entire processing cycle from proposal submission through data delivery is a joint effort between the two countries. In short, the ASCA project must coordinate across continents and languages. To help with the coordination, NASA has agreed to support one person on site at ISAS.

Observation time is allocated 50% Japan only, 25% US-Japan, 15% US only, and 10% ESA-Japan. The ESA time is granted by Japan to encourage international collaboration. A typical observation is for 40 kiloseconds, or about one day of satellite time. Some pointings are as short as 10 ksec; some are as long as 200 ksec. Over the course of a year over 400 observations are made by ASCA, many of these are coordinated with other astronomical platforms.

In order to produce science, these 400 observations must be proposed, coordinated and scheduled, and observed. Then, quicklook and production data must be delivered to the scientists.

2. Announcement and Proposals

The entire process begins with a coordinated announcement by ISAS, NASA, and ESA, 90 days before the deadline. Requests for proposals are made in the vernacular of each country. A proposal has two parts: a cover page and scientific justification. The cover page includes contact information about the proposer and co-investigators, the abstract, target data, and observational requirements. English is used for the cover page while either English or Japanese is acceptable for the scientific justification in Japan.

ASCA is one of the first missions that required electronic submission of the cover page. Electronic submission of proposals allows the database to be populated automatically. The database then serves as the common reference point for each country's efforts, and thus the e-mail proposals are used throughout the observation cycle.

In order to get 100% compliance with electronic submission, we had to provide multiple avenues and make it easy for users. Proposers can construct their cover page via a form on the World Wide Web or a blank e-mail form. About 70% of the Japanese proposals submitted in 1996 used the blank form, while 70% of the US proposers used Web forms. The Web version has many pull-down menus and on-line help. The e-mail method continues to be supported because it still has an active user base.

The Remote Proposal System (RPS) developed at Goddard for preparing, checking, and sending proposals is used. RPS, which is used by other missions including XTE and ROSAT, is available over the Web. RPS is used to check each proposal for format accuracy before it is accepted.

3. Re-order/Review, and Approval

After the proposal deadline, invalid proposals such as duplicates or withdrawn ones are eliminated. Proposals are re-ordered according to category and postal code. A five digit number is assigned to each proposal. This number starts with the AO, and the PI's country code. The proposal number is the key reference through the observation and data distribution.

Formal panels in Japan, America, and Europe review the proposals. A merging meeting convenes to decide the final target list. Email is then used to notify the PIs that their target has been selected. The list of accepted targets is also made public via the Web. This provides both an active and a passive electronic disbursement of the final selection, and reduces miscommunication and communications breakdowns. Selected targets are assigned a sequence number which serves as the main identifier of the observation.

4. Long-term Scheduling

ASCA uses a schedule planning program called SPIKE which was developed at MIT for the Hubble project. The details of SPIKE are beyond the scope of this paper. In general, we plan the time criticals according to the proposal requests and we use a sun angle constraint of +/- 25 degrees.

SPIKE produces a one-year schedule which is divided into weekly periods. One output file is a nice readable ASCII file, which is run through a Perl script to make an HTML file for use on the Web. The long-term schedule displays the list of targets and the planned or observed date. Each target has a hyperlink to its proposal summary.

The proposal summary displays the abstract and the successful targets from the proposal. Comments are listed when they are included with the proposal. There are e-mail hyperlinks to the PI and primary Co-I. This summary serves both the community and the operations team.

5. Co-ordination

Sometimes a PI will request observation time on more than one platform, or with a ground station. Since the observation cycles of each platform are different and in general independent, it is the responsibility of the PI to note that a co-ordinated observation is required.

Then, it becomes necessary to communicate with other schedulers. Usually, an e-mail conveys the list of common targets and a general schedule is discussed. Email has low overhead and fast response time. The primary disadvantage, of course, is that it is ultimately people-based. No mechanism exists (or has ever been proposed) to unify satellite scheduling. Once a coordinated observation is approved, it is solely the responsibility of the schedulers (in consultation, as needed, with project PIs) to make sure the coordination happens.

There is a network of schedulers that was organized by Dave Meriwether when he was at EUVE. Unfortunately, we have not kept the system going after Dave's departure from EUVE. We need to get this network back on track.

6. Short-term Schedule

Short-term schedules of 10-13 days are made about one month in advance. ASCA uses a short-term schedule editor called Needle that was developed in-house at GSFC/ISAS. The primary considerations for a short term schedule are time-critical requests, unofficial time critical requests, bright earth, and star tracker.

After the short term schedule is made, some post processing of the file to be distributed takes place. The latest e-mail address and institute name of the PI are added for the convenience of the contact scientist. Also, relevant comments from the proposal are added. This file is sent to the operations teams in both Japan and the US via e-mail.

After the schedule has been finalized, PIs are notified in English about their upcoming observation. This gives the PI a chance to prepare for the observation in advance of hearing from the contact scientist. By the time the contact scientist notifies the PI, we will have completed several other short term schedules. This means that late notification of a PI could affect the scheduling process. Early notification seems to work in the best interest of both the operations team and the PI.

After the targets have been scheduled and the PIs have been notified, changes are sometimes required. It is important to notify a PI when an observation schedule is changed. ASCA TOOs are subjected to review before approval by two team members in Japan and two in the US. Rapid communications via e-mail or fax means that a gamma ray burst can be reported, discussed, scheduled, and observed by ASCA within 36 hours.

7. Observation and Quick-look Plots

A week or two prior to an observation, the PI is contacted for detailed arrangements. The Japanese contact Japanese PIs, while the US team contacts US and ESA PIs. Email is the primary method of contact.

The ASCA operations ground control center is located at Kagoshima Space Center (KSC) on the south island of Japan. KSC is operated six days a week. When the station is manned, it is possible to do a quick-look plot of data. These plots are distributed via fax.

Electronic delivery would be a useful method to pass the quicklooks along to the PIs. However, the current system is not set up for this, and so a fax machine is used.

8. Data Delivery

After the observation has been made, first reduction files (frf) are sent to ISAS and to Goddard. Japanese PIs are able to process their own data before the CDs are sent out. Finally, the observation data tapes are made and the data are sent to Goddard for storage and distribution. Data are placed on a CD-ROM and mailed to the PI. Because the data have been tagged with the same ID as the schedule products, cross-correlating is trivial. And, the community is automatically aware (by reading the schedule) what products are available in the archive. Data are made public after 1-2 years. This means that the Web-based notification we provide also serves as a "promotional teaser" for upcoming data. The electronic transfer process ensures that data are not lost, misplaced, or wrongly labeled.

The process does require manipulation of the data at different stages, and some manual intervention. A more automated pipeline would be an asset. The most important aspect is that the entire process is transparent to the user. The user sends in an e-mail or Web proposal, and later receives their data (as a CD in the mail, or electronically from the archive.)

9. Summary

ASCA is a very successful mission. Part of the success of the ASCA mission is due to the rapid communication channels that are available to us. We require that all proposers use a standard e-mail format in English. We use the language of each country, when appropriate. English is used as the common language.

When we need to coordinate with other schedulers, we receive timely support. We use the Internet, e-mail, and faxes for communications within the ASCA community. The ultimate strength of the electronic process is that it facilitates better communication, and eliminates the geographic boundaries of this project.

The true reason for ASCA's success is the cooperation of the guest observer teams in Japan, the US, and Europe. We can have the best communications system, but without the cooperation of the operation teams, we would have little.

Astronomical Data Analysis Software and Systems VII
ASP Conference Series, Vol. 145, 1998
R. Albrecht, R. N. Hook and H. A. Bushouse, eds.

Achieving Stable Observing Schedules in an Unstable World

M. Giuliano

Space Telescope Science Institute, 3700 San Martin Drive Baltimore, MD 21218 USA

Abstract. Operations of the Hubble Space Telescope (HST) require the creation of stable and efficient observation schedules in an environment where inputs to the plan can change daily. A process and architecture is presented which supports the creation of efficient and stable plans by dividing scheduling into long term and short term components.

1. Introduction

Operations of the Hubble Space Telescope (HST) require the creation and publication of stable and efficient observation schedules in an environment where inputs to the plan can change daily. Efficient schedules are required to ensure a high scientific return from a costly instrument. Stable schedules are required so that PIs can plan for coordinated observations and data analysis. Several factors complicate creating efficient and stable plans. HST proposals are solicited and executed in multi-year cycles. Nominally, all the accepted proposals in a cycle are submitted in a working form at the beginning of the cycle. However, in practice, most proposals are reworked based on other observations, or updated knowledge of HST capabilities. Another source of instability is due to changes in HST operations. As HST is used the capabilities of some components degrade (e.g., the solar panel rotation engines), and some components perform better than expected (e.g., decreases in acquisition times). Changes in component performance lead to different operation scenarios and different constraints on observations. A plan created with one set of constraints may no longer be valid as operations scenarios are adjusted based on an updated knowledge of HST capabilities. A final source of plan instability is that the precise HST ephemeris is not known more than a few months in advance. As a result highly constrained observations cannot be scheduled with accuracy until the precise ephemeris is known. Given these factors, it is not possible to create a single static schedule of observations. Instead, scheduling is considered as an ongoing process which creates and refines schedules as required.

A process and software architecture is presented which achieves stable and efficient observation schedules by dividing the task into long term and short term components. The process and architecture have helped HST obtain sustained efficiencies of over 50 percent when pre-launch estimates indicated a maximum of 35 percent efficiency. The remainder of the paper is divided as follows. Section 2 presents the architecture. Section 3 discusses observation planning as a process and discusses more details on the long range planner as implemented by

Figure 1. Plan windows for observations 1-6 in weeks 1-8. Week 4 windows are highlighted.

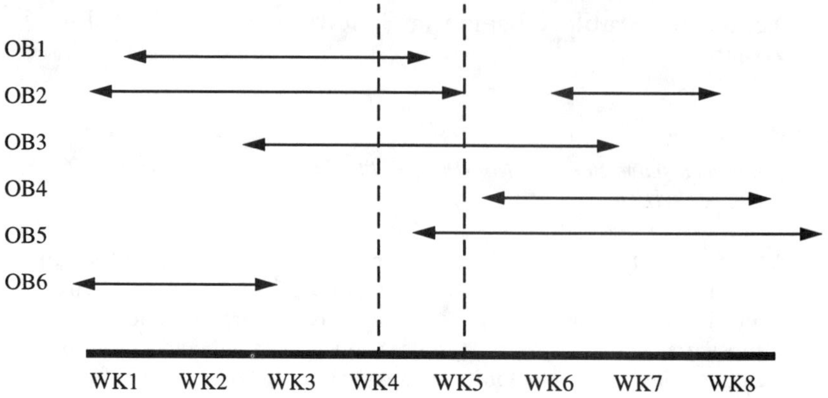

the SPIKE software system (Johnston & Miller 1994). Section 4 evaluates the approach by comparing the system to other approaches.

2. An Architecture for Creating Efficient and Stable Schedules

Efficient and stable observation schedules are created by dividing the scheduling process into long range planning and short term scheduling. The long range planner creates approximate schedules for observations and handles global optimization criteria, and stability issues. The short term scheduler produces precise one week schedules using the approximate schedule produced by the long range planner as input.

The long range planner creates 4-8 week plan windows for observations. A plan window is a subset of an observation's constraint windows, and represents a best effort commitment by the observatory to schedule in the window. Plan windows for different observations can be overlapping. In addition the windows for a single observation can be non-contiguous. Figure 1 shows sample plan windows for six observations. Constraint windows are long term limitation as to when an observation can be executed due to physical and scientific constraints. Constraint windows include sun avoidance, moon avoidance, user window constraints, phase constraints, roll constraints, guide star constraints, and time linkages between observations. The long range planner picks plan windows based on balancing resources (e.g., observing time, observations which can hide the SAA), and optimizing observation criteria (e.g., maximizing orbital visibility, maximizing CVZ opportunities, maximizing plan window size, minimizing stray light, minimizing off-nominal roll).

The short term scheduler builds efficient week long observation schedules by selecting observations which have plan windows open within the week. The short term scheduler is responsible for keeping track of which observations have already been scheduled in past weeks.

Figure 1 shows a schedule with 6 observations with windows in weeks 1-8. The figure shows that observations 1,2,3, and 5 are potential candidates for scheduling in week 4.

Figure 2: Process data flow

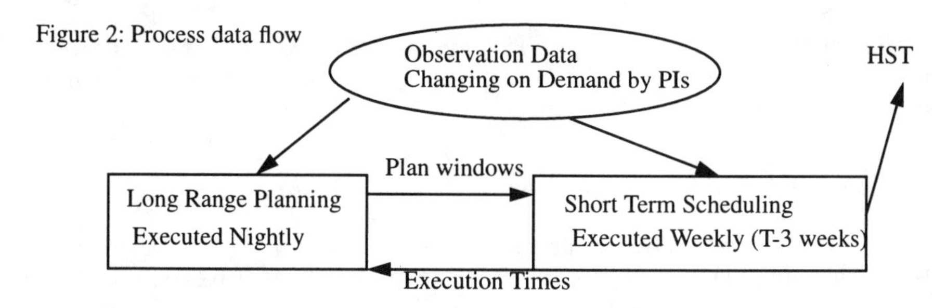

3. Scheduling as a Process

The scheduling process, as illustrated in Figure 2, is an ongoing effort. The long range planner is executed nightly to incorporate the changes made to observing programs during the day. Each week a short term schedule is created which is executed on HST 3 weeks later. Although the long range planner is executed nightly, it must ensure that the resulting plan is stable.

Stability is achieved by considering the long range planner as a function which maps a set of input observing programs, search criteria, and a previous long range plan into a new long range plan. The long range planner partitions observations into those which have windows assigned in the input LRP and those which do not have input windows. In general, the scheduler will assign new windows to those observations which are not scheduled in the input plan and will pass through windows for observations which are scheduled in the input plan. The system assigns windows for observations, which are not scheduled in the input plan, in two steps. First, the system uses user defined criteria to greedily find the best window for each observation. Second, a stochastic search is used to adjust the resource levels.

There are two cases where the plan window for an observation which was planned in the input LRP is changed. First, the system has a special case for linked observations (e.g., OB1 after OB2 by 10-20 days). When the first observation of a link set is executed the system will automatically adjust the plan windows for subsequent observations in the link set based on the actual time the first observation scheduled. A second case occurs when the constraint windows for an observation change. The system measures the percentage of the plan window in the observations current constraint window. If the percentage is below a threshold then, based on a user flag, either the observation is re-planned from scratch or no plan window is written and a status flag is set in the output plan. If the percentage is above a threshold (but some of the plan window is no longer in the constraint window) then the plan window is reduced so that it overlaps with the plan window. The idea is that a minor change in an observations constraint window due to PI adjustments or changes to HST operations will not disrupt the long range plan.

The long range planner must deal with several anomalous cases. First, it may be possible that loading an observation into the system causes a crash or an error. A second anomaly occurs when the input products for an observation are not ready. In either case the problem is caught by the system and any existing plan windows are written to the output plan. A status field is marked so that the anomaly can be investigated. This approach ensures that the LRP process is stable. Problem observations do not drop out of the LRP and an error in one observation does not halt the entire process.

4. Evaluation and Comparison with other Approaches

Using plan windows to communicate between long range planning and short term scheduling has many advantages which are highlighted by contrasting this approach with others. Alternatives are: 1) Short term schedule on demand one week at a time. Do not long range plan; 2) Short term schedule the whole cycle in advance. Do not long range plan; 3) Long range plan to non overlapping bins. Short term schedule using the bin schedule as input. Short term scheduling on demand does not meet the requirement that PIs be informed in advance of the approximate time an observation will be executed. In addition, the approach may run into resource problems as greedy algorithms can schedule all the easy observations first. Short term scheduling the whole cycle in advance does, in principle, tell PIs when observations will be scheduled. However, the schedule would be very brittle. Changing a single observation would require the whole cycle to be rescheduled. The net result would be an unstable plan and lots of rescheduling work for operations staff. Short term scheduling the whole cycle at one time would also result in a combinatorial explosion in the search space effectively preventing optimization of the schedule. Long range planning to week long bins was the approach used in the first four HST observation cycles. The approach was not successful. If the long range planner filled week bins to capacity (or less) then the resulting short term schedules would be inefficient as the short term scheduler would not have the right mixture of observations to select from. If the long range planner oversubscribed week bins then the resulting short term schedules would be efficient. However, the resulting long range plan would be unstable as unscheduled observations in a week bin would have to be rescheduled. This approach would only be feasible if the long range planner knows enough of the short term scheduler constraints to produce week bins with the right mixture of observations.

The plan window approach has advantages over the other approaches. First, it supports the creation of efficient short term schedules without oversubscribing the long range plan. Efficient week long schedules can be created because overlapping plan windows provide a large pool of observations to select from. A second advantage of this approach is stability. It helps to insulate the long range plan from changes in observation specifications and HST operations. A third advantage is that it divides the problem into tractable components. The short term scheduler handles creating efficient schedules while the long range planner handles global resource optimization and stability.

References

Johnston, M., & Miller, G., 1994, in Intelligent Scheduling, eds. Zweben, M., Fox, M. (San Francisco: Morgan Kaufmann), 391, 422

Astronomical Data Analysis Software and Systems VII
ASP Conference Series, Vol. 145, 1998
R. Albrecht, R. N. Hook and H. A. Bushouse, eds.

Data Analysis with ISOCAM Interactive Analysis System — Preparing for the Future

S. Ott[1] and R. Gastaud[2], A. Abergel[3], B. Altieri[1], J-L. Augueres[2], H. Aussel[2], J-P. Bernard[3], A. Biviano[1,4], J. Blommaert[1], O. Boulade[2], F. Boulanger[3], C. Cesarsky[5], D.A. Cesarsky[3], V. Charmandaris[2], A. Claret[2], M. Delaney[1,6], C. Delattre[2], T. Deschamps[2], F-X. Désert[3], P. Didelon[2], D. Elbaz[2], P. Gallais[3], K. Ganga[7], S. Guest[1,8], G. Helou[7], M. Kong[7], F. Lacombe[9], D. Landriu[2], O. Laurent[2], P. Lecoupanec[9], J. Li[7], L. Metcalfe[1], K. Okumura[1], M. Pérault[3], A. Pollock[1], D. Rouan[9], J. Sam-Lone[2], M. Sauvage[2], R. Siebenmorgen[1], J-L. Starck[2], D. Tran[2], D. Van Buren[7], L. Vigroux[2] and F. Vivares[3]

[1]*ISO Science Operations Centre, Astrophysics Division of ESA, Villafranca del Castillo, Spain*

[2]*CEA, DSM/DAPNIA, CE-Saclay, Gif-sur-Yvette, France*

[3]*IAS, CNRS, University of Paris Sud, Orsay, France*

[4]*Istituto TESRE, CNR, Bologna, Italy*

[5]*CEA, DSM, CE-Saclay, Gif-sur-Yvette, France*

[6]*UCD, Belfield, Dublin, Ireland*

[7]*IPAC, JPL and Caltech, Pasadena, CA, USA*

[8]*RAL, Chilton, Didcot, Oxon, England*

[9]*DESPA, Observatoire de Paris, Meudon, France*

Abstract.

This paper presents the latest developments in ISOCAM data analysis with the Interactive Analysis System for ISOCAM (CIA).[10] The main use of the system is now to improve the calibration of ISOCAM, the infrared camera on board the Infrared Space Observatory (ISO) and to perform its astronomical data processing.

We review the algorithms currently implemented in CIA and present some examples. We will also outline foreseen changes to accommodate future improvements for these algorithms.

[10]CIA is a joint development by the ESA Astrophysics Division and the ISOCAM Consortium. The ISOCAM Consortium is led by the ISOCAM PI, C. Cesarsky, Direction des Sciences de la Matiere, C.E.A., France.

1. Introduction

ESA's Infrared Space Observatory (ISO) was successfully launched on November 17th, 1995.[11] ISO is an astronomical, three-axis-stabilized satellite with a 60-cm diameter primary mirror (Kessler et al. 1996). Its four instruments (a camera, ISOCAM, an imaging photo-polarimeter and two spectrometers) operate at wavelengths between 2.5-240 microns at temperatures of 2-8 K.

ISOCAM takes images of the sky in the wavelength range 2.5 to 18 microns (Cesarsky et al. 1996). It features two independent 32 x 32 pixel detectors: the short-wavelength channel (2.5 to 5.5 microns), and the long-wavelength channel (4 to 18 microns). A multitude of filters and lenses enable the observer to perform measurements at different wavelengths, with different fields of view or with polarizers.

The development of CIA was started in 1994, its main goals being to calibrate ISOCAM, to provide the means to perform any sort of investigation requested for problem diagnostics, to assess the quality of ISOCAM pipeline data products, to debug, validate and refine the ISOCAM pipeline and to perform astronomical data-processing of ISOCAM data. The overall system now represents an effort of 20 man-years and comprises over 1300 IDL and 350 FORTRAN, C, and C++ modules and 30 MB of calibration data.

At the moment CIA is distributed to experienced users upon approval; currently it is used at 30 authorized sites. However, this policy will change with the end of ISO operations in summer 1998. Thereafter CIA V3.0 will be made generally available for the astronomical community.

2. Steps for ISOCAM Data Processing

At the moment, the following steps for astronomical ISOCAM data processing are generally performed:

1. data preparation (slicing, generation of CIA prepared data structures)
2. cross-talk correction (for the short-wave detector only)
3. deglitching (removal of cosmic ray hits)
4. dark current subtraction
5. transient correction (compensation for flux-dependent, short timescale variation of the detector response)
6. flat-fielding
7. averaging
8. configuration dependent processing
   (a) raster observations: mosaic generation
   (b) CVF observations: spectra selection and display
   (c) staring and beam-switch observations: image generation
   (d) polarization observations: generation of mosaics for polarizers and hole
9. interfacing with non-ISO data products

---

[11] ISO is an ESA project with instruments funded by ESA member states (especially the PI countries: France, Germany, the Netherlands and the United Kingdom) and with the participation of ISAS and NASA.

3. Review of Selected Steps for ISOCAM Data Processing

3.1. Deglitching and Transient Correction

CIA contains various algorithms for deglitching and transient correction. Multiresolution median and temporal median filtering algorithms are most commonly used for deglitching. Similarly the most commonly used algorithms for transient correction are the inversion-method and exponential fitting. The resulting quality depends on the amount of data available. Therefore we will try in the future to treat a whole revolution instead of single observations.

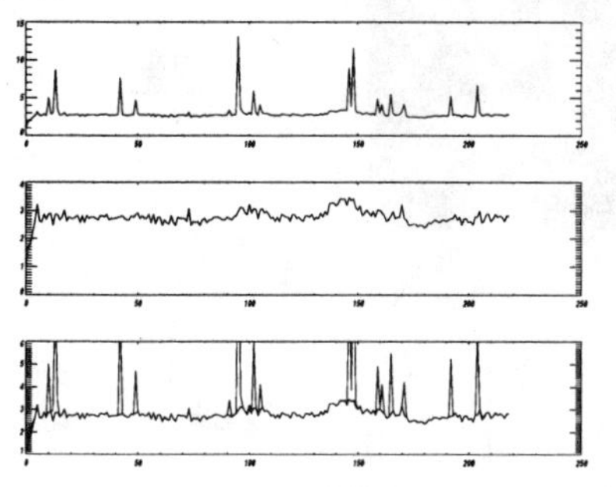

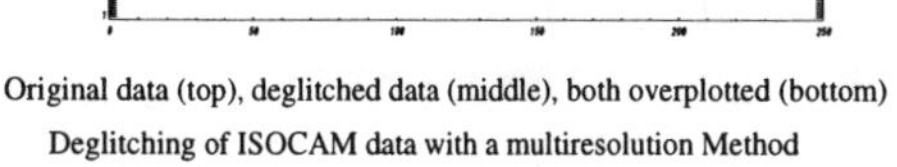

Original data (top), deglitched data (middle), both overplotted (bottom)
Deglitching of ISOCAM data with a multiresolution Method

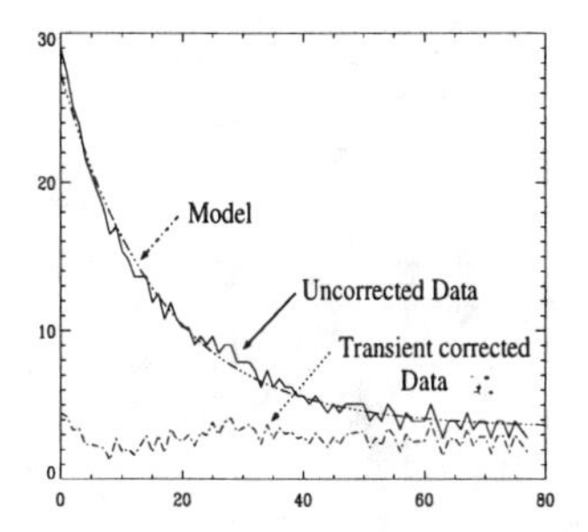

Transient correction of ISOCAM data by exponential fitting

Figure 1. Effects of deglitching and transient correction

3.2. Flat-Fielding

Flat-fielding with library-flats and automatic flat-fielding using the raster observation being analysed are both implemented in CIA. However, the lens wheel does not always come to exactly the same position. When the wheel position for the observation differs from the wheel position reached when the flat-field was obtained, this affects the quality of the final image. Therefore we hope to implement a correction for this effect (lens-wheel jitter) in the flat-field algorithm.

3.3. Dark Current Subtraction

Various algorithms can be used for the dark current subtraction. Currently implemented are the dark correction with library or user supplied darks (e.g., handover darks) as well as a second order dark correction. Studies have shown that the dark can be better predicted as a function of the observation time or by scaling with a reference dark. It is planned to implement these new dark corrections in near future.

3.4. CVF Observations: Spectra Selection and Display

CIA users can play with a 2 dimensional spectrum. From the displayed image a pixel can be selected : the spectrum of this pixel will be displayed. From the spectrum a wavelength can be selected and the image will be displayed at this wavelength.

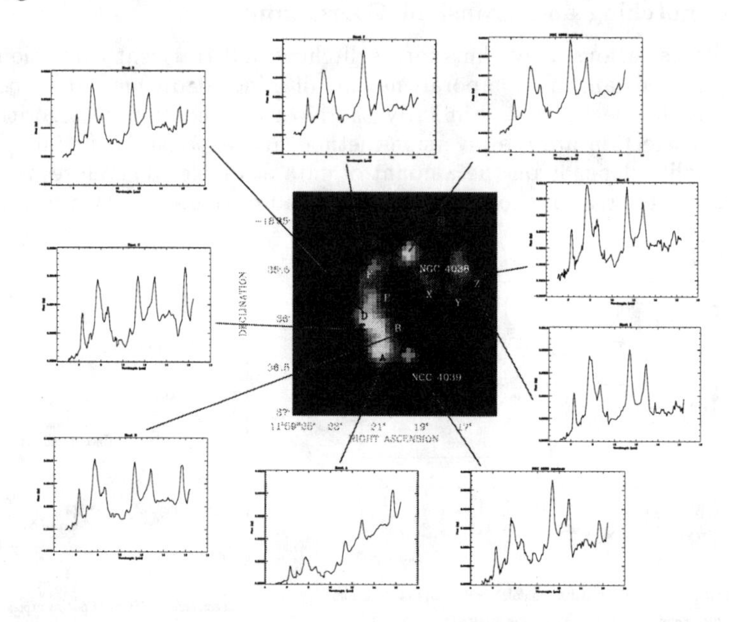

Figure 2. CVF observations: spectra selection and display

3.5. Interfacing with Non-ISO Data Products

CIA users can overlay a contour plot of one image on a colour display of another. The two input images can come from different instruments, have different size and resolution — the only requirement is that there is astrometry information associated with each image and that they overlap spatially. The images can be shifted and annotated.

References

Kessler, M., Steinz, J.A. et al. 1996, A & A, Vol. 315, L27

Cesarsky, C., et al. 1996, A & A, Vol. 315, L32

Delaney, M. ed., ISOCAM Interactive Analysis User's Manual, ESA Document

Ott, S., et al. 1997, "Design and Implementation of CIA, the ISOCAM Interactive Analysis System", in ASP Conf. Ser., Vol. 125, Astronomical Data Analysis Software and Systems VI, ed. Gareth Hunt & H. E. Payne (San Francisco: ASP), 34

Siebenmorgen, R. et al. ISOCAM Data User's Manual, ESA Document, Reference SAI/95-221/DC

Starck, J-L., & Murtagh, F., et al. Image Processing and Data Analysis — The Multiscale Approach, Cambridge University Press

Astronomical Data Analysis Software and Systems VII
ASP Conference Series, Vol. 145, 1998
R. Albrecht, R. N. Hook and H. A. Bushouse, eds.

The Interaction of the ISO-SWS Pipeline Software and the ISO-SWS Interactive Analysis System

E.Wieprecht[1,2], F.Lahuis[1,3], O.H. Bauer[2], D.Boxhoorn[3], R.Huygen[4], D.Kester[3], K.J.Leech[1], P.Roelfsema[3], E. Sturm [2], N.J.Sym [3] and B.Vandenbussche[1,4]

Abstract. We describe the ISO SWS Pipeline software and ISO SWS Interactive Analysis system (IA^3) on different hardware platforms. The IA^2 system design (Roelfsema et al. 1993) has been reviewed and major parts of the concept have been changed for the operating IA^3 system.

The pipeline software is coded in FORTRAN, to work within an environment designed by the European Space Agency (ESA). It is used for bulk processing of ISO data without human interaction, the final product being distributed to the observers. The pipeline s/w is designed in a modular way, with all major steps separated in software sub-modules.

The IA^3 system is set up as a tool box in the Interactive Data Language environment (IDL- a trademark of Research Systems Inc. (RSI)). IA^3 has been developed to fulfill tasks like debugging of pipeline software, monitoring and improve instrument performance and related software, as well for scientific analysis of SWS data. The FORTRAN pipeline modules are included in this system. Thus, it is possible to execute the pipeline step by step within the IA^3 system. The whole IA^3 system is controlled by a configuration control system.

1. ISO and the Short Wavelength Spectrometer SWS

The Infrared Space Observatory (Kessler et al. 1996) is a satellite observatory carrying out astronomical observations in the wavelength range from 2 to 200 μm using a telescope with a primary mirror of 60 cm diameter. The Short Wavelength Spectrometer (SWS) is one of the four instruments of ISO. The spectrometer covers the wavelength range 2-45 μm by two independent gratings and two Fabry Perots. A spectral resolution between 1000 and 20000 can be obtained.

---

[1]ISO Science Operations Centre, ESA SSD, Villafranca,28080 Madrid,Spain

[2]Max Planck Institut für extraterrestrische Physik, Giessenbachstrasse 1, D-85748 Garching, Germany

[3]Space Research Organisation of the Netherlands,PO 800, 9700AV, Groningen, The Netherlands

[4]Katholieke Universiteit Leuven , Institut voor Sterrenkunde, Celestijnenlaan 200 B, B-3001 Heverlee, Belgium

2. The ISO Downlink and overall Data Reduction

The ISO satellite is an observatory on an highly elliptical orbit. The apogee is about 71 000 km, the perigee about 1 000 km. One period is about 24 hours. Thus ISO crosses the radiation belt every day. This limits the actual observation time to 16 hours per day. Because ISO has no on board data storage it is necessary to establish permanent ground contact during observation periods. This has been guaranteed by the ground stations Villafranca de Castillo (Spain) and Goldstone (USA). The raw data are telemetred down and directly stored in an archive. In parallel the data are processed in real time by the RTA (Real Time Assessment)/QLA (Quick Look Analysis) system to monitor the ongoing observation.

After collecting the data from one revolution, where different instruments have been active, the data are processed by the pipeline to produce products which are stored in the archive and distributed to the observer on CD.

After extraction from the archive, it is possible to use the $\texttt{IA}^3$ system to run the SWS specific pipeline step by step for calibration purposes by the Instrument Dedicated Teams (IDTs) at Vilspa. Also the observer might use this s/w by visiting one of the data centers which are running $\texttt{IA}^3$. Here the most recent software is supplied together with the knowledge of the instrument experts.

3. The ISO Pipeline System

The ISO pipeline has been designed for bulk processing. Thus special requirements for robustness had to be accomplished. Due to coding standards it has been required to code the system in FORTRAN. As operation system VAX/VMS has been defined. The overall design of the SWS specific software has been in a modular way to guarantee the step by step processing within the $\texttt{IA}^3$ system.

The ISO pipeline software can be split into two parts. The instrument independent part which has been done by ESA together with the overall design, and the instrument specific part designed and coded by the instrument teams. The pipeline software has been under configuration control and responsibility of ESA. The SWS instrument team developed a special configuration control system which allows the $\texttt{IA}^3$ developers to cooperate in a controlled environment, even they are located at different sites with different h/w (Huygen,R. & Vandenbussche,B.). Because $\texttt{IA}^3$ development and SWS instrument specific software development are related to each other, the SWS instrument specific software has been covered by this system in a special manner.

3.1. SWS Derive Standard Processed Data

After the instrument independent processing, where basically the telemetry file is split into a number of files in a certain time sequence, instrument, for every observation the DERIVE SPD step is following. Here the data are processed in blocks of one reset interval : 1.collect data, 2.determine usable data range, 3.correct reset pulse after effects, 4.measurement linearisation, 5.cross talk removal, 6.glitch recognition/removal, 7.write Trend Analysis glitch table (to database), 8.compute grating/FP positions, 9.slope calculation, 10.convert to voltages, 11.assign wavelength, 12.write product. In almost every step FITS calibration files are used. The products are FITS SPDs which contain for all 52 detectors a wavelength and a slope in μV/reset with some add on informations which are represented in bit code for each reset interval. The second result is a file containing further glitch information.

3.2. DERIVE Auto Analysis Result

This part of the pipeline covers the calibration. The modular design is even more transparent because every step is called with the complete data set of the whole observation to be processed : 1.determine observation specific parameters, 2.dark current subtraction, 3.responsivity calibration, 4.flux conversion, 5.velocity correction, 6.preparation of Auto Analysis Result, 7.Write product. Again we write to database tables during step 2 (dark currents) and step 4 (the result of the internal flux calibrator). The AAR product is the starting point for the Infrared Spectroscopy Analysis Package (Sturm et al. 1998), but within IA^3 all further steps of scientific data processing can be taken as flatfielding, sigma clipping, rebinning, unit conversions, etc.

4. The SWS Interactive Analysis System

Originally it was intended to have a system which fulfils the requirements : - Debugging of pipeline software. Especially for this requirement it was necessary to design an interface between IDL and FORTRAN which considers the characteristics of different hardware platforms/operating systems. - Analyse the performance of SWS. Regular procedures which are executed to check the behaviour of the instrument has been included. - Trouble shooting. - Determine the calibration parameters. An archive of calibration files and their time dependencies is available within IA^3 . The procedures are described and the modules which has been used are included. - Trend Analysis. Certain tools to study the trends of a lot of instrument parameters (as instrument temperatures) and data (as results of internal calibrator checks).

But after the development phase it turned out that IA^3 was used more and more as a very handy tool for Scientific Analysis of SWS data and more effort has been spent to add tools specifically for data analysis. Now it is THE SWS data analysis software.

4.1. Design Aspects and Configuration Control

IA^3 is designed to run on work station class machines (DEC Alpha, Sun Sparc, etc.) under different operating systems (VAX/VMS, Alpha/AXP, HP-UX, SunOS, etc.). It has been decided to set up the IA^3 system on IDL v3.6.1. Because IDL is hardware independent only the specific handling of the FORTRAN code had to be considered.

The SWS configuration control system is not only a configuration control system. It takes also care of system specific compiling, linking, treating object libraries, treating sharable images of the FORTRAN and C code. It is also controls the regular online HELP system update which is an add on to the standard IDL help.

4.2. The IA^3 system as a tool box

IA^3 is set up as a tool box running under the IDL. Although IA^3 is presented as an user-oriented analysis system, it is actually designed as a programmer-oriented environment. Thus programmer freedom and flexibility is woven into the system wherever possible. This way it is possible to use the full power of IDL plus instrument specific features. The major part of IA^3 is keystroke driven. But for many modules which fulfill a certain high level task proper WIDGET GUIs have been coded. IA^3 contains a large number of data structure definitions representing SWS data in a form closely to the SWS standard products/files. As described it is easy to run the pipeline step by step for various reasons. For most of the pipeline steps the system contains at least one alternative which might be highly interactive or simply an improved pipeline module which has been given to the IA^3 system for test. Sometimes even pure IDL might do the job. Pipeline FORTRAN modules are accessed via an IA^3 shell script with the input data structure and necessary calibration file data structures, the structure elements are converted to FORTRAN77 arrays and the parameter are passed to the FORTRAN pipeline module using a special shell procedure which solves the machine dependencies of the IDL CALL_EXTERNAL procedure (library names, entry points, etc.). The user can specify the calibration files they want to use before starting the pipeline command by a GUI select calfile tool. Here the user specifies to use the actual pipeline, test or a self defined calibration file. But the user has also the possibility to specify calibration files on the command line level within the call of the pipeline module.

5. Summary and further plans

The pipeline system has reached version 6.2 and performs well for SWS. Improvements are under development. The DERIVE AAR step contains a few parts where it is recommended to use IA^3 tools. The DERIVE AAR part has been recoded into IDL and it is intended to use this code also within the standard pipeline (version 7.0). It is also planned to incorporate the DERIVE_AAR software together with all relevant data reduction IA^3 tools to a subset IA^3 system which is basing purely on IDL and gives significant support of scientific work on SWS data. This subset will be distributed. Finally it is under investigation to switch to IDL version 5.

References

Huygen.R & Vandenbussche B. 1997, "A Configuration Control and Software Management System for Distributed Multi-platform Software Development", in ASP Conf. Ser., Vol. 125, Astronomical Data Analysis Software and Systems VI, ed. Gareth Hunt & H. E. Payne (San Francisco: ASP), 345

Kessler,M.F. et al. 1996, A&A 15 The Infrared Space Observatory (ISO) mission

Lahuis, F. 1997, "ISO SWS Data Analysis", this volume

Roelfsema,P.R. et al. 1993, "The ISO-SWS Off-Line System", in ASP Conf. Ser., Vol. 52, Astronomical Data Analysis Software and Systems II, ed. R. J. Hanisch, R. J. V. Brissenden & Jeannette Barnes (San Francisco: ASP), 254

Sturm, E. et al. 1998,"The ISO Spectral Analysis Package ISAP", this volume

Astronomical Data Analysis Software and Systems VII
ASP Conference Series, Vol. 145, 1998
R. Albrecht, R. N. Hook and H. A. Bushouse, eds.

The Phase II Language for the Hobby⋆Eberly Telescope

Niall I. Gaffney

Hobby⋆Eberly Telescope, RLM 15.308 University of Texas, Austin, TX 78712, Email: niall@astro.as.utexas.edu

Mark E. Cornell

McDonald Observatory, RLM 15.308 University of Texas, Austin, TX 78712, Email: cornell@puck.as.utexas.edu

Abstract. We describe our recent software effort to enable queued observations on the Hobby⋆Eberly Telescope. For the past year, we have worked on development and implementation of a "Phase II" style observation submission and execution system. We have developed a language in which investigators can describe their observations, a simple method for submission, and tools to execute observations. We discuss the special features of this language that were implemented to accommodate some of the unique characteristics of the Hobby⋆Eberly Telescope.

1. Introduction

The Hobby⋆Eberly Telescope (HET) (Sebring & Ramsey 1994) is a 9.2 meter optical telescope currently undergoing scientific commissioning in west Texas. The telescope is operated by McDonald Observatory on behalf of the HET consortium[1]. Once fully operational, this telescope will operate in a queue scheduled, service observing mode 85% of the time.

What distinguishes the HET from other optical telescopes is its fixed elevation of 55°. The telescope acquires and tracks an object by rotating in azimuth and moving an Arecibo-style tracker mechanism across the focal surface. While this procedure allows the HET to access over 70% of the sky over the course of the year, each object is only observable for approximately 1 hour once or twice a night. We therefore maximize the productivity of this telescope with queue scheduled service observing. To facilitate queue scheduling, we have constructed a scripting language in which investigators can describe their observations in familiar astronomical terms. Using these observing scripts, the HET Operations Staff can build a queue of observations and then execute the observations as required by the investigator.

[1]The HET consortium members are The University of Texas at Austin, The Pennsylvania State University, Stanford University, Ludwig–Maximilians–Universität München, and Georg–August–Universität Göttingen.

2. The Phase II Language

Because of the wide variety of computers in use by our investigators, we needed to develop a platform–independent Phase II system. We chose to use a text-based system of keywords and parameters for the language. These keywords are contained within modular "templates", each of which represent a simple observational concept. These templates are:

- Summary template – Investigators' contact information and projects' scientific goals
- Object/Calibrator template – Information about the object (e.g., position, magnitude)
- Instrument template – Setup information for the instrument
- Constraints template – Conditions acceptable for observations (e.g., sky brightness)
- Lamp/Flat/Dark template– Setup information for unique calibrations required for the data
- Plan template– The sequence of actions to be taken to acquire data

The investigator is able to specify details such as which dates and times are acceptable for the observations, what observing conditions are acceptable, how periodic or phase critical observations are to be carried out, and what the relative priority of different parts of the project are to the overall scientific goal. All of this information is specified in terms familiar to an astronomer with concepts similar to what one would tell an experienced observer unfamiliar with the specifics of the project.

Each template can be called many times from within one plan or from multiple plans, saving the investigator from having to make multiple entries for the same object or instrument setup. In addition to this reusability, we have implemented macros to reduce repetitive typing. Recursive macro calls also allow the user to create complex plans with minimal typing. Details of the language are further described in the Phase II manual[2].

3. Unique Features

Because of the constrained motion of the HET, we needed to develop two unique features for our Phase II language. The first of these is how one requests an observation to be done. One can *do* an observation, which means that the observations are to be executed in the given sequence with no other observations intervening. Alternately one can *schedule* an observation. This allows the HET queue to place that observation at any time in the night that is convenient and allows other observations (possibly from other investigators) to take place

---

[2] http://rhea.as.utexas.edu/HET_documentation

```
begin_object: SAO10
  description: protogalaxy
  ra: 00:04:45
  dec: +22:43:26
  equinox: J2000
  pa: 142.3
  ra_offset: 10.3
  dec_offset: -23.4
  # magnitude of acquisition object
  v: 12.4
  # flux of object
  flux 8300 continuum 21.4
  rv: 430.3
  finding_chart: sao10.fit
end_object:
```

```
begin_plan: SAO10
  instrument: lrs_setup1
  constraint: very_good
  class: required
  # starts on the 18th of November
  # ±3.5 days
  utdate_start: 18/11/1997 delta 3 12:00
  do-flat: 50 5
  do-object: SAO10 800 scaled 2
  do-lamp: ThAr 10 5
  schedule-dark: 500 5
  schedule-calibrator: bluestar 50 fixed 1
  # wait 3 ±1 days
  schedule-wait: 3 delta 1
  do-object: SAO10 800 scaled 2
end_plan:
```

Table 1. An example object template (left) and plan (right). Lines preceded by the "#" character are comments.

during the interim. The plan in the right column of Table 1 demonstrates both of these commands. The sequence of "do" commands require that a flat lamp be taken immediately before and a wavelength calibrator be taken immediately after the observations of the object. The first two "schedule" commands allow the HET Operation staff to make those observations at any time in the night. The schedule-wait command tells the observer to wait for a period of time, allowing other observations to take place during the interim.

The second new feature allows the investigator to give either static or dynamic integration times. As the HET tracks an object across its 11 meter primary mirror, the 9.2 meter beam becomes vignetted. This vignetting is a function of the location of the object on the sky, the time the object was acquired, the length of the track, and the azimuth of the telescope. Because the investigator will typically not know when his/her observation will be scheduled, we allow the investigator to either describe an exposure time that will scaled according to this vignetting function and the observing conditions at the start of the exposure, or to statically set the integration time (e.g., if the exposure is time critical). Exposure time scaling is done to conserve the signal-to-noise ratio of the observation based on the vignetting function and estimates of the atmospheric transmission and seeing losses.

4. Plan Submission and Execution

Once plans are created by the investigator, they are e–mailed to the HET Operations Team in Austin. Upon receipt, the plans are automatically validated using a procmail–type[3] validation system. Approved plans are then written to disk where, at a fixed time each day, any changes and additions are compiled into the HET Observation relational database. This database is a Starbase database (Roll 1996) to which we have added some new HET–related calculations.

---

[3]http://www.ling.helsinki.fi/~reriksso/procmail/mini-faq.html

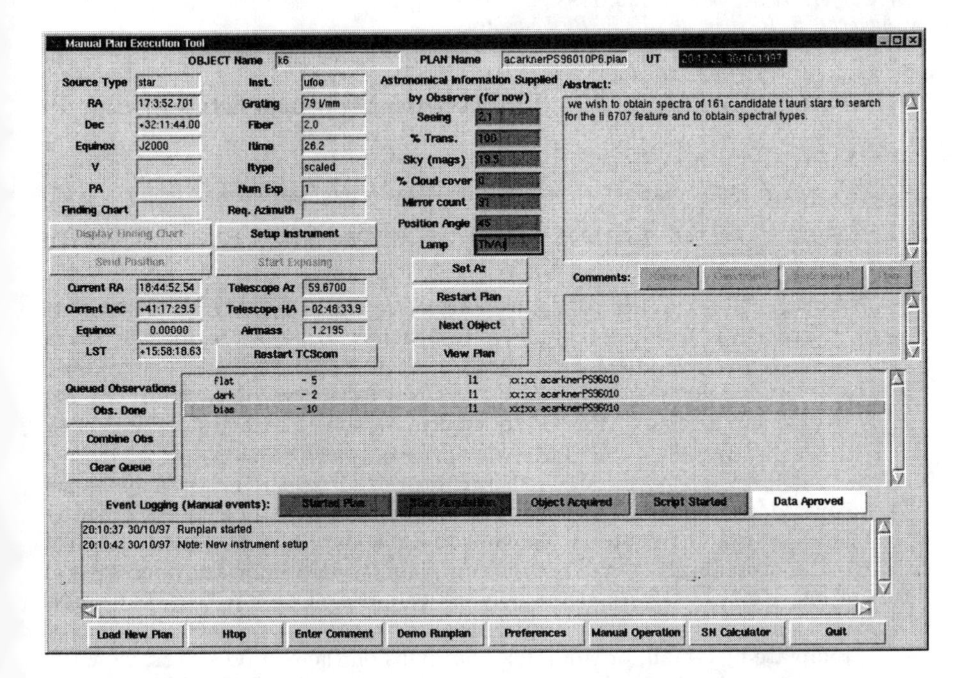

Figure 1. The plan execution tool reads in plan files and coordinates the observations with the other subsystems.

The database is transfered over the Internet to the HET site at the beginning of each night. The database will then be used by the resident astronomer to schedule plans throughout the night (Gaffney & Cornell 1997). When a scheduled target becomes observable, the plan is loaded into the plan execution tool (Figure 1). This tool displays the relational database information in an easy–to–digest GUI format. This GUI sends coordinates to the telescope control system and creates macros to be executed on the instrument control systems (ICS) to acquire the data. When the data is written to disk by the ICS, this tool logs the data as being acquired, appends comments to and fixes up the FITS headers to match the standard header format for the HET, moves the data to the correct directory, and queues the data for backup and transfer to Austin.

References

Gaffney, N. I., & Cornell, M. E. 1997 in ASP Conf. Ser., Vol. 125, Astronomical Data Analysis Software and Systems VI, ed. Gareth Hunt & H. E. Payne (San Francisco: ASP), 379

Roll, J. 1996, in ASP Conf. Ser., Vol. 101, Astronomical Data Analysis Software and Systems V, ed. George H. Jacoby & Jeannette Barnes (San Francisco: ASP), 536

Sebring, T. A., & Ramsey, L. W. 1994, in Advanced Technology Optical Telescopes V, SPIE Tech Conf. 2199

Astronomical Data Analysis Software and Systems VII
ASP Conference Series, Vol. 145, 1998
R. Albrecht, R. N. Hook and H. A. Bushouse, eds.

The NOAO Web-based Observing Proposal System

David J. Bell, Jeannette Barnes and Caty Pilachowski

NOAO[1]*, P.O. Box 26732, Tucson, AZ 85726*

Abstract. A World Wide Web interface to the NOAO observing proposal form is now available. Proposal information is provided by users through a set of HTML forms and submitted to the NOAO server where it is processed by a Perl CGI script. Netscape users may optionally attach locally prepared PostScript figures or ASCII text files for inclusion in their proposals using that browser's file upload feature.

All submitted data is retained on the server so that it may be recovered and modified in later sessions or viewed by collaborators who have shared the proposal password. In addition to the advantages of global availability and interface familiarity, the system provides several other useful options including online verification of LaTeX syntax and a spell-checker. Users can retrieve a filled-in copy of the NOAO proposal template by e-mail, or run latex and dvips on the NOAO server, view the output, and submit the proposal online. The NOAO WWW observing proposal pages can be found at "http://www.noao.edu/noaoprop/".

1. Introduction

Each semester NOAO receives hundreds of requests for observing time in the form of LaTeX documents. An automated system for handling these requests by e-mail has been in use for the past four years (see Bell et al. 1996 for a general description of the original system, which has since been rewritten in Perl). Although the system has always been quite successful, several additional enhancements can be achieved through the use of the World Wide Web, including a friendly and familiar user interface with hypertext help and pull-down menus, online verification and processing, and shared access to proposal information by remote collaborators as the document is being prepared.

By interfacing to the existing form, these new features have been added while retaining the most positive benefits of LaTeX – authors may still include special symbols, equations, figures and tables in their documents and information needed for scheduling can still be automatically extracted from the submitted proposals (Bell 1997).

[1]National Optical Astronomy Observatories, operated by the Association of Universities for Research in Astronomy, Inc. (AURA) under cooperative agreement with the National Science Foundation

2. Discussion

The proposal form is divided into six sections, which can be filled out in any order and revisited at any time. Each HTML page contains hidden fields identifying the proposal ID and section in addition to the standard form fields. A single Perl CGI script is called which saves and retrieves form information on the NOAO server and performs all processing requests. The proposal sections are:

- General Info – Proposal title and other general information. See Figure 1.
- Investigators – Names, addresses, phone numbers, etc.
- Observing Runs – Telescopes are first chosen, then instruments from lists customized for each telescope. Required moon phase, filter information, and desired observing dates are also entered here.
- Justification – The main text sections of the proposal. Text may be edited online or uploaded from prepared files.
- Figures – PostScript figures (using file upload) and captions are controlled from here. Figures may be scaled or rotated.
- Tables – Target object information is entered into one or more HTML table forms. LaTeX tables are built into the proposal document.

The Justification and Figures sections support file upload of ASCII text and PostScript figure files. PostScript bounding boxes are computed for figures and written into the saved files, in addition to optional PostScript commands for rotation.

At any time, the user may choose one of several processing option buttons:

- Run latex – LaTeX proposal is built and processed. If errors occur, line numbers are translated into corresponding form fields. For security, any LaTeX commands capable of probing the host system are removed before processing.
- View PS file – Proposal is built and processed. A hypertext link to a PostScript file is provided for viewing by user.
- Email LaTeX file – A filled-in copy of the LaTeX form is sent to the user.
- Email PS file – A PostScript copy of the proposal is sent to the user.
- Check Proposal – Form data is checked for completeness and basic LaTeX syntax. A spell-checker is run on essay sections.
- Submit Proposal – Proposal is built, checked, and processed. After final user confirmation, proposal and figures are mailed to the NOAO queue and acknowledged by e-mail.

Netscape: NOAO Observing Proposal Form

File Edit View Go Bookmarks Options Directory Window Help

Back | Forward | Home | Reload | Load Images | Open... | Print... | Find... | Stop

Location: http://www.noao.edu/cgi-bin/noaoprop/noaoprop.pl

What's New? | What's Cool? | Destinations | Net Search | People | Software

NOAO Observing Proposal Form

NOAO is operated by the Association of Universities for Research in Astronomy (AURA), Inc. under cooperative agreement with the National Science Foundation

There are six sections or pages to the Web proposal form. This page requests that you enter any general information associated with your proposal. Principal– and co–investigator information is entered in another section. Details of the observing runs requested, along with dates and instrumentation details, are entered in a third section. Essay–type information such as an abstract of the scientific justification, the scientific justification itself, and other background information, is entered on the fourth page. Up to three figures may be included with your proposal and that information can be provided in the fifth section. The last section, for tables, is required for all WIYN proposals but optional for all others. Details about what is needed in each section are provided in the help link at the top of each page. The sections may be completed in any order.

Before starting your proposal please review some general comments about the Proposal Process as well as the Online Help document which provides a field by field description of the proposal form. If you are new to LaTeX please review our LaTeX Pointers page, as well.

DEMONSTRATION ONLY -- no information will be saved.

GENERAL PROPOSAL INFORMATION

Help for this section

Enter the title of the proposal below. Code special symbols as LaTeX symbols or math.

Characteristics of a Well-Crafted Telescope Proposal

TAC Session (KPNO only): Galactic

Is this proposal part of a PhD thesis? Note: If you answer "yes" to this question then you are required to send a letter as per the following instructions.

No

If this is a long-term status request please give details on the line below.

6 nights per semester for 2 years

List dates you cannot use for non-astronomical reasons on the line below.

Please avoid Nov. 4 (Election Day)

Use these buttons to modify or save your entries. The "Undo Changes" button will remove any changes since the last "save" action. The "Clear Form" button will clear all fields but nothing is actually deleted until the next "save" action. The "Save and Verify" button will "save" and "verify" the current form entries.

Undo Changes | Clear Form | Save and Verify

Use these buttons to continue to other sections of the proposal. Other than the figures which are optional and the tables that are only required for WIYN proposals, all of these pages must be completed before the proposal can be submitted. Choosing one of these options will *save* the current form before going on to the requested page.

Enter Investigators | Enter Observing Runs | Enter Justification Topics | Include Figures | Enter Tables

At any time you may choose one of these options. The current form will be saved before any action is taken. If you request a file transfer it will be sent to the email address entered on the proposal form home page.

Run LaTeX | View PS File | Email LaTeX File | Email PS File | Check Proposal | Submit Proposal

National Optical Astronomy Observatories, 950 North Cherry Avenue, P.O. Box 26732, Tucson, Arizona 85726, Phone: (520) 318-8000, FAX: (520) 318-8360

Questions or problems may be directed to noaoprop-help@noao.edu.

Figure 1. A sample NOAO proposal Web form

The LaTeX proposal built on the Web uses the same template as that used in the traditional download/e-mail approach. Users thus always have the option of mailing the proposal to themselves and finishing it with a text editor and sending it in by e-mail (an "import" from the LaTeX form to the Web is planned for the next semester). Proposals completed and submitted by either method are indistinguishable when printed.

3. Conclusion

The NOAO Web-based proposal system went public in August 1997 for the submission period concurrent with this meeting. Initial reviews have been quite favorable and few problems were reported. Currently the only significant limiting aspects of the Web form involve PostScript figures. Although most users had no trouble including figures in their documents, sophisticated customized figure placement is not available. In addition, submitting very large figures becomes inefficient if the user wishes to repeatedly download the PostScript proposal for viewing.

We've found that the system scales well with the experience of the user. Those who know nothing about LaTeX are largely shielded from it and are more comfortable using the Web than the traditional template form. Meanwhile veteran LaTeX users who wish to fill their proposals with special symbols and equations may still do so. Thanks to the online processing and verification, we've found that a much smaller percentage of Web-submitted proposals arrive with problems requiring human attention than those submitted by e-mail. This saves considerable time for the NOAO office staff, as most Web proposals can be automatically filtered for import to observatory databases. User feedback indicates that a substantial time savings in proposal preparation has also been achieved.

Proposal materials at various observatories consist of many types, from simple flat-file templates to sophisticated GUI software tools which each user must download and compile. We've found that the WWW CGI approach described here strikes a good balance by being complex enough to deliver all needed information to the observatory while remaining easy to use.

References

Bell, D. J., Biemesderfer, C. D., Barnes, J., & Massey, P. 1996, in ASP Conf. Ser., Vol. 101, Astronomical Data Analysis Software and Systems V, ed. George H. Jacoby & Jeannette Barnes (San Francisco: ASP), 451

Bell, D. J. 1997, in ASP Conf. Ser., Vol. 125, Astronomical Data Analysis Software and Systems VI, ed. Gareth Hunt & H. E. Payne (San Francisco: ASP), 371

Astronomical Data Analysis Software and Systems VII
ASP Conference Series, Vol. 145, 1998
R. Albrecht, R. N. Hook and H. A. Bushouse, eds.

Observing Control at the UKIRT

A. Bridger and G. Wright

Royal Observatory, Blackford Hill, Edinburgh, EH9 3HJ, United Kingdom

F. Economou

Joint Astronomy Centre, 660 North Aohoku Place, University Park, Hilo, HI 96720

Abstract. Observing with the major instruments at the United Kingdom Infra-Red Telescope (UKIRT) is already semi-automated by using ASCII files to configure the instruments and then sequence a series of exposures and telescope movements to acquire the data. This has been very successful but the emergence of the World Wide Web and other recent software technologies have suggested that it could be developed further to provide a friendlier, more powerful interface to observing at UKIRT. A project is now underway to fully design and implement this system.

1. Introduction

Visiting observers at the United Kingdom Infra-Red Telescope already automate much of their data taking via the use of predefined *Configs* and *EXECs*. Using an off-line preparation system to create these has significantly improved observing efficiency, as also has the automatic, on-line data reduction system in use by one of the main instruments. Experience with the system and also the possibility of changes in the observing practices at UKIRT (an experiment in reactive scheduling is in progress on UKIRT, Davies 1996) make it seem natural to improve the present system, taking advantage of recent developments in software technology. This paper outlines the project (ORAC - "Observatory Reduction and Acquisition Control") that has been formed to implement this evolution. After briefly describing the existing system an overview of the new system is presented and a short description of each component given.

2. Background: The Current System

The UKIRT control system consists of three main elements, the telescope control system, the instrument control system and the on-line data reduction system (Daly, Bridger & Krisciunas 1994 gives more details). The instrument control system controls the instrument via its use of *Configs* and the observing sequence via *EXECs*. Although most high-level telescope control is performed by the telescope operator (TO) the instrument control task also performs some telescope control (*e.g.* "nodding") via the *EXECs*.

A *Config* is a user-defined description of an instrument setup, stored as an ASCII file. It includes items such as the filter required, exposure time, etc for both the target and related calibration observations. An *EXEC* is another ASCII file containing a user-defined sequence of observing commands which control both the instrument and the telescope, *e.g.* configure the instrument, take a dark frame, take an object observation, "nod" the telescope to sky, etc. *EXECs* may also call other *EXECs*, allowing the reuse of standard *EXECs* and the modular building of more complex observing sequences. Both *Configs* and *EXECs* are defined using the *UKIRT Preparation System.* This presents a simple menu-based interface and in the case of *Configs* greatly simplifies the creation of the files, especially by automating some of the selections (though the user may override). Observers are encouraged to prepare *Configs* and *EXECs* before reaching the telescope, usually at the JAC.

For the CGS4 near-infrared spectrometer there is also automatic data reduction - as soon as a new data frame is stored the data reduction system will reduce it, combining it where required with other data frames. This gives a rapid feedback of the true data quality to the observer as well as producing (in many cases) the final reduced data.

Despite its success there are a number of limitations in this system. Bridger and Wright (1996) proposed to evolve this system into a more fully automated version. Since then a project has been created with the stated aim 'To increase the observing efficiency and publication rate of UKIRT'. The approach is based around making it easier, friendlier and more efficient to observe at UKIRT: by providing an easy to use remote preparation system; by providing a more powerful and flexible on-line data reduction system; and by further automating the acquisition of data at the telescope. A long-term goal is to produce a system capable of fully queue-scheduled observing, should UKIRT implement it.

3. ORAC: An Overview

The design is based around the idea that an observation may be defined as consisting of an *Instrument Configuration*, which tells the instrument how it should be setup for an observation, a *Target Description*, to inform the telescope where to point, and an *Observing Sequence*, which describes the sequence of telescope movements and data acquisition commands required to execute the observation. These components are modular and may be used in multiple *Observation Definitions.* Thus *Target Description* **BS623** may be used with *Instrument Configuration* **Std_H_Band** and *Sequence* **Bright_Star** in one observation, and combined with **My_K_Band** and **Std_Jitter** to form a different observation. To this basic definition of an observation are added further components which describe the data reduction *Recipe* to be used on the observation and *Scheduling Information* which may be used in a more sophisticated observation scheduler.

These components, and the full *Observation Definitions*, may be prepared ahead of time (off-line) or at the telescope. In operation the Observatory Control System will read them and distribute the components to the various subsystems, to configure and control the acquisition and the reduction of the data.

3.1. Observation Preparation System

The Observation Preparation System is a replacement for the existing UKIRT Preparation System. Observers must be able to run it from their home institution, without the need for detailed software installations. The system will need to change to reflect changing instrument availability. Both of these suggest a Web-based application. However, speed of the Web, and the loading on the server might be problems. Use of mirror sites could help with this, but the use of automatic software installations and Internet push technology to keep them up to date might be a better solution. This is the approach taken by the Gemini group with their Observing Tool (Wampler et al. 1998).

The output of the preparation system will be one or many files forming an *Observation Definition* that can be stored along with similar definitions from the same or different observing programmes to form the telescope's observation 'database'. Initial versions of the database will probably use the existing combination of ASCII text files and a directory tree structure, for backwards compatibility. Possible future use of a commercial database will be considered.

It is likely that much observation verification will be done by this system - the output of the preparation system should be guaranteed to be able to be performed by the instrument, to eliminate errors at the telescope. The preparation system will present the user with an astronomer-oriented interface, although its output may well be at a lower level. It is anticipated that the system's understanding of the instruments will be codified by a set of rules which will be maintained by the instrument support scientist, adding to the system robustness.

3.2. Observatory Control System

The Observatory Control System performs the sequencing role between the instruments and telescopes and optionally also some of the scheduling of individual observations. The way in which the components of the *Observation Definitions* are to be handled is still to be determined, but the likely approach is to leave the reading and parsing of a component to the system that requires it - for example an *Instrument Configuration* will be read and interpreted by the Instrument Control System. This considerably simplifies the logic of the higher system. Other approaches are possible and will be considered.

The Observatory Control System consists of two main components:

- The Observation Sequencer. This task informs instruments of their required configuration and sequences a set of operations to generate a particular observation. This can include taking Object and Sky sequences, mosaicing, nodding and chopping etc. A key aspect is that it will define a standard, but general, instrument interface to which all UKIRT instruments will need to adhere.

- The Scheduler. This new component in the system is a higher level system that maintains a queue of Observation Definitions prioritised according to their scheduling information. Its goal is to intelligently and automatically handle the correct and efficient scheduling of the observations, informing the observation sequencer of the next observation. It will also provide for monitoring and direct manipulation of the observing queue. The scheduler is likely to be delivered as a later upgrade to the new system.

3.3. Data Reduction System

This performs automatic on-line data reduction after the raw data frames are stored. Its current design is based on the concepts of *Recipes* and *Scripts*:

- A *Recipe* is a series of instructions detailing how to perform a (potentially complex) data reduction operation.
- A *Script* is a series of instructions on how to apply a *Recipe* to reduce an observing sequence using a particular data reduction package.

The user will request a particular reduction sequence by drawing from a series of observatory defined *Scripts* and *Recipes*. Advanced users will be able to provide their own *Recipes* and *Scripts* to reduce particular aspects of their observations, but it is expected that they will all use standard *Recipes* to remove specific instrument dependencies. It is anticipated that the *Scripts* will be coded in a standard scripting language, and it is hoped that the actual data reduction algorithms will be provided by standard data reduction packages. The aim of this is to reduce the support effort at the Observatory, concentrating it on the automatic aspects of the reduction system. As a side effect, all of the actual algorithms used by the system should already be available to the community, and users may be provided with the actual *Scripts* used to reduce their data. The data reduction system is described in more detail in Economou et al. (1997)

4. Conclusions

A constant driving force on the UKIRT control system is the desire of observers for greater efficiency. The ORAC project aims to achieve greater efficiency whilst retaining the flavour and flexibility of current observing practices.

Acknowledgments. We would like to acknowledge the scientists and software groups at UKIRT and ROE, and the UKIRT telescope operators for many helpful discussions, and of course the input of many UKIRT observers.

References

Bridger, A. & Wright, G. S. 1996, in New Observing Modes for the Next Century, ASP Conf. Ser., Vol. 87, eds., T. A. Boroson, J. K. Davies & E. I. Robson (San Francisco, ASP), 162

Daly, P. N., Bridger, A. & Krisciunas, K. 1994, in ASP Conf. Ser., Vol. 61, Astronomical Data Analysis Software and Systems III, ed. Dennis R. Crabtree, R. J. Hanisch & Jeannette Barnes (San Francisco: ASP), 457

Davies, J. K. 1996, in New Observing Modes for the Next Century, ASP Conf. Ser., Vol. 87, eds., T. A. Boroson, J. K. Davies & E. I. Robson (San Francisco, ASP), 76

Economou, F., Bridger, A., Wright, G. S., Rees, N. P. & Jenness T. 1997, this volume

Wampler, S., Gillies, K., Puxley, P. & Walker, S. 1998, in SPIE vol 3112, in press.

Astronomical Data Analysis Software and Systems VII
ASP Conference Series, Vol. 145, 1998
R. Albrecht, R. N. Hook and H. A. Bushouse, eds.

The Ground Support Facilities for the BeppoSAX Mission

Loredana Bruca, Milvia Capalbi and Alessandro Coletta

Telespazio, Via Corcolle 19, I-00133 Roma Italy

Abstract. The BeppoSAX "Satellite per Astronomia X", a program of the Italian (ASI) and the Netherlands (NIVR) Space Agencies, was launched on April 30th into a circular orbit of 590 km altitude and 3.9 degrees inclination.

The Ground Support System components are presented in order to highlight the end-to-end operation approach to the scientific mission. The software systems are described following the forward and backward data flow to/from the satellite: starting from the Observation Proposal reception and archiving, the long term observation scheduling up to the detailed weekly planning and telecommands sequence uplinking, and vice versa the scientific and housekeeping raw telemetry acquisition and archiving, quick look data analysis, data reformatting and distribution as a Final Observation Tape to the Observation Proposal PI. Among these components the Quick-look Data Analysis system is emphasized. The main tasks of this system are both health monitoring of the BeppoSAX instruments and quick-look inspection of the scientific results with the goal of detecting and locating major changes in X-Ray sources and searching for X-Ray transient phenomena up to high energy events as Gamma Ray Bursts. This is carried out by performing both Data Accumulation and Data Presentation and Analysis by using the raw payload telemetry data stored on orbital basis.

1. Introduction

The ground support facilities for the BeppoSAX Mission are split among different ground systems components:

- the Ground Station, located in Malindi (Kenya), for the telecommands uplinking and telemetry retrieving when the satellite is directly visible (approximately only 10 minutes of each 96-100 minute long orbit)
- the Operation Control Centre (BSAX-OCC), which handles spacecraft orbital management and payload monitoring activities
- the Scientific Operation Centre (BSAX-SOC), where the detailed planning of the onboard scientific activities is prepared, the orbit-by-orbit raw telemetry is archived and made immediately available to the quick-look processing
- the Scientific Data Centre-Mission Support Component (BSAX-SDC/MSC), where observation requests from the astronomical community are collected

and checked, long range observation plans are prepared and raw data from the entire mission are archived and delivered to the final users as Final Observation Tapes.

Although all facilities are split into further components (OCC, SOC, SDC/MSC are located in Telespazio, Rome), their design and overall integration allow efficient and successful management of the mission operations.

2. The BeppoSAX forward and backward Data Flow

The aforementioned system components are described following the forward and backward data flow to/from the satellite.

The BeppoSAX Data flow starts at BSAX-SDC/MSC with the reception and management of the Observation Proposals:

- Proposal Reception via e-mail and ordinary mail
- Proposal Checks for completeness, consistency and syntax
- Proposal archiving using database technology
- Proposal Feasibility Checks: target observability and viewing windows, target visibility, bit-rate evaluation

Proposals approved by the Time Allocation Committee and successfully checked contribute to the Mission Plan file preparation. The Mission Plan File is the main input to the Proposal Scheduling steps performed at the BSAX-SOC:

- Long-Term Planning: an optimized timeline designed to maximize satellite performance taking into account the observability and operative constraints and the proposer's specified priorities
- Short-Term Planning: a one-week detailed timeline containing for each requested pointing the attitude sequences, the orbital links, SAGA passages and the on-board instruments set-up
- Urgent Observation (e.g., TOO) Management: TOO feasibility checks, quick replanning of the current short-term schedule

The Short-Term Planning products are the main input to the Spacecraft and Payload command sequence generation. The observation execution involves activities which are performed at the BSAX-OCC:

- Telecommand uplink, Telemetry acquisition
- Spacecraft and Payload Monitoring & Control
- Satellite ephemeris determination: prediction and reconstruction
- Satellite attitude determination and post-facto reconstruction

During the Observation execution the raw telemetry (housekeeping and scientific) is collected at the BSAX-SOC:

- Orbit-by-orbit telemetry acquisition and processing
- Telemetry Data Quality checks for completeness, integrity, consistency and accuracy
- Temporary Telemetry filing (data from the last 60 orbits are always on line)
- Quick-look Data Analysis

Telemetry data, grouped by Observing Period (several orbits time span) and Auxiliary data such as On-Board-Time/Universal-Time conversion Data, Reconstructed Attitude and Satellite ephemeris are then stored in the BeppoSAX Mission Raw Data Archive at the BSAX-SDC/MSC. Its organization and management is done according to the following baseline:

- Optical Disk media as final support for the data archiving
- A relational database to catalogue data and address data files on media
- Optical jukebox in order to keep more media on-line and optimize data retrieval

The last step in the BeppoSAX data flow facilities (BSAX-SDC/MSC) is the data delivery to the final user:

- Observation Data (housekeeping, scientific and auxiliary) retrieving from the BeppoSAX Mission Raw Data Archive, on the basis of the PI proposal
- Data reformatting and storing on the requested media support (e.g., DAT, Exabyte)
- Product delivery to the PI

3. Quick Look Data Analysis System

The Quick Look Analysis (QLA) is mainly conceived as a quick-look system to allow feedback to the BeppoSAX satellite as soon as possible, typically within a few orbital periods from when the relevant events were generated on board.

The QLA is performed by a team of eight Duty Scientists working shifts to ensure 24h data monitoring.

Input data for the QLA are payload telemetry data stored in the BSAX-SOC Archive for a time span encompassing the last 60 orbits.

The main tasks accomplished by QLA are both health monitoring of the BeppoSAX instruments and quick-look inspection of the scientific results with the goal of detecting and locating major changes in X-Ray sources and searching for X-Ray transient phenomena up to high energy events such as Gamma Ray Bursts. These latter events, triggered by the on-board Gamma Ray Burst Monitor, are analysed by the Duty Scientist Team following a well-tested procedure and up to now several real Gamma Ray Bursts have been identified and located, giving input to follow-up observations in different energy bands. In order to accomplish the QLA goals the system provides both Data Accumulation, Data Presentation and Analysis tools. Implemented accumulation functions provide scientific telemetry data retrieving, packet information extraction to perform accumulation of X-Ray events information and integration of on-board accumulated data structures. Data visualization and general analysis are based on a Motif Graphical User Interface developed by Telespazio with the PV-WAVE analysis environment. Based on PV-WAVE Widget Toolbox applications, the ad-hoc software developed using PV-WAVE processing routines provides a data analysis/manipulation system with optimum ease of use.

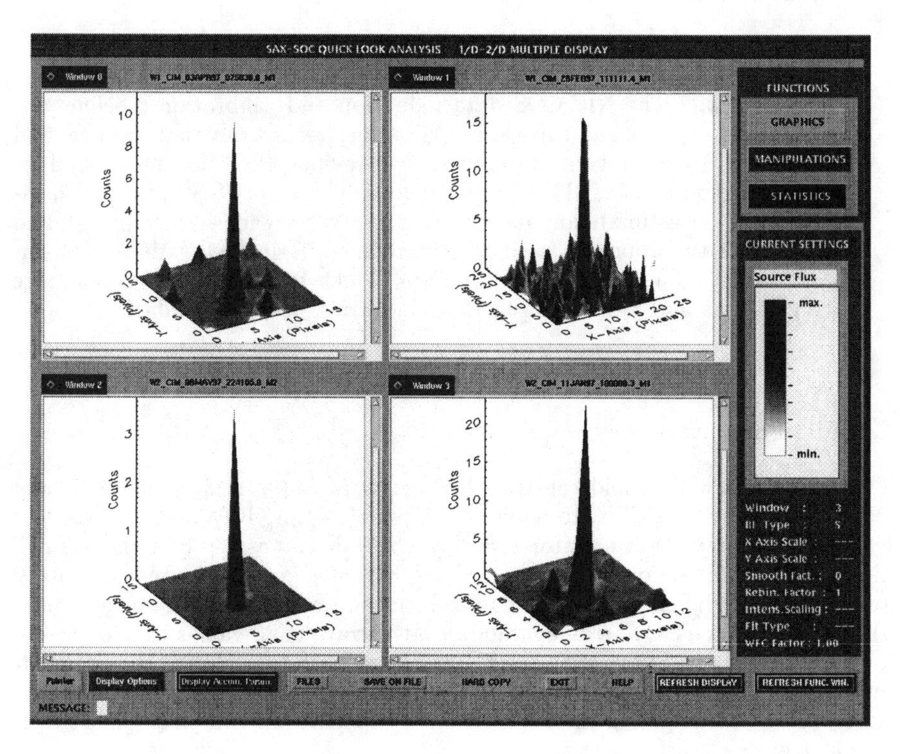

Figure 1. 3D Images of four Gamma Ray Bursts discovered by BeppoSAX (GRB970111, GRB970228, GRB970402 and GRB970508)

Astronomical Data Analysis Software and Systems VII
ASP Conference Series, Vol. 145, 1998
R. Albrecht, R. N. Hook and H. A. Bushouse, eds.

The STScI NICMOS Calibration Pipeline

Howard A. Bushouse

Space Telescope Science Institute, Baltimore, MD 21218, Email: bushouse@stsci.edu

Elizabeth Stobie

University of Arizona, Steward Observatory, Tucson, AZ 85721

Abstract. The NICMOS data reduction and calibration pipeline employs file formats and software architecture techniques that are new and different from what has been used for previous HST instruments. This paper describes the FITS file format used for NICMOS data, which includes error estimate and data quality arrays for each science image, and describes the approach used to associate multiple observations of a single target. The software architecture, which employs ANSI C language algorithms and C bindings to IRAF libraries, is also described.

1. File Formats

The NICMOS archival and run-time file format is FITS, with extensive use of IMAGE and BINTABLE extensions. Each image from NICMOS is represented by five data arrays that are stored as five IMAGE extensions in a single FITS file. The five data arrays are: 1) the science (SCI) image from the instrument; 2) an error (ERR) image giving estimated uncertainties for the SCI image values; 3) a data quality (DQ) array containing bit-encoded flag values that represent known problem conditions with the SCI image values; 4) an array containing the number of samples (SAMP) used to derive the SCI values; and 5) an exposure time (TIME) array giving the total effective exposure time for each pixel in the SCI image.

This group of five data arrays is called an image set, or "imset" for short. NICMOS operation modes that produce multiple readouts of the detector during a single exposure have multiple imsets in a single FITS file; one imset per readout. Multiple observations of a single target are stored in separate FITS files, but the data are logically associated through the use of an "association table", which is a FITS file containing a single BINTABLE extension, listing the names and other relevant information for each observation in the association. One-dimensional spectra that are extracted from NICMOS grism images are also stored in a BINTABLE extension of a FITS file. There is never any image data in the primary header-data unit (HDU) of NICMOS FITS files; all data are stored in IMAGE or BINTABLE extensions. The primary HDUs only contain header keywords with information that is relevant to all extensions in the file.

2. Software Design

All of the NICMOS pipeline calibration software has been written in the ANSI C language. A set of data I/O routines, known as "hstio", has also been written which is used to easily perform the complex mapping between the disk file formats for imsets and their representations as NICMOS-specific C data structures within the pipeline code. The hstio library, in turn, uses a set of C bindings to IRAF image I/O libraries to perform the actual file access and data I/O. C bindings to other IRAF, STSDAS, and TABLES package libraries also provide access to existing mathematical and analysis algorithms, as well as FITS BINTABLE data I/O. The algorithms that apply the actual calibration steps to the science data are completely isolated from and independent of the data I/O routines. The calibration tasks are built in such a way that they can be run from either the host operating system level or as an IRAF native task from within the IRAF cl.

3. Image Associations

Multiple NICMOS observations of a single target field are often obtained for one of several reasons. Multiple exposures at a single sky position can be obtained in order to allow for rejection of cosmic ray hits when the images are later combined by the calibration pipeline. Small-angle dithering of a compact target within the instrument's field of view is useful for removing the effects of defective pixels and for averaging out residual flat fielding uncertainties. For observations at longer wavelengths, where the thermal background signal from the telescope and instrument begins to be significant, it is necessary to alternately "chop" the telescope field of view between the target and nearby blank sky positions in order to obtain measurements of the background signal. For very extended sources an image of the entire target can be created by "mosaicing" individual observations obtained at many sky positions.

These sets of observations are known as image "associations". The associations are a *logical* grouping only, in that the individual observations that make up an association are initially stored and processed separately, but are eventually combined by the pipeline software into a final product. The individual observations are first processed, one at a time, by the `calnica` task, which performs instrumental calibration, and are then processed as a group by the `calnicb` task, which combines the images, measures and removes background signal, and produces final mosaic images. Information about the association is stored in an association table, which is a FITS file with a single BINTABLE extension. The table extension lists the names of the observation datasets contained in the association and their role in the association (target or background image).

4. CALNICA Pipeline

The NICMOS pipeline calibration is divided into two tasks. The first task, `calnica`, applies instrumental calibration to individual exposures, including associated and non-associated observations. The task is completely data driven in that all information necessary to accomplish the calibration is read from image header keywords. These keywords contain instrumental configuration parameters, such as camera number, filter name, and exposure time, as well as "switches" that indicate which calibration steps are to be performed, and the names of reference files (e.g., flat fields, dark current images) that are needed for the calibration.

The major steps performed by `calnica` include the masking (setting DQ image values) of defective pixels in the images, the computation of uncertainties (setting ERR image values) for the image data, which includes both detector readnoise and Poisson noise in the detected signal, dark current subtraction, correction for non-linear detector response, flat fielding, and conversion from detected counts to count rates. Each calibration step that involves the application of reference data (e.g., dark subtraction, flat fielding) propagates estimated uncertainties in the reference data into the ERR images of the data being processed.

Another major step performed by `calnica` is the combination of images generated by the MultiAccum operational mode of the NICMOS detectors, in which multiple non-destructive detector readouts are performed during the course of a single exposure. The images generated by each readout are first individually calibrated and are then combined into a single, final output image. The image combination is accomplished by fitting a straight line to the (cumulative) exposure time vs. detected counts data pairs for each pixel, using a standard linear regression algorithm. Cosmic ray hits in individual readouts are detected and rejected during this processes by searching for and rejecting outliers from the fit. The fit and reject process is iterated for each pixel until no new samples are rejected. The slope of the fit and its estimated uncertainty are stored as the final countrate and error, respectively, for each pixel in the output image SCI and ERR arrays. In addition, the number of non-rejected data samples for each pixel and their total exposure time are stored in the SAMP and TIME arrays, respectively, of the final image.

5. CALNICB Pipeline

The second phase of the NICMOS pipeline is performed by the `calnicb` task, which processes associated observations, each of which must have already received instrumental calibration by `calnica`. This task is also completely data driven, with processing information coming from image header keywords and the association table. The three major steps performed by `calnicb` are: 1) combine multiple images that were obtained at individual sky positions (if any); 2) measure and subtract the thermal background signal from the images; and 3) combine all the images from a given observing pattern into a single mosaic image.

Steps 1 and 3, which both involve combining data from overlapping images, have many common features. First, proper registration of the input images is accomplished by estimating the image-to-image offsets from world coordinate system (WCS) information contained in the image headers, and then applying a cross-correlation technique to further refine the offsets. The cross-correlation uses a "minimum differences" technique which seeks to minimize the difference in pixel values between two images. Second, shifting the images to their proper alignments is done by simple bi-linear interpolation. Third, the combining of pixels in overlap regions uses the input ERR array values to compute a weighted mean, rejects pixels flagged as bad in the DQ arrays, rejects outliers using iterative sigma clipping, and stores the number of non-rejected pixels and their total exposure time in the SAMP and TIME arrays, respectively, of the combined image.

The process of background measurement varies depending on the exact makeup of the image association that is being processed. If observations of background sky locations were obtained by using an observing pattern that includes chopping, then only these images are used to compute the background signal. If background images were not obtained, then an attempt is made to estimate the background from the target images, with an appropriate warning to the user that the background estimate may be biased by signal from the target. A scalar background signal level is measured for each image by computing its median signal level, excluding bad pixels and iteratively rejecting outliers from the computation. The resulting background levels for each image are then averaged together, again using iterative sigma clipping to reject discrepant values for individual images. The resulting average value is then subtracted from all images in the association, both background (if present) and target images. Provisions have also been made for subtracting a background reference image which could include spatial variations in the background signal across the NICMOS field of view, but to date no such spatial variations have been seen.

At the end of `calnicb` processing an updated version of the association table is created which contains information computed during processing, including the offsets for each image (relative to its reference image), and the scalar background level computed for each image. An example of such a table for a "two-chop" pattern is shown in Table 1. This pattern contains one position on the target and two separate background positions. Two images have been obtained at each position. The two images at each position are first combined, the background is measured from the combined images at each background position, the average background is subtracted from all images, and then three final images are produced, one for the target and one for each background position.

Table 1. Output Association Table.

MEMNAME	MEMTYPE	BCKIMAGE	MEANBCK (DN/sec)	XOFFSET (pixels)	YOFFSET (pixels)
n3uw01a1r	EXP-TARG	no	INDEF	0.00	0.00
n3uw01a2r	EXP-TARG	no	INDEF	-0.15	0.00
n3uw01a3r	EXP-BCK1	yes	0.28	0.00	0.00
n3uw01a4r	EXP-BCK1	no	INDEF	0.00	0.20
n3uw01a5r	EXP-TARG	no	INDEF	0.32	0.00
n3uw01a6r	EXP-TARG	no	INDEF	0.00	0.00
n3uw01a7r	EXP-BCK2	yes	0.26	0.00	0.00
n3uw01a8r	EXP-BCK2	no	INDEF	0.00	0.12
n3uw01010	PROD-TARG	no	INDEF	INDEF	INDEF
n3uw01011	PROD-BCK1	no	INDEF	INDEF	INDEF
n3uw01012	PROD-BCK2	no	INDEF	INDEF	INDEF

Astronomical Data Analysis Software and Systems VII
ASP Conference Series, Vol. 145, 1998
R. Albrecht, R. N. Hook and H. A. Bushouse, eds.

Enhanced HST Pointing and Calibration Accuracy: Generating HST Jitter Files at ST-ECF

M. Dolensky, A. Micol, B. Pirenne and M. Rosa

Space Telescope - European Coordinating Facility, Karl-Schwarzschild-Str. 2, D-85748 Garching, Germany

Abstract. After providing on–the–fly re-calibration of HST data to archive users, ST–ECF has embarked on a new project aimed at grouping, cosmic ray cleaning and drizzling images taken by most HST instruments. To perform this task in an unattended way, very precise pointing information is required.

In addition, several other spacecraft parameters contained in the engineering data stream are required as inputs to new, enhanced calibration algorithms. An example is the estimation of the particle induced background count rate in the FOS digicons using actual HST magnetometer readings to scale geomagnetical shielding models. The production of the missing jitter files combined with the extraction of additional engineering parameters will greatly enhance the homogeneity and the scientific value of the entire archive.

We therefore started the generation of the missing jitter files. In this article we explain the various aspects of the task and we give an example of a project which will take advantage of the results.

1. The Pipeline

In order to generate the missing jitter files a pipeline was set up (Figures 1 and 2).

The heart of the pipeline is the Observation Monitoring System (OMS). It correlates the time-tagged engineering telemetry with the Mission Schedule. The Mission Schedule is a list of all commands and events that drive HST activities. The engineering telemetry is stored in the Astrometry and Engineering Data Processing (AEDP) subset files.

Both the Mission Schedule and the AEDP subset files are retrieved from the Data Archive and Distribution System (DADS) at STScI.

In order to reprocess old AEDP subset files it is necessary to set up OMS according to the spacecraft configuration and telemetry format valid at that time. So, configuration files that change over time such as spacecraft characteristic files are retrieved from the Project Data Base PDB.

OMS runs on a VAX/VMS platform. ECF maintains and operates the pipeline remotely using the Space Telescope Operations Support Cluster (STOSC) at STScI, i.e., ECF and STScI are sharing resources.

The AEDP subset files are used in two ways:

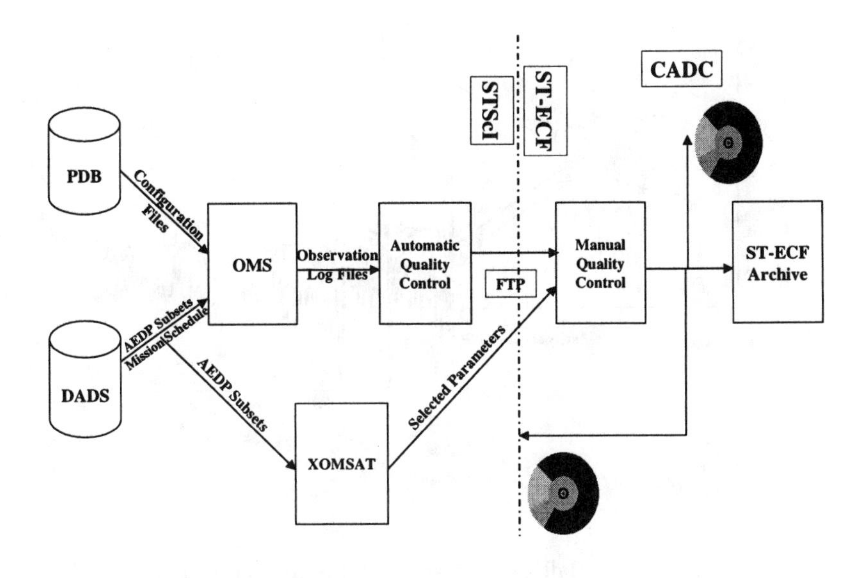

Figure 1. Pipeline reprocessing AEDP Subset Files

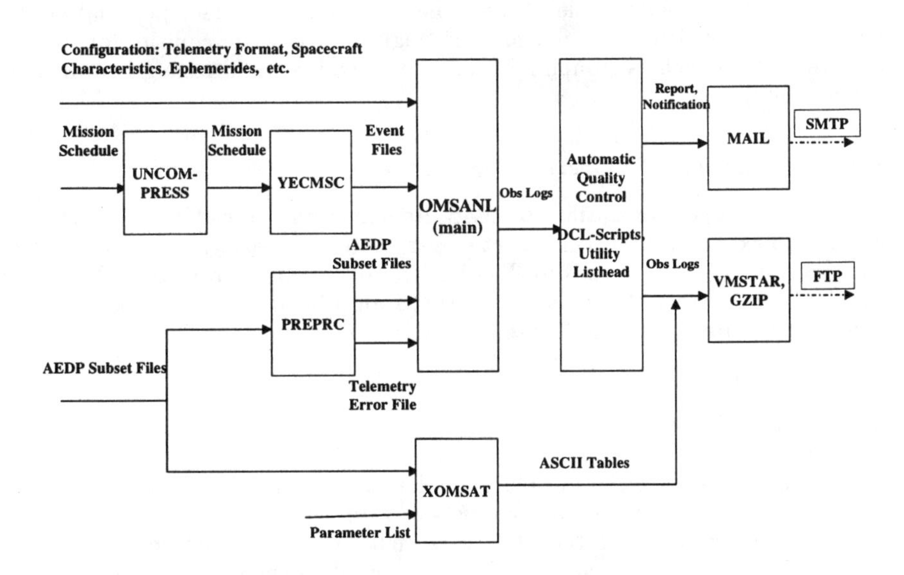

Figure 2. Detailed Data Flow Graph of the Pipeline

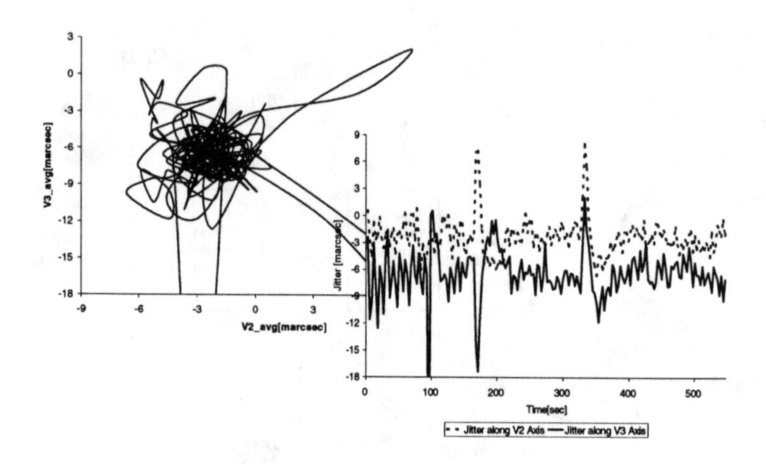

Figure 3. Jitter Ball and Jitter vs. Time for Dataset X22X0103

1.1. Generation of Observation Logs

Observation Logs contain a set of pointing and specialised engineering data (Leitherer et al. 1995; Lupie, Toth & Lallo 1997). The exact pointing of the telescope, also called jitter information (Figure 3), is a prerequisite for further archive services such as grouping, cosmic ray rejection and drizzling HST images (Micol 1998).

1.2. Extraction of Selected Parameters

Xomsat is a stand-alone utility of OMS. Its purpose is to read an AEDP subset file and to extract a user defined set of parameters. Nowadays the new Control Center System (CCS) is used for RT-Analysis of selected parameters. The same capability for historical telemetry is planned for a future release of CCS. It will, however, require more time before it is available.

2. Quality Control

There is a semi-automatic procedure for the quality control of Observation Logs (Figure 3). The automatic part consists of a test series which is performed at STScI. A report is sent to the ECF via e-mail. In order to reduce data volume only the comparatively small jitter files are packed and compressed using vmstar and gzip respectively.

After an additional manual inspection at the ECF the logs finally end up in the archive and become available to the community.

ECF's archive interface includes a Java applet[1] that allows the visualization of any two columns of an obs log as an X/Y plot.

3. Outlook

The pipeline started working in October 97. The first goal is the generation of obs logs for the time period between the first Servicing Mission in Dec. 93 and Oct. 94. In Oct. 94 STScI started producing obs logs on a regular basis. The file format was changed in August 95. For homogeneity the next step is to compute jitter files for the time span until Aug. 95. Finally, the same pipeline could be adapted to get obs logs back to launch in 1990.

Acknowledgments. Setting up this pipeline was only possible with the invaluable help of the DP-, OPUS- and PDB-Teams at STScI. Special thanks to R. Perrine, P. Goldstein, P. Hyde, K. Steuerman and J. Sandoval who sorted out a lot of technical questions as well as DPT group head D. Swade and his deputy M.A. Rose, who arranged for a very productive environment during my visit at STScI. Many thanks also to D. Fraquelli and J. Scott of the OPUS group who showed me a lot of HST's operations and quality control mechanisms. For his collaboration in the early stages of the project many thanks to Joachim Paul of MPE.

References

Leitherer, C. et al. 1995, HST Data Handbook, STScI, 107

Lupie, O., Toth, B. & Lallo M. 1997, Observation Logs, STScI

Micol, A., Bristow, P., & Pirenne, B. 1998, this volume

[1] http://archive.eso.org/archive/jitview.html

Astronomical Data Analysis Software and Systems VII
ASP Conference Series, Vol. 145, 1998
R. Albrecht, R. N. Hook and H. A. Bushouse, eds.

HST Paper Products: A New Way to Look at HST Data

W. Hack
Space Telescope Science Institute, 3700 San Martin Dr., Baltimore, MD, 21218, Email: hack@stsci.edu

J.-C. Hsu
Space Telescope Science Institute, 3700 San Martin Dr., Baltimore, MD, 21218, Email: hsu@stsci.edu

Abstract. Two new instruments, STIS and NICMOS, were added to HST during the SM97 servicing mission, each with new types of data that need to be processed. As a result, a completely new system for generating paper products was created which provides a detailed summary of the observations and of the calibrations applied to the observations in the HST pipeline. This new version of the paper products generating software produces a totally redesigned summary for the older instruments (WFPC2, FOS, GHRS, and FOC) as well as creating paper products for the newly installed STIS and NICMOS. In this paper, we discuss the design and algorithms used to handle the extremely large amounts of data produced by the instruments on board HST, as well as the new task's limitations. This new version of the HST paper product software is now available in STSDAS Release V2.0.

1. Introduction

The Hubble Space Telescope (HST) not only generates a great deal of data in a single day, but a wide variety of data as well - from 1-D and echelle spectra to images and time-tag data. All of this data is processed on the ground to apply basic instrumental calibrations and prepare the data to be ingested into the HST archive and then analyzed by the observer. One final step in this processing is to create a summary for initially judging the quality of the observations and for determining the status of the calibrations applied to the data. This summary, however, must be efficient in its reporting and yet comprehensive enough in its results to be useful. Generating such a report for each instrument's data (WFPC2, FOC, FOS, GHRS, STIS and NICMOS) while retaining a similar enough format for easy comprehension was the primary goal for the software. In addition, the software must run quickly enough to prevent backlogs when ingesting into the archive. These two primary problems guided our effort to create a single software system to provide summaries known as 'paper products' for all HST data. This paper will summarize the major features of this software system.

2. Overview of Data Processing

The paper products software performs the following steps in creating the summaries for each set of observations:

- **Sorts input data**: Input data can be a combination of one or more instrument's data from any number of proposals. The input data can either be in FITS format, as if obtained from the HST DADS system, or GEIS format. All of the observations, or only selected observations, can be processed at one time depending on the user's input. A separate output file with its own cover page and summary will then be created for each logical unit defined by data from each instrument used in each proposal.

- **Summarizes each observation:** The basic exposure information for all observations in the proposal is provided in a tabular form. It includes target data, instrument configuration (such as filters used), and basic data statistics. Observations for each instrument are handled by their own function, allowing the summary to reflect the unique nature of each instrument.

- **Displays the data for each observation or observation set**: Grey scale images or plots of the data are provided for each exposure. These displays are created using SPP functions or CL scripts tailored specifically for the type of data produced by each instrument. In addition to the graphical display of the data, a summary of the exposure information is also provided. The basic forms of displaying the data are given for each instrument:

Table 1.

FOS	Target acquisition position plots, spectral plots, total flux plots
GHRS	spectral plots, jitter plots
FOC	image display
WFPC2	mosaic of CCD images, single chip display
NICMOS	thumbnail index of images, mosaiced images, target acquisition images
STIS	image display, spectral plots, jitter images, target acquisition images

- **Provides a summary of the observations data quality**: The last page(s) of the paper products is a summary of the calibration(s). A summary of pointing information (jitter data), explanations regarding quality flags and some basic image statistics are all reported here with an exhaustive list of calibration files and the origin of each file.

- **Process the IGI scripts**: The results from all the processing are generated in the SPP procedures and CL scripts as IGI commands, then processed by IGI to produce the final output.

This processing will produce a report tailored to each instruments data and contain the following estimated number of pages:

Table 2.

Cover Page and Help Page	1 Page each
Help Page	1 Page
Exposure Summary	1+ Pages
Data Display	1 Page/observation (FOC)
	1- 2 Pages/external observation (WFPC2)
	1-3 Pages/observation (FOS)
	1+ Page/observation (STIS)
	32 dithered obs/page + 1 page/image (NICMOS)
Observation Summary	1-2 pages

HST archive operations currently used Helvetica narrow fonts for all paper products. Use of different fonts may cause overlapping of text in the report.

3. Operation of the Software

The paper products software was designed to be run within the IRAF CL environment. The package consists of one primary task, pp_dads, and many hidden tasks to control each instrument's output. The pp_dads task only uses a small number of parameters; namely, input, device, fits_conv.

The parameter input controls which files are read in for processing, including using wildcards, list files or simply one image name. The parameter device can be set to a PostScript device to create PostScript files or a printer queue to send the output directly to a printer. This relies on using IRAF's printer queues, graphics kernels, and STSDAS's PostScript kernel.

4. Use of FITSIO

All HST data is routinely delivered to observers on tape in FITS format, and archived in the DADS system in FITS format as well. The paper products software, therefore, relies on the latest FITS kernel in IRAF to work with STIS and NICMOS data whilst also working directly with FOC, WFPC2, FOS, and GHRS data that are stored in the GEIS format.

5. SPP, CL and IGI

Most data processing is handled by routines written specifically for each instrument's data. These functions are written in SPP for NICMOS, FOC, WFPC2, and FOS, while CL scripts are used to process STIS and GHRS observations. Modular design can be readily modified for each instrument or even have new code added to support future new instruments. The use of SPP provides efficient access to image headers, the ability to perform statistical calculations while the image remains open, and great control of the output as IGI commands. The CL scripts used for STIS and GHRS are readily edited and can be used without compiling, however, they lack the efficient file I/O that SPP offers.

The results of the calculations and instructions for displaying the data are sent to an output file to be interpreted by IGI, the Interactive Graphics Interface. This provides several advantages for creating the paper products:

- **manages image display**: graphics and images can be displayed with image resizing being done as needed on the fly, enabling creation of thumbnail indices and annotated image

- **machine independent and flexible output**: IGI uses IRAF's graphics kernels and PostScript kernel to create PostScript output for any platform.

6. Computational Efficiency

The paper products software must run fast enough to keep up with the data stream coming from HST in order to avoid backlog. This led to the use of SPP code to generate the paper products for most instruments. In general, the computation of the paper products takes from 30-60 seconds per observation on a Sparc 5 workstation with only one user, and about 2-3 minutes per observation on a DEC Alpha workstation with several users. This is quick enough to avoid most backlog situations. The real bottleneck is the printing of the paper products. A local Printserver 17 printer can print about 15-20 Mb of PostScript output in an hour, however, the paper products code can generate 24Mb of PostScript in about 6 minutes on a Sparc 5, or about 20 minutes on an Alpha workstation. Therefore, a great deal of emphasis has been placed on setting up ways to print these jobs in parallel on multiple printers to keep up with the stream of data coming from HST. A drawback to this is the need for large temporary disk space for block averaged WFPC2/STIS/NICMOS images, PostScript files, and temporary IGI files and an effort is being made to reduce this need.

7. Summary

The new paper products package within STSDAS can process observations from a mix of instruments in one pass resulting in clear and concise documentation. It relies on many of IRAF's built-in functionality as well as many separate IRAF/STSDAS tasks to produce the paper products. The following packages are necessary for the operation of the paper products software: IRAF Version 2.10.4p2, or newer (preferably Version 2.11) and STSDAS/TABLES Version 1.3.5, Version 2.0, or newer. The simple, one task interface makes it easy to run, both interactively by a user after extracting data from DADS and in batch mode from a script, as done by the HST archive. This experience with processing daily observations has proved that this task can function efficiently and effectively with minimal intervention.

Astronomical Data Analysis Software and Systems VII
ASP Conference Series, Vol. 145, 1998
R. Albrecht, R. N. Hook and H. A. Bushouse, eds.

On the Need for Input Data Control in Pipeline Reductions

J.-P. De Cuyper and H. Hensberge

Royal Observatory of Belgium, Ringlaan 3, B-1180 Brussel, Belgium

Abstract.
An analysis of series of flat-field exposures obtained in echelle mode adds support to the view that the performance of an instrument should be monitored closely and described *quantitatively*. This simplifies enormously the data reduction procedure and allows the user to interpret high-quality data correctly.

1. Introduction

Nowadays it is not unusual for the knowledge about an instrument acquired by the design and construction team to be only partially transmitted to those developing the data reduction software or operating the instrument; the feedback to the observer is still smaller. This may result in confusion about basic instrumental parameters, the precision of the set-up, the amount and distribution of calibration frames which are required and finally the assumptions on which the data reduction procedure should rely. The data set then does not deliver the precision aimed at during observing, or observing time was spent inefficiently. The concept of pipeline reductions and on-line controlled operation of the instrument and of the observational procedure offer unprecedented possibilities for delivering data with known error characteristics, provided that the instrumental design, the set-up, the calibration and the data reduction procedure are tuned to each other at a consistent level of accuracy. Our experience with echelle spectroscopy indicates that, as a visiting astronomer, it is impossible to collect the information needed to obtain this goal. Moreover, it is not at all evident to what extent forthcoming pipelines will include realistic error estimates (in addition to random noise estimates).

2. An Example: Series of flat-fields

The results shown here refer to two adjacent series of eight consecutively taken flat-fields (tungsten lamp exposures). The telescope was in the zenith and tracking was off. The analysis method applied is insensitive to global changes in intensity and to small changes in location, width or shape of the cross-order profile (these changes are below 0.5% in intensity and within 0.02 pix for the other quantities in both series). For each pair of frames, a parameter d indicating the lack of similarity in the shape of the blaze profile was computed (Figure

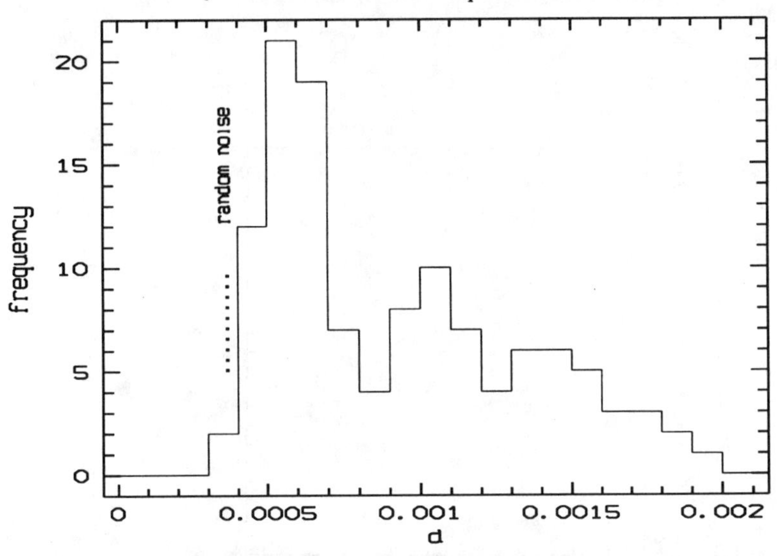

Figure 1. Histogram of the dissimilarity parameter d for all combinations of 2 out of the 16 flat-fields. d^2 scales with the variance due to long-range ($\approx$ 100 pixels) deformations in ratios of extracted flat-field orders, except for an offset due to the contribution of random noise. The value for pure random noise is indicated

1). It turns out that the excess of d over its value calculated from the case of random noise can be modelled as a "distance" between frames (Figure 2).

The lack of repeatability in subsequent frames is due to instabilities that grow and disappear, rather than to a slow, continuous change with time. Consecutively taken frames differ more than the ones separated by an intermediate frame, and the global pattern of changes repeats in the two independent series (Figure 2). Relative deformations in the shape of the blaze profile appear over several spectral orders in the same set of rows of the detector (see e.g., Figure 3 near $x \approx 300$).

It is not our intention to show that things *can* sometimes go very wrong, but merely that the accuracy is *generally* limited by systematic errors. The example shown above does not refer to an exceptional malfunctioning, but to a common situation. Notice that it is not uncommon to detect stronger effects when comparing exposures taken with longer time delays and/or in different telescope positions. The detectability of such systematic effects sets a natural limit on the precision of order merging (since the intensity ratio of the wavelength overlap region of consecutive orders is affected), on the level up to which faint, shallow spectral features can be trusted and on the precision of the continuum placement.

3. Data Reduction Strategy

Experience with echelle spectroscopy confirms that the previous example is not an isolated case of bias. Systematic errors are detectable in almost all types of frames and they influence directly the applicability and the accuracy of the

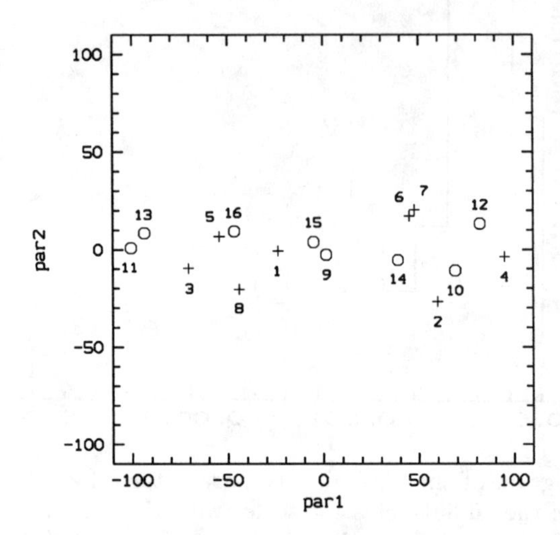

Figure 2. The dissimilarity of flat-fields represented as an Euclidian distance between frames situated in a plane (*par1, par2*). The numbering indicates the order of the exposures. Notice the usually large distance between consecutive frames, and the similarity between the series 1-8 and 9-16

data reduction algorithms. Depending on the ratio of systematic to random noise, algorithms that are based on the dominance of random noise (such as the detection of radiation events, the application of optimal extraction and any reduction step involving non-robust least-squares fitting) may need refinements.

Rather than commenting on specific sources of bias, we like to outline a procedure, interconnecting the different phases from instrument development and testing to data reduction software development, that in our opinion would permit the user to evaluate properly the quality of the final data with regard to the particular aspects of interest to his/her specific purpose:

- Characterize detector and instrument using the most stable mode of operation. Specify the maximum precision that will be supported in the set-up, calibration and data reduction flow.
- Specify the target set-up accuracies (consistent with the data reduction requirements) and check during operation whether the calibration and science frames fall within these specifications. Specify how frequently extensive stability checks are needed.
- Use robust techniques. Exploit knowledge about the instrument and environment, at least in a differential if not in an absolute way. A data frame

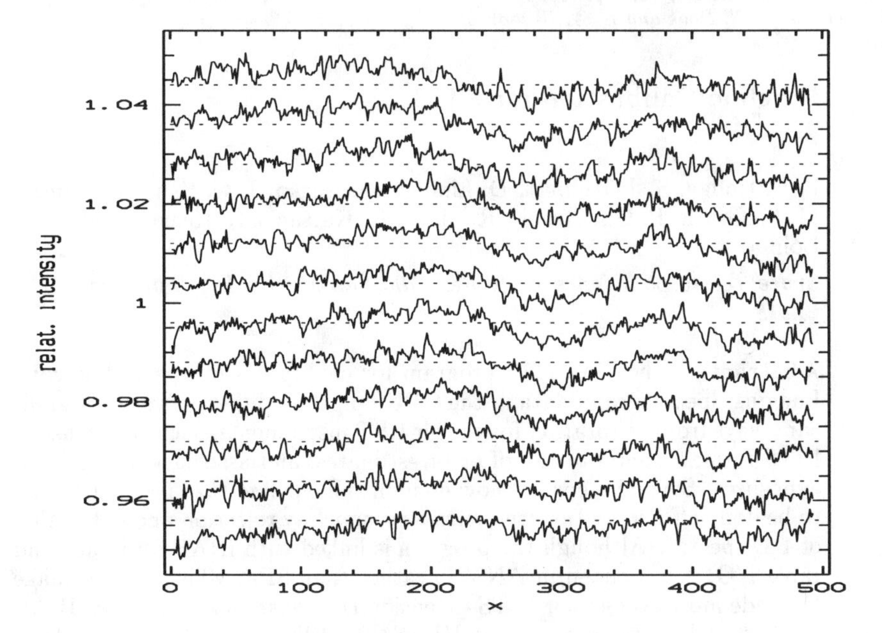

Figure 3. *Relative* deformation of the blaze profile expressed as the intensity ratio of the sum of frames #11 and #13 to the sum of #10 and #12. This ratio is shown along 12 consecutive orders, shifted vertically for clarity (the dashed lines indicate the relative intensity level 1.0 for each order). Instrument: CASPEC, detector: ESO CCD #32

with its specific calibration frames should not be treated independently from the global observing run. Several factors vary systematically with quantities that are known or measurable and do not require ill-determined free parameters.

- Quantify the extent to which assumptions made during the data reduction are invalid and archive them with the final data.
- Use the experience gained during the reduction process to improve at convenient times (e.g., when important instrumental interventions are unavoidable) the observing and data reduction strategy, and, ultimately, the development of new instruments.

Acknowledgments. This research was carried out in the framework of the project 'IUAP P4/05' financed by the Belgian Federal Scientific Services (DWTC/SSTC). We thank W. Verschueren (RUCA, University of Antwerp), who obtained the calibration spectra discussed in this paper, and H. Van Diest for help with the data handling. This work is based on observations obtained at the European Southern Observatory (ESO), La Silla, Chile.

Astronomical Data Analysis Software and Systems VII
ASP Conference Series, Vol. 145, 1998
R. Albrecht, R. N. Hook and H. A. Bushouse, eds.

Pipeline Calibration for STIS

P. E. Hodge, S. J. Hulbert, D. Lindler[1], I. Busko, J. C. Hsu, S. Baum, M. McGrath, P. Goudfrooij, R. Shaw, R. Katsanis, S. Keener[2] and R. Bohlin

Space Telescope Science Institute, 3700 San Martin Dr, Baltimore, MD, 21218

Abstract. The CALSTIS program for calibration of Space Telescope Imaging Spectrograph data in the OPUS pipeline differs in several significant ways from calibration for earlier HST instruments, such as the use of FITS format, computation of error estimates, and association of related exposures. Several steps are now done in the pipeline that previously had to be done off-line by the user, such as cosmic ray rejection and extraction of 1-D spectra. Although the program is linked with IRAF for image and table I/O, it is written in ANSI C rather than SPP, which should make the code more accessible. FITS extension I/O makes use of the new IRAF FITS kernel for images and the HEASARC FITSIO package for tables.

1. Introduction

The Space Telescope Imaging Spectrograph (STIS) is a complex instrument containing both CCD and MAMA detectors. STIS can observe in both spectroscopic and imaging modes, using either first order or echelle gratings, and taking data in either accumulate or time-tag mode. STIS has many apertures, some filtered, including large apertures for imaging as well as long and short slits for spectroscopy. On-board processing options include binning and restricting the image to a subset of the full detector.

CALSTIS is the program that performs the "pipeline" calibration of STIS data, the processing that can be done non-interactively and with table-driven parameters.

See the STIS Instrument Science Reports (ISRs) and HST Data Handbook for further information. These can be reached via the URL
http://www.stsci.edu/documents/data-handbook.html

2. Discussion

The previous pipeline calibration tasks read and write so-called GEIS image format, known in IRAF as Space Telescope Format (STF). The new programs,

---

[1]Advanced Computer Concepts, Inc.

[2]University of Illinois, Urbana-Champaign, Urbana, IL 61801

on the other hand, use FITS format. Global keywords are written to the primary header, and the images are written as IMAGE extensions. The primary data unit is null, i.e. is not present. Three IMAGE extensions are used for a STIS image, the science data, corresponding error estimates, and data quality flags; this is called an "image set," or imset. Reference images (e.g. flat fields) use the same format, while reference tables are in FITS BINTABLE extensions. Multiple exposures at the same pointing (e.g. cosmic-ray splits or repeat observations) are stored in the same file, one image set per exposure.

Previous pipeline routines were tasks under the IRAF CL and were written in either SPP or FORTRAN. The new routines are host level programs written in ANSI C, although they are still linked with IRAF for I/O. The I/O interface makes use of HSTIO and CVOS, written by Allen Farris. HSTIO is the higher level interface for images; this can be used to read or write an entire image set in one call. HSTIO is independent of IRAF at the interface level, although the current implementation calls CVOS routines. CVOS is a C-callable interface to the standard IRAF subroutines, in particular for image and table I/O. Image I/O uses the IRAF FITS image kernel written by Nelson Zarate. Table I/O uses the HEASARC FITSIO package written by Bill Pence. For both images and tables, an input or output file with filename extension ".fits" will automatically be treated as a FITS file.

There are several major components to CALSTIS: cosmic-ray rejection for CCD data, basic 2-D reduction (bias, dark, flat, etc.), wavecal processing, 2-D spectral rectification, and 1-D spectral extraction. These can be run either combined into a single executable or as several individual executables (eight, actually). The former is the pipeline version, but both options are available off-line. When running the single executable (cs0.e), the calibration switches are read from the input primary header, as with previous calibration pipeline routines. The individual executables, however, take the calibration switches from the command line; the default is to perform all relevant steps.

Although the CALSTIS executables are host level programs, CL scripts have been provided in the STIS package for running CALSTIS from IRAF, for the convenience of IRAF users. `calstis` is the single executable, the pipeline version; `basic2d` does basic 2-D reduction; `ocrreject` is the cosmic-ray rejection task; `wavecal` does wavecal processing; `x2d` does 2-D rectification; and `x1d` does 1-D spectral extraction. These scripts have parameters for input and output file names, for options, and–except for the `calstis` task itself–for the calibration switches. They construct a command line string beginning with "!" (to escape from IRAF to Unix), and they invoke the program using the syntax: `print (cmdline) | cl`.

3. Further Details

Processing starts with populating the data quality array from a reference table and assigning initial values for the error estimates based on pixel values, gain, and read noise.

For CCD data, the bias level from the overscan will be subtracted and the overscan regions removed. If the file contains multiple exposures, the separate exposures will then be combined to reject cosmic rays. The algorithm is very similar to the one used in the `crrej` task in the `wfpc` package. Parameters controlling the cosmic ray rejection are read from a reference table; the row in that table is selected based on exposure time and number of exposures. The individual exposures must have been taken at the same pointing.

The usual bias, dark, and flat field corrections are then applied. MAMA data are corrected for nonlinearity. For CCD data, code is in place for shutter shading and analog-to-digital correction, but these steps are not performed because they do not appear to be required.

For medium or high resolution spectroscopic MAMA data, the reference files need to be convolved to account for Doppler shift, because a correction for spacecraft motion during the exposure is applied on-board. That is, whenever a photon is detected, the on-board software reads the detector coordinates of the event, adjusts the pixel number in the dispersion direction by the amount of the Doppler shift at that moment, then the image buffer is incremented at the adjusted location. During calibration, therefore, the data quality initialization table and the dark and flat field reference images need to be shifted by the same amount before being applied. This amounts to a convolution by the various Doppler shifts throughout the exposure.

For spectroscopic data, a wavecal (line lamp) observation is used for determining the location of the spectrum on the detector. The location is needed for accurate assignment of wavelengths, and for position information in the case of long-slit data. Flat fielding and 2-D rectification are first applied to the wavecal, and cosmic rays (in CCD data) are rejected by looking for outliers in the cross dispersion direction. The offset in the dispersion direction between the expected location of the spectrum and the actual location is found by cross correlating the observed spectrum with a template. In the cross dispersion direction, the offset is found either by finding edges or by cross correlation, depending on the length of the slit. The long slits have two occulting bars, and it is the edges of these bars that are used for finding the location. Edges are found by convolving with [-1, 0, +1]. The location is found to subpixel level by fitting a quadratic to the three pixels nearest the maximum (or minimum, depending on the edge).

2-D rectification is performed for long-slit data; it can can also be done for imaging mode but currently is not. For each pixel in the output rectified image, the corresponding point is found in the input distorted image, and bilinear interpolation is used on the four nearest pixels to determine the value to assign to the output. No correction for flux conservation is applied, as this is accounted for in the flat field. Mapping from an output pixel back into the input image makes use of the dispersion relation and 1-D trace. The dispersion relation gives the pixel number as a function of wavelength and spectral order number. The 1-D trace is the displacement in the cross dispersion direction at each pixel in the dispersion direction. Both of these can vary along the slit, so the dispersion coefficients and the 1-D trace are linearly interpolated for each image line. Corrections are applied to account for image offset, binning, and subarray. For imaging mode, the mapping is a 2-D polynomial (Chebyshev) representing the optical distortion, and the IRAF gsurfit package is used to evaluate the function.

Extraction of 1-D spectra is performed in the pipeline for echelle data, and it can be performed for long-slit data. When running the off-line `x1d` task, the target location on the slit can be specified. When running the pipeline version, however, the brightest object on the slit is assumed to be the target, and the extraction will be centered on that object. The same corrections for image offset are applied for 1-D extraction that were done in 2-D rectification. The location of the spectrum is then refined by cross correlating the data with the 1-D trace. At each pixel in the dispersion direction, the data are summed over some range of pixels for the on-source region and background regions. The displacement of these extraction regions in the cross dispersion direction is taken from the same 1-D trace used in 2-D rectification. No resampling is done in the dispersion direction; the associated wavelengths are therefore nonlinear, and they are computed using the dispersion relation. The output is written as a FITS BINTABLE extension, with one row per spectral order. For long-slit data, then, there will be just one row, while there can be 50 rows for echelle data.

Astronomical Data Analysis Software and Systems VII
ASP Conference Series, Vol. 145, 1998
R. Albrecht, R. N. Hook and H. A. Bushouse, eds.

Data-flow for the ESO Imaging Survey (EIS)

R. N. Hook[1], L. N. da Costa, W. Freudling[1] and A. Wicenec

European Southern Observatory, Karl Schwarzschild Str-2, D-85748, Garching, Germany, e-mail: rhook@eso.org, ldacosta@eso.org, wfreudli@eso.org, awicenec@eso.org

E. Bertin

Institute d'Astrophysique de Paris, 98bis Boulevard Arago, 75014 Paris, France, e-mail: bertin@iap.fr

E. Deul

Sterrewacht Leiden, P.O. Box 9513, 2300 RA Leiden, The Netherlands, e-mail: Erik.Deul@strw.LeidenUniv.nl

M. Nonino

Osservatorio Astronomico di Trieste, Via G.B. Tiepolo 11, I-34131, Trieste, Italy, e-mail: nonino@ts.astro.it

Abstract. The ESO Imaging Survey (EIS) uses the ESO New Technology Telescope (NTT) and the EMMI camera to image several square degrees of sky to moderate depth through B,V and I filters. In later parts of the survey smaller areas will be imaged more deeply in a wider selection of spectral bands. The primary aim of the project is to provide a statistically useful selection of interesting faint objects as targets for the ESO Very Large Telescope (VLT) which will see first light in 1998. This paper gives an overview of the data flow for this project, some comments on the methodology employed and examples of the initial data. Some techniques for attacking the generic problem of mosaicing inhomogeneous overlapping images and extracting catalogues from the resulting coadded frames are described. More information can be found at the EIS Web page: http://www.eso.org/eis/.

1. Introduction

During the second half of 1997 and much of 1998 ESO is conducting a deep, multicolour imaging survey using the EMMI camera on the NTT. The ESO Imaging Survey (EIS, Renzini & da Costa 1997), will generate suitable statistical samples for a variety of astronomical applications, ranging from candidate objects at the outer edge of the Solar System all the way to galaxies and quasars

[1]Space Telescope — European Coordinating Facility

at extremely high redshift. EIS data should provide a suitable database from which targets can be drawn for observations with the VLT, in its early phase of scientific operation (from the third quarter of 1998 to approximately the end of 2000). EIS has been conceived as a service to the ESO community and all the data will become public immediately after its completion.

In the initial (EIS Wide) phase of the project many exposures are being taken using a stepping pattern on the sky such that each step is about half of the frame size in both X and Y and on average each point is observed twice. This strategy allows accurate astrometric association between frames and gives an overlap pattern which shows clearly on Figure 3.

Figure 1 gives an overview of the data flow of the project starting with the preparation of "observation blocks" which are the basic unit of telescope execution and working through to the preparation and archiving of catalogues from the final coadded images.

2. Overview of Software Development

The data volume of EIS is considerable and requires a fully automated pipeline for both observation planning and execution and subsequent data handling. There were insufficient resources within ESO to develop the complex software system required for EIS starting from scratch. Instead funds were used to employ visitors who had extensive experience from other, similar, projects to adapt software and integrate it into a complete EIS pipeline. This was run on the ESO "Science Archive Research Environment" (a multi-processor Ultra Sparc server) and is acting as a test case for this system.

There were four areas of software re-use (excluding subsequent science software):

- Standard IRAF tools were used for the initial calibration of each input image which were taken from the ESO archive.

- The DENIS (DEep Near IR Survey of the Southern Sky, Epchtein, 1994) data pipeline, was extensively modified for EIS and used to perform photometric and astrometric calibration.

- The SExtractor object detection and classification code (Bertin & Arnouts, 1996) was modified for EIS and integrated into the pipeline. It was used both to find objects during the calibration steps and also to prepare final science catalogues.

- The Drizzle image coaddition software (Hook & Fruchter 1997; Fruchter & Hook 1997), originally developed for HST, was modified to efficiently handle the EIS geometry and create coadded final output images from the many, overlapping, input frames.

A major aim of this software is to handle the generic problem of mosaicing overlapping images with widely varying characteristics and the extraction of information from the resulting inhomogeneous coadded frames.

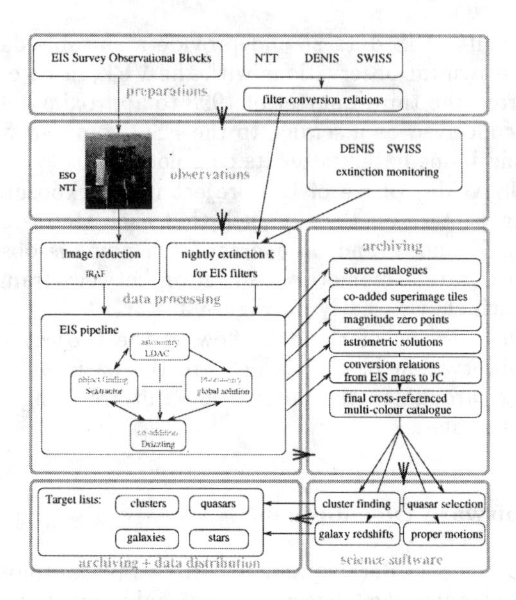

Figure 1. An overview of the EIS dataflow

3. Example Data

At the time of writing approximately half of the EIS survey has been completed. Conditions have not been uniform and hence the resulting coadded images vary considerably in seeing, transparency and depth. This variation poses many problem to automatic data analysis, particularly on the coadded images which may have discontinuities in PSF and noise levels. This problem has been addressed by creating not just a final data image but also an image with the weight of each output pixel and another in which the "context" is encoded. The context is essentially a list of input images which have contributed to a given output pixel. This allows the PSF and other characteristics of sections of the coadded image to be reconstructed at each pixel when needed.

Figure 2 gives an example of a subset of an output coadded EIS image. It covers about 18 arcminutes and is the result of combining about 10 EMMI images. Figure 3 shows how the weights vary across this image. Image defects such as dead columns and saturated regions show clearly as does the variation of weight among the frames.

4. Future Plans

The primary aim of EIS is to provide data products, both images and derived catalogues, which will enhance the scientific returns from the forthcoming 8m telescopes, in particular the ESO VLT. An important secondary aim is to act as a test-bed for the VLT data flow concept, in particularly the observation preparation stage and the exploitation of the data within the Science Archive Research Environment. The software developed for EIS will act as the basis for

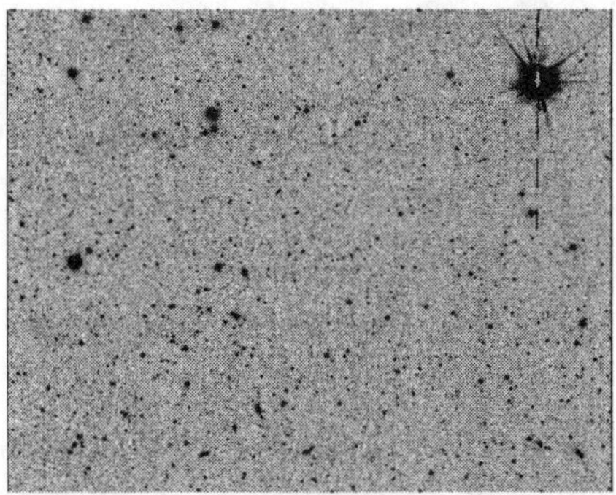

Figure 2. An example of part of a coadded image from EIS Wide.

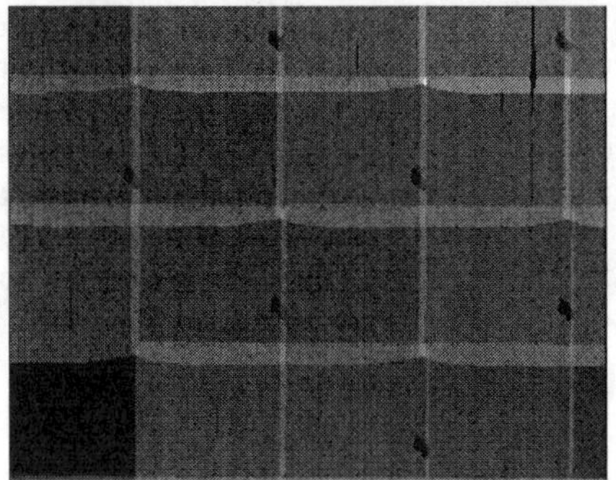

Figure 3. The weighting image associated with Figure 2.

a similar imaging survey to be conducted using the ESO 2.2m telescope with a large mosaiced CCD camera and may later also lead into a dedicated imaging survey telescope on Paranal with the ESO VLT.

References

Bertin, E. & Arnouts, S., 1996, Astronomy and Astrophysics Supplement, 117, 393

Epchtein, N., 1994, Astrophysics and Space Science, 217, 3.

Fruchter, A. S. & Hook, R. N., 1997, in Applications of Digital Image Processing XX, ed. A. Tescher, Proc. S.P.I.E. vol. 3164, 120

Hook, R. N. & Fruchter, A. S., 1997, in ASP Conf. Ser., Vol. 125, Astronomical Data Analysis Software and Systems VI, ed. Gareth Hunt & H. E. Payne (San Francisco: ASP), 147

Renzini, A. & da Costa, L. N., 1997, The Messenger, 87, 23

Astronomical Data Analysis Software and Systems VII
ASP Conference Series, Vol. 145, 1998
R. Albrecht, R. N. Hook and H. A. Bushouse, eds.

System Interfaces to the STIS Calibration Pipeline

S. Hulbert

Space Telescope Science Institute, 3700 San Martin Drive, Baltimore, MD 21218

Abstract. The routine processing of STIS data in the calibration "pipeline" requires an interrelated set of software tasks and databases developed and maintained by many groups at STScI. We present the systems-level design of this calibration pipeline from the perspective of the STIS instrument group at STScI. We describe the multiple interfaces to the pipeline processes that the STIS group works across in an effort to provide for the routine calibration of STIS data. We provide a description of the pipeline processing while treating the actual processing software as a collection of black boxes. We describe the systems engineering requirements levied by and against the STIS group needed to carry out day-to-day calibration of STIS observations.

1. Functional View of STIS Calibration Pipeline

The functional view of the STIS calibration pipeline (Figure 1) illustrates the interactions between the external world and the pipeline itself. In this case, there are two sets of external users: observers (including GO and GTO) and the STIS group at STScI. The two groups of users are similar in that they both receive data products consisting of raw and calibrated STIS data from the calibration pipeline system. Additionally, the STIS instrument scientists are responsible for calibrating STIS and feeding these calibration parameters back to the pipeline system. (Note: the arrows that are used in all three figures indicate the flow of information between groups.)

2. Top Level Object Interaction Diagram of STIS Calibration Pipeline

To demonstrate the number of details that must be managed to run a calibration pipeline for STIS we identify distinct entities within the Space Telescope Science Institute with which the STIS group interacts in the course of keeping the STIS calibration pipeline up and running. Borrowing from the object-oriented paradigm, we designate these entities as "objects". The Object Interaction Diagram (OID) of the STIS calibration pipeline (Figure 2) shows in detail the "objects" within STScI that the STIS group interacts with in the process of ensuring that STIS data are properly calibrated. These objects consist of three distinct classes: Operations, Database and Tools. The Operations class consists of subsystems (usually controlled by distinct management groups) that actually handle STIS data on a daily basis. The Database class consists of databases con-

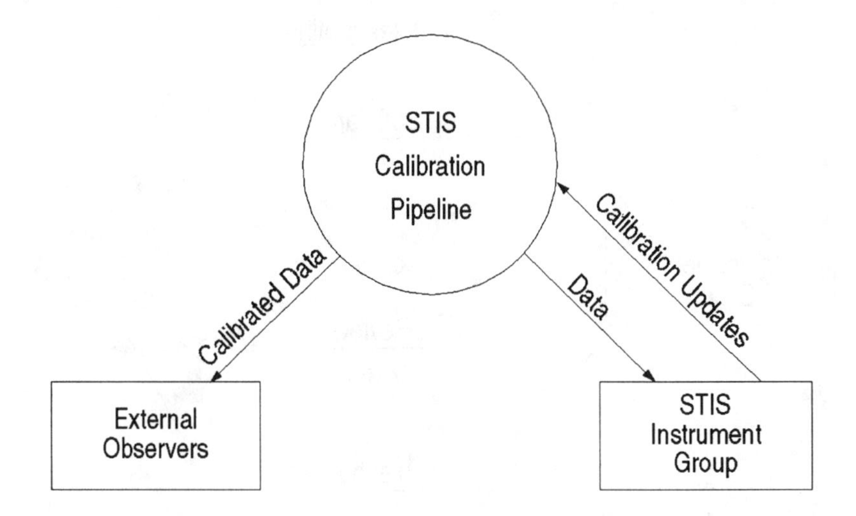

Figure 1. Functional View of STIS Calibration Pipeline.

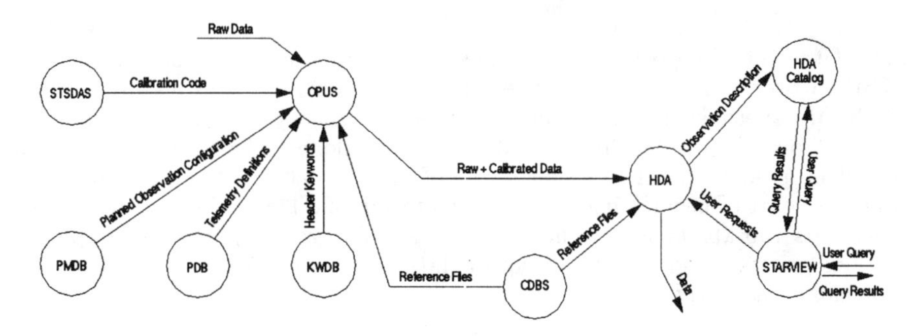

Figure 2. Top Level Object Interaction Diagram of STIS Calibration Pipeline

taining myriad details of STIS operations which are cared for by the database designers and operators. The third class of object in this model, Tools, consists of groups of programmers with the responsibility of producing specialized software tools. The OID concepts used in this description of the calibration pipeline have been adapted from Morris, et al. (1996).

3. Object Interaction Diagram (OID) of the STIS Group Interface with the STSDAS Group

Missing from the OID of the STIS calibration pipeline is the explicit interaction between the STIS group and each object in the model of the STIS calibration pipeline. The OID shown in Figure 3 gives, as an example, the detailed descrip-

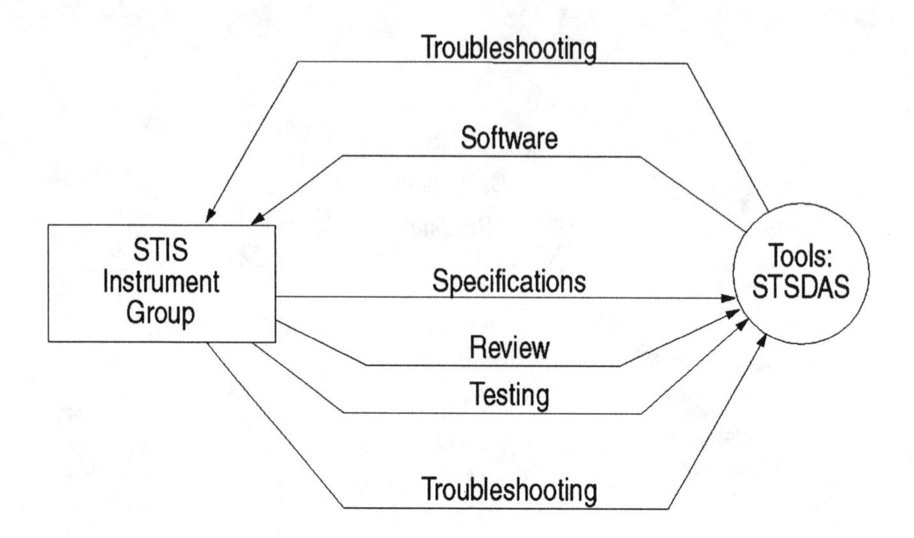

Figure 3. Object Interaction Diagram of the STIS Group Interface with the STSDAS Group

tion of the interface between the STIS group and the Tools class object, the STSDAS programming group. The STSDAS group is responsible for crafting the highly specialized calibration code used to process STIS data. This code runs not only in the pipeline but also as stand-alone tasks. Special attention should be paid to the interactions (pair of arrows) labeled "Troubleshooting". A critical part of negotiating the implementation of STISspecific features of the pipeline is the ability and willingness of the STIS group and the relevant operations group, programming group or database group to identify and resolve problems in a timely manner.

4. Systems Engineering Details

To assist in the management of the STIS portion of the calibration pipeline, the STIS group is subject to constraints imposed by the systems engineering of the pipeline process. For example, the STIS group:

- maintains ICD-19 which is a description of the data formats for all STIS data including specifications for the header keywords. This information is stored in the Keyword Database (KWDB).
- maintains ICD-47 which is a description of the STIS calibration reference file formats. This information is stored in the Calibration Database System (CDBS).
- performs cross-subsystem testing of software changes affecting STIS.
- participates in the biweekly Pipeline Working Group consisting of representatives from all of the calibration pipeline "objects" (groups).

- provides documented specifications via instrument science reports (ISRs), technical instrument reports (TIRs) and the problem reporting (OPR) system.

References

Morris D., Evans, G., Green, P. & Theaker, C. 1996. Object Oriented Computer Systems Engineering. (New York: Springer), 70

Astronomical Data Analysis Software and Systems VII
ASP Conference Series, Vol. 145, 1998
R. Albrecht, R. N. Hook and H. A. Bushouse, eds.

REMOT: A Design for Multiple Site Remote Observing

P. Linde

Lund Observatory, Box 43, SE-221 00 Lund, Sweden, Email: peter@astro.lu.se

F. Pasian and M. Pucillo

Osservatorio Astronomico, Via G.B.Tiepolo 11, I 34131 Trieste, Italy , Email: {pasian,pucillo}@ts.astro.it

J. D. Ponz

ESA Villafranca P. O. Box 50727, 28080 Madrid, Spain, Email:jdp@vilspa.esa.es

Abstract. The REMOT project objective is to develop and validate a generic approach to allow remote control of scientific experiments and facilities that require real time operation and multimedia information feedback. The project is funded through the European Union Telematics initiative and is a collaboration, involving representatives from both the astro- and plasma physics communities. Standard communications infrastructure and software solutions will be used wherever possible. In the first step, requirements have been collected and analysed, resulting in a set of service definitions. These have been partly implemented to perform a set of demonstrations, showing the feasibility of the design.

1. Introduction

As a response to the increasing need for remote control of large scientific installations, the REMOT (Remote Experiment MOnitoring and conTrol) project was initiated in 1995. The final goal of the project is to develop and validate a generic approach to allow real time operation and multimedia information feedback, using communications infrastructure available to European researchers.

Funded through the European Union Telematics Organisation, the project involves representatives from the astro- and plasma physics communities and from several teleoperation and software design institutes and companies. The partners from the astrophysical community include OAT (Italy), IAC (Spain), LAEFF (Spain), Nordic Optical Telescope and CDS (France).

2. Requirements and Operational Scenario

The design of the generic system and communications architecture started by creating a large database of user requirements, collected from each participating partner. The requirements were derived from the following operational scenario:

- The scientist generates the observing proposal. Upon approval of the proposal, a detailed observing plan is prepared, using the instrument simulation facility to allow optimal usage of the instrument. The plan is then submitted to the scheduling subsystem to prepare the actual time allocation plan.
- Several time allocation plans are generated using a scheduling subsystem, defining a primary session and several alternate sessions. The actual operational session is designed in a flexible form, allowing for observing conditions and priority requirements.
- When a session is allocated for operations, the end user node is contacted, and the client capabilities are validated to define the supported services and hand-over takes place.
- After hand-over, the actual operational session takes place, controlled by the instrument server, with minimal input from end user clients.
- During the session, several clients can work with a given instrument server. One of the clients is the primary, operational node, the others acting as passive, mirror nodes. A given client can work with several different instrument servers, allowing simultaneous observations.
- Instrument servers include an on-line archive facility, to keep record of the real-time activity and to store the observed data.
- Observed data and related archive data are transferred from the server to the clients upon request, for quick-look, calibration and analysis.

3. Services Provided

The services that REMOT will provide can be roughly divided into four categories:

- Monitoring activities, to follow basic operations performed at the telescope. No control operation is performed at this level, therefore no real-time activity is really required and no guaranteed bandwidth is required.
- Remote control activities, controlling the activities of telescope and instruments, including performing real-time operations. A minimal guaranteed bandwidth for the support of real-time is an absolute requirement.
- Remote observing activities, including monitoring and some of the control activities, such as the setup of basic modes and functions of telescope and instruments, download of compressed (lossy) data for quick-look purposes, etc. Guaranteed bandwidth for the support of real-time is required together with intermediate transfer rates.

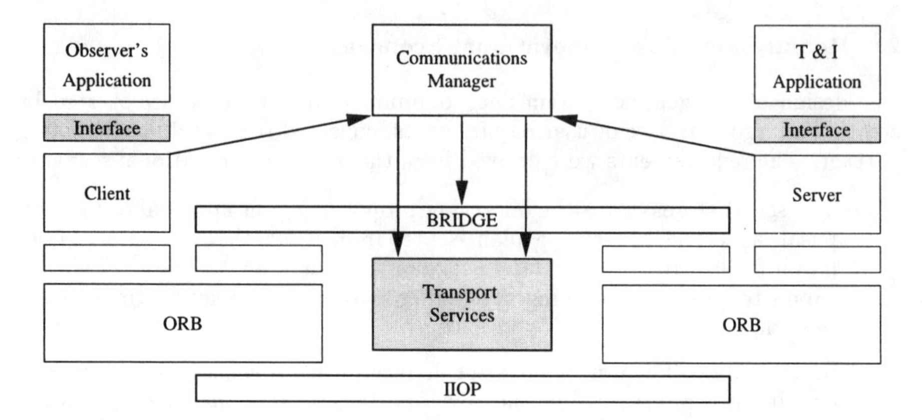

Figure 1. The architecture of the REMOT system. The observer's application is interfaced to a CORBA client and, through the Object Request Broker (ORB) mechanism, communicates with the Telescope and Instruments application. A Communications Manager arbitrates if the communication is to take place through the standard IP-based channel (IIOP), through a bridge supporting Quality of Service (QoS), or using directly the transport services.

- Data transfer activities,for the transfer of scientific data files (lossless compressed). Guaranteed bandwidth is not a requirement but high transfer rates are desirable.

4. Implementation of Services

Analysis of the collected database of requirements resulted in a set of service definitions. These services will be implemented by using two different conceptual planes (see Figure 1). These are (1) the *System Plane*, supporting Quality of Service (QoS), (2) the *Management Plane*, supporting reliable communications. Operational services will be performed in the System Plane and managed over the Management Plane, while auxiliary services will be supported over the Management Plane.

For the communications architecture we use a CORBA compliant mechanism: the Management Plane is implemented by using CORBA and its IP-based layer (IIOP) directly. When QoS is required (System Plane), a bridging mechanism bypassing IIOP can be used, or alternatively there is the possibility of using the transport services directly. The presence of a Communications Manager to arbitrate communication resources is necessary in both cases.

5. An Example: The Galileo telescope (TNG)

Galileo's high-level control system is based on a distributed computer architecture in which both the code and the data are shared by all the workstations supporting the systems (Pucillo 1994). Remote control is implicitly embedded in the design, since the control workstation can either be located or be connected via a bridge or a router.

REMOT improves this remote access philosophy since, besides using a common user interface for different telescopes, it does not require TNG-specific hardware, but will allow remote observations to be run "at the astronomer's own desk", on a simple PC.

Integrating the control system of the Galileo telescope in the REMOT project is simple:

- The IP protocol will be implemented on top of ISDN, so to allow testing the control system concept on ISDN-based connections.
- An appropriate software interface will be built between the control system and the REMOT-provided communications architecture. Direct use of the communications infrastructure and the bridging mechanism guaranteeing QoS will be provided.
- The use of ATM services may be considered in a possible extension of the project.

The software interface between the control system and REMOT consists of an ancillary process running under the same environment. This process is also responsible for retrieving from the TNG on-line database the updated status of the system, and for downloading observational data in original (FITS) or compressed (quick-look) format to the remote user.

6. Project Status and Future Activities

A prototype of the complete system is under development, including the basic services. Since REMOT is limited in scope, only a subset of the design will be currently implemented. This will, however, allow a practical demonstration to be performed using the new Galileo telescope located at La Palma and the IAC 80 cm telescope located on Tenerife.

REMOT will be extended in a continuation project DYNACORE, where a complete implementation is targeted. The design will then be upgraded to include real time access to data bases, flexible scheduling and dynamic reconfiguration. Additional information on the project can be found at `http://www.laeff.esa.es/~tcpsi/`.

References

Pucillo, M. 1994, *Handling and Archiving Data from Ground-Based Telescopes*, M.Albrecht & F.Pasian eds., ESO Conferences and Workshop Proceedings, Vol. no. 50, 77

Astronomical Data Analysis Software and Systems VII
ASP Conference Series, Vol. 145, 1998
R. Albrecht, R. N. Hook and H. A. Bushouse, eds.

The Distributed Analysis System Hierarchy (*DASH*) for the SUBARU Telescope

Y. Mizumoto, Y. Chikada, G. Kosugi[1], E. Nishihara[2], T. Takata[1] and M. Yoshida[2]

National Astronomical Observatory, Osawa, Mitaka, Tokyo 181, Japan

Y. Ishihara and H. Yanaka

Fujitsu Limited, Nakase, Mihama, Chiba 261, Japan

Y. Morita and H. Nakamoto

SEC Co. Ltd. 22-14 Sakuragaoka, Shibuya, Tokyo 150, Japan

Abstract. We are developing a data reduction and analysis system *DASH* (Distributed Analysis Software Hierarchy) for efficient data processing for the SUBARU telescope. We adopted CORBA as a distributed object environment and Java for the user interface in the prototype of *DASH*. Moreover, we introduced a data reduction procedure cube (*PROCube*) as a kind of visual procedure script.

1. Introduction

The purpose of *DASH* is efficient data processing for the SUBARU telescope which will produce up to 30TB of data year. This data production rate requires high computational power and huge mass storage capacity. *DASH* is, therefore, designed as an observatory system for astronomical data reduction and analysis, which cooperates with the SUBARU observation software system and the data archival system. *DASH* is, of necessity, designed for a distributed heterogeneous computer environment; considering role sharing and joint work. We adopt CORBA (Common Object Request Broker Architecture) for a distributed environment platform and Java for the user interface objects. As the first step of the development, which is a three year project, we made a prototype of *DASH* for trial of the new computer technology such as CORBA and Java.

---

[1]Subaru Telescope, National Astronomical Observatory of Japan, Hilo, HI 96720, USA

[2]Okayama Astrophysical Observatory, Kamogata-cho, Okayama 719-02, JAPAN

2. Design Concept and Requirement of *DASH*

The requirements of the data reduction system for SUBARU telescope are the following;

1. *DASH* has an enriched function of data management. The data management cooperates with SUBARU data archives and database (Ichikawa 1995, Yoshida 1997). It can collect all kind of data necessary for data reduction from the SUBARU data archives using Observation Dataset, which is produced by the SUBARU observation control systems (Kosugi 1996). The Observation Dataset is a set of information about acquired object frames, related calibration frames such as a bias frame of CCD, and others. The data management also keeps track of temporary files that are created during data reduction process.

2. *DASH* can be used on a distributed heterogeneous computer environment. In particular, the console or operation section of *DASH* must work on many kinds of computers (workstations and PCs).

3. It is easy to assemble a suitable processing pipeline with the steps chosen through a series of data processing trials.

4. *DASH* aims at open architecture. We intend to use applications from widely used astronomical data reduction packages such as IRAF and Eclipse within *DASH*.

According to the above requirements, we chose the following line;

1. We adopt CORBA as a distributed object environment (Takata 1996).

2. We adopt Java for the operation section or user interface of *DASH*. The user interface is defined on the character basis. A Graphical User Interface (GUI) is built on it.

3. We use object oriented methodology. We acquire new experience in object oriented analysis and design of software development.

3. DASH System Architecture Model

We analyzed the data reduction system of astronomical CCD images and made a restaurant model shown in Figure 1. The model has 3 tier structure. The first tier is "Chef" which corresponds to the user interface or console section. The second tier is "Kitchen" and "Books". The third tier is composed of "Ware House" which is the data server or data archives, "Cooks" which is analysis task or CPU server, and "Books" which is the database. These components are on the network computers. Each tier is linked through the Object Request Broker.

4. PROCube

PROCube is an object that unifies the whole process of the data reduction of an astronomical object frame (CCD image). It handles processing flow together

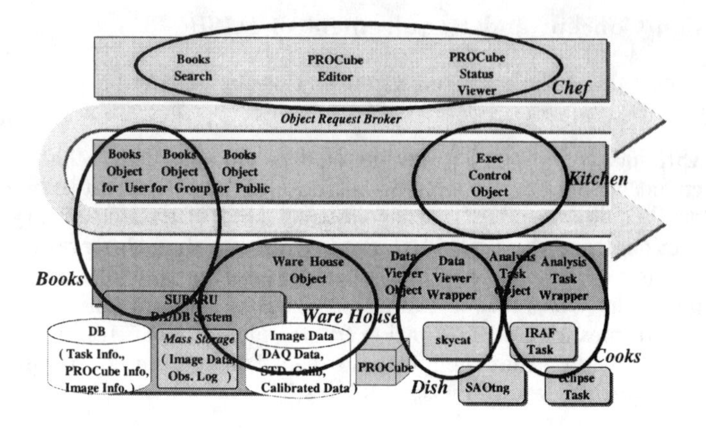

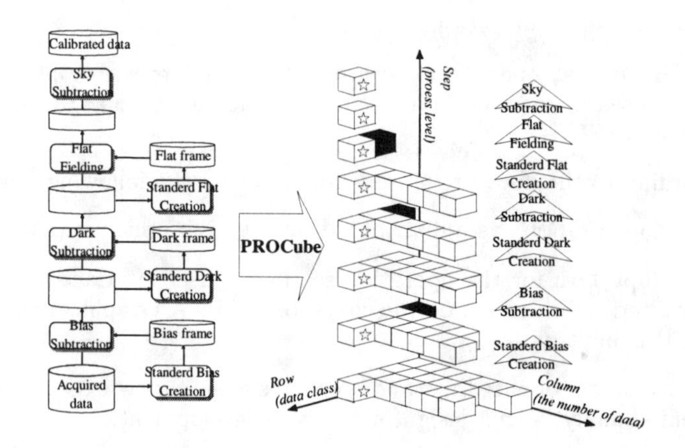

Figure 2. *PROCube*

with image data. Figure 2 shows a *PROCube* like a 3 dimensional flow chart. The X-axis is the kind of image frame, the Y-axis is the number of image frames, and the Z-axis is the flow of reduction. The base bricks of the cube stand for raw data files of an object frame and related calibration frames to be used in the reduction process. The apex of the cube corresponds to the resultant file. *PROCube* is a kind of visual procedure script which can be executed by "Kitchen". A log of its execution is also recorded in the *PROCube*. Abstract *PROCube* that is extracted from the executed *PROCube* works as a procedure script with a variable of an object frame. The abstract *PROCube* can be used for the pipeline.

5. Prototype of *DASH*

We made a prototype for a feasibility study of *DASH* in a distributed heterogeneous computer system using CORBA (ObjectDirector2.0.1), Java (JDK1.0.2), and Web. Some tasks of IRAF 2.11b and Eclipse 1.3.4, and Skycat 1.1 are used as an analysis engine and an image viewer using a wrapping technique. *PROCube* is also implemented in the prototype.

The prototype is tested and evaluated on a small scale system. We used two workstations (Solaris 2.5.1 J) connected through FibreChannel switch (266Mbps) and two personal computers (Windows 95 J). The four computers composed a distributed computer system using Ethernet (10baseT). This computer system is isolated from the other networks.

6. Results from the Prototype

A data reduction system implemented the *PROCube* is realized on the network computers using CORBA as a platform for the distributed software environment. The prototype works well on a system composed of 2 server workstations and 2 client PCs. From a client PC, 2 server workstations look like one network computer.

The wrapping of an existing software package as a CORBA object is not easy and it takes a long time to study details of the package. Eclipse and Skycat were easier to wrap than IRAF. A series of data reduction processes can be done with a *PROCube* that uses IRAF tasks and Eclipse tasks together. The *PROCube* is able to work as a data reduction pipeline. Parallel processing of the tasks is realized in one *PROCube*. It is confirmed that parallel processing is effective for I/O bound tasks.

Problems or disadvantages of the prototype also become clear. CORBA and Java are still under development. So, the products CORBA and Java may have some unsatisfactory features as well as bugs. Tasks for image data reduction are I/O bound in most cases. It is important in data transfer to harmonize performance of the network with that of disk I/O. NFS through 10Mbps Ethernet is not satisfactory for data transfer of large image files.

Data reduction with a *PROCube* may create a lot of temporary files, which eat a lot of disk space. Yet it is hard to manage lifetimes of temporary files in a *PROCube*.

References

Ichikawa, S., et al. 1995, in ASP Conf. Ser., Vol. 77, Astronomical Data Analysis Software and Systems IV, ed. R. A. Shaw, H. E. Payne & J. J. E. Hayes (San Francisco: ASP), 173

Yoshida, M., et al. 1997, in ASP Conf. Ser., Vol. 125, Astronomical Data Analysis Software and Systems VI, ed. Gareth Hunt & H. E. Payne (San Francisco: ASP), 302

Kosugi, G., et al. 1996, in ASP Conf. Ser., Vol. 101, Astronomical Data Analysis Software and Systems V, ed. George H. Jacoby & Jeannette Barnes (San Francisco: ASP), 404

Takata, T., et al. 1996, in ASP Conf. Ser., Vol. 101, Astronomical Data Analysis Software and Systems V, ed. George H. Jacoby & Jeannette Barnes (San Francisco: ASP), 251

Astronomical Data Analysis Software and Systems VII
ASP Conference Series, Vol. 145, 1998
R. Albrecht, R. N. Hook and H. A. Bushouse, eds.

Pipeline Processing and Quality Control for Echelle Data

G. Morgante and F. Pasian

Osservatorio Astronomico di Trieste, 34131 Trieste, Italy
Email: pasian@ts.astro.it

P. Ballester

European Southern Observatory, D-85748 Garching, Germany

Abstract. In the framework of a collaboration between ESO and OAT concerning the development of a data reduction pipeline and the related quality control subsystems software for UVES/VLT, a detailed feasibility study is being performed on the basis of data extracted from the EMMI/NTT archive.

The pipeline reduction is based on an accurate set of "pre-calibrated solutions" and will also result in a powerful tool for the release and the distribution of the scientific data, from a simple quick look at the data just collected to a complete echelle reduction with the highest possible accuracy.

1. Overview

It is ESO policy to perform data reduction for VLT instruments in "pipeline" mode, either at the observing premises or at headquarters. The result of the pipeline will be a "standard quality" data product, certified by the Observatory.

In order to verify if a pipeline was feasible for echelle data, in particular for UVES[1] (Ultraviolet-Visible Echelle Spectrograph), a test on EMMI[2] (ESO Multi-Mode Instrument) data acquired in echelle mode and extracted from the NTT archive was performed:

- a preliminary processing pipeline was defined using the MIDAS echelle package;
- a set of pre-calibrated solutions was available, and other have been defined, to optimize the processing speed while achieving results within an acceptable level of accuracy;
- archive data were processed in pipeline mode;

[1] http://www.eso.org/observing/vlt/instruments/uves/

[2] http://www.ls.eso.org/lasilla/Telescopes/NEWNTT/NTT-MAIN.html

- quality control on the data was performed, with the purpose of determining the instrumental stability.

It is important to note that the approach followed in the development of the work has been to minimize assumptions, or *a priori* knowledge of the EMMI instrument, of its data and parameters; the decision was made to rely uniquely on the contents of the archive and of the keywords of the files stored therein. As a consequence, an evaluation of the completeness of the archive of EMMI echelle data was also performed.

2. Key Items

The key points of the work were identified to be the following:

- test of the performance of the preliminary version of the automatic reduction chain compared with the interactive calibration;

- adaptability of the pipeline procedures on data collected with the R4 grating;

- verification of instrument stability over time with particular reference to geometrical and photometric stability;

- analysis of the accuracies expected from wavelength and flux calibrations for the eventual definition of new calibration strategies;

- evaluation of different effects, such as atmospheric variations, differential refraction, the impact of mechanical deformations on optical performance and instrument maintenance;

- definition of procedures for quality control tests for EMMI data;

- estimation of CPU and hardware requirements to run the final pipeline;

- the completeness of the information stored in the archive, and the corresponding requirements to be imposed to operations to ensure such completeness;

- requirements to be imposed to data processing to ensure that the inevitable instrument instabilities are correctly recovered by the data reduction software, both in pipeline and in interactive modes.

3. Pipeline Processing and Pre-calibrated Solutions

The MIDAS environment, and in particular its Echelle Package, include proper tools for building pipeline DRS for generic echelle instruments, and in the future for UVES, with some minor improvements. In the pipeline built for EMMI, pre-calibrated solutions (i.e., tables containing the results of interactive processing on selected "reference" spectra) were used for order detection and wavelength calibration as starting points.

The `UPDATE/ORDER` MIDAS command was used to re-define the positions of the orders starting from the pre-calibrated order table.

The wavelength calibration has been performed using `IDENT/ECHELLE` with the `GUESS` method (i.e., searching for the optimal solution starting from a pre-defined position of calibration lines). The result was accepted only after checking the plot of residuals and the "percentage of identification among the half brighter lines", that should be the highest, and in any case above 50%.

The analysis of applicability for the pre-calibrated solutions has verified that, for a stable instrumental configuration and performance, the current procedures are quite adequate. Minor software upgrades could be made to take care of extreme situations; e.g., for EMMI a non-rigid shift in order positions due to a changed field lens was corrected with a custom modification of the `UPDATE/ORDER` command. A new set of pre-calibrated solutions shall however be computed by staff whenever a manual intervention on the instrument is made; such interventions shall be appropriately reported in the archive. New pre-calibrated solutions should be computed during every test period, to minimize *a-posteriori* software recovery.

4. Quality Control

For an efficient quality control to take place, the setup of proper operational procedures is necessary:

- **at the telescope:**
  - test frames must be regularly acquired (once during every test period):
    - bias and long dark frames are useful to monitor the detector parameters and features;
    - flat fields for different instrumental configurations provide additional information on the instrument stability;
  - scientific data sets must contain all calibration frames necessary for re-using the data and assessing their quality:
    - the optical performance of the instrument can be monitored by the analysis of the flat field stability and standard stars fluxes;
    - a list of standard stars of a very early spectral type or with a featureless spectrum should be selected in order to evaluate and subtract the atmospheric contribution;
    - the ThAr spectra allow the geometrical stability and the dispersion relation accuracy over time to be checked;
  - information on manual interventions must be regularly logged and archived;
  - instruments should always be in a stable and documented state.
- **at headquarters:**
  - whenever a modification to an instrument occurs, the related processing script (or parameters) should be updated;

– previous versions of processing scripts should be archived;
– the test data should be promptly processed to produce a proper trend analysis on instrument behaviour.

5. Archive Completeness

In this work, it was decided to rely uniquely on the contents of the archive and of the keywords of the files, because this is the standard approach an ESO user would take in accessing data he/she would have not acquired directly. In accordance with the results of Péron et al. (1994), some incompleteness in the archive of EMMI echelle data has been identified. In the future, for EMMI but especially for UVES, the following are key items for archive completeness:

- routine acquisition (e.g., in service mode) of all frames necessary for proper reduction of the scientific data;
- indication in the archive or in the header of the scientific frame, or in a dedicated Observation Summary Table (OST) of the calibration files considered by ESO as the ones to be used for an "observatory standard" reduction of data;
- creation of *data sets* (one per scientific frame), including science data, calibration frames (acquired with the science data, or extracted from the archive), and the OST;
- correctness of the keywords in the frame headers.

Acknowledgments. The work has been carried out under ESO Contract No. 47980/DMD/96/7513 to OAT. Useful information and valuable suggestions were provided during meetings at ESO by H.Dekker, S.d'Odorico, H.Hernsberge, R.Gilmozzi, P.Grosbøl, L.Kaper, B.Leibundgut, P.Molaro, P.Nielsen, and L.Pasquini. Useful discussions with P.Bonifacio and S.Monai of OAT are gratefully acknowledged.

References

Ballester, P. 1992, in Proc. 4th ESO/ ST-ECF Data Analysis Workshop, ESO Conf. and Workshop Proc. No. 41, 177

Ballester, P., Rosa, M. R. 1997, Astr. Astroph., in press, ESO Scientific Preprint No. 1220

Morgante, G., Pasian, F. 1997, Feasibility Study of Pipeline and Quality Control Software for Echelle Data - Final Report, OAT Technical Report

Péron, M., Albrecht, M. A., Grosbol, P. 1994, in Proc. ESO/OAT Workshop Handling and Archiving Data from Ground-based Telescopes, M.A.Albrecht & F.Pasian eds., ESO Conf. and Workshop Proc. No. 50, 57

Verschueren, W., Hernsberge, H. 1990, in Proc. 2nd ESO/ST-ECF Data Analysis Workshop, ESO Conf. and Workshop Proc. No. 34, 143

Astronomical Data Analysis Software and Systems VII
ASP Conference Series, Vol. 145, 1998
R. Albrecht, R. N. Hook and H. A. Bushouse, eds.

On-The-Fly Re-Calibration of HST Observations

B. Pirenne and A. Micol

Space Telescope – European Coordinating Facility, ESO, Garching, D-85748

D. Durand and S. Gaudet

Canadian Astronomy Data Centre, DAO, Victoria, BC

Abstract. On-the-fly re-calibration of HST data (OTF) has now been available at the CADC and ST-ECF for about 18 months. With the installation of STIS and NICMOS on the spacecraft, a new breed of data types and data organisations was challenging our existing data retrieval and processing service. We briefly describe the OTF method below.

1. What is On-the-fly re-calibration?

HST observations are normally calibrated and stored in the archive immediately after reception by the ground system. The calibration can only be executed using the calibration reference files which are available at that time. Some may be missing or may not be the most appropriate ones as new calibration observations might be obtained later on. Moreover, the calibration software of those instruments hardly ever stabilizes. In other words, the longer one waits before calibrating an observation, the better the results should be.

This is the concept that we decided to implement. The recipe is simple in principle: recalibrate only at the time of the data delivery to the user. This is "Just-in-time" HST data!

2. Implementation

The implementation is best explained by considering the data flow model presented in Figure 1.

In this model, users perform the traditional catalogue browsing activities: selection of data according to search criteria, examination of the results, refinement of the selection etc., until the proper set of observations has been identified from the science database, perhaps helped by looking at quick-look samples of the data. Subsequently, users will mark those records for retrieval and will typically select the on-the-fly reprocessing of the datasets.

After proper identification, selection of the output media etc, the request is registered in the archive database. Then, a first automatic process reads the files required from the data repository, while a second process starts the actual re-calibration.

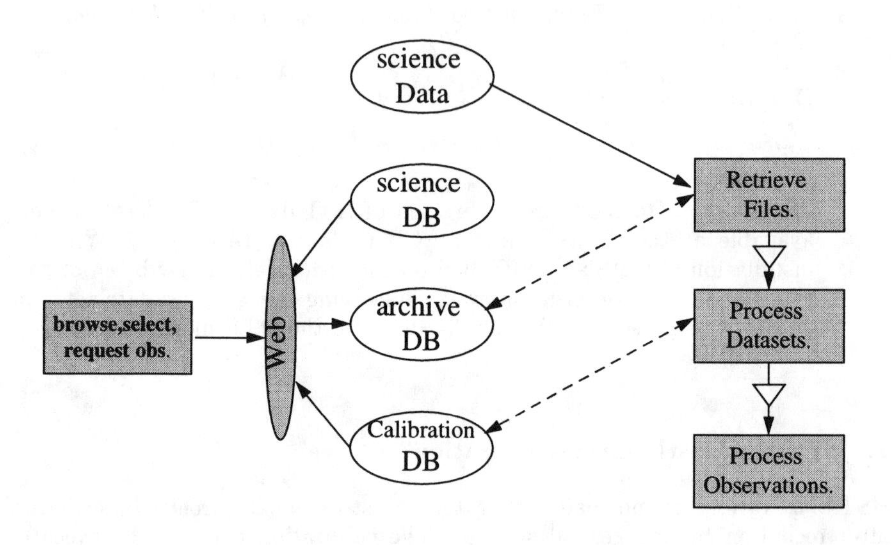

Figure 1. On-the-fly re-calibration data flow: from request submission to data delivery.

Currently, this processing step involves the selection of the best calibration reference files using a specialized database. This database relates the current best calibration reference files to any given HST observation. Using this information, the actual calibration reference files (flat, bias, dark, etc.) are retrieved from magnetic disks and applied to the science data being re-calibrated.

As soon as all exposures belonging to a request have been reprocessed, the data is made available to the user.

On-the-fly re-calibration has another advantage for our archive sites: we only need to keep the raw data and calibration reference files on-line for the process to be automatic. As a matter of fact, original calibration, engineering data and other auxiliary data need not be available to the re-calibration process and can remain on secondary storage.

3. Recent Improvements

The ST-ECF and CADC are currently working towards improving this system through two major new facilities.

- First, the integration of the calibration pipeline for the two new HST instruments (NICMOS and STIS) complements our existing service.
- The science catalogue is now showing associations of WFPC2 exposures that can be retrieved and re-calibrated as one single entity, enabling further processing steps such as cosmic ray rejection and image co-addition. These latest improvements are described at length in Alberto Micol's article in this volume.

4. Conclusions

The mere fact that the service is used a lot is for us both proof of the usefulness of the OTF concept and an encouragement to develop it further: The integration of more advanced processing steps (cosmic-ray removal, co-addition of frames) pushes OTF even further by allowing users to concentrate more on data analysis and leave the reduction tasks to automated procedures.

References

Micol A., Bristow P., & Pirenne B. 1998, this volume

Crabtree, D., Durand, D., Gaudet, S., & Pirenne, B., 1996,"The CADC/ST–ECF Archives of HST data: Less is More", in ASP Conf. Ser., Vol. 101, Astronomical Data Analysis Software and Systems V, ed. George H. Jacoby & Jeannette Barnes (San Francisco: ASP), 505

Astronomical Data Analysis Software and Systems VII
ASP Conference Series, Vol. 145, 1998
R. Albrecht, R. N. Hook and H. A. Bushouse, eds.

OPUS-97: A Generalized Operational Pipeline System

J. Rose

CSC: Computer Sciences Corporation, Inc.

Abstract. OPUS is the platform on which the telemetry pipeline at the Hubble Space Telescope Science Institute is running currently. OPUS was developed both to repair the mistakes of the past, and to build a system which could meet the challenges of the future. The production pipeline inherited at the Space Telescope Science Institute was designed a decade earlier, and made assumptions about the environment which were unsustainable.

While OPUS was developed in an environment that required a great deal of attention to throughput, speed, efficiency, flexibility, robustness and extensibility, it is not just a "big science" machine. The OPUS platform, our baseline product, is a small compact system designed to solve a specific problem in a robust way.

The OPUS platform handles communication with the OPUS blackboard; individual processes within this pipeline need have no knowledge of OPUS, of the blackboard, or of the pipeline itself. The OPUS API is an intermediate pipeline product. In addition to the pipeline platform and the GUI managers, the OPUS object libraries can give your mission finer control over pipeline processing.

The OPUS platform, including a sample pipeline, is now available on CD-ROM. That package, designed to be run on the Solaris operating system, can help you decide whether OPUS can be used for your own mission.

OPUS was developed in an environment which demanded attention to productivity. A ten-fold increase in the volume of data was anticipated with the new instrumentation to be flown on the Hubble, and the ground system required a pipeline which could accommodate that load. OPUS has met these challenges.

A distributed pipeline system which allows multiple instances of multiple processes to run on multiple nodes over multiple paths may seem like an operational nightmare. To resolve this, OPUS includes two pipeline managers: Motif GUI applications which assist operations staff in monitoring the system. The Process Manager not only assists with the task of configuring the system, but monitors what processes are running on which nodes, and what they are doing currently. The Observation Manager provides a different view of the pipeline activities, monitoring which datasets are in which step in the pipeline and alerting the operator when observations are unable to complete the pipeline.

[1]rose@stsci.edu

The success of OPUS can be attributed in part to adopting a blackboard architecture of interprocess communication. This technique effectively decouples the communication process and automatically makes the entire system more robust. Based upon the standard file system, OPUS inherits a simple, robust and well-tested blackboard.

OPUS has been operational at the Space Telescope Science Institute since December 1995, and has now been packaged[2] so that other missions can take advantage of this robust, extensible pipeline system.

1. Fully distributed processing...

OPUS supports a fully distributed processing system. This means multiple instances of a process are able to be run simultaneously without interference from one another. In addition it can support a variety of processes, each a step in the pipeline.

Moreover multiple pipelines, or paths, are supported. For example, at the Space Telescope Science Institute it is necessary to operate a real-time pipeline at the same time that a production pipeline is processing. And a reprocessing pipeline may be simultaneously converting science images in the background.

In addition to several pipelines with identical processing steps, OPUS supports any number of distinct pipelines all running on the same set of processors. Thus, in addition to the science pipelines, OPUS accommodates an engineering data pipeline, a separate pipeline for other non-science data, as well as an interface to Goddard Space Flight Center for data receipt.

All pipelines are defined in simple text files. The pipeline path file defines a set of network-visible directories on the shared disks. While one set of disks is being used to process one kind of data, another set can be employed to process a different type of data. ASCII text files are the basis for configuring any component of the OPUS environment. Adding an additional machine to the set of nodes is accomplished by editing a text file: that node will immediately be available to share the load.

The OPUS managers are uniquely suited to keep track of the environment. The Process Manager keeps track of what is running where, while the Observation Manager is monitoring the progress of observations being processed in any one of the pipelines. Multiple Observation Managers can each monitor their own pipelines without interference from one another.

2. ...or a simple data pipeline

Even though OPUS was developed in an environment that required a great deal of attention to throughput, speed, efficiency, flexibility, robustness and extensibility, it is not just a "big science" machine. The OPUS platform, our

---

[2]http://www.stsci.edu/opus/

baseline product, is a compact[3] system designed to solve a specific problem in a robust way.

OPUS is implemented as an automated pipeline, one which can start up automatically, send exposures from task to task automatically, and monitor how things are proceeding automatically. OPUS is designed to achieve a "lights-out" operation: data can enter the pipeline, be processed, then archived without intervention.

The OPUS pipeline managers monitor the status of each exposure in the system: how far it got in the pipeline, whether it failed and where it failed. The GUI interfaces provide convenient tools to investigate problems, examine log files, view trailers, and restart troubled exposures at any step in the pipeline.

The FUSE (Far Ultraviolet Spectrographic Explorer) team selected OPUS for their pipeline even though they will be receiving only a moderate amount of data that will processed on a single workstation. OPUS was chosen because it frees the FUSE team to concentrate on the science and calibration issues which are unique to their instrument.

By handling the mechanics of data processing, OPUS frees the scientists to do science.

3. OPUS-97 platform

The OPUS platform is the baseline pipeline product. All communication with the OPUS blackboard is handled by the platform; an individual process within this pipeline need have no knowledge of OPUS, of the blackboard, or of the pipeline itself.

OPUS can accommodate any non-interactive shell script. When there is work to be performed by that script, OPUS will pass the name of the dataset to be processed, the location of that dataset and other auxiliary datasets, as well as other parameters required by the script.

Similarly the OPUS platform can wrap any stand-alone, non-interactive executable which takes the name of the input dataset as a single argument. All other information for that task is either passed by OPUS through environment variables (symbols) or is obtained from the dataset itself.

OPUS is fully table-driven. To add another process to the OPUS pipeline only requires the development of an ASCII text file describing the command line arguments, the pipeline triggers, subsequent process triggers, and other such control information.

Processes (or scripts) can be triggered in three ways: the most common way is to allow the completion of one or more previous pipeline steps to act as the process trigger mechanism. Another useful technique is to use the existence of a file as the trigger. Alternatively one can use a time event to trigger an OPUS process (eg: wake up once an hour).

---

[3]The OPUS baseline system is less than 20,000 lines of code.

The OPUS platform is being distributed now on CD-ROM [4] for the Solaris platform. This distribution comes with a sample pipeline which shows how to set up the system and how to modify it to reflect your own needs.

4. An OPUS Pipeline

Where the OPUS platform is a generalized pipeline environment, it does not provide the applications which handle mission-specific data. The OPUS team at the STScI has a variety of tools and packages at hand to help build functional telemetry pipelines. Packages to handle FITS files, keyword dictionaries, database access, keyword population, message handling, and the like, form the building blocks of a standard astronomical data processing pipeline.

Certainly the specific applications are not portable, but the experience of the OPUS team in developing complete pipelines for Hubble and for FUSE can be easily applied to other missions.

Is OPUS overkill for a small mission? No. First, OPUS is not a large system. It is compact, designed to solve a specific problem in a robust way: distributed processing with controlled monitoring. Second, OPUS does the controlled monitoring. Telemetry processing does not have to be a labor intensive task. OPUS relieves your talented engineering and science staff to do more interesting work. Third, OPUS exists. Your mission is to understand the science, not to build pipelines.

Acknowledgments. The entire OPUS team at the Space Telescope Science Institute was involved in the development of OPUS-97: Daryl Swade (CSC), Mary Alice Rose (AURA), Chris Heller, Warren Miller, Mike Swam and Steve Slowinski are all to be congratulated.

References

Rose, J., Choo, T.H., & Rose, M.A. 1996, "The OPUS Pipeline Managers", in ASP Conf. Ser., Vol. 101, Astronomical Data Analysis Software and Systems V, ed. George H. Jacoby & Jeannette Barnes (San Francisco: ASP), 311

Rose, J. et al. 1994, "The OPUS Pipeline: A Partially Object-Oriented Pipeline System", in ASP Conf. Ser., Vol. 77, Astronomical Data Analysis Software and Systems IV, ed. R. A. Shaw, H. E. Payne & J. J. E. Hayes (San Francisco: ASP), 429

Nii, H.P. 1989, "Introduction" in *Blackboard Architectures and Applications*, Jagannathan, V., Dodhiawala, R., Baum, L., editors, Academic Press, San Diego, CA, xix

[4]http://www.stsci.edu/opus/opusfaq.html

Astronomical Data Analysis Software and Systems VII
ASP Conference Series, Vol. 145, 1998
R. Albrecht, R. N. Hook and H. A. Bushouse, eds.

Remote Observing with the Keck Telescope: ATM Networks and Satellite Systems

P.L. Shopbell, J.G. Cohen, and L. Bergman[1]

California Institute of Technology, Pasadena, CA. 91125

Abstract. As a technical demonstration project for the NASA Advanced Communications Technology Satellite (ACTS), we have implemented remote observing on the 10-meter Keck II telescope on Mauna Kea from the Caltech campus in Pasadena. The data connection consists of ATM networks in Hawaii and California, running at OC-1 speeds (45 Mbit/sec), and high data rate (HDR) satellite antennae at JPL in Pasadena and Tripler Army Medical Center in Honolulu. The ACTS network is being used to enable true remote observing, as well as remote eavesdropping. The software environment is identical to that used for on-site observing at the Keck telescope, with the added benefit of the software, personnel, and other facilities provided by observing in a local environment. In this paper, we describe our high-speed remote observing network, assess the network's level of performance, and summarize the benefits and difficulties encountered in this project.

1. Introduction

Remote use of astronomical telescopes has been a topic of interest for many years, even before space-based observing platforms (e.g., IUE) began to demonstrate total remote operation out of sheer necessity. However, only very recently are optical telescopes beginning to realize the benefits of true remote observing: for example, observations with modest size detectors at Apache Point Observatory are being carried out remotely using the Internet (York 1995, BAAS, 186, 44.04). For this project, we have established remote interactive observing capabilities for Keck Observatory on Mauna Kea for observers at Caltech, in Pasadena, California. The recently commissioned twin 10-meter Keck Telescopes are the largest optical/infrared telescopes in the world and thereby typify the data and network requirements of a modern observatory. In undertaking this project, we were motivated by several operational and scientific advantages that remote observing would offer, including alleviating altitude-related difficulties, saving time and money due to travel, enabling remote diagnosis of hardware and software problems, and simplifying access to the telescopes for educational use.

[1]Jet Propulsion Laboratory, California Institute of Technology

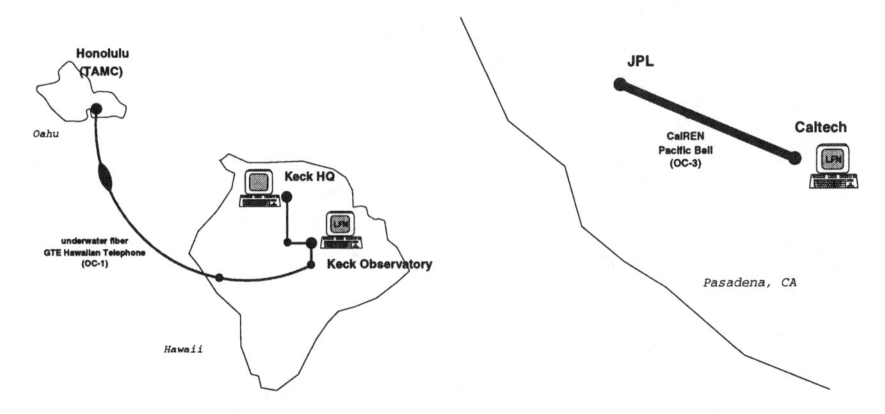

Figure 1. Schematics of the final terrestrial networks in Hawaii and California, as used for the Keck remote observing project. These networks were connected via NASA's ACTS satellite.

2. Network Architecture

Our network consists of three major segments: the ground network in California, the satellite link across the Pacific Ocean, and the ground network in Hawaii.

The ground network in California connects Caltech with JPL, the site of the ACTS ground station. This segment was established as part of Pacific Bell's extant CalREN fiber optic network and has proved to be the most reliable portion of our network.

The satellite connection was made available to us through a grant from NASA as part of the Gigabit Satellite Network (GSN) testbed program. NASA's Advanced Communications Technology Satellite (ACTS) was built to explore new modes of high speed data transmission at rates up to OC-12 (622 Mbit/sec). The 20–30 GHz frequency band has been employed for the first time by a communications satellite, with extensive rain fade compensation.

The ground network in Hawaii, which connects Keck observatory with the other ACTS ground station at Tripler Army Medical Center in Honolulu has been somewhat more complex in its evolution. This was primarily due to the relative inexperience of GTE Hawaiian Telephone and a lack of prior infrastructure in Hawaii. This segment initially consisted of a combination of underwater fiber, microwave antennae, and buried fiber. The higher bit error rates (BER) of the non-fiber segment produced noticeable instability in the end-to-end network. Fortunately, in January of 1997 this portion of the ground network in Hawaii was upgraded to optical fiber. The improved performance for high-speed data transfers of the final all-fiber network was immediately apparent.

In order to support standard higher-level (IP) networking protocols, we installed an Asynchronous Transfer Mode (ATM) network over this infrastructure. The transfer of 53-byte ATM cells is performed by hardware switches throughout the network, at speeds of OC-1 (51 Mbit/sec) and above. Several vendors have supplied the ATM switches and Network Interface Cards (NICs), providing a stringent test of compatibility in the relatively new ATM environment. Al-

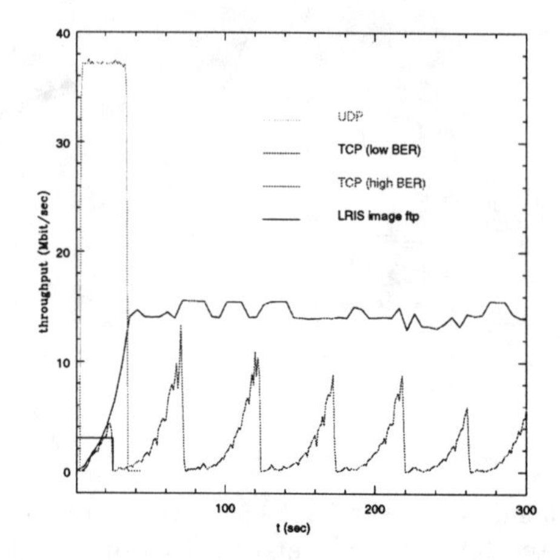

Figure 2. Bandwidth test results between Keck Observatory and the Caltech campus in Pasadena, California, over the ACTS satellite network. TCP exhibits a remarkable dependence on the bit error rate.

though we have encountered several interoperability problems, none have been serious, and the ATM and telephone vendors have been extremely helpful.

In order to facilitate reliable data transfer, as well as to allow the use of the wealth of software tools already available, we are running the standard IP protocols over ATM using a pseudo-standard implementation known as "Classical IP". This enables the use of the standard network-based applications that are in widespread use on the Internet. Tools such as `ftp` and `telnet` are part of every observing run, as are additional special-purpose applications, such as an audio conferencing tool (`rat`) and a shared whiteboard tool (`wb`).

3. Network Performance

The most important impact of a satellite component on a high-speed network is the relatively large delay introduced by the round-trip signal travel time to the satellite. In our network, this travel time is approximately 0.55 seconds, which corresponds to 3.5 Mbytes of data at OC-1 speeds (51 Mbit/sec). The problem has to do with the connection-oriented nature of TCP/IP: TCP sends a very specific amount of data, known as a "window", after which time it expects an acknowledgment from the other end of the connection. However, this window size is often very small; the default value for workstations running the SunOS 4.1.4 operating system is only 4 Kbytes.

Fortunately, this problem is well-known in the high-speed networking community. Networks such as ours are known as "long fat networks" (LFN; see RFC 1323). In the case of the SunOS operating system (to which we are constrained by legacy control software at Keck), we obtained the TCP-LFN package from Sun Consulting, which purports to support the RFC 1323 extensions. Unfortunately, a number of limitations of SunOS 4.1.4 conspire to prohibit one from obtaining extremely large window sizes, regardless of the TCP-LFN software. In our case, the compiled-in kernel limit of 2 Mbytes of Mbuf memory (i.e., IP packet wrappers) turned out to be the major constraint, limiting our window size to no more than 1 Mbyte. Indeed, our final tuned network delivered the expected maximum TCP/IP performance of approximately 15 Mbit/sec ($\sim \frac{1}{3}$ of OC-1). Although perhaps disappointing in a relative sense, this bandwidth is far in excess of T1 Ethernet speed (1.44 Mbit/sec) and allows an 8 Mbyte image to be transferred in approximately 5 seconds. As a further comparison, this bandwidth exceeds by 50% that which is available on the local area Ethernet network at the Keck Telescope itself. Figure 2 illustrates typical bandwidth measurements of our network for UDP and TCP, the latter before and after the network was upgraded to fiber in Hawaii.

While network performance was perhaps not at the level desired, due to developing infrastructure in Hawaii and idiosyncrasies within the operating system, issues of network reliability had far greater impact on our remote observing operation. The experimental and limited nature of the ACTS program created a number of difficulties which one would almost certainly not face if using a more developed and/or commercial satellite system. The impact of the reliability issue is that at least one observer must be sent to Hawaii to use the telescope, in case of ACTS-related problems.

4. Conclusions

This experiment has explored the data requirements of remote observing with a modern research telescope and large-format detector arrays. While the maximum data rates are lower than those required for many other applications (e.g., HDTV), the network reliability and data integrity requirements are critical. The former issue in particular may be the greatest challenge for satellite networks for this class of application. We have also experimented with the portability of standard TCP/IP applications to satellite networks, demonstrating the need for alternative TCP congestion algorithms and minimization of bit error rates.

Reliability issues aside, we have demonstrated that true remote observing over high-speed networks provides several important advantages over standard observing paradigms. Technical advantages include more rapid download of data and the opportunity for alternative communication facilities, such as audio- and videoconferencing. Scientific benefits include involving more members of observing teams while decreasing expenses, enhancing real-time data analysis of observations by persons not subject to altitude-related conditions, and providing facilities, expertise, and personnel not normally available at the observing site.

Due to the limited scope of the ACTS project, future work from the standpoint of Keck Observatory will be concerned with establishing a more permanent remote observing facility via a ground-based network. At least two projects are under way in this direction: remote observing from the Keck Headquarters in Waimea, from where up to 75% of observing is now performed every month, and remote observing from multiple sites on the U.S. mainland using a slower T1 connection (Conrad et al. 1997, SPIE Proc. 3112). Trial tests of this latter approach over the Internet have been extremely promising.

Astronomical Data Analysis Software and Systems VII
ASP Conference Series, Vol. 145, 1998
R. Albrecht, R. N. Hook and H. A. Bushouse, eds.

NICMOS Software: An Observation Case Study

E. Stobie, D. Lytle, A. Ferro and I. Barg

University of Arizona, Steward Observatory, NICMOS, Tucson, AZ 85721

Abstract. The Near Infrared Camera and MultiObject Spectrometer (NICMOS) was installed on the Hubble Space Telescope during the Second Servicing Mission in February, 1997. NICMOS Proposal 7111, the imaging of Orion OMC-1, was selected as an Early Release Observation (ERO) to be executed in the Science Mission Observatory Verification (SMOV) period shortly afterwards. Calibration and analysis of this observation posed very interesting challenges for the authors, in particular dealing with an exciting new instrument whose on-orbit calibration was not yet well defined.

1. Introduction

NICMOS contains three cameras with adjacent, but not spatially contiguous fields-of-view of 11, 19.2, and 51.2 arcseconds which provide infrared imaging and spectroscopic observations that cover the spectral range of 0.8 to 2.5 microns. Proposal 7111 imaged the core of OMC-1 using the NICMOS high and intermediate resolution cameras with seven 160-second exposures, composed of 13 non-destructive reads (multiaccum mode) for both the F212N and F215N filters. Five mosaic positions were observed during five consecutive orbits as well as observations for dark and flat calibration. The purpose of the observation was to resolve structure in the molecular hydrogen outflow close to the outflow source (or sources) in OMC-1.

Many steps were required to process the raw OMC-1 data into a final calibrated mosaiced image. Darks and flats had to be created from contemporaneous reference observations and problems within the calibration process had to be resolved. Mosaicing software had to be developed to create the final product.

2. Reference Files

2.1. Introduction

The NICMOS chips present some interesting differences in the reference files which are used to generate calibrated data. Because the ability to do non-destructive reads of the chip and the associated electronics, as well as the materials used in construction of the chips, there are many differences between reference data in the NICMOS chips and those found in normal CCD reductions.

2.2. Darks

Special Considerations: In addition to the dark current, similar to that found in CCDs, NICMOS dark frames are also used to remove several other systematic effects. These include the signature of the electronics, known as "shading", and the contamination from the heat generated by the detector's readout amplifiers, known as amp glow. Figure 1 shows a dark frame in which these signatures can be seen.

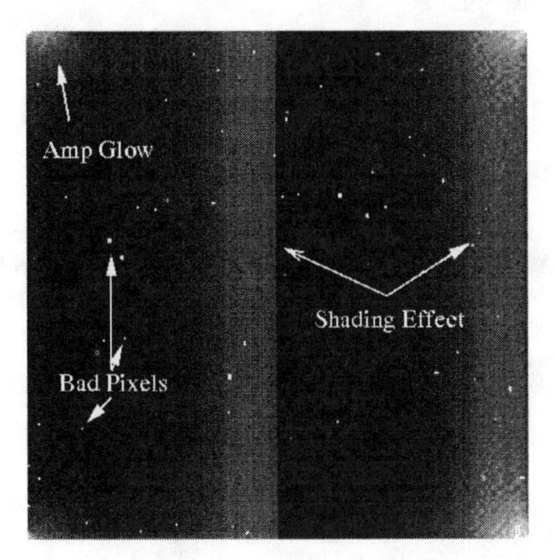

Figure 1. Dark with problem areas.

How we generated the dark: Application of the darks in calibration requires that the darks be treated as counts as opposed to count-rates. Also, darks must be applied to individual readouts of the detector. Each readout from a data sequence must have a corresponding dark readout.

Because the darks are combined in terms of individual readouts, normal NICMOS cosmic ray detection (noting a rapid increase in charge in a pixel), cannot be used. Instead a number of darks were taken and median combined. This produced a final dark series in which very few if any cosmic rays should be present. The final step in producing the dark used for calibration of the data was to recombine the individual readouts into a single file in the format required by the *CALNICA* program, an IDL procedure called *BuildRef*, software written by the authors.

2.3. Flats

Special Considerations: The NICMOS detectors have a large change in quantum efficiency across the chip. This results in a flat field which has a great deal of structure. One problem with this data set is that the filters used, 212N and 215N are quite narrow, which means that a well exposed flat frame was difficult to obtain. In the NICMOS instrument, flat fields are generated by observing a blank part of the sky and illuminating one of the optical elements with a flat field illuminator. Care must be taken that the observed sky does not contain any strong sources.

How we generated the flats: The flat fields were generated by running the data through normal processing with *CALNICA* up to the point of removing the flat field. This included subtraction of the zeroth read, dark subtraction and cosmic ray rejection. The images were median combined to increase signal to noise and to obtain a single flat image.

3. Calibrating the Data

3.1. Introduction

The calibration process using *CALNICA* was straightforward once the headers of the raw images were edited to use our newly created reference darks and flats. However, it became obvious that there were problems in the the linearity correction and in the detection of saturated pixels that led to poor cosmic ray detection.

3.2. Repairing the Zeroth Read

When observing very bright sources with NICMOS (the central object, BN, is approximately 5th magnitude), signal from the object may appear in the zeroth read (bias frame). This results in too much bias being subtracted from all of the subsequent reads causing pixels to fall on an incorrect location on the nonlinearity curve. In extreme cases the data may be saturated by the first readout.

We compared the zeroth read of each observation with a super bias frame from a large set of dark observations. Where the value in the zeroth read deviated by more than 5 sigma from the super bias value, the pixel value was replaced with the super bias value (procedure *satfix*). No pixels were shown to be approaching saturation by the first read, i.e., the difference between a value in the zeroth read and the super bias image is greater than 12,000 ADU.

3.3. Detecting Saturated Pixels

During the cosmic ray rejection step of the calibration process many pixels were flagged as containing cosmic rays when they were actually saturated. This caused a bad fit across the readouts and a poor value for the pixel in the final image. The problem was that a substantial number of pixels in the saturation node of the nonlinearity reference file were defined too high. (Pixels that depart by more than 6% from linearity are considered saturated and unusable.) Since very little on-orbit linearity/saturation calibration data had been taken to date, we used the solution suggested by STScI, i.e., adjust all values in the saturation node downward by 9%. With this change to the saturation node values all saturated pixels were properly detected.

4. Creating the Mosaiced Image

The five positions surrounding BN were observed without the use of a dither pattern and the orientation varied from the exposure on one position to another so that they could not be processed through the STScI CALNICB step of the pipeline. We developed an interactive mosaic task, *NICMosaic*, to combine the images together. Figure 2 shows the mosaic image created by *NICMosaic*.

Figure 2. Mosaic of OMC-1.

5. Summary:

The primary obstacles in the reduction were due to the unique aspects of the NICMOS detectors, as well as the new, more complex, format of the data from HST. Once these factors are understood, the reduction and analysis of NICMOS data is straightforward.

Acknowledgments. We would like to thank Ed Erickson and Susan Stolovy for allowing us to use their ERO data in this paper. We also appreciate many discussions with Susan regarding the analysis of the data.

Astronomical Data Analysis Software and Systems VII
ASP Conference Series, Vol. 145, 1998
R. Albrecht, R. N. Hook and H. A. Bushouse, eds.

The *INTEGRAL* Science Data Centre

Roland Walter, Alain Aubord, Paul Bartholdi, Jurek Borkowski, Pierre Bratschi, Tomaso Contessi, Thierry Courvoisier, Davide Cremonesi, Pierre Dubath, Donnald Jennings, Peter Kretschmar, Tim Lock, Stéphane Paltani, Reiner Rohlfs and Julian Sternberg[1]

INTEGRAL Science Data Centre, Chemin d'Écogia 16, CH–1290 Versoix, Switzerland

Abstract. *INTEGRAL* is a gamma-ray observatory mission of the European Space Agency to be launched in 2001. The *INTEGRAL* Science Data Centre, implemented by a consortium of scientific institutions, is the part of the ground segment which deals with the interface between the scientific community and the *INTEGRAL* data.

1. The *INTEGRAL* Mission

The International Gamma-Ray Astrophysics Laboratory is a "medium mission" of the scientific programme of the European Space Agency (ESA) to be launched in 2001. It was selected to provide the astronomical community with a gamma ray observatory capable of the best possible spectral and angular resolution in order to extend, in the early years of next century, the set of astronomical tools covering the electromagnetic spectrum well into the difficult gamma ray domain.

INTEGRAL will provide both excellent (in terms of gamma ray astronomy) spectral and imaging resolution. The three high energy detectors (the imager IBIS, the spectrometer SPI and the X-ray monitor JEM-X) all use coded mask technology. *INTEGRAL* will therefore have an angular resolution of $\simeq$ 13'. The precision of location of a bright point source will be about one arc minute. The spectral resolution will be 2 keV at 500 keV. This will allow astronomers to measure the profiles of gamma ray lines and to measure Doppler shifts within the galaxy. An optical monitoring camera completes the scientific payload.

Gamma ray astronomy is now one of the tools necessary for the understanding of a wide range of cosmic phenomena, from the study of the interstellar medium to that of active galactic nuclei. It is therefore important to open the observing program to astronomers in the whole community and not to restrict this access to those teams that build the instruments. This is not only necessary for the community in general, but it also ensures that the instruments are used to study the most relevant problems and thus increases the scientific output

[1]Astrophysics Division, Space Science Department of ESA, ESTEC, 2200 AG Noordwijk, The Netherlands

of the mission. *INTEGRAL* was therefore conceived from the beginning as an observatory mission.

2. The *INTEGRAL* Science Data Centre

A gamma ray mission open to the astronomical community at large requires that the data will be calibrated and prepared so as to be understood by non-specialists. This has led to the concept of the *INTEGRAL* Science Data Centre (ISDC). This centre is the interface between the *INTEGRAL* data and the users' community. The ISDC is provided by the scientific community (the ISDC consortium). It is hosted by the Observatory of Geneva and started its activities in 1996.

Gamma-ray instruments are complex and the data reduction and calibration rest on a detailed knowledge of the instruments. The *INTEGRAL* data reduction will therefore be based on instrument specific software modules written by the teams developing and building the instruments.

The main ISDC responsibilities are (a) to receive the telemetry in real time from the Mission Operation Centre, (b) detect gamma-ray bursts within seconds and alert the community, (c) monitor in near real time the health of the scientific instruments and investigate solutions with the instrument teams to resolve problems, (d) perform a quick-look analysis of the data within 10 hours to detect unexpected features and events (TOO), (e) convert the raw data products into physical units, (f) taking into account the instrument characteristics, (g) deduce source properties (images, spectra, light curves) from the observed data, (h) archive the final data product, (i) distribute data and some software and support the users.

3. The ISDC Software System

The ISDC software system is made of data structures, analysis executables, pipelines and applications.

Data structures are tables and multidimensional arrays stored in FITS files. A typical *INTEGRAL* observation will be split into a few hundred short (20 minutes) pointings and slews. Each pointing will result in

- 10^5 to 10^6 events and several images for the high energy instruments
- 600 images for the optical camera
- 200 housekeeping records per instrument
- auxiliary data (attitude, orbit, timing, planning, radiation monitoring data).

The data structures, raw to processed, that correspond to a pointing will be grouped together. All data corresponding to an observation will form a group of about 5000 data structures for the high energy instruments and 10^5 data structures in total. All those data structures will be grouped in a hierarchy that

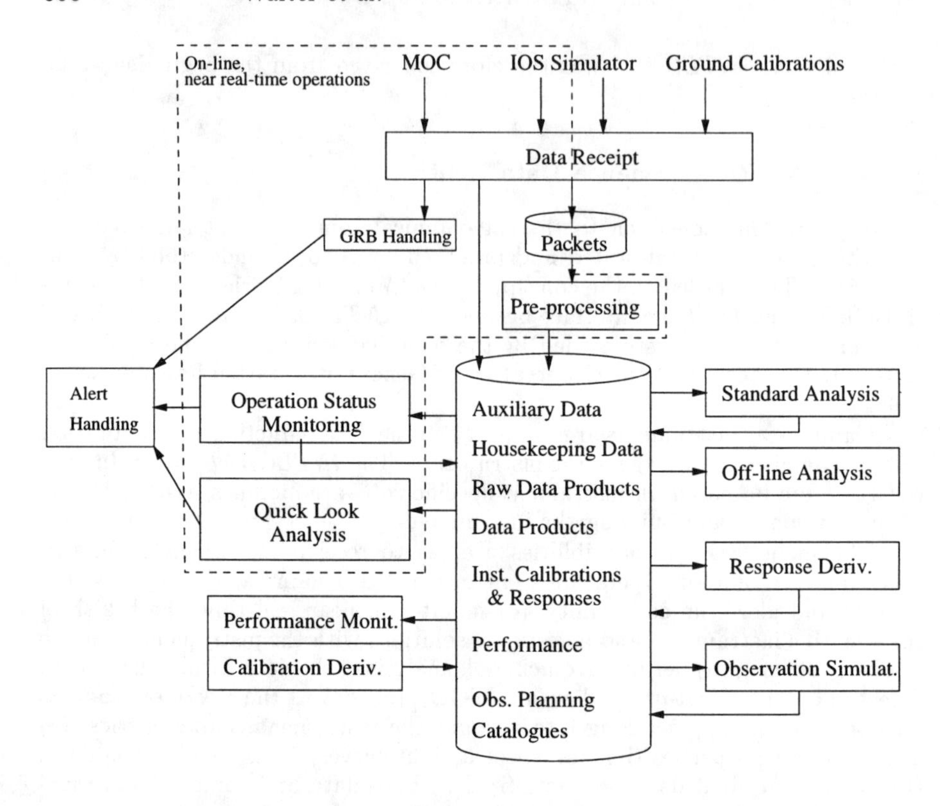

is described in the data themselves. The data and their hierarchy are available through the ISDC Data Access Layer (see Jennings et al. this proceedings). In addition to the standard groups created for the proposed observations, any group of pointings can be created for archival studies.

Analysis executables are C or F90 programs reading their inputs from and writing their outputs to the FITS files and reading their parameters from IRAF parameter files. The analysis executables are compatible with the FTOOLS standards, with some extensions. Analysis executables are standalone and do not require any other software to run.

Analysis pipelines link analysis executables together through flow control instructions. The parameters of each executable are defined at the pipeline level from pipeline parameters, data or database queries. Data structures are not modified at the pipeline level but always through analysis executables. The current baseline for pipeline scheduling is to use the OPUS system developed for the HST. The data access layer supports data storage in shared memory to accelerate data transmission when necessary.

Analysis applications are written to support interactive analysis of the data as well as automatic display of the data. Interactive applications call analysis executables for data manipulation and use native functions of the environment to generate displays and interact with the users. The object oriented framework ROOT developed at CERN is currently under study for writing analysis applications.

4. Development Phases

After a phase of requirement definition, system libraries development and interface specification, the architecture of the ISDC system is being built. Several subsystems will be under development starting early in 1998 and will be integrated and system tested during 1999. Overall tests involving the full ground segment will be conducted during the year 2000.

Additional information on the different elements of the ISDC can be found on the World Wide Web pages of the ISDC (`http://obswww.unige.ch/isdc/`).

Part 7. Archives and Information Services

Astronomical Data Analysis Software and Systems VII
ASP Conference Series, Vol. 145, 1998
R. Albrecht, R. N. Hook and H. A. Bushouse, eds.

The VLT Science Archive System

M. A. Albrecht, E. Angeloni, A. Brighton, J. Girvan, F. Sogni, A. J. Wicenec and H. Ziaeepour

European Southern Observatory, send e-mail to: malbrech@eso.org

Abstract. The ESO[1] Very Large Telescope (VLT) will deliver a Science Archive of astronomical observations well exceeding the 100 Terabytes mark already within its first five years of operations. ESO is undertaking the design and development of both On-Line and Off-Line Archive Facilities. This paper reviews the current planning and development state of the VLT Science Archive project.

1. Introduction

The VLT Archive System goals can be summarized as follows: i) record the history of VLT observations in the long term; ii) provide a research tool - make the Science Archive another VLT instrument; iii) help VLT operations to be predictable by providing traceability of instrument performance; iv) support observation preparation and analysis.

		1999	2000	2001	2002	2003	2004
UT1	isaac	4	4	4	4	4	4
	fors1	0.5	0.5	0.5	0.5	0.5	0.5
	conica		1.5	1.5	1.5	1.5	1.5
	conica (speckle)		40	40	40	40	40
UT2	TestCam	0.5	0.5				
	uves	2.5	2.5	2.5	2.5	2.5	2.5
	fuegos			2	2	2	2
	fors2		0.5	0.5	0.5	0.5	0.5
UT3	TestCam		0.5	0.5			
	vimos		20	20	20	20	20
	visir			1	1	1	1
UT4	TestCam		0.5	0.5			
	nirmos			48	48	48	48
	Typical mix (GB/night)	3.0	19.1	55.6	55.6	55.6	55.6
	TB/Year	1.07	6.80	19.81	19.81	19.81	19.81
	TB cumulative	**1.07**	**7.87**	**27.68**	**47.49**	**67.30**	**87.11**

The data volume expected from the different instruments over the next years is listed in Table 1 Figures are given in gigabytes for a typical night during steady state operations. Estimated total rates per night are derived by making assumptions on a mixture of instrument usage for a typical night.

[1] http://www.eso.org/

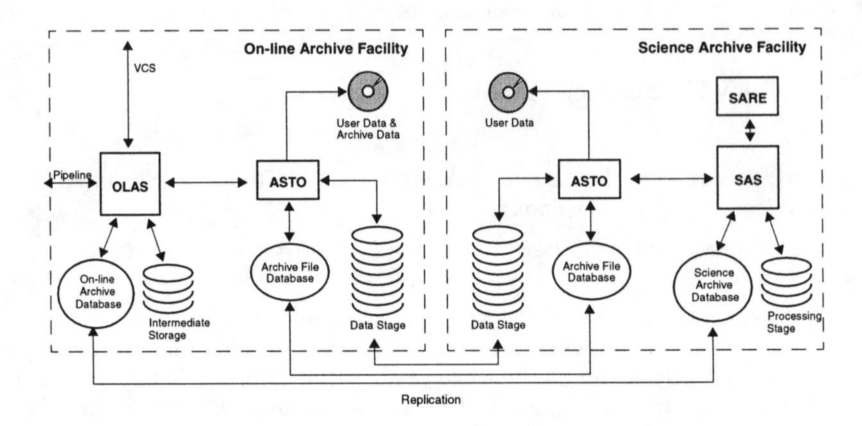

Figure 1. Overview of the VLT Archive System Architecture.

In order to achieve the goals listed above, a system is being built that will include innovative features both in the areas of technology and functionality. Among its most distinct features, the system a) will be scalable through quasi on-line data storage with DVD Jukeboxes and on-line storage with RAID arrays and HFS; b) will include transparent replication across sites; c) will be data mining-aware through meta-databases of extracted features and derived parameters.

2. System Architecture

The main components of the VLT Archive System are (see figure 2): the On-Line Archive Facility (OLAF) and the off-line Science Archive Facility (SAF). The On-Line Archive System (OLAS) takes care of receiving the data and creates the Observations Catalog while the Archive Storage system (ASTO) saves the data products onto safe, long-term archive media. The SAF includes a copy of ASTO used mainly for retrieval and user request handling, the Science Archive System (SAS) and the Science Archive Research Environment (SARE). The SAS stores the Observations Catalog in its Science Archive Database. All the data is described in an observations catalog which typically describes the instrument setup that was used for the exposure. Other information included in the catalog summarize ambient conditions, engineering data and the operations log entries made during the exposure. In addition to the raw science data, all calibration files will be available from the calibration database. The calibration database includes the best suitable data for calibrating an observation at any given time.

The Science Archive Research Environment (SARE) provides the infrastructure to support research programmes on archive data. Figure 3 shows an overview of the SARE setup. Archive Research Programmes are either user defined or ESO standard processing chains that are applied to the raw data. Each of the processing steps is called a Reduction Block (RB). Typically the first reduction block would be the re-calibration of data according to the standard calibration pipeline. A reduction block consist of one or more processes which are treated by the system as black boxes, i.e., without any knowledge of its im-

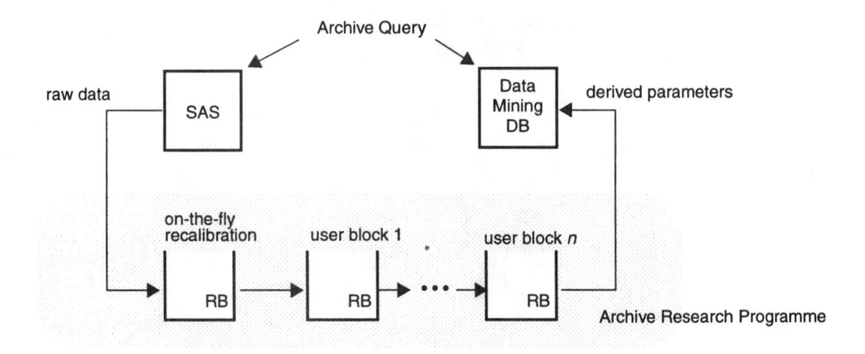

Figure 2. Overview of the VLT Science Archive Research Environment.

plementation. However, the reduction block interface (input and output data) do comply to a well defined specification. This feature allows any reduction module to become part of the chain. In fact, this flexible architecture also allows the research programme to analyze different kinds of data from images and spectra to catalogs and tables of physical quantities. The output of an archive research programme will be derived parameters that are fed into the data mining database.

3. Data Mining in the Science Archive Research Environment

Observation data will be stored within the VLT Science Archive Facility and will be available to Science Archive Research programmes one year after the observation was made.

However, in face of the very large data amounts, the selection of data for a particular archive research project becomes quickly an unmanageable task. This is due to the fact that even though the observations catalog gives a precise description of the conditions under which the observation was made, it doesn't tell anything about the scientific contents of the data. Hence, archive researchers have to first do a pre-selection of the possibly interesting data sets on the basis of the catalog, then assess each observation by possibly looking at it (preview) and/or by running some automated task to determine its suitability. Such procedure is currently used for archive research with the HST Science Archive and is acceptable when the data volume is limited (e.g., 270 GB of WFPC2 science data within the last 3.5 years of HST operations).

Already after the first year of UT1 operations, the VLT will be delivering data quantities that make it not feasible to follow the same procedure for archive research. New tools and data management facilities are required. The ESO/CDS Data Mining Project aims at closing the gap and develop methods and techniques that will allow a thorough exploitation of the VLT Science Archive.

One approach at tackling this problem is to extract parameters from the raw data that can be easily correlated with other information. The main idea

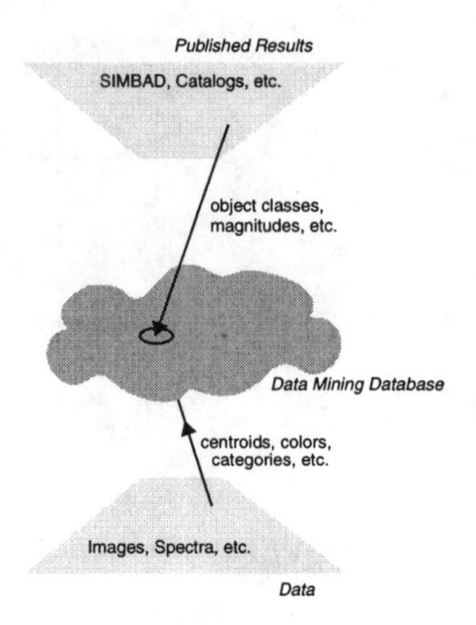

Figure 3. Overview of the data mining environment.

here is to create an environment that contains both extracted parametric information from the data plus references to existing databases and catalogs. In its own way, this environment then establishes a link between the raw data and the published knowledge with the immediate result of having the possibility to derive classification and other statistical samples. Figure 3 illustrates the general concept.

An example of a semi-automatic parameter extraction is the object detection pipeline used by the ESO Imaging Survey (EIS) Project. Every image in the survey is subject of a set of reduction steps that aim at extracting object parameters such as 2-D Gaussian fitted centroids, integrated magnitudes, etc. The cross-correlation of parameters of this kind with selected databases and catalogs (e.g., eccentric centroids with galaxy catalogs) would provide a powerful tool for a number of science support activities from proposal preparation to archive research.

4. Conclusions

The VLT Archive System being developed will provide the infrastructure needed to offer the Science Archive as an additional instrument of the VLT. The main capabilities of the system will be a) handling of very large data volume, b) routine computer aided feature extraction from raw data, c) data mining environment on both data and extracted parameters and d) an Archive Research Programme to support user defined projects.

Astronomical Data Analysis Software and Systems VII
ASP Conference Series, Vol. 145, 1998
R. Albrecht, R. N. Hook and H. A. Bushouse, eds.

A Queriable Repository for HST Telemetry Data, a Case Study in using Data Warehousing for Science and Engineering

Joseph A. Pollizzi, III and Karen Lezon

Space Telescope Science Institute, 3700 San Martin Drive, Baltimore, MD 21218, Email: pollizzi@stsci.edu

Abstract. The Hubble Space Telescope (HST) generates on the order of 7,000 telemetry values, many of which are sampled at 1Hz, and with several hundred parameters being sampled at 40Hz. Such data volumes would quickly tax even the largest of processing facilities. Yet the ability to access the telemetry data in a variety of ways, and in particular, using ad hoc (i.e., no a priori fixed) queries, is essential to assuring the long term viability and usefulness of this instrument. As part of the recent NASA initiative to re-engineer HST's ground control systems, a concept arose to apply newly available data warehousing technologies to this problem. The Space Telescope Science Institute was engaged to develop a pilot to investigate the technology and to create a proof-of-concept testbed that could be demonstrated and evaluated for operational use. This paper describes this effort and its results.

1. HST as a Telemetry Source

The Hubble Space Telescope (HST) is well a known source of significant and substantial amounts of Astronomy data. Less known, however, is that the HST is also one of the most highly instrumented non-manned platforms ever launched. Over 6,000 telemetry points are monitored on the HST. These "monitors" cover practically every aspect of the platform and of the various instrument environmental and state conditions.

In addition to routine study and problem analysis, we use telemetry to look for long term trends in how the platform is behaving. By carefully studying such trends, we hope to uncover potential problems before they arise. In this way, we plan to extend the scientific utility of this unique instrument as far as possible through its planned lifetime (now scheduled through to 2010).

1.1. Complications in Querying Telemetry Values

Since telemetry values are sampled at widely different rates, looking for a "cause-effect" relationship between monitors can rarely be found by identifying time matches between records. Rather, the queries tend to look for overlapping time windows when the monitors acquire some state. We have coined the term, "Fuzzy Query" to describe this kind of query. Using a stylized SQL, a fuzzy query typically appears as:

Select $ParamA \ldots ParamN$ where $Param1 > somelimit$ AND $Param2 > somelimit$ AND $\underline{Time(Param2) \leq Time(Param1) + someDelta}$

The underlined portion in the above query highlights this 'fuzzy' aspect. An SQL designer will recognize the complexity such a query requires.

1.2. Dealing with the Telemetry Volume

Then there's the shear volume of the telemetry data. At its nominal format and rate, the HST generates over 3,000 monitored samples per second. Tracking each sample as a separate record would generate over 95 giga-records/year, or assuming a 16 year Life-of-Mission (LOM), 1.5 tera-records/LOM. Assuming a minimal 20 byte record per transaction yields 1.9 terabytes/year or 30 terabytes/LOM. Such volumes are supported by only the most exotic and expensive custom database systems made.

By careful study of the data, we discovered two properties that could significantly reduce this volume. First, instead of capturing each telemetry measurement, by only capturing when the measurement changed value - we could reduce the volume by almost 3-to-1. Second, we recognized that roughly 100 parameters changed most often (i.e., high frequency parameters) and caused the largest volume of the "change" records. By averaging these parameters over some time period, we could still achieve the necessary engineering accuracy while again reducing the volume of records. In total, we reduced the volume of data down to a reasonable 250 records/sec or approximately 2.5 terabytes/LOM.

2. Data Warehousing as a Potential Solution

The complexities and expected volumes in dealing with telemetry data naturally lead us to consider Data Warehousing. Beyond its data handling abilities, Data Warehouses are designed to be balanced in their approach to the data. That is, they expect to handle the ad-hoc type queries with little or no pre-knowledge of what will be of interest.

2.1. What is Data Warehousing

From a User's perspective, a data warehouse looks very similar to a typical relational database management system (RDBMS). A user will find a query language very similar to (if not the same as) SQL. Typically, the warehouse will support one or more programming interfaces, such as ODBC, which allows the warehouse to be accessed by familiar reporting or analysis tools (such as Microsoft's Access, IDL, PV-Wave,)

To the database designer, a different perspective is shown. There is no transaction or update facility for the warehouse. The warehouse operates either with many readers or a single writer, and the warehouse is "loaded" as opposed to being updated. In laying out the warehouse, the designer quickly learns that their logical definition (i.e., the layout of the tables and attributes of the warehouse) is more intimately tied to the physical definition (how the warehouse is laid-out on the physical i/o subsystem). Choices in one will often significantly affect the other.

In giving up the flexibility in transactions and by having the closer linkage between the physical and logical views, the warehouse provides a number of new features particularly in supporting efficient indices for very large data volumes.

2.2. The Star Index

One of the most common indexing methods that characterizes a data warehouse is called the Star. In using a Star index, the designer starts by designing a few number of *fact* tables.

Generally, a fact table can be viewed as a flat file version of an entire collection of typical database tables. The goal here is not to normalize data. Instead, a fact table attempts to bring together as many related attributes as can be expressed in a single table with (effectively) an unlimited number of columns.

While a fact table holds the records of interest to be searched, *dimension* tables provide the meta-data that describes aspects of a fact table and supports the rapid indexing of data. A dimension table can be formed for any column within a fact table, when all possible values for that column can be taken from a pre-defined set of values. For example, a column holding a person's age can be reasonably limited to the set of whole integers from $\{1 \ldots 150\}$; sex from { "Male", "Female" }. Even an arbitrary set of values is appropriate. Consider the social security numbers for all the employees of a company. While the individual numbers themselves may be an arbitrary string of digits, all the numbers are known and can be listed within a dimension table.

The star index then relates a fact table to one or more corresponding dimension tables.

3. Resulting Fact Tables

Applying this strategy to the HST telemetry problem produced the following Fact Tables:

- Telemetry Facts with the columns: Mnemonic, Format-type, Flags, Raw Value, Start Time, Stop Time, Start Millsec, Discrete Value (NULL if the telemetry parameter is not a discrete), and Engineering Units (EU) Value (NULL if the telemetry parameter is a discrete).

- Averaged Facts with the columns: Mnemonic, Format-type, Flags, Start Time, Stop Time, Start Millsec, Averaged, Maximum, Minimum EU Values, and Number of Samples in period.

and with the dimension tables:

- Mnemonics - enumerated all Mnemonics, and other meta-data

- Time - enumerated as years, months, days, hours, minutes, seconds

- Milliseconds - enumerated as 0 - 999

- Format - enumerated as the defined HST telemetry format codes

- Discretes - enumerated as the list of all possible discrete text strings (i.e., 'on', 'off', 'set clear', 'set on', 'state level 1', 'state level 2', ...)

4. Benchmarking

The apparent simplicity of the above tables belies the subtlety and the iterations necessary to converge to these tables. Key to this refinement was the use of benchmarking against a reasonable set data. The importance of benchmarking against a sufficiently sized dataset cannot be understated. With the traversal power of data warehouse engines, even the most poorly defined warehouse structure will be quickly searched when the datasets are of typical RDBMS testing sizes. For the nuances of a warehouse design to come out, the dataset must reach a reasonable fraction of the total expected size of the holding. In our case, we pushed our test warehouse up to 100 gigabytes in testing for a planned size of 1 to 2 terabytes.

In doing the benchmarks, we constructed a family of queries, each meant to push some aspect typical of the queries we expected the warehouse to handle. The queries were specifically meant to validate the efficiency of the indexing as it related to the size of the warehouse. In terms of the telemetry data, we ran the query suite against the warehouse with 12 through 36 weeks of data.

It was only through this process that we were able to understand the implications of our design choices, and then refine the warehouse scheme to that shown above.

5. Lessons Learned

This prototype effort demonstrated both the capabilities and limitations of warehousing technology. On the plus side, warehousing technology shows a lot of promise. Once the design was settled, we were impressed with the performance and shear data handling capability of the warehouse product. It is clear that this technology can have significant benefit for those working with reams of discrete information.

As a weakness, this technology is still quite young. The products are only beginning to stabilize and one must be prepared for a number of false starts. For the scientific/engineering user, it is important to realize that warehouse technology is being driven by the commercial sector. There is little experience on the part of the vendors in scientific data issues, and in many cases the warehouse product might have a rich set of functions and primitives for commercial or financial use - but be missing rudimentary scientific ones. In particular, be aware of the use of time, high precision real numbers, scientific notation and functions. Most importantly, it must be remembered that a data warehouse is not designed as one would design a relational database. The warehouse designer has new indexing tools that tend to drive the design more, and the tight linkage between the logical and physical designs must also be reckoned with.

Finally, the critical lesson is the absolute need for benchmarking a sizable dataset. It was only through the actual trial-and-error of manipulating the design, and pushing the technology against a fairly sized dataset, that the real power and limitations of the warehouse are exposed.

Astronomical Data Analysis Software and Systems VII
ASP Conference Series, Vol. 145, 1998
R. Albrecht, R. N. Hook and H. A. Bushouse, eds.

Accessing Astronomical Data over the WWW using *dat*OZ

Patricio F. Ortiz

Department of Astronomy, University of Chile, Casilla 36-D, Santiago, Chile, Email: ortiz@das.uchile.cl

Abstract.
The Department of Astronomy of the University of Chile hosts a number of astronomical databases created with *dat***OZ** (Ortiz, 1997). This is a site in which databases are accessed interactively by the use of an HTML interface and CGI database engines. Data can be retrieved by the users in a highly flexible way, from lists of user-selected quantities, to customized plots, including any additional multimedia information available for the database elements. The latest additions to the system point in two directions: a) to allow access to combinations of database variables for different purposes, and b)to allow the retrieval and further correlation of information stored in different *dat***OZ**'s databases, by means of a second tool named **Cat***MERGER*. Another tool lets users search for objects in all catalogs created with *dat***OZ** at the same time.

The most recent development is the creation of a new version of *dat***OZ** capable of handling catalogs such as USNO's A1.0 with 488 million objects. Specially designed indexing and data storage techniques allow the catalog size to be reduced by half.

1. Introduction

Electronic sharing of scientific data is a concept which has been around for at least the last fifteen years. The first efforts pointed in the direction of creating sites accessible via ftp. Pioneer work at the Strasbourg CDS (SIMBAD), and by NASA (IPAC's NED, GSFC's ADC ftp site) amongst others, have set the ground to share data in a more efficient way amongst astronomers than the traditional printed, journals.

The WEB gives scientists a much richer way of sharing information. We can now have quite a number of things done by transferring just a few bytes between our computers and host computers running httpd (the HTTP daemon) by invoking CGI (common gateway interface) "scripts". The applications are almost unlimited, and we see more and more of them at astronomical sites around the world.

Scientists are usually interested in just a minor percentage of the data kept in a catalog to pursue their research. We might need a few "columns" of a catalog, and maybe just a small subset of the catalog's elements. There are currently a few efforts in the world pointing in that direction, VizieR at CDS, Wizard at NASA, Pizazz at NCSA Skycat at STScI, Starlink at the UK, *dat***OZ**, at the University of Chile, and possibly others. All of them should point to create a uniform method of data retrieval.

In the following sections an overview of the features of databases created with *dat*OZ will be presented. Detailed information can be found in the "user's manual" for the system, where static and dynamic databases are discussed.

2. What is *dat*OZ and what does *dat*OZ offer?

*dat*OZ is a computer tool which creates the interface and the source code capable of handling a specific catalog (Ortiz 1997). The source code is written in C, with some specific routines written in Perl. The data is stored in machine format, with some degree of indexation for fast access, a key for large catalogs. The modified version of *dat*OZ for large catalogues fully indexes the data-file, introduces compressed variables and reduces the number of significant figures for RA and Dec without degradation of the positions.

Each database element can have associated several "multimedia" files (.gif, .jpg, .ps, spectra-like data files, and a note-pad for extra annotations), and supports a kind of variable which allows hyper-links to other sources of information. The system was built with the capability of accessing interfaces in more than one language; something particularly useful if this system is to be accessed by people in countries where English is not the official language.

*dat*OZ offers the user a database with fast and flexible data retrieval and visualization capabilities; it also offers uniformity, as all databases look alike and behave in the same way. These databases can receive requests from related tools developed with the purpose of combining information amongst databases created with *dat*OZ. Links to these tools are found in the homepage.

*dat*OZ's home page is: **http://146.83.9.18/datoz_t.html**

3. *dat*OZ Retrieval Modes

*dat*OZ' databases have a flexible retrieval access mode also known as Advanced Access Mode (**AAM**), which lets the user define the quantities he/she wants to retrieve amongst the "natural" database variables or by using mathematical expressions formed with these variables. The opposite of the **AAM** is the Basic Access Mode (**BAM**) which restricts operations to the values of the database variables.

The reason for implementing the flexible access mode has to do with the fact that catalogs are usually created with one purpose in mind, and only certain quantities become the catalog's variables. A typical case is when the catalog contains "variables" for **V**, **(B-V)**, **(V-R)**, and **(V-I)**, but if the user needs the magnitude of an object in the **I** band he/she would normally have to retrieve **V** and **V-I** and then perform the difference by him/herself. The **AAM** lets the user specify the quantity to get as **V - (V-I)**, i.e., **I**. The way to express this quantity is using Reverse Polish Notation (RPN) (complete details are found on the manual and HTML help pages).

Another advantage of the **AAM** is that the user can impose constraints on math expressions formed with "numerical" variables. Let's assume that, from

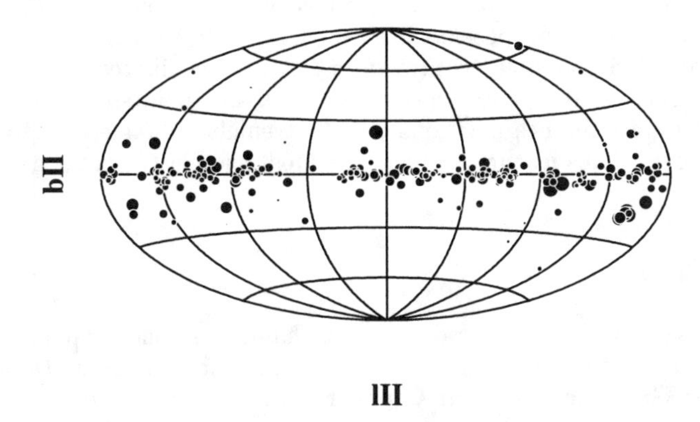

Figure 1. Distribution of O stars in the galaxy.

the same catalog, we need sources brighter than 17 in the **I** band, but as we saw, **I** is not one of the catalog variables. We can define the expression to use as a constraint as: `V V-I -` and impose it to be < 17.

One of the most important pluses of the **AAM** is that it lets the user submit a file with a list of the names of the objects which he/she needs to get information for, and get whatever data is stored in the database for those objects only. This is particularly useful when we deal with catalogs with a large number of objects, such as Hipparcos, or ROSAT for example. The files are transfered from the computer where the user is running **Netscape** to the database's host computer and then analyzed locally.

4. *dat*OZ Visualization Tools

Visualization is the key to getting most of the science out of the data stored in a catalog. It allows quick checking of determined properties, and it is also a valuable teaching tool. ***dat*OZ** provides its databases with the capability to use any numerical variable to generate fully customized PostScript plots.

There are several types of plots that can be obtained: **histograms**, **scatter plots**, **scatter plots with variable symbol size**, **scatter plots with error bars**, **annotated scatter plots** (where the user prints the value of a variable instead of a symbol), **All-sky, equal area projection plots**, and **line plots**.

Plots can be viewed on the spot with `ghostview`, `ghostscript`, or any other PostScript previewer, and/or saved to disk using any of the ways Netscape provides.

5. Searching for Neighbors with *dat*OZ

The word "neighbor" must be understood in a broad sense here. It could mean, of course, objects "close" in the sky, or it could apply to closeness in any numerical variable or expression.

The neighbor search operation is extremely valuable to explore the content of a catalog for diverse purposes. Besides using elements in the catalog as reference points, it is possible to provide a list of coordinates and ask which objects in the catalog are located near these reference points. The submission of files with lists of reference positions is an extremely valuable task to quickly search for matches. This feature is one of the most used in the catalogs installed in our system.

6. Conclusions

*dat***OZ** has proved to be a very powerful tool to handle astronomical information, at the present time, the following catalogs are implemented in our Department (Cerro Calán): **Galactic Globular Clusters**, Harris W.E. Astron. J. 112, 1487 (1996). **QSO's and AGNs** Veron-Cetty M.P., et al. (1996). **ROSAT Bright Sources Catalog** Max-Planck-Institut fuer extraterrestrische Physik, Garching (1996). **Catalog of Principal Galaxies**, Paturel et al. 1989 **Third Revised Catalog (RC3)**, de Vaucouleurs G. et. al. (1991). **AAT Equatorial photometric calibrators**, Boyle et al.'s (1995). **A Catalogue of Rich Clusters of Galaxies**, Abell G.O., Corwin Jr. H.G., Olowin R.P. (1989) **IRAS Catalog of Point Sources**, V2.0 Joint IRAS Science W.G. IPAC (1986). **Hipparcos Main Catalog** ESA, (1997). **Spectrophotometric Standards used at CTIO MK Spectral Classifications** - Twelfth General Catalog Buscombe W., and Foster B. E. (1995). **Northern Proper Motions, Positions and Proper Motions, SOUTH**, Bastian U., and Roeser S. (1993). **Catalog of 558 Pulsars**, Taylor J.H. et al. (1983). **Line Spectra of the Elements**, Reader J., and Corliss Ch.H. (1981). **USNO's A1.0 – 488 million stars**, Monet, D. (1996).

The near future will bring big survey works (Sloan, 2DF, 2MASS, AXAF, and EROS, to name a few). Making the data available to the community will be crucial to increase these projects' science impact. The current trend of having data available on CD's or ftp sites may discourage an important number of scientists to explore or analyze the data, tools like *dat***OZ** will make the data readily available to anyone.

7. Acknowledgments

I thank the support of project FONDECYT 1950573 lead by Dr. José Maza. To the organizers of ADASS 97 my most sincere thanks for allowing me to assist to the conference. I am also very thankful of Professors Luis Campusano and Claudio Anguita for valuable comments and support to mount some databases on their machines; and to Sandra Scott for installing a demonstration database at the University of Toronto during the ADASS 97 conference.

References

Ortiz, P. F. 1997, in Proceedings of the Fifth workshop on Astronomical Data Analysis, Erice, 1996.

Astronomical Data Analysis Software and Systems VII
ASP Conference Series, Vol. 145, 1998
R. Albrecht, R. N. Hook and H. A. Bushouse, eds.

Distributed Searching of Astronomical Databases with Pizazz

K. Gamiel, R. McGrath and R. Plante

National Center for Supercomputing Applications, University of Illinois, Urbana, IL 61801

Abstract. Robust, distributed searching of networked-based astronomical databases requires an investment not only by individual data providers, but also by the astronomical community as a whole. We describe both these investments and introduce a supporting software project called Pizazz. It is our hope that the community will appreciate the social requirement placed on them to work together and to participate in efforts towards a globally integrated, distributed astronomical search and retrieval system.

1. Background

Establishing the infrastructure required for robust, distributed searching of databases is not a simple proposition. In our view, there are four major steps involved; analysis of existing data sources, choosing a model, profiling, and implementation.

In this case, analysis means finding various astronomical database sources around the net of interest to the community and analyzing the interactive capabilities and content. For example, what is the protocol used, what is the query format, what is the response format, etc.

Next, we must choose a model for interacting with the identified data sources in parallel. We narrow these models to two. The first model is the "hands-off" model, that is, we do not impose any changes on the original data sources whatsoever. In such a model, one would build a distributed searching gateway that would offer a single user interface, but would map the user's query to each and every original data source in its native protocol and interaction style. While this is obviously an appealing option for the original data providers, we find that the wide range of protocols and interactive paradigms that must be supported severely limit the overall usage and value of response. We feel this approach is short-sighted and ultimately of little value. The second and preferred model is to agree on a single protocol for information retrieval. By agreeing on a single protocol, we ensure a consistent and highly extensible interface to each and every original data provider. If we agree, for example, on a single protocol and profile, one can easily imagine autonomous user agents interacting with the system on behalf of users.

Profiling is a step that must be taken regardless of which model is chosen. Profiling, in the current context, means agreeing on general features of a distributed searching system. For example, a profile for such a system might state that all original data providers must provide access to title, author, and abstract fields. It may state that a response must include the same information and maybe a URL. In other words, a profile is a document that ensures some level of consistency when interacting across data sources.

The last step is implementation. This involves work depending on which model is chosen. If the "hands-off" model is chosen, one must implement the distributed searching gateway and for each original data source, a driver must be written in order to interoperate with that source, mapping to the user interface. In this model, the programmer must be available at a moment's notice to alter any of the possibly hundreds of drivers when the original data provider alters their interface. Since no open standards are imposed, it is certain that frequent changes will occur, particularl since most of these systems will probably be HTML/CGI-based systems and will change for purely superficial reasons. If the second model is chosen, a communications server must be written and installed at the original data provider's site. A gateway application that need only speak a single protocol in parallel is then written. In this model, the gateway need only point to new data providers as they come online, not requiring any changing of source code.

2. Pizazz

We chose a model based on a single information retrieval protocol, namely ANSI/NISO Z39.50. Z39.50 is a well-defined international standard used in academic, government, and commercial institutions. It defines a standard for interactive search and retrieval of database records from a data source over the Internet. We here announce a software distribution called Pizazz that includes a Z39.50 server toolkit. More information on Pizazz and other project details can be found at http://webstar.ncsa.uiuc.edu/Project30/.

The server toolkit included with the Pizazz distribution builds on an existing application called pizazzd. The typical scenario is that a data provider who wants to participate in the distributed searching system will download Pizazz and build the default pizazzd server. The provider will then alter a single C file that includes callbacks for the four major information retrieval functions, namely initialize, search, present, and close. In those four functions, the provider will make calls to his own native database system, building the response as appropriate with pizazz library calls we provide. Once that interface is written, the provider then notifies NCSA and a pointer to their server is added to the distributed searching gateway.

It would be a disservice to say that creating this interface is simple. However, we are committed to making it as easy as possible with helper tools and, more importantly, feedback from users ("user" in this sense is the programmer who builds the interface between pizazzd and the native database). By far the most difficult part of implementing the interface is in support of the nested RPN query structure. The user has access to function calls for walking the query tree structure and while doing so, must generate a query based on terms and attributes suitable for his native system.

The information retrieval model used by pizazzd is as such. The server receives a single initialize request from the client. The initialize callback is invoked where any native database initialization is performed. Typically, the server then receives a search request which includes the database name to search and a potentially complex, but standard, RPN query, along with other search parameters. The search callback is invoked where the query is translated and passed to the native database system. The search results in the formation of a logical result set. From that result set, the client may then request to have records presented. In that case, the server receives a present request asking, for example, records 1 through 5 from the Default result set in HTML format. The present callback is invoked and the requested records are retrieved from the native database and returned to the client. Finally, the client sends a close request, whereas the close callback is invoked, releasing any resources used by the native database.

More information on the Pizazz software distribution can be found at the project home page, http://webstar.ncsa.uiuc.edu/Project30/. The project team recognizes and appreciates the complexity of interfacing a communications server with a native database system and, accordingly, are happy to help with the effort in any way possible. Contact information may also be found on that Web page.

3. Conclusion

Designing a distributed, networked-based astronomical information system requires a four-step approach including original data source analysis, modeling the system, profiling the system, and implementation of the system. We described each step and proposed our solution. We have a project underway to implement such an information system and are committed to helping the astronomical community realize the dream of a robust, single interface to any and all relevant data sources on the Internet.

Astronomical Data Analysis Software and Systems VII
ASP Conference Series, Vol. 145, 1998
R. Albrecht, R. N. Hook and H. A. Bushouse, eds.

New Capabilities of the ADS Abstract and Article Service

G. Eichhorn, A. Accomazzi, C.S. Grant, M.J. Kurtz and S.S. Murray

Smithsonian Astrophysical Observatory, 60 Garden Street, Cambridge, MA 02138, Email: gei@cfa.harvard.edu

Abstract. The ADS abstract service at: http://adswww.harvard.edu has been updated considerably in the last year. New capabilities in the search engine include searching for multi-word phrases and searching for various logical combinations of search terms. Through optimization of the custom built search software, the search times were decreased by a factor of 4 in the last year.

The WWW interface now uses WWW cookies to store and retrieve individual user preferences. This allows our users to set preferences for printing, accessing mirror sites, fonts, colors, etc. Information about most recently accessed references allows customized retrieval of the most recent unread volume of selected journals. The information stored in these preferences is kept completely confidential and is not used for any other purposes.

Two mirror sites (at the CDS in Strasbourg, France and at NAO in Tokyo, Japan) provide faster access for our European and Asian users.

To include new information in the ADS as fast as possible, new indexing and search software was developed to allow updating the index data files within minutes of receipt of time critical information (e.g., IAU Circulars which report on supernova and comet discoveries).

The ADS is currently used by over 10,000 users per month, which retrieve over 4.5 million references and over 250,000 full article pages each month.

1. Introduction

The Astrophysics Data System (ADS[1]) provides access to almost 1 million references and 250,000 scanned journal pages (Eichhorn 1997). These can be accessed from the World Wide Web (WWW) through a sophisticated search engine (Kurtz et al. 1993), as well as directly from other data centers through hyperlinks or Perl scripts (Eichhorn 1996). Our references in turn link to other data and information sources. This cross-linking between different data systems provides the user with the means to find comprehensive information about a given subject.

---

[1]http://adswww.harvard.edu

2. New Search Features

1. Complex Query Logic

   The search system allows the user to specify complex queries in two forms:

   (a) Simple logic: This allows the user to specify that certain words must appear in the selected reference (+word) or must not appear in the select reference (-word). Phrases of multiple words can be specified by enclosing the words in double quotes. An example is in Figure 1 in the title field:

   +"black hole" -galaxies +=unstable

   This query searches for references that contain the phrase "black hole", but not the word "galaxies" or its synonyms (like "galaxy" or "galactic"). They must also contain the word "unstable". The '=' before "unstable" turns off the automatic synonym replacement. It is not sufficient for a reference to contain a synonym of "unstable".

   (b) Full boolean logic: This allows the user to use "AND", "OR", "NOT", and parentheses for grouping to build complex logical expressions. An example is in Figure 1 in the abstract text field:

   ("black hole" or "neutron star") and ("globular cluster" or binary) and not ("cataclysmic variable" or CV)

   This expression searches for references that contain one of the expressions "black hole" or "neutron star" as well as either the expression "globular cluster" or the word binary, but neither the expression "cataclysmic variable" nor the word CV.

2. Object Queries

   (a) Extra-Solar system objects: By selecting SIMBAD and/or NED above the input field for object names, the user can query for objects outside the solar system.

   (b) Solar system objects: By selecting LPI, a small database of meteorite and lunar sample names can be queried.

   (c) IAU objects: By selecting IAU, the database of names that have appeared in IAU Circulars is searched.

3. Indexing

1. Preprint Database: We are now indexing the preprints from the Los Alamos preprint server on a regular basis. Every night after the preprint server is updated, we automatically retrieve the articles from the preprint server and index them into a separate database. This database can be searched through the same interface as our other databases (Astronomy, Instrumentation, and Physics/Geophysics).

2. Mirror Sites: The mirror sites and the associated updating procedures are described by Accomazzi et al. (this volume).

3. Quick Updates: The quick updates allow us to quickly enter new data into the database. This was mainly developed to enable us to include IAU Circulars into the ADS within minutes after publication. Normally a full indexing of our database takes more than one day. Quick updates append new index sections to the original ones and link these new sections to the existing ones. The searches are not noticeably slowed down by these additional links. Every two to three weeks we re-index the complete database to include these additional sections in the main index.

4. New User Interface Features

We implemented WWW cookies in the ADS user interface. WWW cookies are a system that allows us to identify individual users uniquely, even without knowing who they are. This allows us to customize our response to individual users. Users can for instance select which mirror sites they prefer for some external services, how to print articles by default, and how the pages of the system should look (font sizes, colors, etc). The system also remembers through the cookie mechanism which tables of contents have been retrieved by the user for several different journals. Lastly, we can send out one-time messages to users. The ADS remembers which message has already been sent to a user. The user database is of course completely confidential and will not be made available to any other system.

5. Future Plans

Two major projects that will require software development in the future are the OCRing (Optical Character Recognition) and indexing of the historical literature once we have scanned it, and the parsing of the scanned articles for references.

In the next year we plan to scan several major journals back to volume 1. We will OCR these scans and make the OCR'd text available to selected researchers.

The other major project is the parsing of the references from the scanned literature. This will allow us to update and expand the reference and citation lists that are already available. This will be a very difficult task and there is no time line yet as to when we will be able to get useful data from that project.

Acknowledgments. This work was funded by NASA under grant NCCW 00254.

References

Eichhorn, G., 1997, Astroph. & Space Sci., 247, 189

Kurtz, M.J., et al. 1993, in ASP Conf. Ser., Vol. 52, Astronomical Data Analysis Software and Systems II, ed. R. J. Hanisch, R. J. V. Brissenden & Jeannette Barnes (San Francisco: ASP), 132

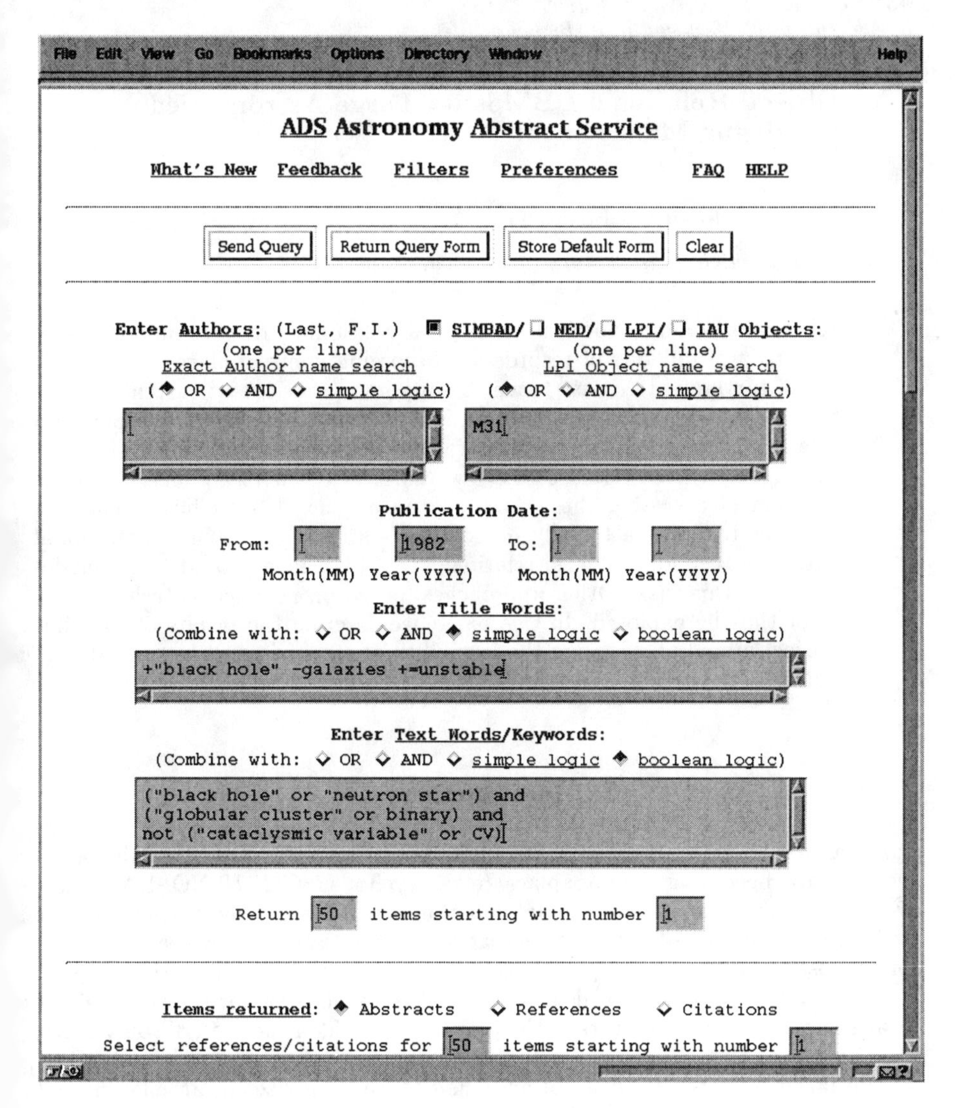

Figure 1. ADS Query page with examples of logical query constructs.

Eichhorn, G., et al. 1996, in ASP Conf. Ser., Vol. 101, Astronomical Data Analysis Software and Systems V, ed. George H. Jacoby & Jeannette Barnes (San Francisco: ASP), 569

Accomazzi, A., et al., 1998, this volume

Astronomical Data Analysis Software and Systems VII
ASP Conference Series, Vol. 145, 1998
R. Albrecht, R. N. Hook and H. A. Bushouse, eds.

Object–Relational DBMSs for Large Astronomical Catalogue Management

A. Baruffolo and L. Benacchio

Astronomical Observatory of Padova, Italy

Abstract. Astronomical catalogues containing from a million up to hundreds of millions of records are becoming commonplace. While they are of fundamental importance to support operations of current and future large telescopes and space missions, they appear also as powerful research tools for galactic and extragalactic astronomy. Since even larger catalogues will be released in a few years, researchers are faced with the problem of accessing these databases in a general but efficient manner, in order to be able to fully exploit their scientific content. Traditional database technologies (i.e. relational DBMSs) have proved to be inadequate for this task. Other approaches, based on new access technologies, must thus be explored. In this paper we describe the results of our pilot project aimed at assessing the feasibility of employing Object–Relational DBMSs for the management of large astronomical catalogues.

1. Introduction

Large astronomical catalogues, with one million up to hundreds of millions of records, are becoming commonplace (e.g., Tycho, GSC I, USNO–1.A). They have an obvious operational use, in that they will be employed throughout the cycle of observations of the next generation large telescopes and space missions for proposal and observation preparation, telescope scheduling, selection of guide stars. However, they appear also as powerful research tools for the study of the local and grand scale structure of the Galaxy, cross–identification of sources etc.

Since even larger catalogues will be available in the near future (e.g., GSC II), we are faced with the problem of accessing these databases in an efficient but general manner. If the scientific content of these catalogues is to be fully exploited, astronomers must be allowed to issue almost any query on the database without being hampered by excessively long execution times.

2. The Large Astronomical Catalogues Management Problem

There are many possible approaches to the problem of managing very large astronomical catalogues.

Data can be organized in a catalogue specific file structure and accessed by means of programs. This approach allows for fast access for a defined set of queries, on the other side new queries require writing of programs and access is limited to one catalogue only.

One can then consider the use of "custom", astronomical, DBMSs, like, e.g., DIRA (Benacchio 1992) or Starbase (Roll 1996). They support astronomical data and queries, and are freely available. However they typically do not support large DBs, since data is often stored in flat ASCII files and secondary access methods are usually not provided.

Commercial Relational DBMSs have also been used in the past: they are robust systems, widely used in the industry, whose data model is close to the structure of astronomical catalogues (Page & Davenhall 1993). On the other side, they have limited data modeling capabilities, and their access methods support indexing on simple data types and predefined set of query predicates. Their use with large astronomical catalogues has proved to be problematic (Pirenne & Ochsenbein 1991).

Another possible approach is to use an Object–Oriented DBMS. These systems feature a powerful data model, which allows data and operations to be modelled. However, they do not provide an efficient query processing engine, such facility must be implemented on the top of the DBMS (Brunner et al. 1994).

3. An Object–Relational Approach

From the discussion above it is apparent that, in order to give astronomers a general and efficient access to new databases, a DBMS must be employed that support astronomical data and queries, and that is able to efficiently execute them.

Recently, a new class of DBMSs has emerged, Object–Relational DBMSs, that provide:

- user–defined data types and functions which allow to define methods to create, manipulate and access new data types;
- user–defined index structures that can speed up the execution of queries with predicates that are *natural* to the new data types;
- an extensible optimizer that can determine the most efficient way to execute user queries.

In a word, they provide powerful data modeling and efficient query processing capabilities. We thus conducted a pilot project aimed at assessing the feasibility of employing ORDBMSs in managing large astronomical catalogues.

We built a prototype catalogue management system on a Sun Sparc Ultra 1/140, equipped with 128 MB RAM and 10 GB HD, using PostgreSQL 6.0 (Yu & Chen 1995), as the Object–Relational DBMS. Software was developed, in the C language, for the custom data types and functions, to extend the DB B–tree index to support astronomical coordinates, and to implement a two dimensional R–tree (Guttman 1984) index on coordinates on top of the DB GiST (Hellerstein et al. 1995) secondary access methods.

We defined in the DBMS typical astronomical data types (e.g., coordinates) and implemented functions acting on them. The DB query language was then extended by bounding these functions to user–defined operators so that they could be employed in formulating queries. For example, typical astronomical queries that were supported in this way are:

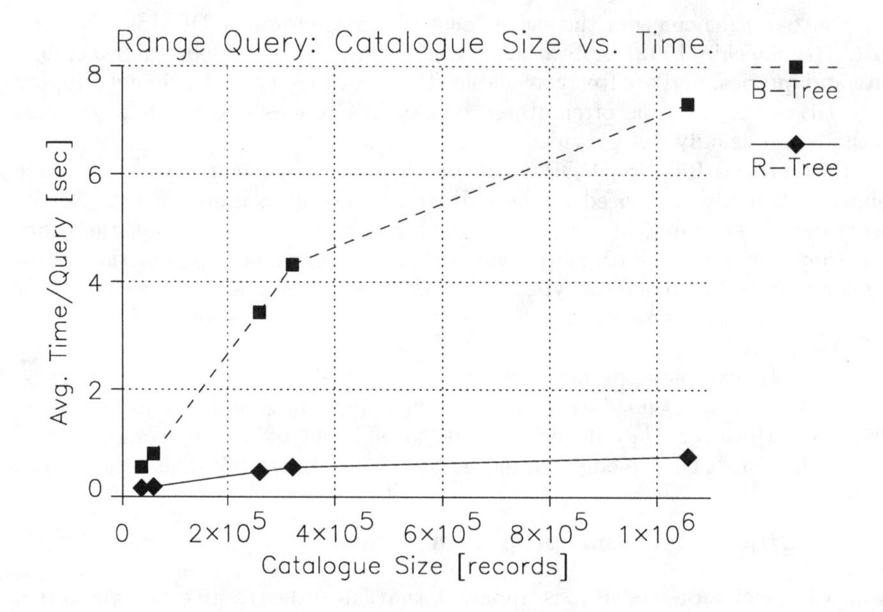

Figure 1. Execution times for range queries, covering an area of $40^{\circ 2}$ in the sky, over coordinates on catalogues of various sizes indexed using B–Trees (dashed line) and R–Trees (solid line). Times were averaged over 100 queries and include actual retrieval of the data from disk.

- Range query on coordinates:

```
SELECT * FROM AGK3 WHERE POS -> '(0:4,-55:16,0:8,-54:16)';
```

- Search–by–cone:

```
SELECT * FROM AGK3 WHERE POS @> '(0:4,-55:16,0.1)';
```

- Cross–correlation of catalogues based on sky position:

```
SELECT A.ID,B.ID,A.POS,B.POS FROM AGK3 A, TYCHO B WHERE
  A.POS @> SkyPosToSkyCone(B.POS,0.1);
```

In order to evaluate the performance improvement that can be obtained by employing multi–dimensional indexes, we created and populated a database with data from five catalogues, ranging in size from $\sim$ 35000 records (IRS) up to a million records (Tycho). From these catalogues we extracted: ID, α, δ, μ_α, μ_δ, σ_α, σ_δ, σ_{μ_α}, σ_{μ_δ}, magnitude and spectral type. All catalogues were then indexed on coordinates using both B–Trees and R–Trees, and a series of tests were run to measure the performance of these access methods.

Results for one of these tests are shown in Figure 1. From this graph it is apparent that, even for a simple range query over coordinates, execution times are greatly reduced when using an R–tree index with respect to a B–tree index, which is the access method commonly employed in relational DBMSs.

It is to be noted that absolute query execution times are only indicative of the DB performance, because they depend on the actual content of the catalogues, system hardware, etc. Relative performance of the R–tree based indexes with respect to B–trees is more significant, because all other conditions are identical. Another important point is that other typical astronomical queries besides the simple range query (e.g., search–by–cone) can take advantage from the presence of an R–tree based index, while their execution can't be speeded up using B–tree indexes.

4. Conclusions

Our experience in employing an ORDBMS to manage astronomical catalogues has been positive. The data modeling capabilities of this DBMS allow to define typical astronomical data in the DB. We verified that it is possible to extend the DB query language with astronomical functionalities and to formulate queries with astronomical predicates. Further, the execution of these queries is speeded up by the use of multidimensional indexes. Performance improvements, with respect to traditional access methods, are apparent even with small catalogues.

On the minus side, it should be noted that substantial effort is required to add new index structures to the DB, however some commercial ORDBMSs already supporting R–Trees and other third party "extensions" (with access methods) are also available. They can be customized to support astronomical data and predicates.

We also experienced long data loading and index building times, this is an architectural issue though, it is not a fundamental limitation due to the specific data model of the DBMS. In fact, commercial ORDBMS usually provide parallel operations for data loading, index creation and query execution.

The bottom line is that in our experience ORDBMS provide the basic building blocks for creating systems for an efficient and general access to large astronomical catalogues. We think that this technology should be seriously taken into account by those planning to build such systems.

References

Benacchio, L. 1992, ESO Conf. and Workshop Proc. 43, 201

Brunner, R. J., Ramaiyer, K., Szalay, A., Connolly, A. J., & Lupton, R. H. 1995, in ASP Conf. Ser., Vol. 77, Astronomical Data Analysis Software and Systems IV, ed. R. A. Shaw, H. E. Payne & J. J. E. Hayes (San Francisco: ASP), 169

Guttman, A. 1984, Proc. ACM SIGMOD, 47

Hellerstein. J. M., Naughton, J. F., & Pfeffer, A. 1995, Proc. Int. Conf. on VLDBs, 562

Pirenne, B., & Ochsenbein, F. 1991, ST–ECF Newsletter, 15, 17

Page, C. G., & Davenhall, A. C. 1993, in ASP Conf. Ser., Vol. 52, Astronomical Data Analysis Software and Systems II, ed. R. J. Hanisch, R. J. V. Brissenden & Jeannette Barnes (San Francisco: ASP), 77

Yu, A., & Chen, J. 1995, PostgreSQL User Manual

Roll, J. 1996, in ASP Conf. Ser., Vol. 101, Astronomical Data Analysis Software and Systems V, ed. George H. Jacoby & Jeannette Barnes (San Francisco: ASP), 536

Astronomical Data Analysis Software and Systems VII
ASP Conference Series, Vol. 145, 1998
R. Albrecht, R. N. Hook and H. A. Bushouse, eds.

The VizieR System for Accessing Astronomical Data

François Ochsenbein

Centre de Données Astronomiques, Strasbourg Observatory, France

Abstract. The recently reshaped VizieR[1] system, a unified query interface to an increasing number of catalogs (presently ~ 1500), is presented.

1. Historical Background

VizieR was first presented at the AAS meeting in early 1996 (Ochsenbein et al. 1995), as the result of a joint effort between CDS[2] and ESA-ESRIN[3] (the European Space Agency's Information Systems division) in order to provide the astronomical community with a dedicated tool for retrieving astronomical data listed in published catalogs and tables — a follow-up of the ESIS (European Space Information System) project.

Shortly after this first version, which has been effectively accessible since February 1996, new needs for performance and for standardisation led to basic changes in the system: the ASU[4] (Astronomical Standardized URL), resulting from discussions between several institutes, was adopted as the way to specify constraints in the new version of VizieR, which was introduced on 29 May 1997 — just in time for the distribution of the results of *Hipparcos* catalogs. The basic concept of ASU is a standardized way of specifying catalogs (as `-source=`*catalog_designation*), target positions (as `-c=`*name_or_position*`,rm=`*radius_in_arcmin*), output format (as `-mime=`*type*), and general constraints on parameters (as *column_name*=*constraint*).

Besides the adoption of this new protocol, the most visible changes in this new version of VizieR are an easy access to notes, and possibilities of navigation between the tables of a catalog.

The quantitative daily usage of VizieR is presently (September 1997) about 1,000 external requests from 75 different nodes; 1,000 different nodes effectively submitted queries to VizieR during the first 3 months of the new installation (June to August 1997); among all queries, about 40% of the hits concern the recent results of the *Hipparcos* and *Tycho* missions.

---

[1] http://vizier.u-strasbg.fr

[2] http://cdsweb.u-strasbg.fr

[3] http://www.esrin.esa.it/esrin/esrin.html

[4] http://vizier.u-strasbg.fr/doc/asu.html

2. How to Query in VizieR

The "standard query" in VizieR consists in a few steps:

1. Locate the interesting catalogs in the VizieR Service[5]. This page presents various ways of finding out the interesting catalog among this large set:
   (a) from one of their usual *acronyms*, like **GSC**, **HD**, **HIC**, etc...
   (b) from a set of words (author's names and/or words from the title of the catalog), and/or keywords attached to each catalog;
   (c) or by clicking in a Kohonen Self-Organizing Map[6], a map created by neural network techniques which tends to group in nearby locations those catalogs having similar sets of keywords. This technique is the same as the one used by Lesteven et al. (1996) to index the bibliography.

2. Once a catalog table – or a small set of catalog tables — is located (for instance the *Hipparcos* Catalog[7] resulting from the *Hipparcos* mission), *constraints* about the input and/or output can be specified, as:
   - constraints based on the celestial coordinates (location in the neighbourhood of a target specified by its actual coordinates in the sky, or its name, using SIMBAD[8] as a name resolver);
   - any other qualification on any of the columns of the table(s); the standard comparison and logical operators are available, detailed in the VizieR help pages[9];
   - which columns are to be displayed, and in which order the matching rows are to be presented.

   By pushing the appropriate buttons, it is for instance easy to get the list of *Hipparcos* stars closer than 5 parsecs to the Sun, ordered by their increasing distance[10].

3. Obtaining full details about one row is achieved by simply clicking in the first column of the result: for instance, the first row of the search for nearby stars described above leads to the VizieR Detailed Page with *Hipparcos* parameters and their explanations concerning Proxima Centauri[11].

---

[5]http://vizier.u-strasbg.fr/cgi-bin/VizieR

[6]http://vizier.u-strasbg.fr/cgi-bin/VizieR#Qkmap

[7]http://vizier.u-strasbg.fr/cgi-bin/VizieR?-source=I/239/hip_main

[8]http://simbad.u-strasbg.fr/Simbad

[9]http://vizier.u-strasbg.fr/cgi-bin/Help?VizieR/intro

[10]http://vizier.u-strasbg.fr/cgi-bin/VizieR?-source=I/239/hip_main&-sort=-Plx&Plx=%3e=200

[11]http://vizier.u-strasbg.fr/cgi-bin/VizieR-5?-source=I/239/hip_main&HIP=70890

4. Finally, there may be correlated data, like notes or remarks, references, etc.... In our example, Proxima Centauri is related to the α Cen system, which components can be viewed from the CCDM link appearing in the detailed page.

It should be noted that the usage of the ASU protocol allows to write anywhere in a text (like it is done in this short article) a hyperlink to the result of a query: for instance, all parameters from the *Hipparcos* catalog for the star HIP 12345 can be pointed to by a call to VizieR with parameters: `-source=I/239/hip_main&HIP=12345`[12]; or a pointer to all *Tycho* stars closer than 0.5° to Sirius, ordered by their increasing distance to the brightest star, can be written by a call to VizieR with parameters:
`-source=I/239/tyc_main&-c=Sirius,rm=30&-sort=_r`[13]

3. VizieR Structure

VizieR is based on the usage of a relational DBMS: the data tables are stored as relational tables, and a set of *META* tables — a structure which was called *Reference Directory* in the previous version of VizieR — contains the necessary descriptions of all items:

- *METAcat* is a table providing the description of *catalogs* (a *catalog* is a set of related tables, like a table of observations, a table of mean values, a table of references, ...); *METAcat* details the authors, reference, title, etc...of each stored catalog. There are presently ~ 1500 rows in this table.
- *METAtab* is a table providing the description of each *table* stored in VizieR: title, number of rows, how to access the actual data, the equinox and epoch of the coordinates, etc...There are presently ~ 3500 rows in this table, *i.e.* an average of $2\frac{1}{3}$ tables per catalog.
- *METAcol* is a table providing the description of each *column* stored in VizieR: labels, units, datatypes, etc...There are presently ~ 45000 rows in this table, *i.e.* an average of 13 columns per table.
- about 10 more tables exist in the system to detail other parameters, like the definitions of keywords, the acronyms associated with the catalogs, the notes, etc...

4. The VizieR Feeding Pipeline

It is of course not possible to enter all details describing the ~ 45000 columns by hand: the VizieR feeding pipe-line is fully automatic using as input the standardized description[14] of the catalogs shared by the Astronomical Data Centers,

---

[12] http://vizier.u-strasbg.fr/cgi-bin/VizieR?-source=I/239/hip_main&HIP=12345

[13] http://vizier.u-strasbg.fr/cgi-bin/VizieR?-source=I/239/tyc_main&-c=Sirius,rm=30&-sort=_r

[14] http://vizier.u-strasbg.fr/doc/catstd.htx

and used for the description of the electronic tables published by a large fraction of the major astronomical journals (A&A, ApJ, AJ, PASP).

The addition of a new catalog into VizieR consists in two steps: **(1)** adding into the *META* tables all items describing the new catalog: catalog title, authors, table captions, etc...and details about every column of each table; and **(2)** converting the data of each original (ASCII) file making up the catalog into a relational table. In this step, the existence of data (the so-called *NULL values*) is carefully tested, and some important parameters like astronomical positions, magnitudes and color indexes are converted to uniform units. We however take care of storing, as much as possible, the catalogs in their original form, and for instance the coordinates are stored in their original equinox and epoch.

5. Access to the Very Large Catalogs

Very Large Catalogs are defined here as catalogs made of more than 10^7 objects — a size which can hardly be managed by the existing DBMs. Dave Monet's (1977) USNO-A1.0 catalog[15] gathering 488,006,860 sources is a typical example: it consists originally in a set of 10 CDROMs (about 6Gbytes) with 12-bytes binary records (positions, magnitudes, and a couple of flags).

This catalog was losslessly compressed by grouping small regions of the sky, allowing to store only position offsets instead of full-range values positions: the resulting catalog occupies only 3.4Gbytes, allowing therefore faster access since the queries are heavily i/o-limited: on a Sparc-20 (72MHz), the average search time (radius of 2.5') is less than 0.1s, and the whole 488×10^6 objects are tested in about 40 minutes (i.e., $5\mu s$ per object).

6. VizieR developments

The current new developments include: *(a)* more *connectivity* between catalogs, with Simbad, and more *remote connectivity* with external databases, Observatory Archives, and other search engines; the *GLU* (Fernique et al. 1998) will most likely be extensively used for this purpose; *(b)* the creation of shared indexes on fundamental parameters like celestial positions in order to allow queries directed to a large number of tables; *(c)* a possibility to submit large lists for queries; and *(d)* a facility to present the results in graphical form.

References

Fernique, P., Ochsenbein, F., & Wenger M. 1998, this volume

Lesteven, S., Poinçot, P., Murtagh, F. 1996, Vistas in Astronomy 40, 395

Monet, D. 1997, "The USNO A1.0 catalog", http://psyche.usno.navy.mil/pmm

Ochsenbein, F., Genova, F., Egret, D., Bourekeb, I., Sadat, R., Ansari, S.G., & Simonsen, E. 1996, Bull. American Astron. Soc. 187 #9103

[15] http://vizier.u-strasbg.fr/cgi-bin/VizieR?-source=USNO-A1.0

Astronomical Data Analysis Software and Systems VII
ASP Conference Series, Vol. 145, 1998
R. Albrecht, R. N. Hook and H. A. Bushouse, eds.

The ASC Data Archive for the AXAF Ground Calibration

P. Zografou, S. Chary, K. DuPrie, A. Estes, P. Harbo and K. Pak

Smithsonian Astrophysical Observatory, Cambridge, MA 02138

Abstract. A data archive is near completion at the ASC to store and provide access to AXAF data. The archive is a distributed Client/Server system. It consists of a number of different servers which handle flat data files, relational data, replication across multiple sites and the interface to the WWW. There is a 4GL client interface for each type of data server, C++ and Java API and a number of standard clients to archive and retrieve data. The architecture is scalable and configurable in order to accommodate future data types and increasing data volumes. The first release of the system became available in August 1996 and has been successfully operated since then in support of the AXAF calibration at MSFC. This paper presents the overall archive architecture and the design of client and server components as it was used during ground calibration.

1. Introduction

The ASC archive is projected to contain terabytes of data including ground and on orbit raw data and data products. The archive stores the data following requirements for data distribution across sites, secure access, flexible searches, performance, easy administration, recovery from failures, interface to other components of the ASC Data System and a user interface through the WWW. The architecture is extensible in order to accommodate new data products, new functions and a growing number of users.

2. Data Design

Data such as event lists and images need to be kept in files as they are received. They also need to be correlated with engineering and other ancillary data which arrive as a continuous time stream and to be associated with a calibration test or an observation ID. A level of isolation between the data and users is desirable for security, performance and ease of administration. The following design was chosen. Files are kept in simple directory structures. Metadata about the files, extracted from file headers or supplied by the archiving process, is stored in databases. This allows file searches on the basis of their contents. Continuous time data extracted from engineering files is also stored in databases so the correct values can be easily associated with an image or an event list with defined time boundaries. In addition to partial or entire file contents, file

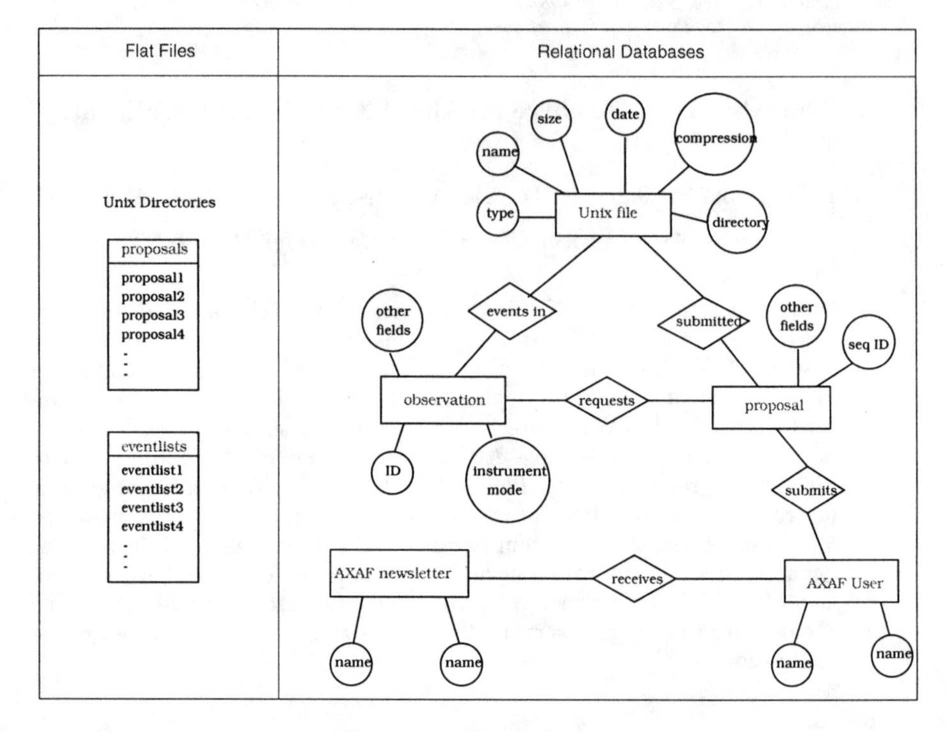

Figure 1. Data Design.

external characteristics such as its location in the archive, size, compression, creation time are also stored in databases for archive internal use. In addition to databases with contents originating in files, there are also databases which are updated independently, such as the AXAF observing catalog and the AXAF users database.

A simplistic example of the data design is shown on Figure 1. The archive contains a number of proposal files submitted by users. It also contains a number of event files, products of observed data processing. A table in a database contains the characteristics of each file. The proposal table contains a record for each proposal which points to the associated file in the file table. The observation table contains a record for each observed target and has a pointer to the associated proposal. An observation is also associated with files in the file table which contain observed events. Related to the proposal is the AXAF user who submitted it and for whom there is an entry in the AXAF user table. An AXAF user may have a subscription to the AXAF newsletter.

3. Software Design

The data is managed by a number of different servers. A Relational Database server stores and manages the databases. It is implemented using the Sybase SQL Server. An archive server was developed to manage the data files. The

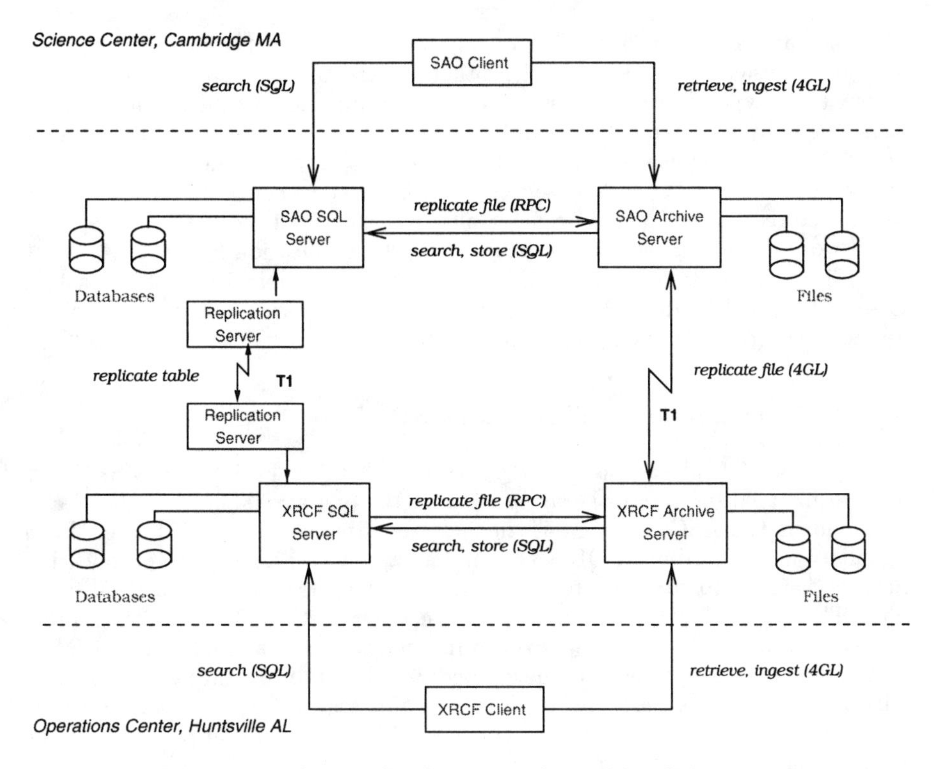

Figure 2. Server Configuration for XRCF Calibration.

archive server organizes files on devices and directories. It keeps track of their location, size, compression and other external characteristics by inserting information in a table in the SQL Server when the file is ingested. It also has data specific modules which parse incoming files and store in databases their contents or information about their contents. The server supports file browse and retrieve operations. A browse or retrieve request may specify a filename or enter values for a number of supported keywords such as observation or test ID, instrument, level of processing, start and stop time of contained data. Browse searches the database and returns a list of files, their size and date. Retrieve uses the same method to locate the files in the server's storage area and return a copy to the client. The archive server responds to language commands and remote procedure calls. Language commands are used by interactive users or processes in order to archive or retrieve data. A custom 4GL was developed in the form of a "keyword = value" template which is sent by clients and is interpreted at the server. The remote procedure call capability is used for automated file transfer between two remote servers.

The server infrastructure uses the Sybase Open Server libraries which support communications, multi-threading, different types of events and call-backs and communications with the SQL server. A C++ class layer was developed to interface the libraries with the rest of the system (Zografou 1997). File transfer uses the same communications protocol as the SQL server which is optimized for data transfer and integrates with other server features such as security.

A third type of server was needed in order to automatically maintain more than one copy of the data at two different locations. The Sybase Replication Server is used to replicate designated databases. Via triggers in the database at the target site the local archive server is notified to connect to its mirror archive server at the source site and transfer files. Queuing and recovery from system or network down-time is handled entirely by the Replication Server.

Client applications use the Sybase Open Client libraries with a custom C++ interface (Zografou 1997). The same client libraries are used for client applications to either the SQL or the archive server.

4. Configuration for Calibration at XRCF

During ground Calibration at the X-Ray Calibration Facility at MSFC two archive installations were operating, one at the operations site at XRCF and a second at the ASC. Communications across sites were via a T1 line. Each installation consisted of a SQL Server and an archive server. A set of replication Servers were setup to replicate all databases which triggered replication of all files. The system layout is shown on Figure 2. Data the in form of files entered the system at XRCF, which was the primary site, and was replicated at SAO. With some tuning to adjust to unexpectedly high data rates the system kept up with ingestion, replication and retrievals by processing pipelines at XRCF and users at the ASC. There were no incidents of data corruption or loss and the overall system was very successful.

5. Conclusion

At the end of the XRCF calibration the system was adapted to support ASC operations at the SAO and AXAF OCC sites connected with a T3 line. In the new configuration only critical data is being replicated. All other data is distributed according to user access. A new server component, the Sybase JConnect Server, and a new Java/JDBC client interface have been added to support WWW access (Chary 1997). The second release of the system, including the WWW interface, is currently operational in support of proposal submission.

References

Zografou, P. 1997, Sybase Server Journal, 1st Quarter 1997, 9

Chary, S., Zografou, P. 1997, this volume

Astronomical Data Analysis Software and Systems VII
ASP Conference Series, Vol. 145, 1998
R. Albrecht, R. N. Hook and H. A. Bushouse, eds.

Mirroring the ADS Bibliographic Databases

Alberto Accomazzi, Guenther Eichhorn, Michael J. Kurtz, Carolyn S. Grant and Stephen S. Murray

Smithsonian Astrophysical Observatory, 60 Garden Street, Cambridge, MA 02138, USA

Abstract.
During the past year the Astrophysics Data System has set up two mirror sites for the benefit of users around the world with a slow network connection to the main ADS server in Cambridge, MA. In order to clone the ADS abstract and article services on the mirror sites, the structure of the bibliographic databases, query forms and search scripts has been made both site- and platform-independent by creating a set of configuration parameters that define the characteristics of each mirror site and by modifying the database management software to use such parameters. Regular updates to the databases are performed on the main ADS server and then mirrored on the remote sites using a modular set of scripts capable of performing both incremental and full updates. The use of software packages capable of authentication, as well as data compression and encryption, permits secure and fast data transfers over the network, making it possible to run the mirroring procedures in an unsupervised fashion.

1. Introduction

Due to the widespread use of its abstract and article services by astronomers worldwide, the NASA Astrophysics Data System (ADS) has set up two mirror sites in Europe and Asia. The European mirror site is hosted by the Centre De Données Stellaires (CDS) in Strasbourg, while the Asian mirror is hosted by the National Astronomical Observatory of Japan (NAO) in Tokyo.

The creation of the ADS mirrors allows users in different parts of the world to select the most convenient site when using ADS services, making best use of bandwidth available to them. For many users outside the USA this has meant an increase in throughput of orders of magnitude. For instance, Japanese users have seen typical data transfer rates going from 10 bytes/sec to 10K bytes/sec. In addition, the existence of replicas of the ADS services has taken some load off of the main ADS site at the Smithsonian Astrophysical Observatory, allowing the server to respond better to incoming queries.

The cloning of databases on remote sites does however present new challenges to the data providers. First of all, in order to make it possible to replicate a complex database system elsewhere, the database management system and the underlying data sets have to be independent of the local file structure, operating system, hardware architecture, etc. Additionally, networked services which rely on links with both internal and external Web resources (possibly available on different mirror sites) need to have procedures capable of deciding how the links should be created, possibly giving users the option to review and modify the system's linking strategy. Finally, a reliable and efficient mechanism should be in place to allow unsupervised database updates, especially for those applications involving the publication of time-critical data.

2. System Independence

The database management software and the search engine used for the ADS bibliographic services have been written to be system-independent.

Hardware independence is made possible by writing portable software that can be either compiled under a standard compiler and environment framework (e.g., GNU gcc) or interpreted by a standard language (e.g., Perl5). All the software used by the ADS mirrors is first compiled and tested for the different hardware platforms on the main ADS server, and then the appropriate binary distributions are mirrored to the remote sites.

Operating System independence is achieved by using a standard set of Unix tools which abiding to a well-defined standard (e.g., POSIX.2). Any additional enhancements to the standard Unix system tools are achieved by cloning more advanced software utilities (e.g., GNU shell-utils) and using them when necessary.

File-system independence is made possible by organizing the data files for a specific database under a single directory tree, and creating configuration files with parameters pointing to the location of these top-level directories. Similarly, host name independence is achieved by storing the host names of ADS servers in configuration files.

3. Resolution of Hyperlinks

The strategy used to generate links to networked services external to the ADS which are available on more than one site follows a two-tiered approach. First, a "default" mirror can be specified in a configuration file by the ADS administrator. This configuration file is site-specific, so that appropriate defaults can be chosen for each of the ADS mirror sites depending on their location. Then, ADS users are allowed to override these defaults by using a "Preference Settings" page to have the final say as to which site should be used for each link category (see Figure 1). The use of preferences is implemented using HTTP "cookies" (Kristol & Montulli, 1997). The URLs relative to external links associated with a particular bibliographic references are looked up in a hash table and variable substitution is done if necessary to resolve those URLs containing mirror site variables, as shown in the examples below.

1997Icar..126..241S ⇒ $IDEAL$/cgi-bin/links/citation/0019-1035/126/241
⇒ http://www.idealibrary.com/cgi-bin/links/citation/0019-1035/126/241
1997astro.ph..8232H ⇒ $PREPRINTS$/abs/astro-ph/9708232
⇒ http://xxx.lanl.gov/abs/astro-ph/9708232

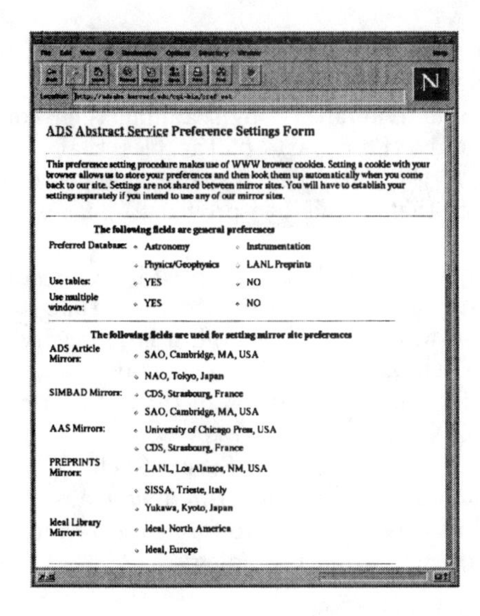


```
# $Id: ads_sites.config,v 1.1 1997/09/08 19:52:16 ads Exp ads $
# Sets possible values for mirror sites used by the ADS
# bibliographic services, with default values for the SAO site.

# ADS article sites
SET_1=ARTICLE
DESC_1=ADS Article Mirrors
TAGS_1=ARTICLE
DEF_1=1
ARTICLE_DESC_1=SAO, Cambridge, MA, USA
ARTICLE_DESC_2=NAO, Tokyo, Japan
ARTICLE_1=http://adsbit.harvard.edu
ARTICLE_2=http://ads.nao.ac.jp

# SIMBAD DB mirrors
SET_2=SIMBAD
DESC_2=SIMBAD Mirrors
TAGS_2=SIMBAD
DEF_2=2
SIMBAD_DESC_1=CDS, Strasbourg, France
SIMBAD_DESC_2=SAO, Cambridge, MA, USA
SIMBAD_1=http://simbad.u-strasbg.fr/simbo.pl
SIMBAD_2=http://simbad.harvard.edu/simbo.pl
                                       :
# AAS publications (University Chicago Press)
SET_3=AAS
DESC_3=AAS Mirrors
TAGS_3=AAS_APJ AAS_AJ
DEF_3=1
AAS_DESC_1=University of Chicago Press, USA
AAS_DESC_2=CDS, Strasbourg, France
AAS_APJ_1=http://www.journals.uchicago.edu/ApJ/cgi-bin/resolve
AAS_APJ_2=http://cdsaas.u-strasbg.fr:2001/ApJ/cgi-bin/resolve
AAS_AJ_1=http://www.journals.uchicago.edu/AJ/cgi-bin/resolve
AAS_AJ_2=http://cdsaas.u-strasbg.fr:2001/AJ/cgi-bin/resolve

# Preprint Servers
SET_5=PREPRINTS
DESC_5=PREPRINTS Mirrors
TAGS_5=PREPRINTS
DEF_5=1
PREPRINTS_DESC_1=LANL, Los Alamos, NM, USA
PREPRINTS_1=http://xxx.lanl.gov
PREPRINTS_DESC_2=SISSA, Trieste, Italy
PREPRINTS_2=http://xxx.sissa.it
PREPRINTS_DESC_3=Yukawa, Kyoto, Japan
PREPRINTS_3=http://xxx.yukawa.kyoto-u.ac.jp

# IDEAL library (Academic Press publications)
SET_6=IDEAL
DESC_6=Ideal Library Mirrors
TAGS_6=IDEAL
DEF_6=1
IDEAL_DESC_1=Ideal, North America
IDEAL_DESC_2=Ideal, Europe
IDEAL_1=http://www.idealibrary.com
IDEAL_2=http://www.europe.idealibrary.com
```

Figure 1. Left: the Preference Setting form allows users to select which mirror sites should be used when following links on ADS pages. Right: mirror sites configuration file for the ADS server at SAO.

1997ApJ...486L..75F ⇒ AAS_APJ?1997ApJ...486L..75FCHK
⇒ http://www.journals.uchicago.edu/ApJ/cgi-bin/resolve?1997ApJ...486L..75FCHK

While more sophisticated ways to create dynamic links are being used by other institutions (Fernique et al. 1998), there is currently no reliable way to automatically choose the "best" mirror site for a particular user. By saving these settings in a user preference database indexed on the cookie ID, users only need to define their preferences once and our interface will retrieve and use the appropriate settings as necessary.

4. Mirroring Software

The software used to perform the actual mirroring of the databases consists of a main program running on the ADS master site initiating the mirroring procedure, and a number of scripts, run on the mirror sites, which perform the transfer of files and software necessary to update the database. The main program, which can be run either from the command line or as a CGI script, is an Expect/Tcl script that performs a login on the mirror site to be updated, sets up the environment by evaluating the mirror site and master site's configuration files, and then initiates the updating process.

The updating procedures are specialized scripts which check and update different parts of the database and database management software (including the procedures themselves). The actual updating of the database files is done by using a public domain implementation of the rsync algorithm (Tridgell &

Mackerras, 1996), with local modifications. The advantages of using rsync to update data files rather than performing complete transfers are:

Incremental updates: rsync updates individual files by scanning their contents and copying across the network only those parts of the files that have changed. Since only a small fraction of the data files actually changes during our updates (usually less than 5% of them), this has proved to be a great advantage.

Data integrity: should the updating procedure be interrupted by a network error or human intervention, the update can be resumed at a later time and rsync will pick up transferring data from where it had left off. File integrity is checked by comparing file attributes and via a 128-bit MD4 checksum.

Data compression: rsync supports internal compression of the data stream by use of the zlib library (also used by GNU gzip).

Encryption: rsync can be used in conjunction with the Secure Shell package (Ylonen 1997) to transfer the data for added security. Unfortunately, transfer of encrypted data could not be performed at this point due to foreign government restrictions and regulations on the use of encryption technology.

5. Conclusions

The approach we followed in the implementation of automated mirroring procedures for the ADS bibliographic services has proved to be very effective and flexible. The use of the rsync algorithm makes it practical to update portions of the database and have only such portions automatically transferred to the mirror sites, without requiring us to keep track of what individual files have been modified. Because of the reduced amount of data that needs to be transferred over the network, we typically achieve speed gains from 1 to 2 orders of magnitude, which makes the updating process feasible despite poor network connections. We plan to improve the reliability of the individual transfers (which occasionally are interrupted by temporary network dropouts) by using sensible time-outs and adding appropriate error handlers in the main transfer procedure.

As a result of the proliferation of mirror sites, we have provided a user-friendly interface which allows our users to conveniently select the best possible mirror site given their local network topology. This model, currently based on HTTP cookies, can be easily adapted by other data providers for the benefit of the user. An issue which still needs to be resolved concerns providing a fall-back mechanism allowing users to retrieve a particular document from a backup mirror site should the default site not be available. It is possible that new developments in the area of URN definition and management will help us to find a solution to this problem.

Acknowledgments. This work is funded by the NASA Astrophysics Program under grant NCCW-0024.

References

Fernique, P., Ochsenbein, & F., Wenger, M. 1998, this volume

Kristol, D., & Montulli, L. 1997, HTTP State Management Mechanism, RFC2109, Internet Official Protocol Standards, Network Working Group.

Tridgell, A., & Mackerras, P. 1996, The rsync algorithm, Joint Computer Science Technical Report Series TR-CS-96-05, Australian National University.

Ylonen, T. 1997, SSH (Secure Shell) Remote Login Program, Helsinki University of Technology, Finland.

Astronomical Data Analysis Software and Systems VII
ASP Conference Series, Vol. 145, 1998
R. Albrecht, R. N. Hook and H. A. Bushouse, eds.

Archiving Activities at the Astronomical Observatory of Asiago

A. Baruffolo and R. Falomo

Astronomical Observatory of Padova, Italy

Abstract. Since July 1993 all observations collected by the 1.82m telescope at Cima Ekar (Asiago) are systematically archived on DAT cartridges and a listing of collected data, including all parameters of the observations, is maintained in a relational DBMS. In order to promote the scientific reuse of observations stored in this archive, we have developed a WWW interface based on the WDB software, that enable users to perform complex searches in the observation log. In this paper we give an overview of the archive, briefly describe the WWW–based search interface as well as planned future enhancements to the archiving system.

1. Introduction

In the past few years astronomers which observed with the Copernico 1.82m telescope have asked for an archiving facility that could serve as a repository of all observational data and provide astronomers with a backup copy of their observations in case of data loss.

Following this request, since July 1993 an archiving facility has been set up at the Astronomical Observatory of Asiago (Mt. Ekar) where all observations are routinely stored on DAT cartridges and catalogued using a relational DBMS. In a first phase, the archive facility did not foresee network access: archive searches could be performed only at the Mt. Ekar Observatory archive system, or at a twin system located at the Astronomical Observatory of Padova.

In order to give access to the Copernico Archive to the widest possible community of users, a WWW–based search facility was developed and installed at the Astronomical Observatory of Padova, which is better connected to the Internet.

In this report we give an overview of the archiving system and describe the main features of the WWW–based search interface.

2. The Archiving System

The Copernico Telescope Archiving System is made up of a *master archive*, where the observation list is generated and maintained, and a *public archive*, where a copy of the observation list is accessible through a WWW–based interface.

The *master* archive software consists of a collection of programs that run in the Windows environment and that use the FoxPro relational database management system (RDBMS) for data handling. It allows for:

- daily ingestion of the list of new observations;
- correction/insertion of incorrect/missing data;
- local archive searching.

The observation list maintained in the *master* archive is named the *master catalogue*: corrections and updates are made to this list only, they eventually propagate to the *public* archive during the periodic transfer of data from the *master* to the *public* archive.

The *public* archive software consists of a collection of programs that run on a Linux system and that use the MiniSQL RDBMS (Hughes Technologies 1997). It is responsible for:

- periodic synchronization of the *public* archive's content with the *master* archive;
- handling of queries from archive users.

3. Overview of the Archiving Process

During the observations all images and spectra taken by the astronomer are stored on disk in FITS files by the data acquisition system (Baruffolo & D'Alessandro 1993). At the end of each observing night these files are transferred on DAT cartridges: during this step a listing of observations is generated and this list is *ingested* into the archive. For each data file parameters that characterize the observation are extracted or derived from the FITS header and the data are displayed and checked for consistency; three other parameters which give the location of the data on the archive medium are then added to those describing the observation.

Periodically the most recent part of the observation catalogue is dumped to ASCII files and transferred to the *public* archive. In the public archive machine a procedure automatically starts at fixed intervals, checks for new data coming from the master archive and loads them into a miniSQL database. During this pass some additional checks are performed on the data and a report is generated signaling errors and/or missing data so that the *master* catalogue can be corrected further.

4. Archive Content

At the time of writing, the Asiago archive contains more than 16000 frames. This figure includes all science exposures (images and spectra) and all calibration data (bias, dark, flat field and comparison spectra). Table 1 gives a summary of all archived data by image type and by instrument.

Table 1. Summary of the archive content by image type and by instrument

Image type	Nr. of Frames
tvcamera	11
test	143
bias	1069
calib	3339
dark	153
flat	2248
focus	244
object	8821

Instrument	Nr. Of Frames
Boller & Chivens	7105
Echelle	5498
Imaging camera	3425

5. WWW Access to the Archive

WWW access to the archive is available from the Asiago Observatory Home page[1] and is realized using the WDB software version 1.3a (Rasmussen 1995), modified to work with miniSQL and to allow ordering of results on a user-selected field. When the user accesses the archive by opening its URL in a Web browser, a query form is generated on–the–fly by WDB on the basis of a form definition file (FDF). Searches can be constrained by filling in the form and then clicking on the `search` button; archive last modification time can be retrieved by following an *hyperlink* located towards the top of the query form.

The query form consists of *action buttons*, which initiate an action when clicked, *query fields*, which are used to constrain searches, *additional parameters*, which can be added to the query output, and *output controls*, which control how query output is formatted and how many data rows are retrieved by a query. These items have been discussed in detail in another paper (Baruffolo, Falomo, & Contri 1997).

6. Requesting Archived Data

At present there is no facility for automatically requesting data that have been selected using the archive query form. A user willing to retrieve data from the archive must send the list of requested observations (that must include the keyword "`CODICE`") to the archive maintainer who will forward it to the technical staff at the Asiago Observatory. Observations will then be retrieved from the archive media and delivered to the requester on DAT cartridges or stored on disk for ftp transfer. Further information on how to request archived data and contact points for the archive maintainer can be found on the archive WWW page.

The actual policy of the Observatory on data distribution is that the observations become public domain two years after the observation date.

---

[1]http://www.pd.astro.it/asiago/asiago.html

7. Conclusions and Future Work

In 1993 we started to systematically store all observations collected at the 182 cm. telescope in FITS format and to produce listings containing the parameters that characterize them. At that time the development of the Web was still in its infancy and network access to the archive was not foreseen, since this would have implied the development of a custom client/server user interface. However, having maintained the observation list in a relational DBMS has allowed us to upgrade the archive system with Web access with relatively little effort.

The WWW interface to the archive has been available for more than one year now, no major problems were reported during this initial operational period.

Future work on the public archive is planned to provide preview images, storage of data on optical disks (recordable CD–ROMs) and to automate archive data requests.

Finally, it is to be noted that provision has been made in the public archive software for inclusion of observations coming from the other two telescopes working at the Asiago sites: the 67/92 cm. Schmidt and the 122 cm. telescope. For the first of these, which still operates using photographic plates, a preliminary database (only working on a PC) is available, it contains data collected since 1965. The 122 cm. telescope has been recently equipped with a CCD detector and it is planned to only archive these recent data.

References

Baruffolo, A., & D'Alessandro, M. 1993, Padova and Asiago Observatories Tech. Rep. No. 3

Rasmussen, B. F. 1995, in ASP Conf. Ser., Vol. 77, Astronomical Data Analysis Software and Systems IV, ed. R. A. Shaw, H. E. Payne & J. J. E. Hayes (San Francisco: ASP), 72

Hughes Technologies, 1997, "Mini SQL 2.0. User Guide."

Baruffolo, A., Falomo, R., & Contri, L. 1997, Padova and Asiago Observatories Tech. Rep. No. 12

Astronomical Data Analysis Software and Systems VII
ASP Conference Series, Vol. 145, 1998
R. Albrecht, R. N. Hook and H. A. Bushouse, eds.

The IUE Archive at Villafranca

M. Barylak and J. D. Ponz

ESA Villafranca
P.O. Box 50727, 28080 Madrid, Spain

Abstract.
The International Ultraviolet Explorer (IUE) mission has produced a large collection of spectroscopic data containing about 104,000 spectra of ca. 9,600 different objects. The IUE Final Archive (IUEFA) project will generate a high quality and uniform spectral archive during the final phase of the mission (when specialized knowledge on the instrument and calibration procedures are still available), and maintain it so that it is accessible to the scientific community.

This contribution describes the IUEFA project at Villafranca, and the plans to make the archive available to the scientific community under through the INES (IUE New Extracted Spectra) project.

1. Introduction

There is no doubt that scientific archives represent an invaluable resource. The IUEFA will provide important reference information for current and future UV space missions. No maintenance and support of a data archive was foreseen in IUE's original mission plan. Hence, in the context of the IUEFA project, a special effort has been dedicated to the definition of the mission archive and its distribution to the scientific community world-wide. It has been said and truly written that "IUE has led the way in promoting archiving services".

2. The Final Archive

The IUE Final Archive (IUEFA) was defined as a project between NASA, ESA and PPARC, according to the following main requirements:

- Produce a *uniformly processed* archive. Through the years of the mission, IUESIPS has experienced several modifications and different calibrations have been used, so that it has been difficult to compare data processed at different epochs.
- Apply new image processing algorithms to *improve the photometric accuracy* and *signal-to-noise ratio* of the reduced spectra.
- New calibrations have been derived to estimate *absolute fluxes* and new blaze models have been incorporated to better correct for the ripple effect in high-dispersion spectra.

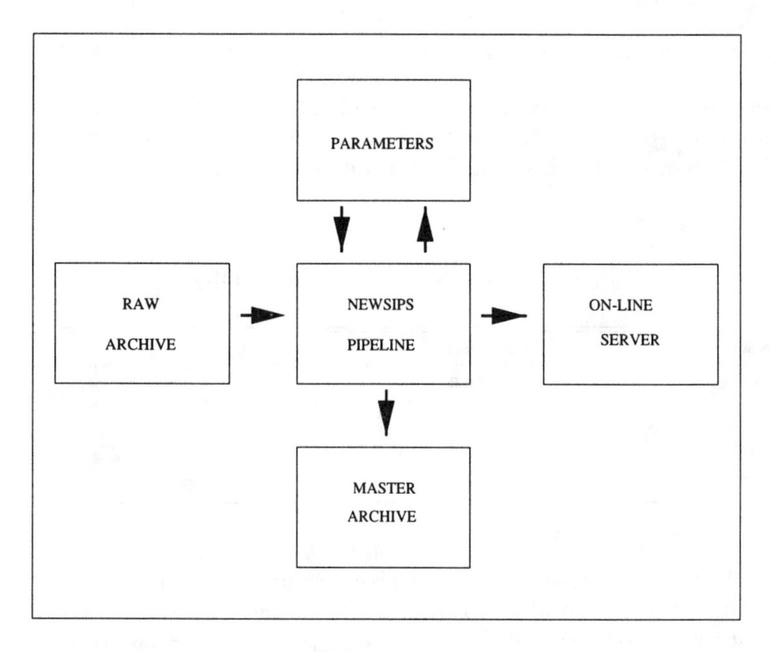

Figure 1. Final Archive data flow.

- Quality control on instrument and observational parameters. Special attention was devoted to the compilation and verification of a set of image parameters, identified as *Core Data Items*, that are required to adequately re-process the raw image and are essential for scientific analysis.

- Archived spectra written in standard FITS format (Ponz et al. 1994) on high density optical disks.

- Compatibility of output products generated at both observing stations.

3. The Production at Villafranca

Based upon the previous requirements, NEWSIPS (Garhart et al. 1997) was developed and implemented at both GSFC and VILSPA observing stations. The main elements for the production system, depicted in Figure 1, are: (1) Raw archive stored on optical disks, provides the input images to the NEWSIPS production pipeline. (2) Image parameters, defining the main object characteristics and instrumental setup. These parameters determine the calibration options, provide quality control information and are maintained under a relational data base management system (DBMS) (Barylak, 1996). (3) Pipeline (NEWSIPS), implemented under MIDAS. (4) Master archive, stored on optical disks. (5) On-line data server with extracted spectra – both two-dimensional and absolutely calibrated fluxes – available on-line.

4. Project Status

The number of observations in the archive is about 70,000 low and 30,000 high resolution images. The volume of the information of the master archive is shown in Table 1, indicating the size in Gbytes for each file type.

Table 1. Archive volume (sizes in Gbytes).

Data set	Low res.	High res.	Total
Raw images	42.0	18.0	60.0
Linearized	168.0	72.0	240.0
2D spectra	16.8	72.0	88.8
1D spectra	4.2	36.0	40.2
Total	231.0	198.0	429.0

At the time of writing, the raw archive is complete and all image parameters have been verified. Local processing at Villafranca and data exchange with GSFC are well advanced, so that the low-dispersion set is nearly finished and more than 50 % of the high-dispersion spectra have been processed and are available on the data server. The master archive will be completed by December 1997.

5. The Data Server

Access to the ESA IUE data server (Yurrita & Barylak, 1997) is available at the address `http://iuearc.vilspa.esa.es/`. This server implements the following basic features:

User identification The user logs into the data server to identify a new session. This accounting system allows users to recall queries and retrieve data from previous sessions and supports detailed usage statistics.

Query by form A simple form allows searching for objects by name, position, object type and observing date. Instrument parameters such as camera name, dispersion or aperture can also be specified.

List of observations The result of the search is a list of observations that can be used to plot selected spectra or to transfer the selected data to the local node, using different compression methods.

6. INES: The IUEFA Data Distribution System

After completion of IUEFA, data distribution is planned following the original, low-cost, distributed archive concept of ULDA/USSP (Wamsteker, et al. 1989).

The **INES** (*IUE New Extracted Spectra*) project will contain (1) the re-extracted set of low-dispersion spectra together with all high-dispersion observations re-sampled to the low-dispersion domain, (2) the two dimensional line-by-line spectra, and (3) a new version of high-dispersion spectra, with orders concatenated.

INES will be based on the following structure:

Principal center: Master archive repository, containing all the data items indicated above. Distributes the compressed versions of the archive to the National hosts via CD-ROMs.

National hosts: Located in different countries, containing the INES access catalogue and serving the absolute flux calibrated 1D spectra via HTTP.

End users: Unlimited number of nodes that access the archive at the National hosts via standard WWW browsers.

The suggested configuration for the National host data server is based upon a PC running the Linux operating system. Our prototype is running under the free DBMS MySQL employing Perl to interface with both the APACHE HTTP server and the WWW browser. This data server also implements the syntax proposed for information exchange among remote astronomical services (see ASU `http://vizier.u-strasbg.fr/doc/asu.html`). Furthermore, the access catalogue includes references to IUE spectra included in scientific publications which can be retrieved via the standard `BIBCODE` as defined in the NASA Astrophysics Data System (ADS) (Eichhorn et al. 1998).

References

Barylak, M. 1996, ESA IUE Newsletter No. 46, February 1996, 19

Eichhorn, G., Accomazzi, A., Grant, C.S., Kurtz, M.J., & Murray, S.S. 1998, this volume

Garhart, M.P., Smith, M.A., Levay, K.L., & Thompson, R.W. 1997, *NEWSIPS Information Manual*, version 2.0

González-Riestra, R., Cassatella, A., & de la Fuente, A. 1992, *Record of the IUE Three Agency Coordination Meeting*, November 1992, D-78

Kinney, A.L., Bohlin R.C., & Neil, J.D. 1991, PASP, 103, 694

Linde, P., & Dravins, D. 1990, ESA SP-310, 605

NASA/Science Office of Standards and Technology 1995, *Definition of the Flexible Image Transport System (FITS)*, NOST 100-1.1

Ponz, J.D., Thompson, R.W., & Muñoz, J.R. 1994, A&AS, 105, 53

Wamsteker, W., Driessen, C., Muñoz, J.R., Hassall, B.J.M., Pasian, F., Barylak, M., Russo, G., Egret, D., Murray, J., Talavera, A., & Heck, A. 1989, A&A, 79, 1

Yurrita, I., & Barylak, M. 1997, *Access to the ESA IUE Data Server*, ESA IUE Newsletter No. 47, January 1997, 34.

Astronomical Data Analysis Software and Systems VII
ASP Conference Series, Vol. 145, 1998
R. Albrecht, R. N. Hook and H. A. Bushouse, eds.

Search and Retrieval of the AXAF Data Archive on the Web using Java

Sumitra Chary and Panagoula Zografou

Smithsonian Astrophysical Observatory, AXAF Science Center, Boston, MA 02138

Abstract. An important component of the AXAF Data Archive is the interface through the WWW. Internet applications are moving beyond delivering static content to more complex dynamic content and transaction processing. While HTML and CGI based interfaces have been possible for some time, it was not until the advent of JDBC (Java[1] Database Connectivity) that a uniform and powerful interface to the different types of servers in the archive, with secure access to proprietary data, could become a reality.
This paper presents a multi-tier architecture which integrates the data servers, Web server and a HTTP/JDBC gateway and enables data on any of the data servers in the AXAF archive to be available to clients using a Web browser. This Web-based solution to data browsing is currently available for the contents of the AXAF observing catalog and other catalogs in the archive. A Java API for other database applications to access the data is presented. Performance issues and security limitations and workarounds are discussed.

1. Introduction

The AXAF archive architecture involves a number of servers, RDBMS servers for managing relational data and archive servers that manage datafiles. Access to datafiles is provided by existing client applications which send data ingest, search and retrieval requests in a 4gl language to an archive server (Zografou et al. 1997). An architecture was developed to build a powerful and uniform interface to the various servers through the World Wide Web. The advent of JDBC a specification for an API that allows Java applications to access various DBMS using SQL became the obvious choice to build this interface.

2. System Architecture

Figure 1 shows both the client-server architecture developed using the Database vendor API (Zografou et al. 1997) where both the client and the data servers communicate using the same protocol and the 3-tier architecture which allows

[1]Trademark of Sun Microsystems

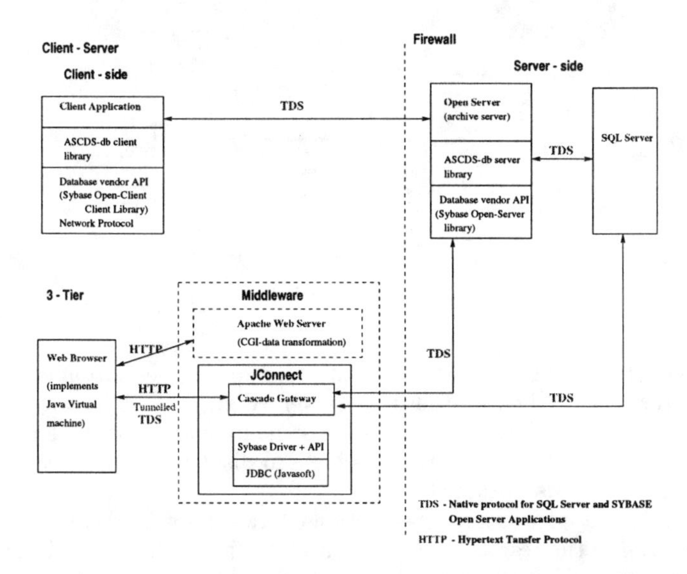

Figure 1. System Architecture [Client-Server and 3-Tier]

a Java client applet to communicate with the data servers behind a firewall. The middle tier consists of Sybase jConnect[2] for JDBC package. This includes the Cascade HTTP Gateway[3] which is the gateway to the RDBMS and archive servers.

According to Java security restrictions, applets can only connect back to the host they were downloaded from. If the Web server and database server were on the same machine then the Cascade gateway is not required. But in this case the data server hosts are behind a firewall hence the Cascade gateway acts as a proxy and provides the path to the data servers. The HTML documents and Java code are served from the Apache Web Server[4] tree. The Apache Server is also used to execute a cgi-script which performs transformations of input and output data when required. Hence the Apache and the Cascade run on the same host on different ports so that the applet downloaded from the Apache server tree can connect back to the same host at the port of the Cascade gateway.

3. Search Operation

Searching involves querying databases on the SQL Server and viewing the results on the client screen or saving them in a file. Both command-line client applications and Java applets have been written to send browse requests to the archive and the database servers. The primary difference between the above

[2]http://www.sybase.com/

[3]http://www.cascade.org.uk/

[4]http://www.apache.org/

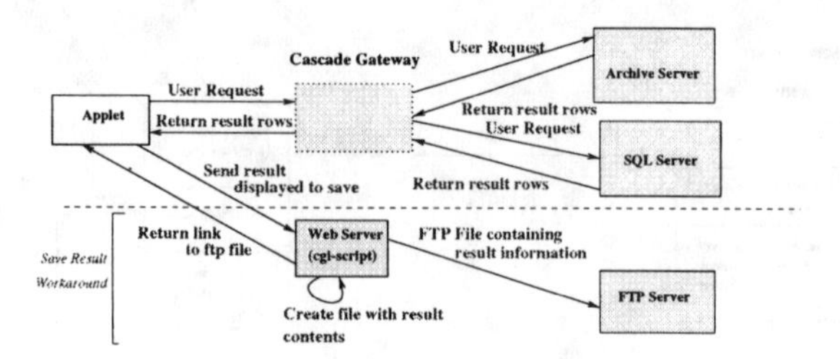

Figure 2. Search Element

two types of client requests is that the latter's request goes through the Cascade HTTP/gateway and then to the data server (Figure 1). The way the browse request is handled on the server-side is the same for either client. The results are received in the same way on the client-side and displayed. One major difference is in saving the results obtained.

Due to Java applet security restrictions, files cannot be created on the client filesystem. Hence the result contents are sent by the applet to a cgi-script on the Web Server host which creates a file on its filesystem from the binary stream. This file is then FTP'ed to the FTP server in a previously created location. The cgi-script informs the client of this path which is displayed to the user.
In the future with JDK1.1 and a compatible Web browser it may be possible to re-create files on the client side depending on authentication and access control schemes. This would eliminate the extra step of passing data to a cgi-script.

4. Retrieve Operation

This involves retrieving files stored in an archive from the archive server via metadata browsing in the SQL server. As shown in Figure 3 the ideal retrieve scenario would function the same way for either client type. The client would receive the filenames and file contents as a binary stream, re-create the files on the client filesystem. Since this is not possible with applets due to security restrictions, two methods are presented to resolve this issue.

1. The file contents are forwarded to the middle-tier and the files are re-created on the Web server's filesystem. These files are then FTP'ed to the FTP Server and the client is notified of the link to access the files. This is similar to the "save results" method outlined above.

2. The second method is much more efficient both from the performance point of view and because it avoids of excessive network activity. The applet in this case only serves as the GUI and passes the user input to the Web server. The Web server now behaves as the client to the archive server. The CGI-script directly executes the "arc4gl" client application which is the command-line client application. Hence it has the ability to send and receive data from the archive server. The files retrieved are re-created on the Web server filesystem. Then they are FTP'ed over to the FTP server. The link to the files is returned and displayed at the client applet end. If the files are of a format supported for viewing in the Web Browser e.g.,

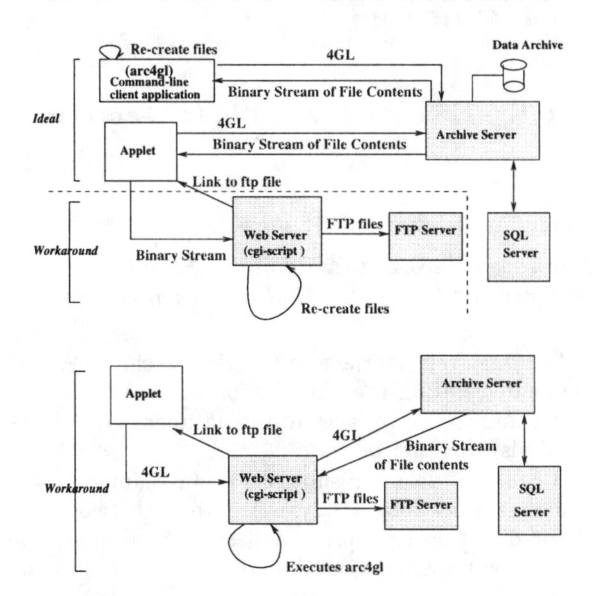

Figure 3. Retrieve Element

HTML, GIF, PS then the files, if placed under the Web Server tree, can be directly viewed by the user.

To port this setup to various environments such as test and production, a cgi-script is used to dynamically create the HTML file presented below with the appropriate information.

```
<HTML>
<applet
       code = "ocat.class"
       codebase = "base URL of applet to be displayed"
       archive = ocat.zip
       width=700 height = 650>
       <param name=proxy value="IP address:port of Cascade Gateway">
       <param name=host value="IP address of DataServer">
       <param name=port value="Port of DataServer">
       <param name=uid value=""> <param name=pass value="">
       <param name=cgiserver value="IP address of Web Server">
       <param name=cgiserverport value="Port of Web Server">
</applet>
</HTML>
```

References

Zografou P., Chary, S., DuPrie, K., Harbo, P., & Pak K. 1998, "The ASC Data Archive for AXAF Ground Calibration", this volume

Astronomical Data Analysis Software and Systems VII
ASP Conference Series, Vol. 145, 1998
R. Albrecht, R. N. Hook and H. A. Bushouse, eds.

Browsing the HST Archive with Java-enriched Database Access

M. Dolensky, A. Micol and B. Pirenne

Space Telescope - European Coordinating Facility,
Karl-Schwarzschild-Str. 2, D-85748 Garching, Germany

Abstract. Our Web interface to the HST archive (WDB) was substantially augmented during the past few months: A seamless integration of Java applets and CGI interfaces to the database now allows basic manipulation and visualization of data.

WDB can now delegate subtasks to Java applets. We use this ability to visualize quick-look (Preview) data with two applets: JIPA and Spectral for direct FITS images and spectra visualization respectively. They both allow basic data manipulation such as contrast enhancement, zooming or X/Y plots.

Another utility, the Skymap applet, can display the location of HST observations on a sky projection.

These applets only require a common Web browser and offer very helpful features to facilitate archive data selection.

1. Browsing the HST Archive

There are already a number of ESO and ECF archive services[1] and projects that make use of the new Java technology. We will briefly discuss some applets used for the following purposes:

Jplot	Plot Contents of HST Observation Log Files
Spectral	Plot Spectra - Preview of HST Observations
JIPA	Display FITS Images of HST Observations
Skymap	Present Maps of Parallel HST Observations

1.1. JPlot

ECF's HST archive uses WDB to implement query forms for its Web interface. A few months ago the only way to retrieve data was to invoke a CGI script. Now we also use WDB to fire up applets. In this way it is possible to replace component by component on demand, without changing the whole archive system at once. Jplot is an example of this strategy. It allows visualization of ASCII tables and the production of basic X/Y plots (Figure 1). The same is true for the Spectrali applet.

[1] http://archive.eso.org/

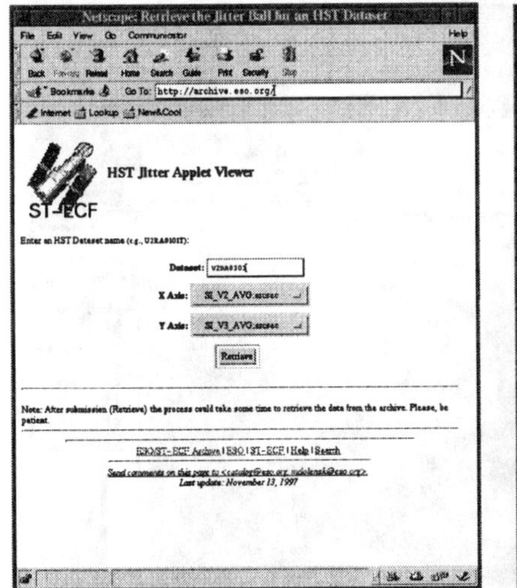

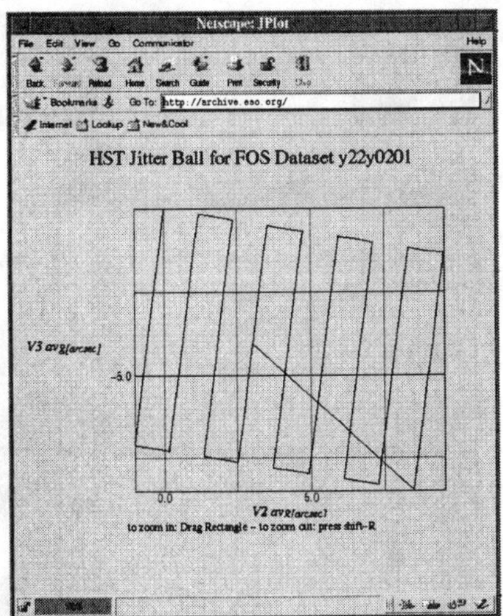

Figure 1. Requesting and Displaying Pointing Information using Jplot

1.2. Spectral

This applet is a previewer for HST spectra. It is quite similar to Jplot and is integrated into the WDB Web interface (Figure 3). The user has various options for inspecting plots with the mouse and by means of hotkeys. As the screenshots in this paper show, a standard Web browser supporting either JDK 1.0 or JDK 1.1 is sufficient to run this applet. No special software is required, in particular no appletviewer or plug-ins for the Web browser have to be installed on the client side. Micol et al. (1997) discussed this issue in more detail.

1.3. Java Image Preview Application (JIPA)

While Spectral presents plots of spectra, JIPA's task is to visualize images of HST and to allow basic image processing. The input data format is compressed FITS which makes JIPA the ideal previewer for the HST archive. There are several options for contrast enhancement, zooming and displaying header keywords (Figure 2). Another feature is the conversion of the mouse position from pixel space into the World Coordinate System (WCS). JIPA is written in pure Java like the other applets presented here and hence is platform independent.

1.4. Skymap

Figure 3 shows the Skymap applet. It graphically displays locations on the sky, where observations took place. It is possible to select observations taken with one or more specific instruments and to transform the mouse position into right ascension and declination.

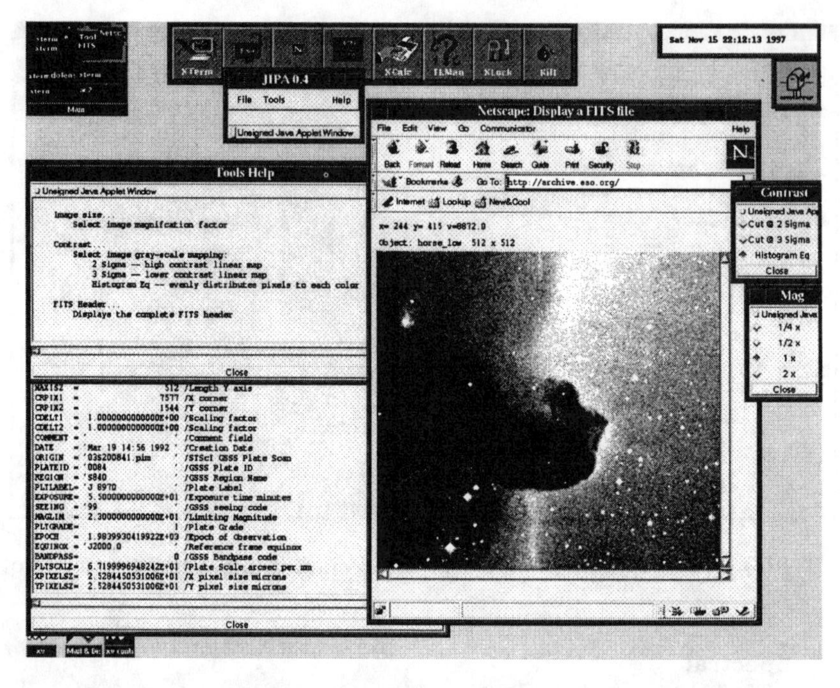

Figure 2. Inspecting FITS Image with the Java Applet JIPA

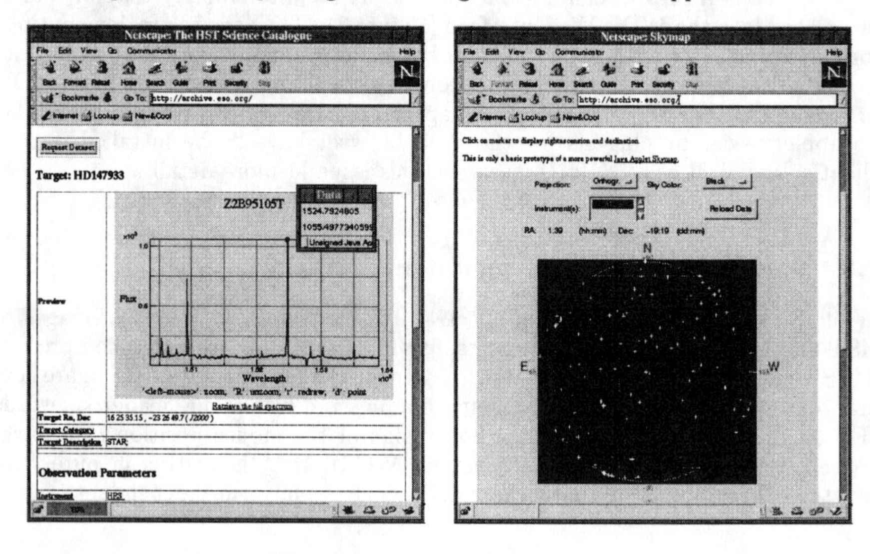

Figure 3. Applets Spectral (left) and Skymap (right)

2. Outlook

In the future more and more components of the archive user interface will be replaced by Java applets. In this way not only data, but also programs are provided to the user. It also means that the CPU on the client side will become more and more busy manipulating data instead of running idle while waiting for the response from some archive server.

References

Micol, A., Albrecht, R., & Pirenne, B. 1997, in ASP Conf. Ser., Vol. 125, Astronomical Data Analysis Software and Systems VI, ed. Gareth Hunt & H. E. Payne (San Francisco: ASP), 104

Astronomical Data Analysis Software and Systems VII
ASP Conference Series, Vol. 145, 1998
R. Albrecht, R. N. Hook and H. A. Bushouse, eds.

Prototype of a Discovery Tool for Querying Heterogeneous Services

D. Egret, P. Fernique and F. Genova

CDS, Strasbourg, France, E-mail: egret@astro.u-strasbg.fr

Abstract. The CDS has recently developed a tool for managing URLs in a context of distributed heterogeneous services (GLU, Uniform Link Generator: Fernique et al. 1998). This includes the development of a 'URL dictionary' maintained by the data providers contributing to the system.

Based on such a system, it becomes possible to create automatically a homogeneous interface to the services described in the dictionary. This tool is currently available as a prototype, under the name of AstroGLU, with the aims of demonstrating the feasibility of the concept, and helping orienting future developments in the domain.

Being a flexible, easily maintainable tool, AstroGLU is a strong incentive for an increased cooperation of all astronomical data providers.

1. Introduction

How to help the user find his path through the jungle of information services is a question which has been raised during the past years (see e.g., Egret 1994), when it became clear that a big centralised system was not the efficient way to go.

Obviously the World-Wide Web brought a very interesting medium for solving this question: on the one hand the WWW provides a common language for all information providers (but flexible enough so that it does not place unbearable constraints on existing databases); on the other hand the distributed hypertextual approach opens the way to navigation between services (provided a minimum of coordinating spirit can be achieved). Let us note that it has been already widely demonstrated that coordinating spirit is not out of reach in a (small) community such as astronomy, which also remains largely sheltered from commercial influence.

The CDS (Centre de Données astronomiques de Strasbourg) has recently developed a tool for managing remote links in a context of distributed heterogeneous services (GLU, Générateur de Liens Uniformes, i.e., Uniform Link Generator; Fernique et al. 1998). First developed for ensuring efficient interoperability of the several services existing at CDS (Astronomer's Bazaar, VizieR, SIMBAD, bibliography, documentation, etc. — see Genova et al. 1996), this tool has also been designed for maintaining addresses (URLs) of remote services (ADS, NED, etc.).

Capacity of leaving the management of these addresses to the remote data provider (rather than having to centralize all the knowledge of the distributed services) is an essential feature of the GLU.

A key element of the system is the 'GLU dictionary' maintained by the data providers contributing to the system, and distributed to all machines of a given domain. This dictionary contains knowledge about the participating services (URLs, syntax and semantics of input fields, descriptions, etc.), so that it is possible to generate automatically a correct query to be submitted.

A typical scenario is the following: we want to get from service S, a piece of information J, corresponding to the data D: the GLU system will ensure that the correct query is generated, i.e., in WWW syntax:

`address_of_S/query_for_J?field_D`

In fact, the remote user has no visibility of the GLU: GLU is essentially a tool for the data providers. The user will simply see a series of features accessible in a sequence, such as, for example, once an author name is provided (by the user, or retrieved from a database in a previous step), it becomes possible to press a button and to obtain the list of all papers published by this author (from ADS), or the list of recent preprints (in astro-ph) mentioning his name. How the button leading to this information is generated remains transparent to the user, who probably does not care to know (even if, hopefully, he or she appreciates the opportunity which is being offered).

The service provider (data center, archive manager, or Webmaster of an astronomical institute) has used the GLU for coding the query, making use of the easy update of the system: knowing which service to call, and which answer to expect from this service, the programmer does not have to worry about the precise address of the remote service at a given time, nor of the detailed syntax of the query (expected format of the equatorial coordinates, etc.).

2. What can we find about ... ?

Let us imagine, now, another scenario: we have the data D (for example an author's name, position or name of an astronomical object, bibliographical reference, etc.), and we would like to know more about it, but we do not know which service S to contact, and what are the different types of information J which can be requested. While the first scenario was typical of an information provider (who knows the astronomical landscape of information services, and has developed contacts with the managers of interesting remote databases), this latter scenario is typical of a scientist, exploring new domains as part of a research procedure.

2.1. A Reference Directory

The GLU dictionary can also be used for helping to solve this question: the dictionary can be considered as a reference directory, storing the knowledge about all services accepting data D as input, for retrieving information types J, or K. For example, we can easily obtain from such a dictionary the list of all services accepting an author's name as input : information which can be accessed, in return, may be an abstract (service ADS), a preprint (LANL/astro-ph), the author's address (RGO e-mail directory) or his personal Web page (StarHeads), etc.

Based on such a system, it becomes possible to create automatically a simple interface guiding the user towards any of the services described in the dictionary.

2.2. AstroGLU

This idea has been developed as a prototype tool, under the name of AstroGLU[1], in order to demonstrate the feasibility of the concept, convince more data providers to collaborate, and help orienting future developments in the domain.

The current steps of a usage scenario are the following:

1. Data Type Selection: First, the user can select among the available data types those which correspond to the data D for which additional information is needed. The principal data types already available in the current prototype version (3.0, September 1997) are the following: name of an astronomical object, celestial position, last name of a person (e.g., an author's name), keyword (in English natural language), reference code (bibcode used by NED/SIMBAD/ADS), catalog name, dataset number in HST or CFHT archive, etc.

At this stage, the user can already input the data D itself, assuming it is a simple character string (e.g., an astronomer's name). But the user does not need to have an *a priori* knowledge of the existing services, or even of the information types corresponding to potential query results.

2. Service List: Based on this data type, AstroGLU scans the dictionary and selects all services (among those known from the system) that support queries involving this data type; for example, if the data type is 'astronomical object name', SIMBAD, NED, or the ALADIN sky atlas (among others) are listed.

This list is supposed to answer simultaneously two questions: *what ?* (i.e. which type of information can be found) and *where ?* (i.e. which service can provide it). But the focus is made on the first aspect, the second one being generally kept implicit until the selection of the information type is done.

3. Query Submission: The user can finally select one of the proposed services, and will receive a form for submitting the query to the remote service, to which it is finally submitted for processing. These forms frequently imply giving additional parameters in complement to the data D (e.g., epoch of a position; year limits for a bibliographical query, etc.).

Where can we find ... ? Alternatively, the user can specify what he "looks for", and according to the service qualifications contained in the dictionary, the user will be presented with a selection of services able to answer his query.

---

[1]http://simbad.u-strasbg.fr/demo/cgi-bin/astroglu-m1.pl

3. Current AstroGLU Functionality

AstroGLU functionality is constructed around the main dictionary. In step 1, all data types listed are those occuring at least once in the dictionary (or, more specifically in the subset of the dictionary related to the specific domain on which AstroGLU is working). These data types may be sorted according to eventual conversions (using remote resolvers, or local rules). At the end, they are sorted in alphabetical order.

In step 2, the list of 'actions' (i.e. possible queries) using the data type selected in step 1 is displayed, and, if a data string has been given, some tests can be performed, when a test method has been implemented in the dictionary (e.g., compliance of a refcode with the corresponding dataset). Some of these actions may imply use of one, or more, intermediate resolution (e.g., call SIMBAD for finding the celestial position of a given object, before sending a query to an archive).

Step 3 may include, in the future, examples and default values of additional parameters.

The complete list of actions can be displayed on request.

AstroGLU can be automatically implemented for all or part of the domains cooperating to the GLU system. All the forms are generated from the GLU Dictionary information. The dictionary being very easily and efficiently maintained, this is a strong incentive for an increased cooperation of astronomical data providers.

4. Final Remarks

A major aim of this tool is to help the user find his way among several dozens (for the moment) of possible actions or services. A number of compromises have to be made between providing the user with the full information (which would be too abundant and thus unusable), and preparing digest lists (which imply hiding a number of key pieces of auxiliary information, and making subjective choices).

A resulting issue is the fact that the system puts on the same line services which have very different quantitative or qualitative characteristics. Heck (1997) has frequently advocated that high quality databases should be given preference, with respect to poorly updated or documented datasets. We do not provide, with AstroGLU, efficient ways to provide the user with a hierarchy of services, as a gastronomic guide would do for restaurants. This might come to be a necessity in the future, as more and more services become (and remain) available.

References

Egret, D. 1994, in ASP Conf. Ser., Vol. 61, Astronomical Data Analysis Software and Systems III, ed. Dennis R. Crabtree, R. J. Hanisch & Jeannette Barnes (San Francisco: ASP), 14

Fernique, P. et al. 1998, this volume

Genova, F. et al. 1996, Vistas in Astronomy 40, 429

Heck, A. 1997, in "Electronic Publishing for Physics and Astronomy", Astrophys. Space Science 247, Kluwer, Dordrecht, 1 (ISBN 0-7923-4820-6)

Astronomical Data Analysis Software and Systems VII
ASP Conference Series, Vol. 145, 1998
R. Albrecht, R. N. Hook and H. A. Bushouse, eds.

Hubble Space Telescope Telemetry Access using the Vision 2000 Control Center System (CCS)

M. Miebach

European Space Agency, Space Telescope Science Institute, 3700 San Martin Drive, Baltimore, Maryland USA 21218

M. Dolensky

Space Telescope - European Coordinating Facility, Karl-Schwarzschild-Str. 2, D-85748 Garching, Germany

Abstract. Major changes to the Space Telescope Ground Systems are presently in progress. The main objectives of the re-engineering effort, Vision 2000[1] , are to reduce development and operation costs for the remaining years of Space Telescope's lifetime. Costs are reduced by the use of commercial off the shelf (COTS) products wherever possible.

Part of CCS is a Space Telescope Engineering Data Store, the design of which is based on modern Data Warehouse technology. The purpose of this data store is to provide a common data source for telemetry data for all HST subsystems. This data store will become the engineering data archive and will provide a query-able DB for the user to analyze HST telemetry. The access to the engineering data in the Data Warehouse is platform-independent from an office environment using commercial standards (Unix, Windows/NT, Win95). Latest Internet technology is used to reach the HST community. A WEB-based user interface allows easy access to the archives.

Some of the capabilities of CCS will be illustrated: sample of real-time data pages and plots of selected historical telemetry points.

1. CCS Architecture

A Web browser with Java support and an account to establish a secure Internet HTTP connection to the Goddard Space Flight Center (GSFC), is everything needed in order to use the new Control Center System (CCS). Then full access to the telemetry of the Hubble Space Telescope (HST) is given (Figure 2). A public version of CCS[2] is also available.

How does that work? The telemetry data stream of HST is transfered to the Front End Processor (FEP) via Nascom (Figure 1). FEP provides a communication interface between the vehicle and ground control. It also captures all the

[1] http://vision.hst.nasa.gov/

[2] http://v2mb20.hst.nasa.gov:4041/demo.html

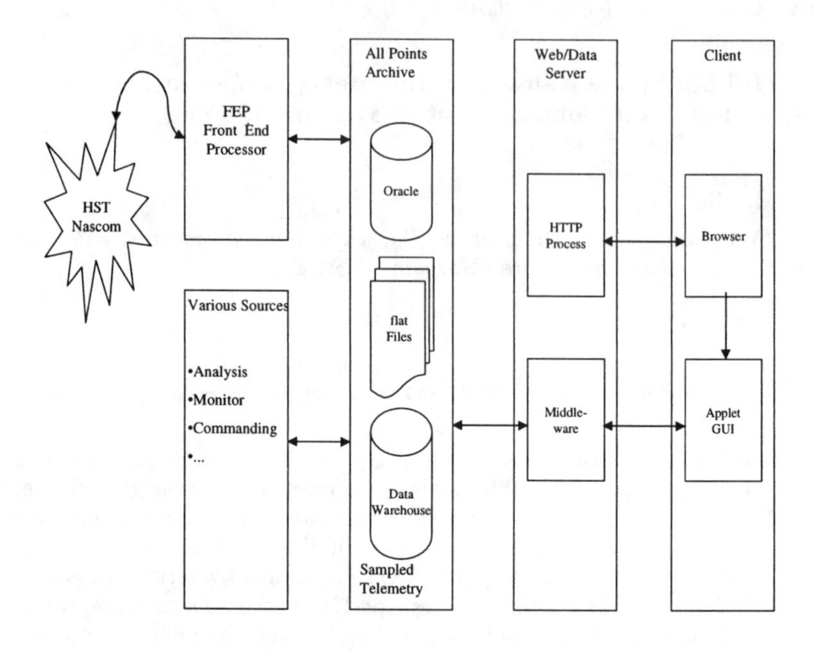

Figure 1. CCS Archive Architecture

downlinked data and forwards them to the attached All Points Archive. Besides the All Points Archive there is also the Red Brick Data Warehouse for sampled data. Without the warehouse the All Points Database containing several terabytes of telemetry parameters would be of limited use only. The warehouse samples data to reduce their size. In this way it is possible to further analyze them on standard platforms like Pentium PCs. The warehouse also provides derived engineering parameters and advanced query options. The warehouse is still under construction, but a Java applet prototype[3] is already available at the home page of the Science & Engineering Systems Division (SESD) at STScI.

In order to support its heterogeneous clients and DB servers a three-tiered solution was chosen (Mickey, 1997). A Java applet on the client side speaks to a middleware component on the data server. The middleware then takes care of the DB access. Rifkin (1997) gives a more comprehensive description.

2. Applications

CCS's main task is not scientific data reduction. Nonetheless, it provides the means to develop e.g., new calibration algorithms. This is possible, because for the first time, one can have access to all engineering parameters via the Internet. So far, only a very limited subset of telemetry parameters is available through

[3] http://www.sesd.stsci.edu/

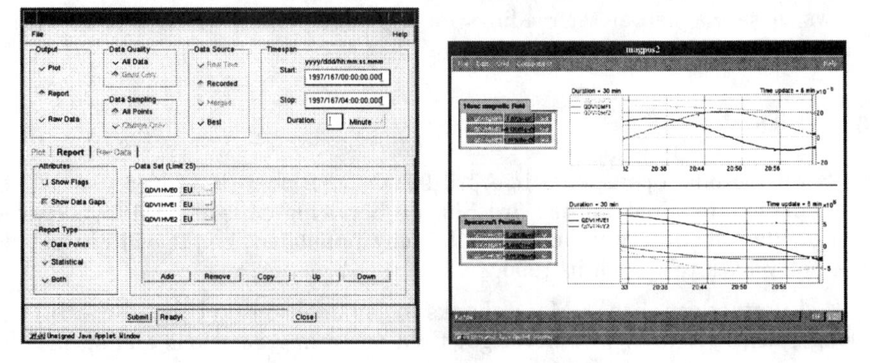

Figure 2. CCS Archive Query Form (l.) and Real-Time Data Page (r.)

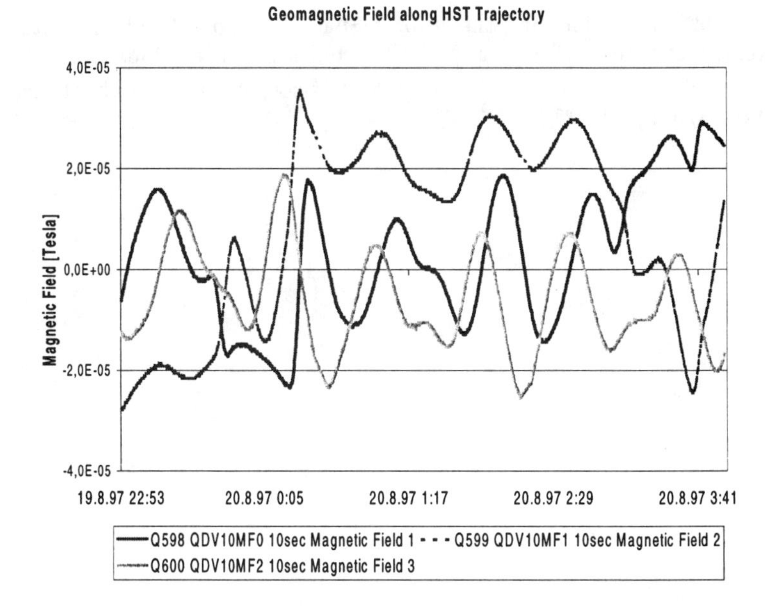

Figure 3. Geom. Field (3 Components) over a Period of 3 Orbits

observation log files. One possible application is mentioned here to illustrate this.

The Faint Object Spectrometer (FOS) of the Hubble Telescope is an instrument that was replaced during the 2nd Shuttle Servicing Mission in February 97. As part of the final archive for this instrument, ST-ECF would like to apply an improved model of the geomagnetic field to the calibration pipeline. The required magnetometer readings from the onboard sensors are not part of the observation log files, but CCS could now provide the missing values. Figure 3 shows these magnetometer readings for a 5-hour-period, which equals roughly three HST orbits.

3. Outlook

CCS will become operational in April 98. At this point it will provide all the functionality required for the 3rd Shuttle Servicing Mission in 1999. By the end of 1999 the data warehouse will be fully populated, i.e., it will contain the telemetry back to launch in 1990.

Acknowledgments. Many thanks to Jeff Johnson (Lockheed Martin) and Doug Spiegel (NASA) who never got tired answering questions and spent a lot of time on user support for CCS.

References

Rifkin, A. 1997, Reengineering the Hubble Space Telescope Control Center System, Institute of Electrical and Electronics Engineers Inc.

Mickey, S. 1997, Internet Java & ActiveX Advisor, Advisor Publications Inc., Vol. 4/94, p. 16-21

Astronomical Data Analysis Software and Systems VII
ASP Conference Series, Vol. 145, 1998
R. Albrecht, R. N. Hook and H. A. Bushouse, eds.

An Archival System for Observational Data Obtained at the Okayama and Kiso Observatories. III

Eiji Nishihara and Michitoshi Yoshida

Okayama Astrophysical Observatory, National Astronomical Observatory of Japan, Kamogata, Okayama 719-02, Japan

Shin-ichi Ichikawa, Kentaro Aoki and Masaru Watanabe

Astronomical Data Analysis Center, National Astronomical Observatory of Japan, Mitaka, Tokyo 181, Japan

Toshihiro Horaguchi

National Science Museum, Shinjuku, Tokyo 169, Japan

Shigeomi Yoshida

Kiso Observatory, University of Tokyo, Mitake, Nagano 397-01, Japan

Masaru Hamabe

Institute of Astronomy, University of Tokyo, Mitaka, Tokyo 181, Japan

Abstract. We present the newly developed version of the Mitaka-Okayama-Kiso data Archival system (MOKA3). MOKA3 consists of three parts: 1) a distributed database system which manages the observational data archive of Okayama and Kiso observatories, 2) a Web-based data search and request system, and 3) a delivery system for requested data. The client system of MOKA3 is implemented as a Java Applet so that many functions, e.g., GUI arrangement, display of preview images, and conversion of coordinates, can be processed locally on the client side without accessing the server system. Moreover, environmental data, such as meteorological data and all-sky images, are also archived in the database system of MOKA3 in order to increase the utility of its archival observational data.

1. Introduction

Observational data is one of the most important fundamentals in astronomy, and so the importance of archiving and reusing observational data is widely recognized. The original version of MOKA (Mitaka-Okayama-Kiso data Archival system; Horaguchi et al. 1994; Takata et al. 1995; & Ichikawa et al. 1995) is the first fully-fledged observational data archival system in Japan. It was developed for the observational data taken with the Spectro-NebulaGraph (SNG; Kosugi et al. 1995) of the 188cm telescope at the Okayama Astrophysical Observatory, and those taken with the prime focus CCD camera of the 105cm Schmidt telescope at the Kiso Observatory.

Basically, the original MOKA was not a client-server system. Therefore, the sites where users can operate MOKA were considerably restricted. In order to solve this problem, we developed the second version of MOKA (MOKA2[1]; Yoshida 1997) which is based on the World Wide Web (WWW). MOKA2 has been in operation since September 1996. Through the operation of MOKA2, several points to be improved were clarified. These were with respect to displaying preview images, managing database systems, and integrating environmental data. Then we started to develop the third version of MOKA (MOKA3), which will be in operation late in 1997. In this paper, we present a technical overview of MOKA3.

2. System Overview

MOKA3 consists of three parts: 1) a distributed database system which manages the observational data archive of Okayama and Kiso observatories, 2) a Web-based data search and request system, and 3) a delivery system of requested data. The data flow of MOKA3 is shown in Figure 1.

2.1. Distributed Database System

The following three kinds of data files are managed in the MOKA3 distributed database system. 1) Header information database: Requisite header items are extracted from header sections of original CCD frames and stored in database tables. These tables are managed by an ORACLE DBMS (DataBase Management System) and used for a search of CCD frames. 2) Header files: In addition to the header database above mentioned, we stored header files, which are duplicate copies of header sections of original CCD frames. These files are plain text files and used to show detailed information about CCD frames. 3) Preview image files: By binning and resampling, reduced-size preview images are generated from original CCD frames. They are stored as gzipped FITS files.

These data files are stored in the server machines at Mitaka, Okayama, and Kiso respectively. Newly produced data from night-by-night observations are processed into these three kinds of data at their source observatory. These new processed data are sent to Mitaka, then other observatories copy them from Mitaka. In MOKA3, this data transfer is achieved by distributed database mechanism in ORACLE for database files and by ftp-mirroring for header and preview image files. This mechanism efficiently copies only new data and keeps the database at each site consistent.

2.2. Data Search & Request System

The data search and request system of MOKA3 is implemented as a Java applet. Hence, users can operate MOKA3 with a Java-enabled WWW browser on any computer connected to the Internet. Searching constraints input in this

---

[1]http://moka.nao.ac.jp/

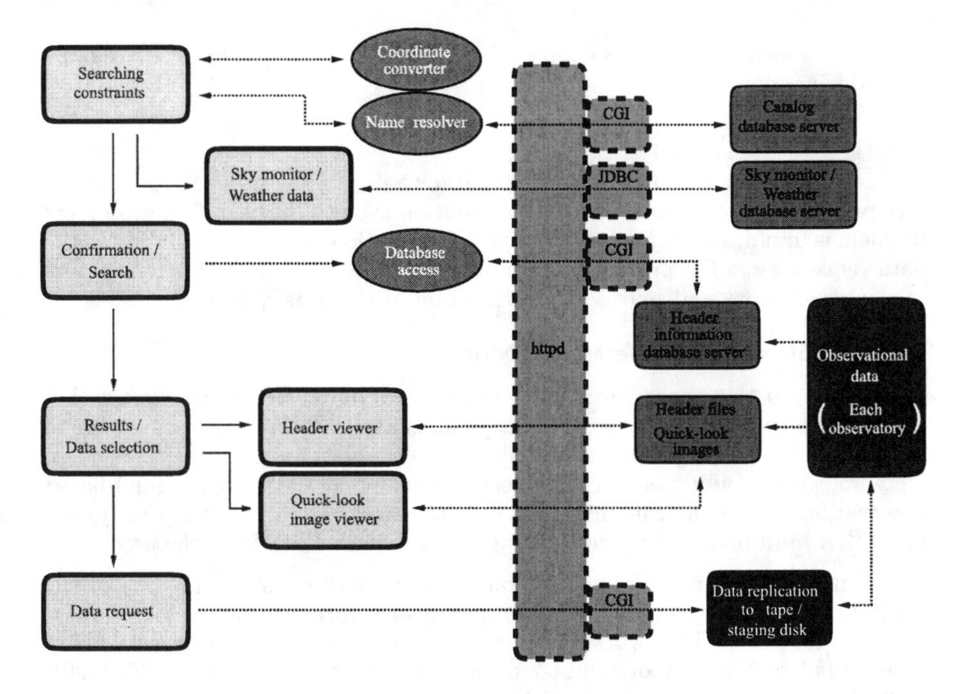

Figure 1. Data flow of MOKA3.

applet are passed to the DBMS server computer through CGI. SQL statements corresponding to these constraints are created, and a search is performed by ORACLE. The results of the search are displayed on the client computer as a list of frames appropriate to the constraints specified. Users can also display the detailed header information and/or preview images of selected frames on the client computer to examine whether these frames are appropriate for their purpose.

In the above procedure, many functions are processed locally on the client computer without accessing the server computer. For example, GUI components are dynamically switched over in compliance with the user's selection of searching strategies (e.g., by name or by coordinate). This local processing by the Java applet reduces load the server computer and the network. Moreover, as a part of the applet, the viewer of preview FITS images is also implemented in Java. Therefore, users can flexibly display the preprocessed images with their favorite display levels, and plot them at any position. In the earlier MOKA2, such a viewer was prepared as an external viewer of the WWW browser. Hence, available platforms are restricted. In addition, users must first install the viewer on their client machine.

The original CCD frames are stored in the observatory where they were produced. Registered users can utilize these data. MOKA3 has a window to request for these original data. From this window, the request for selected frames is automatically generated and sent to the MOKA3 administrator at the source observatory as an e-mail. Registration for MOKA3 data request is also available from this window.

2.3. Data Delivery System

The administrator who receives the e-mail for a data request checks the propriety period of the data, and sends a distribution tape to the user if no proprietry problem is found. A mechanism to make a distribution tape corresponding to the data request e-mail is prepared. This decreases the load on the administrators at the observatory and makes the distribution of the data quicker.

2.4. Weather Database & Sky Monitor

In addition to the above three main body of MOKA3, access to the weather database and the all-sky images at Okayama is available from the home page of MOKA3. They are implemented as independent applets for the present. Because meteorological conditions severely affect the quality of data from ground-based observations, these environmental data are also archived in the database system of MOKA3 in order to help users to sift out adequate data from the archive.

Acknowledgments. The study on the data archival system is one of the projects promoted by the Japan Association for Information Processing in Astronomy (JAIPA). This work was supported in part under the Scientific Research fund of Ministry of Education, Science, Sports and Culture (06554001, 07304024 and 08228223), and JAIPA Nishimura foundation. The authors thank S. Nishimura for his continuous encouragement.

References

Horaguchi, T., et al. 1994, Publ. Natl. Astron. Obs. Japan, 4, 1

Ichikawa, S., et al. 1995, in ASP Conf. Ser., Vol. 77, Astronomical Data Analysis Software and Systems IV, ed. R. A. Shaw, H. E. Payne & J. J. E. Hayes (San Francisco: ASP), 173

Kosugi, G., et al. 1995, PASP, 107, 474

Takata, T., et al. 1995, Publ. Natl. Astron. Obs. Japan, 4, 9

Yoshida, M. 1997, in ASP Conf. Ser., Vol. 125, Astronomical Data Analysis Software and Systems VI, ed. Gareth Hunt & H. E. Payne (San Francisco: ASP), 302

Astronomical Data Analysis Software and Systems VII
ASP Conference Series, Vol. 145, 1998
R. Albrecht, R. N. Hook and H. A. Bushouse, eds.

Querying by Example Astronomical Archives

F. Pasian and R. Smareglia

Osservatorio Astronomico di Trieste, 34131 Trieste, Italy
Email: pasian@ts.astro.it

Abstract. In this paper, a project aiming to allow access to data archives by means of a query-by-example mechanism is described. The basic user query would simply be the submission of an input image, either 1-D or 2-D; the system should allow retrieval of archived images similar to an input one on the basis of some "resemblance parameter".

1. Introduction

With the exponential expansion of information systems, accessing data archives plays a role which is becoming increasingly important for the dissemination of scientific knowledge at all levels. The need is felt to make access to data archives simpler for the community of users at large. Queries on the data, regardless of their complexity, should be easier to specify, more intuitive, simplified with respect to those currently in use. In other words, users should be able to express queries in an intuitive way, without knowing the detailed physical structure and naming of data. Content-based information retrieval is receiving increasing attention from scientists, e.g., see Ardizzone et al. (1996), and Lesteven et al. (1996).

This project aims to allow access to data archives by means of a query-by-example mechanism: the user should be put in a position to provide a "user object", a template of what he/she wishes to retrieve from the archive. The system managing the archive should provide the user with archived data resembling the template on the basis of some desired characteristic, and with some "resemblance parameter". Data understanding (classification and recognition of descriptive features) is deemed to be an essential step of the query-by-example mechanism.

2. Project Objectives

The objective of this project is to integrate database, mass storage, classification - feature recognition, and networking aspects into a unique system allowing a "query-by-example" approach to the retrieval of data from remotely-accessible large image archives exploiting the algorithms up to their performance limits.

The basic idea, from the user's point of view, is to be able to submit a query to a remote archive in the form of an image (2-D or 1-D), telling the system: *"get me all the images/plots in the archive looking like this one, or having this specific feature"*. The basic user query would therefore simply be the submission of an image, either previously extracted from the very same archive, or owned by the user, or built via a modelling software. This operation must be feasible while connected remotely to the archive.

Several problems have been already identified. In the "global" approach to the query-by-example mechanism, the various characteristics of images have the same weight. Difficulties here include coping with different image resolutions, being able to search for images having a specific feature in them by just submitting a small image template containing the feature itself, building models compatible with the archived images, *etc.* A more sophisticated approach would be to add a level of interaction with the user, allowing him/her to know which are the features of the images the system is able to recognize, and allow a choice of the features to be searched for.

3. QUBE - Overall Features

A system called QUBE is currently being designed to support the QUery-By-Example paradigm in accessing astronomical data archives. A preliminary list of features is as follows:

- Network-based interface, for backward-compatibility with already existing image archives: standard queries on metadata (data descriptions) should always be possible through the known interface.
- Extensions to standard SQL, to allow handling of image templates by a relational database: in principle, specifying a query in this extended SQL should always be possible for an expert user. As a first order approximation, SQL extensions can be handled by an interpreter.
- Ingestion of "user objects" in the system in different formats, with priority given to astronomy-specific and commercial standards (FITS, but also GIF, JPEG, etc.).
- "Global" approach to query-by-example based on classification methods such as artificial neural networks: unsupervised in the general case, supervised for specific applications.
- Feature recognition algorithms may be used, able to give different weights to different characteristics of the "user object" (the template being compared with the archived data).
- Transmission of data (both "user objects" and retrieved data) via the standard TCP/IP mechanisms.
- An updating capability should be available: in the case classification is required on an image subset (e.g., wavelength range in a spectral archive) the system should be able to re-compute classification parameters for all relevant image subsets in the archive.
- A feedback mechanism should be built in the system, to allow detection and correction of misclassified data.

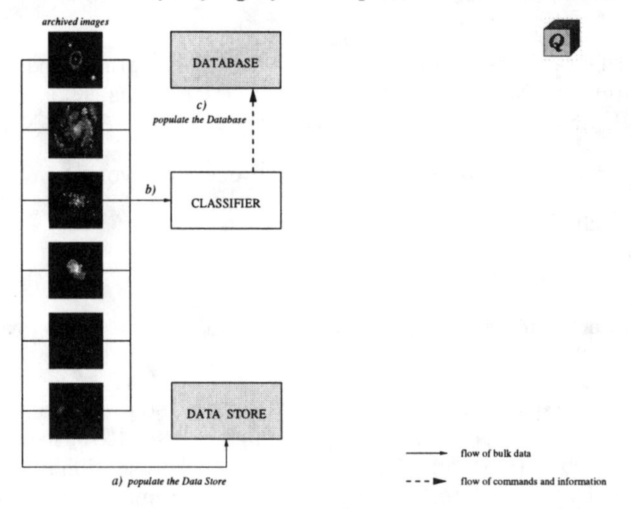

Figure 1. Operational scenario for the data ingest phase of the QUBE system

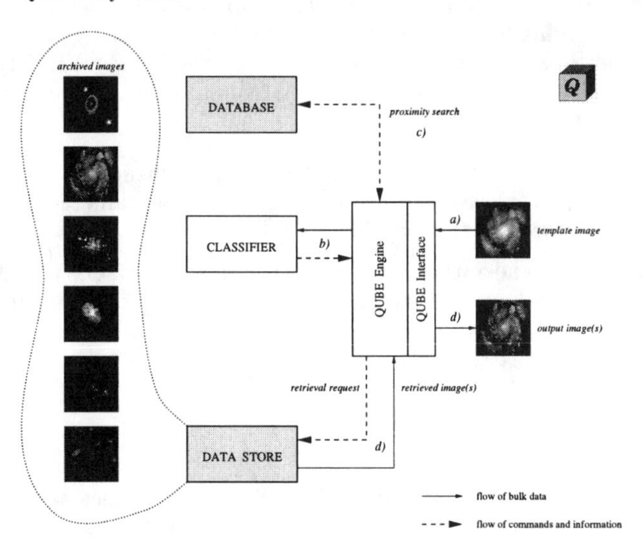

Figure 2. Operational scenario for the data retrieval phase of the QUBE system

4. QUBE - Structure and Operational scenario

The QUBE system is composed of:

- the Archive (a Data Store containing the images, using optical media and a jukebox system; a Database containing their descriptions, managed by a relational DBMS);
- a Classifier (parametric or based on Artificial Neural Networks);
- the QUBE Search Engine;
- the QUBE User Interface.

The following operational scenario is envisaged, and is represented graphically in Figures 1 and 2, respectively:

Data ingest phase : When ingested in the Data Store (step *a*), the images are also analyzed by a Classifier, either parametric or ANN-based, as in Pasian et al. (1997) (step *b*). The result of the classification phase is ingested in the Database (step *c*), together with the image parameters, e.g., extracted from the FITS headers.

Data retrieval phase : A template image is given by the user as input to the system (step *a*). The QUBE Search Engine re-scales it according to the resolution of the archived images, possibly subsets it, and submits it to the Classifier, obtaining the classification parameters (step *b*). A proximity search is then performed on the Database (step *c*); the image(s) satisfying the desired conditions are identified, retrieved from the Data Store and fed to the user as output (step *d*).

More than one classifier may be envisaged to make the mechanism more flexible: in this case, more than one table containing the classification parameters for the archived images will be stored in the Database.

Acknowledgments. The authors are grateful to H.M.Adorf, O.Yu.Malkov, J.D.Ponz and M.Pucillo for having discussions about the concepts of querying image archives by example. The images used for the figures in this paper are thumbnails of Hubble Space Telescope PR images.

References

Ardizzone, E., Di Gesù, V., & Maccarone, M. C. 1996, in Strategies and Techniques of Information in Astronomy, Vistas in Astronomy, 40, 401

Lesteven, S., Poinçot, P., & Murtagh, F. 1996, in Strategies and Techniques of Information in Astronomy, Vistas in Astronomy, 40, 395

Pasian, F., Smareglia, R., Hantzios, P., Dapergolas, A., & Bellas-Velidis, I. 1997, in: Wide-Field Spectroscopy, E.Kontizas, M.Kontizas, D.H.Morgan, G. Vettolani eds., Kluwer Academic Publishers, 103

Astronomical Data Analysis Software and Systems VII
ASP Conference Series, Vol. 145, 1998
R. Albrecht, R. N. Hook and H. A. Bushouse, eds.

Integrating the ZGSC and PPM at the Galileo Telescope for On-line Control of Instrumentation

F. Pasian, P. Marcucci, M. Pucillo and C. Vuerli

Osservatorio Astronomico di Trieste, 34131 Trieste, Italy
Email: pasian@ts.astro.it

O. Yu. Malkov and O. M. Smirnov

Institute of Astronomy (INASAN), Moskow 109017, Russia

S. Monai[1], P. Conconi and E. Molinari

Osservatorio Astronomico di Brera, 22055 Merate, Italy

Abstract. The usage of catalogs is extremely important for efficiently observing with up-to-date instrumentation. In the work described in this paper, GSC sources are cross-correlated with entries in the PPM catalog, and used as an input to the control software of the Galileo Telescope (TNG). A graphical user interface (GUI) based on IDL has also been built. The system will be used during observing time at the TNG.

1. Background

The Guide Star Catalog (GSC) was created by STScI to support HST observations. With positions and magnitudes for about 20 million objects, it is by far the biggest all-sky catalog to date. The GSC is a unique tool for many astronomical applications, however, its utility is somewhat hampered by its format. The two CD-ROMs make for relatively slow access times. Maintaining the catalog on-line requires either two dedicated CD-ROM drives (or a jukebox), or over 1GB of hard disk space. Furthermore, the actual data in the catalog is not easily accessible. The sky is divided into regions of a complex organization, with the data for each region stored in a separate FITS table. The coordinates are given in one standard system (J2000.0).

The GSC sources can be cross-correlated with entries in a number of astrometric catalogs. One of the best of them is a family of PPM catalogs, namely: the Catalog of Positions and Proper Motions; the Catalog of Positions and Proper Motions - South; the Bright Stars Supplement to the PPM and PPM South Catalog, Revised Edition; and the 90000 Stars Supplement to the PPM Star Catalog (Roeser & Bastian, 1988; Roeser & Bastian, 1993; Roeser et al. 1993). These four PPM catalogs give J2000 positions and proper motions for 468861 stars altogether. Their main purpose was to provide a convenient, dense, and

[1] and Osservatorio Astronomico di Trieste, Italy

accurate net of astrometric reference stars that represents the IAU coordinate system on the sky.

The Galileo telescope (TNG) is currently in its testing phase. Some efforts are currently being directed towards using specific GSC2 fields for commissioning purposes; this paper deals instead with the development of tools to be used at the telescope when observing with TNG instruments, and in particular with the Low Resolution Spectrograph, also known as DOLORES (Device Optimized for LOw RESolution).

2. GUIDARES and the ZGSC

To help solve the problem of GSC data retrieval, the Guide Star Catalog Data Retrieval Software, or GUIDARES (Malkov & Smirnov 1995), has been created. This is a user-friendly program which lets one easily produce text samplings of the catalog and sky maps in Aitoff or celestial projections, given a rectangular or circular region in any standard coordinate system. Originally developed for the PC, the low-level GSC access routines of GUIDARES have since been ported to a variety of Unixes, and equipped with a GUI developed using IDL widgets.

We have created a compressed version of the GSC, called ZGSC (Smirnov & Malkov 1997a). By using a binary format and an adaptive compression algorithm, the GSC was losslessly compressed by a factor of 6, giving the ZGSC a total size of about 200 Mb. This makes it entirely practical to keep the ZGSC on-line on a hard disk and gain a dramatic improvement in access times.

An extensive software package, derived from the GUIDARES project, was developed to work with the ZGSC. This includes a suite of IDL routines that retrieve data from the ZGSC into IDL arrays, and supporting C libraries for on-the-fly decompression of the catalog. The software facilitates retrieval of circular regions, specified by center and size. Four coordinate systems are supported: equatorial and ecliptic (any equinox), galactic and supergalactic. The software also allows retrieval of objects of a particular type and/or in a particular magnitude range.

3. The XSKYMAP application

The XSKYMAP software is an IDL widget application for retrieval, visualization and hard copy of ZGSC samplings. The applications of the XSKYMAP are finder charts, GSC studies (Malkov & Smirnov 1997; Smirnov & Malkov 1997b), *etc.* XSKYMAP is fully integrated with ZGSC and provides easy access to all retrieval options of the ZGSC. It also allows for mouse-based catalog feedback (i.e., objects may be selected with the mouse, directly on the map, to view their corresponding full catalog entries). The software provides mouse operations for zoom in/out and recenter region, click-and-drag for computing angular separation and positional angles, and tracking of mouse coordinates with dynamic display of sky coordinates and separation/positional angle relative to center of area. The user can interactively change the map legend (i.e., symbol and color used for each type of object), and selectively display and label particular types of objects.

XSKYMAP uses a custom map projection routine, one which allows plotting of truly rectangular areas even in polar regions, and supports arbitrary rotation of the map relative to North on the sky. An instrument's field-of-view box may also be plotted on the map. XSKYMAP can also display a 2D image (e.g., directly from the instrument), and overplot the catalog map on top. Hard copy in PostScript format is also provided, both in map-only mode, and in image+map overplot mode. Another useful feature of XSKYMAP is built-in coordinate system conversion. For example, the user can start exploring the catalogs using the galactic system, and once the necessary sky area is obtained, the coordinates may be switched to a different system (e.g., the native system of the telescope), while preserving the same area on the screen. The necessary map rotation (to compensate for orientation of different systems relative to each other) is computed and introduced automatically.

Recently, XSKYMAP has been extended with a module to access the PPM catalog. For this purpose the PPM catalogs were reformatted to have the same file structure as the GSC. It is now possible to execute a query on both catalogs (ZGSC and PPM) simultaneously, and view the objects plotted on the same field. Objects from the PPM are plotted using a different symbol or color; their complete catalog entries can also be accessed by clicking the mouse over them. Thus, GSC's depth of field can be combined with PPM's extremely high astrometric accuracy, all within one plot. This development has led to a restructuring of the XSKYMAP software. Data access will now be handled by generic modules with a well-defined interface. Thus it will be possible to easily add capabilities for access to other catalogs in addition to the ZGSC and PPM. To this end, XSKYMAP is being overhauled to take advantage of the new object-oriented features of IDL 5.0.

4. Integrating with the Galileo instrument software

The integration of a new instrument with Galileo is a rather smooth effort, since WSS, the Workstation Software System (Balestra et al. 1991), takes care of all communications with subsystems, information and data management and handling, and its tabular structure simplifies any addition of new configurations. Newer versions of WSS support complete integration with IDL (Balestra 1997), thus guaranteeing the possibility of binding XSKYMAP with the control of TNG instruments. The Observation Software (OS) for the instrument, if properly integrated in WSS, is guaranteed access to all facilities related to TNG control and information handling.

This work is dedicated to the Observation Software for DOLORES, the TNG low resolution spectrograph and imager. The ZGSC and PPM catalogs will be accessed by means of the GUIDARES and XSKYMAP for a number of different purposes including:

- choice of field to be observed;
- assistance to the observer, both in "blind" mode (direct access to the catalog) and in "guided" mode (comparison with observational data taken in imaging mode);
- setup of the MOS systems:

– positioning of the slitlets;

– commands to be sent to the punching machine.

The compactness of the compressed versions of the catalogs and the efficiency of the access software allows the system to be installed within WSS and to perform operations while observing, with no need for local or external network connections to databases or data centers.

Access to ZGSC and PPM is supported by tools and procedures integrated with IDL and WSS. Therefore the system, although built specifically for DOLORES, can be used with instrument-specific modifications for any other TNG instrument controlled by WSS.

The system is planned to be available within the Archives At the Telescope (Pasian 1996), so as to guarantee a reference version on the mountain. In the case of problems, the system can be downlinked from the Archive Server to the Instrument Workstation and installed by means of an automatic procedure at WSS re-start time.

The system is not a general-purpose one, and is dedicated to on-line use during observing time. It is planned to also make it available within the DOLORES Observation Support Software, and will be distributable via the network. At the level of observation preparation, however, network connections with other facilities and/or data centers (e.g., VizieR at the CDS) may be also advisable and supported by the Galileo Observatory.

Acknowledgments. Two of the authors (OYM and OMS) visited the Osservatorio Astronomico di Trieste supported by the INTAS grant no. 94-4069 and by a TNG contract; SM was supported by a grant of the Osservatorio Astronomico di Brera-Merate.

References

Balestra A. 1997, OAT Technical Report no. 24

Balestra, A., Marcucci, P., Pasian, F., Pucillo, M., Smareglia, R., & Vuerli, C. 1991, Galileo Project, Technical Report no. 9

Conconi P. et al. 1997, Mem. SAIt, in press

Malkov, O. Yu., Smirnov, O. M. 1995, in ASP Conf. Ser., Vol. 77, Astronomical Data Analysis Software and Systems IV, ed. R. A. Shaw, H. E. Payne & J. J. E. Hayes (San Francisco: ASP), 182

Malkov, O. Yu., Smirnov, O. M. 1997, in ASP Conf. Ser., Vol. 125, Astronomical Data Analysis Software and Systems VI, ed. Gareth Hunt & H. E. Payne (San Francisco: ASP), 298

Pasian, F. 1996, in ASP Conf. Ser., Vol. 101, Astronomical Data Analysis Software and Systems V, ed. George H. Jacoby & Jeannette Barnes (San Francisco: ASP), 479

Smirnov, O. M., Malkov, O. Yu. 1997a, in ASP Conf. Ser., Vol. 125, Astronomical Data Analysis Software and Systems VI, ed. Gareth Hunt & H. E. Payne (San Francisco: ASP), 429

Smirnov, O. M., Malkov, O. Yu. 1997b, in ASP Conf. Ser., Vol. 125, Astronomical Data Analysis Software and Systems VI, ed. Gareth Hunt & H. E. Payne (San Francisco: ASP), 426

Roeser, S., Bastian, U. 1988, Astron. Astrophys. Suppl. Ser., 74, 449

Roeser, S., Bastian, U. 1993, Bull. Inform. CDS, 42, 11

Roeser, S., Bastian, U., Kuzmin, A. 1993, Astron. Astrophys. Suppl. Ser., 105, 301

Astronomical Data Analysis Software and Systems VII
ASP Conference Series, Vol. 145, 1998
R. Albrecht, R. N. Hook and H. A. Bushouse, eds.

The ISO Post-Mission Archive

R.D. Saxton, C. Arviset, J. Dowson, R. Carr, C. Todd, M.F. Kessler, J-L. Hernandez, R.N. Jenkins, P. Osuna and A. Plug

ESA, ISO Science Operations Centre, Villafranca del Castillo, Apartado 50727, 28080 Madrid, Spain, Email: rsaxton@iso.vilspa.esa.es

Abstract.
ISO (the Infrared Space Observatory) was launched in November 1995 and is now expected to have an operational lifetime of 2.5 years, as compared to the design requirement of 18 months. It is performing astronomical observations at wavelengths from 2.4-240 microns with four instruments; a camera (2.5-18 microns), a photometer (2.5-240 microns), a short-wavelength spectrometer (2.4-45.2 microns) and a long-wavelength spectrometer (43-197 microns). By the end of the mission it will have made in excess of 25,000 high quality Infrared observations which will leave an important astronomical archive.

The ISO post-mission archive (PMA) is being designed to fully exploit this legacy by providing access to the data across the Internet. It will provide an on-line service to the data, supporting documentation and software through a WWW interface which has the goals of being friendly for the novice user, flexible for the expert and rapid for everybody.

1. Introduction

ISO will make its final measurement on April 10 1998 (±2.5 weeks) when the liquid Helium coolant is expected to run out. All of the data will then be bulk reprocessed with the latest version of the ISO processing pipeline to produce an INTERIM archive; which will be stored on CD in a set of jukeboxes to give easy near-line access to the entire dataset.

The ISO PMA[1] will be hosted at the ESA site, VILSPA, in Villafranca del Castillo, Spain and will become available to the public three months after the end of the mission (i.e., it is expected to go live in July 1998).

The processing pipeline will continue to be regularly updated in the post-mission phase as instrument knowledge and techniques improve. An On-The-Fly-Reprocessing (OFRP) option will be made available to archive users to allow them to retrieve observations processed by the latest software and using the latest calibrations.

Three and a half years after Helium-loss the pipeline will be frozen and all observations reprocessed to form a LEGACY archive. Data will become public

[1] See http://isowww.estec.esa.nl/science/pub/isopma/isopma.html

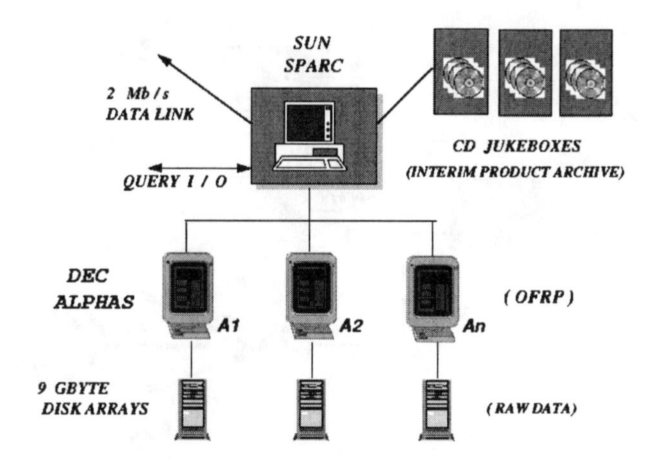

Figure 1. Hardware configuration

one year after they have been shipped to the observer in fully-calibrated form, which means that all data are likely to be public by April 1999. Until this point is reached proprietary data rights will be respected.

2. High-Level Design

The INTERIM product archive will be stored in a near-line system on ~ 700 CDs spread between three jukeboxes. Each jukebox will contain four readers and the retrieval software will optimise the performance by scheduling CDs to be read in parallel.

The archive and associated *Sybase* database will be hosted on a Sun Sparcstation connected by a 2Mb/s line to REDIRIS, which has a fast link to Europe through the *Ten 34* capital city network and is linked to the USA by 2 further 2Mb/s lines.

The user interface will be HTML and Java based and will issue SQL queries to the database via *JCONNECT*. The basic interface will allow observations to be selected based on astronomical considerations such as coordinates, source category, exposure time, wavelength etc. An 'expert' interface will also be provided to allow calibration scientists to query the database using engineering, trend and housekeeping parameters.

OFRP will be performed on a cluster of Alpha machines running OpenVMS. The raw data will be stored locally on an array of 9 Gbyte disks and the processing load will be shared between the machines and between disks so that a configurable number of observations (initially 10) can be processed simultaneously.

Files will be returned in FITS format, optionally compressed, into a transfer directory. For large datasets, or when requested, data will be sent on CD to a

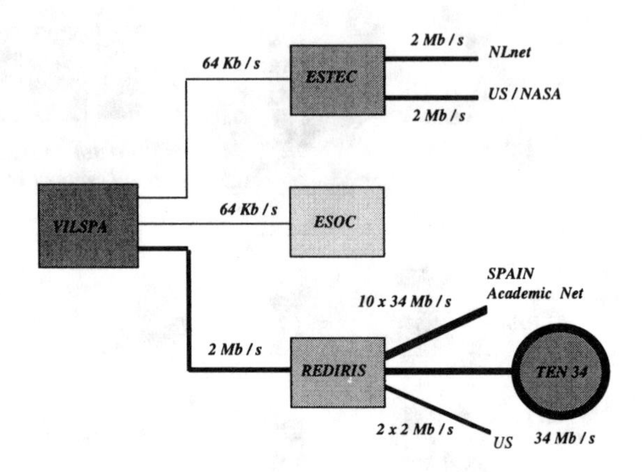

Figure 2. Network access

mail address. In this case data will be copied into a CD holding area and a batch of CDs written once a week. To improve throughput a cache area of 30Gbytes ($\sim$ 10% of the archive) will be maintained on magnetic disk to avoid retrieving from CD, or reprocessing, popular observations many times.

3. User Interface

This will be a suite of HTML pages and Java applets providing on-line access to the archive. The observation catalogue searching and product request functionality will be controlled by a single applet, working in a *modulable load* configuration whereby parts of the applet are downloaded only when necessary.

3.1. Browse form

- allows the user to select observations based upon a wide range of astronomical and engineering parameters
- a Java applet will run on the client machine to translate the submitted request into SQL and send it to the database via *JCONNECT*
- the observation list will be displayed in a *Results form* which will include a link to a viewable postage stamp product
- each observation may be moved into a SHOPPING BASKET

3.2. Shopping basket form

For each observation the user specifies:

- Archived product or OFRP

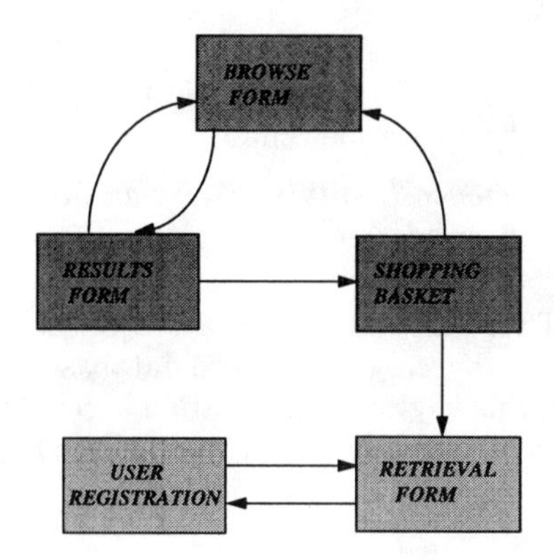

Figure 3. Schematic search/retrieve transition diagram

- Level of products required (Raw data, basic science data etc)

When the shopping basket is complete the data may be retrieved

3.3. Retrieval form

- summarises request and displays estimated retrieval time
- asks for transfer medium (FTP or CD)

Before the data is retrieved the user is asked to register

3.4. Registration form

First time users are asked to provide a Username,Password,E-mail address, Mail address and Preferred transfer medium. Observers who own observations will be sent a username and password by E-mail at the start of the archive phase.

Astronomical Data Analysis Software and Systems VII
ASP Conference Series, Vol. 145, 1998
R. Albrecht, R. N. Hook and H. A. Bushouse, eds.

HST Keyword Dictionary

Daryl A. Swade [1], Lisa Gardner, Ed Hopkins, Tim Kimball, Karen Lezon, Jim Rose [1] and Bernie Shiao

Space Telescope Science Institute, 3700 San Martin Drive, Baltimore, MD 21218, Email: swade@stsci.edu

Abstract. STScI has undertaken a project to place all HST keyword information in one source, the keyword database, and to provide a mechanism for making this keyword information accessible to all HST users, the keyword dictionary, which is a WWW interface to the keyword database.

1. Overview

The HST OPUS[2] data processing pipeline receives the telemetry stream from the spacecraft and formats the data as either GEIS or standard FITS files. In both cases header keywords are used to characterize the data. Along with forming the headers for the science data, most keyword values are ingested into HST archive catalog fields where they are accessible for archive searches and data characterization. One goal of this project is to strive for unity in keyword definitions across all HST science instrument data. In order to establish standards for HST keywords and to provide a central location for keyword information, STScI has undertaken a project to place all keyword information in one source - the keyword database.

In addition, a Web-based interface to the keyword database, the keyword dictionary[3], has been developed as a mechanism for making this keyword information accessible to all HST users.

2. Keyword Flow

The keyword database provides the capability to trace the flow of a keyword value from source, through headers, to the archive catalog.

Keyword values for the HST science data are derived from three main sources: the Proposal Management Database (PMDB), spacecraft telemetry as defined in the Project Database (PDB), and calculations performed during data processing. The PMDB contains information derived from the submitted proposals and their scheduling process. The PDB contains mnemonic definitions

---

[1] Computer Sciences Corporation

[2] http://www.stsci.edu/software/OPUS/

[3] http://archive.stsci.edu/keyword/

for downlinked telemetry values and aperture positions. A table in the keyword database links a keyword to the source of its value.

Headers are constructed and the keyword values are populated in the OPUS (science) data processing pipeline and the OMS (engineering and FGS) data processing pipeline. All science data keywords are inserted into the headers during a processing step called Generic Conversion. Some of the keyword values may be left blank at this stage and populated in subsequent pipeline steps. For example, the OPUS pipeline includes the instrument specific STSDAS calibration step that will update some keyword values.

Along with forming the headers for the data, most keyword values are cataloged into HST archive database fields where they are accessible for archive searches and data characterization. Tables in the keyword database associate the header keywords with the archive catalog field into which they are ingested.

3. Keyword Database

The keyword database is a relational database that contains all relevant information about a keyword. It is the definitive source of keyword information for HST data processing and archiving software systems. This information includes the keyword source, datatype, default value, possible values, units, header comment, and textual long description with detailed information. The keyword database also contains all the necessary information on how to build headers from the individual keywords, which archive catalog fields are populated by each keyword, and information necessary to build DDL for StarView screens. A detailed description of the keyword database schema is available through the keyword dictionary Web site.

4. Information Contained in the Keyword Database

The following keyword related information is contained within the keyword database fields.

- Keyword name - name of the keyword following FITS standards
- HST instrument - the instrument onboard HST to which the keyword applies (the same keyword can be used for multiple instruments)
- Datatype - datatype of the keyword value: character, I2, I4, R4, R8, or L1
- Units - the units of the keyword value
- Default value - a default value for each keyword may be specified
- Possible values - a list of all possible values for a given keyword. This may be an enumerated list in which case all values are stored in the database, or a range in which case the minimum and maximum values are stored.
- Source - the source of the keyword value (PMDB, PDB, or calculated; see above)

- Long description - an arbitrarily long text field which describes any relevant information about a keyword or how its value is derived
- Header structure information - all the information necessary to construct each HST data header including which keywords to include and their order
- Optional flag - a flag to identify a keyword as optional. Optional keywords will not appear in the header if their value is blank.
- Short comment - a short (<48 character) description of the keyword for inclusion in the headers
- Archive catalog table - the archive catalog table to which the keyword is ingested
- Archive catalog field - the field in the archive catalog table to which the keyword is ingested
- StarView DDL - information necessary to construct StarView screens

5. Keyword Database Products

The following products are currently generated from the keyword database.

- ICD-19: The HST project document that defines the headers for HST science data files. This document now resides on-line within the keyword dictionary.
- OPUS load files: The OPUS data processing system uses database load files generated by the keyword database to populate tables in the operational database which define the keywords contained in the data headers, the order those keywords appear in the headers, and the source of the keyword values.
- OMS keyword file: In a manner similar to OPUS, the OMS engineering data processing system uses the keyword database to define the contents and order of keywords in observation logs.
- Archive catalog verification file: The archive catalog is verified to insure all keywords ingested into the catalog correspond to the keyword value in the header. This file contains archive catalog keyword/fieldname mapping.
- StarView DDL files: StarView is the interface to the HST archive catalog. DDL files provide the database attributes defined on StarView screens.

6. Keyword Database Configuration Process

A formal configuration management procedure has been established to control modifications to the database. There are three versions of the keyword database active at any one time. Development of new keywords and headers is performed in keyworddev. After development the new keywords enter a test phase and the contents of keyworddev are copied to keywordtst. New development can continue in keyworddev as the previously developed keywords are tested. After successfully completing the test phase keywordtst is copied into the operational keyword database. In addition to the three versions of the keyword database, the latest version of the OPUS load files derived from the keyword database are configured in a CMS software archive.

7. Keyword Update Process

STScI has set up a Keyword Review Team (KRT) to consider all proposed changes to HST headers. Proposals to change a keyword or any keyword attribute are submitted to the KRT coordinator. Proposals have originated from a number of sources such as science instrument teams, software developers, system engineers, user support personnel, etc. Once the proposed changes are clearly documented, the proposal is forwarded to the rest of the KRT for consideration and evaluation of any impacts. When approved the change is implemented in the development version of the keyword database, and the revised headers enter into the formal software test cycle.

During development of headers for new science instruments placed aboard HST during servicing missions, the process is streamlined in that entire headers are considered by the KRT instead of individual keywords.

8. Keyword Dictionary

The HST Keyword Dictionary Web Interface is written in Perl using the Sybperl module. For Javascript-enabled browsers, a small window can be popped up that allows quick searches "on the side". Frames were used where possible in the initial design, but they encumber the interface (and the CGI script) somewhat; the next generation will design them out.

An administration tool was written in Tcl/Tk. (At the time, Web security was judged to be insufficient for an administrative interface.) The keyword database administrator can use this tool to load new or edit existing database records.

9. Keyword Dictionary Capabilities

The HST keyword dictionary allows for access to keyword information through a number of different approaches. It is possible to search for a specific keyword, browse all the keywords in alphabetical order for a specific HST instrument, or generate schematic headers for each science file generated from HST data. All information about each keyword contained within the keyword database is accessible through any of these routes. It is also possible to look in the development, test, or operational versions of the keyword database.

Since the keyword dictionary accesses the same database that provides the information and instructions on how to construct the headers in the data processing pipeline, this documentation can never be out of date with respect to current data processing!

Part 8. Astrostatistics and Databases

Astronomical Data Analysis Software and Systems VII
ASP Conference Series, Vol. 145, 1998
R. Albrecht, R. N. Hook and H. A. Bushouse, eds.

Noise Detection and Filtering using Multiresolution Transform Methods

Fionn Murtagh (1,2) and Jean-Luc Starck (3)

(1) Faculty of Informatics, University of Ulster, BT48 7JL Derry, Northern Ireland (e-mail fd.murtagh@ulst.ac.uk)
(2) Observatoire Astronomique, 11 rue de l'Université, 67000 Strasbourg Cedex, France (e-mail fmurtagh@astro.u-strasbg.fr)
(3) CEA, DSM/DAPNIA, CE-Saclay, 91191 Gif-sur-Yvette Cedex, France (e-mail jstarck@cea.fr)

Abstract. A new and powerful methodology for treating noise and clutter in astronomical images is described. This is based on the use of the redundant à trous wavelet transform, and the multiresolution support data structure. Variance stabilization and noise modeling are discussed. A number of the examples used in the presentation are shown.

1. Introduction

Noise is of fundamental importance in astronomy. In practice we must model it. In this article we present a powerful framework for doing this.

Information on the detector's noise properties may be available (Snyder et al. 1993, Tekalp and Pavlović 1991). In the case of CCD detectors, or digitized photographic images, additive Gaussian and/or Poisson distributions often provide the most appropriate model.

If the noise model is unknown, e.g., due to calibration and preprocessing of the image data being difficult to model, or having images of unknown detector provenance, or in a case where we simply need to check up on our preconceived ideas about the data we are handling, then we need to estimate noise. An algorithm for automated noise estimation is discussed in Starck and Murtagh (1997).

2. Multiresolution Support

This describes in a logical or boolean way if an image I contains information at a given scale j and at a given position (x, y). If $M^{(I)}(j, x, y) = 1$ (or $=$ *true*), then I contains information at scale j and at the position (x, y).

M depends on several parameters:

- The input image.
- The algorithm used for the multiresolution decomposition.
- The noise.

- All constraints which we want the support additionally to satisfy.

The multiresolution support of an image is computed in several steps:

- Step 1 is to compute the wavelet transform of the image.
- Booleanization of each scale leads to the multiresolution support.
- A priori knowledge can be introduced by modifying the support.

The last step depends on the knowledge we have of our images. E.g., if no interesting object smaller or larger than a given size in our image, we can suppress, in the support, anything which is due to that kind of object.

3. Multiresolution Support from the Wavelet Transform

The à trous algorithm is described as follows. A pixel at position x, y can be expressed as the sum all the wavelet coefficients at this position, plus the smoothed array:

$$c_0(x,y) = c_p(x,y) + \sum_{j=1}^{p} w_j(x,y) \tag{1}$$

The multiresolution support is defined by:

$$M(j,x,y) = \begin{cases} 1 & \text{if } w_j(x,y) \text{ is significant} \\ 0 & \text{if } w_j(x,y) \text{ is not significant} \end{cases} \tag{2}$$

Given stationary Gaussian noise,

$$\begin{array}{ll} \text{if } \mid w_j \mid \geq k\sigma_j & \text{then } w_j \text{ is s significant} \\ \text{if } \mid w_j \mid < k\sigma_j & \text{then } w_j \text{ is not significant} \end{array} \tag{3}$$

4. Variance Stabilization

We need the noise standard deviation at each scale for iid (independent identically distributed) Gaussian pixel values. In the case of additive Poisson or Poisson plus Gaussian noise, we transform the image so that this is the case. The transform we use is one for variance stabilization.

To compute the standard deviation σ_j^e at each scale, simulate an image containing Gaussian noise with a standard deviation equal to 1, and take the wavelet transform of this image.

The standard deviation of the noise at a scale j of the image, $\sigma_j = \sigma_I \sigma_j^e$, where σ_I is the standard deviation of the noise in the original image.

If the noise in the data I is Poisson, the transform

$$t(I(x,y)) = 2\sqrt{I(x,y) + \frac{3}{8}} \tag{4}$$

acts as if the data arose from a Gaussian white noise model, with $\sigma = 1$, under the assumption that the mean value of I is large.

For combined Gaussian and Poisson noise:

$$t(I) = \frac{2}{\alpha}\sqrt{\alpha I(x,y) + \frac{3}{8}\alpha^2 + \sigma^2 - \alpha g} \qquad (5)$$

where α is the gain, σ and g the standard deviation and the mean of the read-out noise.

5. Algorithm for Noise Filtering

Reconstruction of the image, after setting non-significant wavelet coefficients to zero, at full resolution provides an approach for adaptive filtering. A satisfactory filtering implies that the error image $E = I - \tilde{I}$, obtained as the difference between the original image and the filtered image, contains only noise and no 'structure'. We can easily arrive at this objective by keeping significant wavelet coefficients and by iterating. The following algorithm (Starck et al. 1997) converges quickly to a solution which protects all signal in the image, and assesses noise with great accuracy.

1. $n \leftarrow 0$.
2. Initialize the solution, $I^{(0)}$, to zero.
3. Determine the multiresolution support of the image.
4. Estimate the significance level (e.g., 3-sigma) at each scale.
5. Keep significant wavelet coefficients. Reconstruct the image, $I^{(n)}$, from these significant coefficients.
6. Determine the error, $E^{(n)} = I - I^{(n)}$ (where I is the input image, to be filtered).
7. Determine the multiresolution transform of $E^{(n)}$.
8. Threshold: only retain the coefficients which belong to the support.
9. Reconstruct the thresholded error image. This yields the image $\tilde{E}^{(n)}$ containing the significant residuals of the error image.
10. Add this residual to the solution: $I^{(n)} \leftarrow I^{(n)} + \tilde{E}^{(n)}$.
11. If $\mid (\sigma_{E^{(n-1)}} - \sigma_{E^{(n)}})/\sigma_{E^{(n)}} \mid > \ \epsilon$ then $n \leftarrow n + 1$ and go to step 4.

6. Example 1: Simulation

A simulated image containing stars and galaxies is shown in Figure 1 (top left). The simulated noisy image, the filtered image and the residual image are respectively shown in Figure 1 top right, bottom left, and bottom right. We can see that there is no structure in the residual image. The filtering was carried out using the multiresolution support.

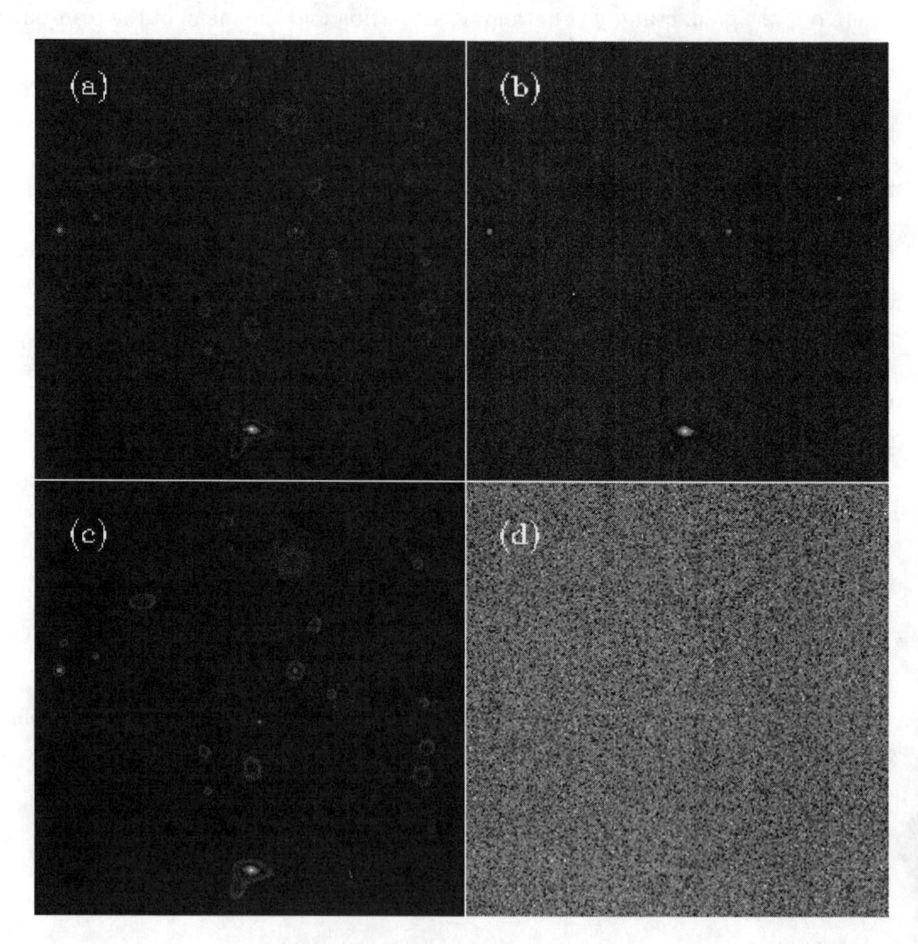

Figure 1. (a) Simulated image, (b) simulated image and Gaussian noise, (c) filtered image, and (d) residual image.

7. Example 2: Spectrum Filtering

Figure 2 shows a noisy spectrum (upper left, repeated lower right). For the astronomer, the spectral lines – here mainly absorption lines extending downwards

– are of interest. The continuum may also be of interest, i.e. the overall spectral tendency.

The spectral lines are unchanged in the filtered version (upper center, and upper right). The lower center (and lower right) version shows the result of applying Daubechies' coefficient 8, a compactly-supported orthonormal wavelet. This was followed by thresholding based on estimated variance of the coefficients, as proposed by Donoho (1990), but not taking into account the image's noise properties as we have done. One sees immediately that a problem- (or image-) driven choice of wavelet and filtering strategy is indispensable.

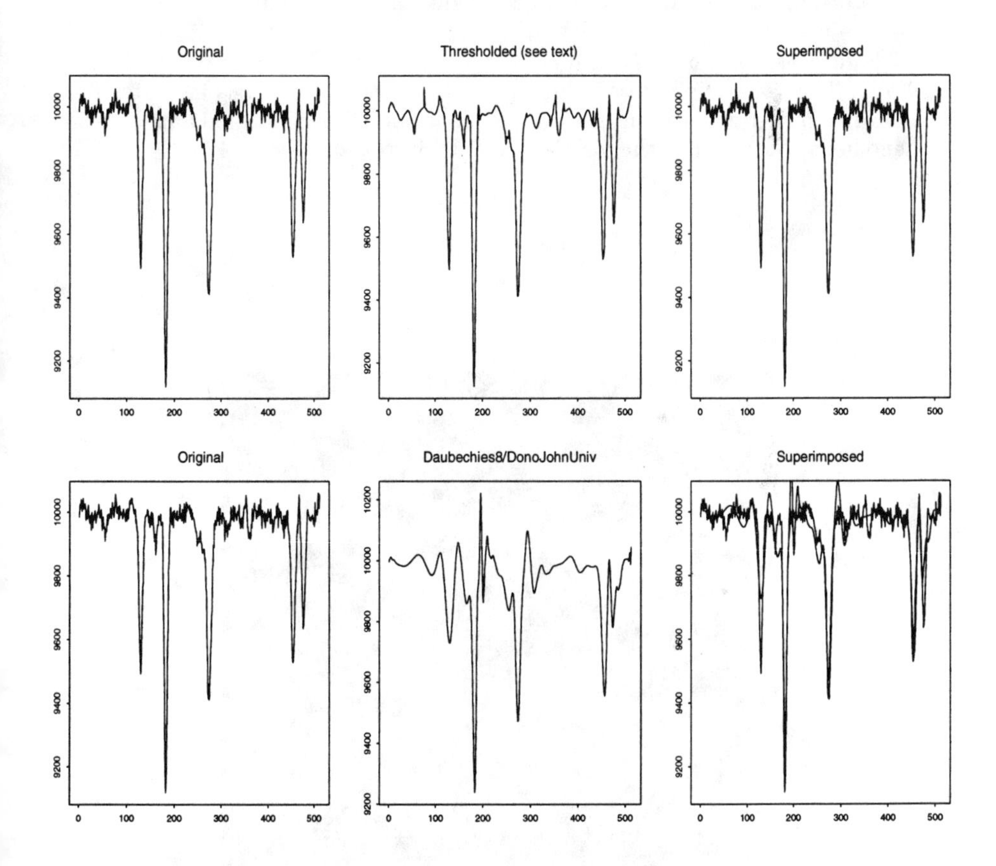

Figure 2. Top row: original noisy spectrum; filtered spectrum using the multiresolution support method; both superimposed. Bottom row: original; filtered (using Daubechies coefficient 8, and Donoho and Johnstone "universal" thresholding); both superimposed.

8. Example 3: X-Ray Image Filtering

The galaxy cluster A2390 is at a redshift of 0.231. Figure 3 shows an image of this cluster, obtained by the ROSAT X-ray spacecraft. The resolution is one arc second per pixel, with a total number of photons equal to 13506 for an integration time of 8.5 hours. The background level is about 0.04 photons per pixel.

It is obvious that this image cannot be used directly, and some treatment must be performed before any analysis. The standard method consists of convolving the image with a Gaussian. Figure 4 shows the result of such processing. (The Gaussian used had full-width at half maximum equal to 5", which is approximately that of the point spread function). The smoothed image shows some structures, but also residual noise, and it is difficult to give any meaning to them.

Figure 5 shows an image filtered by the wavelet transform (Starck and Pierre 1997a). The noise has been eliminated, and the wavelet analysis indicates faint structures in X-ray emission, allowing explanation of gravitational amplification phenomena, observed in the visible domain (Starck and Pierre 1997b).

Figure 3. ROSAT Image of the cluster A2390.

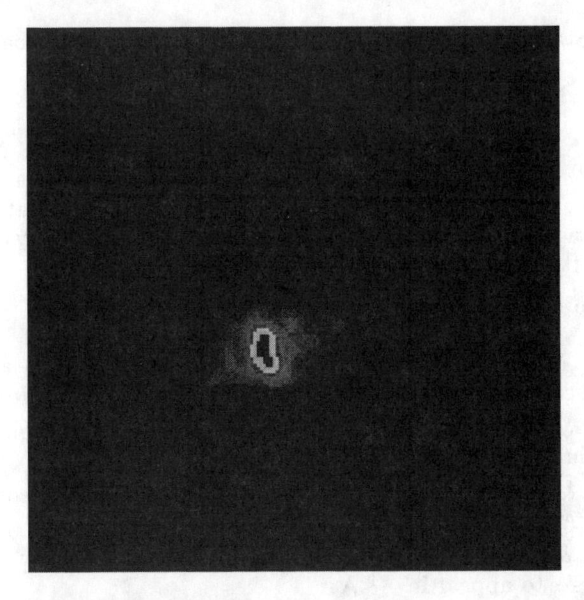

Figure 4. ROSAT image of the cluster A2390 filtered by the standard method (convolution by a Gaussian).

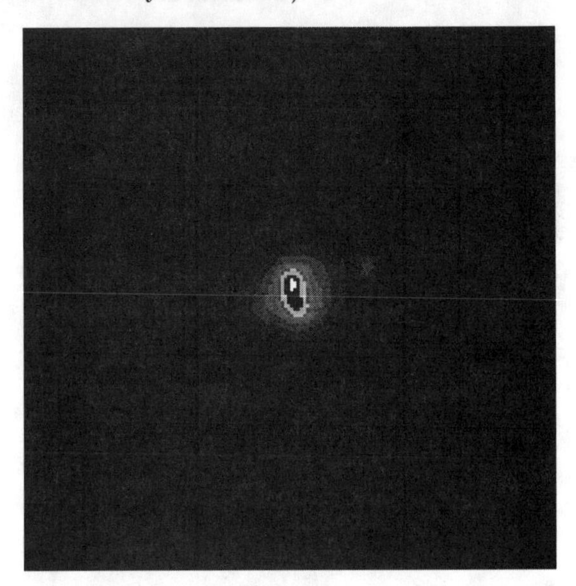

Figure 5. ROSAT image of the cluster A2390 filtered by the method based on wavelet coefficients.

9. Conclusion

The methods described in this article have been applied to many practical problems over the last few years. A comprehensive treatment of these methods can be found in Starck et al. (1998). A large software package will be available from early 1998.

References

Donoho, D.L. & Johnstone, I.M. 1993, "Ideal spatial adaptation by wavelet shrinkage", Stanford University, Technical Report 400 (available by anonymous ftp from playfair.stanford.edu: /pub/donoho)

Snyder, D.L., Hammoud, A.M. and White, R.L. 1993, J. Optical Soc. Am. 10, 1014

Starck, J.-L. Murtagh, F. & Bijaoui, A. 1998, Image and Data Analysis: the Multiscale Approach, Cambridge University Press, in press

Starck, J.-L. & Murtagh, F. 1997, "Automatic noise estimation from the multiresolution support", PASP, in press

Starck, J.-L. & Pierre, M. 1997a, "X-ray structures in galaxy cores", to appear in A&A

Starck, J.-L. & Pierre, M. 1997b, "Structure detection in low intensity X-ray images", to appear in A&A

Tekalp, A.M. and Pavlović, G. 1991, in A.K. Katsaggelos, ed., Digital Image Restoration, Springer-Verlag, New York, 209

Astronomical Data Analysis Software and Systems VII
ASP Conference Series, Vol. 145, 1998
R. Albrecht, R. N. Hook and H. A. Bushouse, eds.

LINNÉ, a Software System for Automatic Classification

N. Christlieb[1] and L. Wisotzki
Hamburger Sternwarte, Gojenbergsweg 112, D-21029 Hamburg, Germany

G. Graßhoff
Institut für Wissenschaftsgeschichte, Georg-August-Universität Göttingen

A. Nelke and A. Schlemminger
Philosophisches Seminar, Universität Hamburg

Abstract. We report on the software system LINNÉ, which has been designed for the development and evaluation of classification models. LINNÉ is used for the exploitation of the Hamburg/ESO survey (HES), an objective prism survey covering the entire southern extragalactic sky.

1. Introduction

The Hamburg/ESO survey (HES) was originally conceived as a wide angle objective prism survey for bright quasars. It is carried out with the ESO Schmidt telescope and its 4° prism and covers the total southern extragalactic sky. For a description of the survey see Wisotzki et al. (1996).

A few years ago we started to develop methods for the systematic exploitation of the *stellar* content of the survey by means of automatic spectral classification. A short overview of the scientific objectives is given in Christlieb et al. (1997) and Christlieb et al. (1998), where also a detailed description of the classification techniques can be found. In this paper we report on the software system LINNÉ, that has been designed for the development and evaluation of classification models.

2. Classification models

A *classification model* (CM) consists of the following components:

Class definitions are given by means of a learning sample, implicitly including the number N and names of the defined classes.

[1]E-mail: nchristlieb@hs.uni-hamburg.de

Class parameters include the *a priori* probabilities $p(\omega_i)$, $i = 1 \dots N$, of the N defined classes and the parameters of the multivariate normal distribution of the class-conditional probabilities $p(\vec{x}|\omega_i)$.

Classification aim can be one of the following items:

(1) Perform a "simple", i. e. Bayes-rule classification.

(2) Compile a complete sample of class $\omega_{\text{target}} \in \{\omega_1, \dots, \omega_N\}$ with minimum cost rule classification.

(3) Detect "un-classifiable" spectra, i. e. spectra to which the reject option (Christlieb et al. 1998) applies. Note that e. g. quasar spectra belong to this class.

Feature space The space of features in which the search for the optimal subset is carried out. Note that in certain cases one may want to exclude available features beforehand to avoid biases, so that the feature space is not necessarily identical to the *total* set of available features.

Optimal feature set for the given classification aim.

Optimal loss factors In case of classification aim (3) a set of three optimal loss factors – weights for different kinds of misclassifications – has to be stated (Christlieb et al. 1998).

Once a CM is established, it is straightforward to derive from it a *classification rule* for the assignment of objects of unknown classes to one of the defined classes.

The aim of LINNÉ is to permit easy and well controlled access to the variation of the model components and effective means to evaluate the resulting quality of classification. The performance of a model with classification aim (1) can be evaluated by e. g. the total number of misclassifications, estimated with the *leaving-one-out* method (Hand 1981); in case of aim (3) the model is usually assessed by the number of misclassifications between the target class and the other classes.

3. Description of the system

The core of LINNÉ was implemented in an object-oriented extension to Prolog, with the numerical routines – e. g. for estimation of the parameters of the multivariate normal distributions – written in C. To facilitate user interaction and to ensure effective control over the model components and performance, a graphical user interface (GUI) for LINNÉ was developed (see Figure 1). After the first implementation, using Prolog's own graphical library (SWI-Prolog plus XPCE), we recently started to switch to Java for reasons of system independence and remote access via WWW. At present, LINNÉ has a server-client architecture, the Prolog server communicating with a Java client through TCP/IP sockets. The server keeps the learning sample data, read in from MIDAS via an interface and converted into Prolog readable terms. It is not yet possible to select *all* model components interactively via the GUI, so that partly pre-designed models have to be used. They are also provided from the server side.

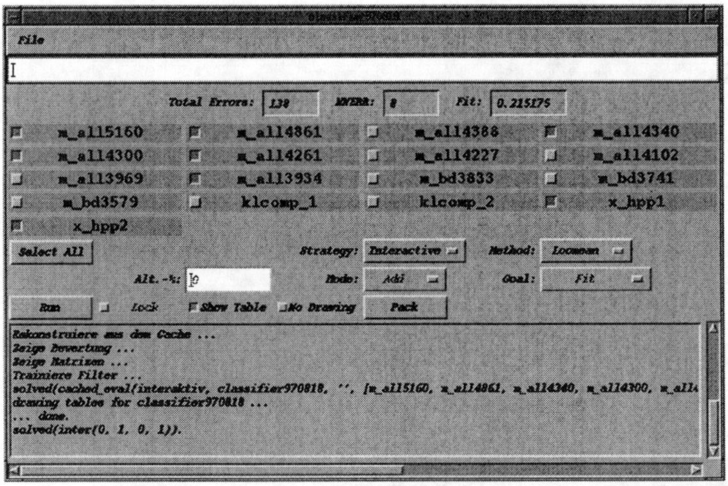

Figure 1. Main control panel of LINNÉ. The three upper text fields show model performance parameters. Below them the feature selection area is placed (`m_al15160...x_hpp2`). The automatic feature selection is controlled by the menus above the Prolog server messages window.

The results of the classification model evaluation are presented on the client. A confusion matrix and loss matrix window assists the user in the analysis of the model. The user may then alter components and repeat the evaluation to improve the model step by step.

The search for the optimal feature set can also be done automatically. Since the set of available features may easily become too large to perform exhaustive search among all possible combinations, apart from the exhaustive search a hill-climbing like, stepwise search has been implemented. It can be controlled from the client side, using different strategies and branching parameters.

LINNÉ also provides a tool for the systematic and efficient adjustment of loss factors (see Figure 2).

4. Application of classification models

Once a CM has been established, evaluated, and the evaluation has pleased the user, its parameters can be exported to MIDAS tables. The classification of spectra of unknown classes can then be carried out under MIDAS. The typical computing time for the classification of all spectra on one HES plate – mapping $5^\circ \times 5^\circ$ of the sky and yielding typically $\sim 10,000$ non-disturbed spectra with $S/N > 10$ – is less than 5 min on a Linux PC with a Pentium 133 MHz processor.

So far LINNÉ has been used for some first test applications, i. e. compilation of a sample of extremely metal poor halo stars and a search for FHB/A stars. It

Errors Target -> Other

	0	0.1	0.2	0.3	0.4	0.5	0.6	0.7	0.8	0.9	1
0	93	0	0	0	0	0	0	0	0	0	0
0.1	106	4	3	3	3	3	3	2	2	2	2
0.2	106	6	4	4	3	3	3	3	3	3	3
0.3	106	9	6	4	4	4	3	3	3	3	3
0.4	106	9	6	5	4	4	4	4	3	3	3
0.5	106	9	6	6	5	4	4	4	4	4	3
0.6	106	9	9	6	6	4	4	4	4	4	4
0.7	106	9	9	6	6	5	4	4	4	4	4
0.8	106	10	9	8	6	6	5	4	4	4	4
0.9	106	10	9	9	6	6	6	5	4	4	4
1	106	11	9	9	6	6	6	5	5	4	4

t->o
o->t

Errors Other -> Target

	0	0.1	0.2	0.3	0.4	0.5	0.6	0.7	0.8	0.9	1
0	548	548	548	548	548	548	548	548	548	548	548
0.1	0	4	7	9	12	14	16	18	18	18	20
0.2	0	3	4	5	7	9	9	11	12	13	14
0.3	0	2	4	4	5	5	7	9	9	9	11
0.4	0	2	3	4	4	5	5	5	7	8	9
0.5	0	2	3	4	4	4	5	5	5	6	7
0.6	0	2	2	3	4	4	4	5	5	5	5
0.7	0	2	2	3	4	4	4	4	5	5	5
0.8	0	2	2	2	3	4	4	4	4	4	5
0.9	0	2	2	2	3	4	4	4	4	4	4
1	0	2	2	2	3	3	4	4	4	4	4

t->o
o->t

Run New OK

Figure 2. Tool for interactive adjustment of loss factors. The upper half of the window shows the number of target class spectra which have been erroneously assigned to one of the other classes in dependence of the loss factors $c_{\mathrm{target}\to\omega_i}$ (abscissa) and $c_{\omega_i\to\mathrm{target}}$ (ordinate). The lower half shows the same for the target class contamination. The third loss factor, $c_{\omega_i\to\omega_j}$, does not have to be adjusted but can be held constant at a small value.

will be developed further and extended in functionality and will be applied to the exploitation of the huge HES data base, which will finally consist of $\sim 5,000,000$ digitised objective prism spectra.

Acknowledgments. N.C. acknowledges an accommodation grant by the conference organizers. This work was supported by the Deutsche Forschungsgemeinschaft under grants Re 353/40-1 and Gr 968/3-1.

References

Christlieb, N. et al. 1997, in Wide-Field Spectroscopy, ed. Kontizas, E. et al., Kluwer, Dordrecht, 109

Christlieb, N. et al. 1998, to appear in Data Highways and Information Flooding, a Challenge for Classification and Data Analysis, ed. Balderjahn, I. et al., Springer, Berlin.

Hand, D. 1981, Discrimination and Classification, Wiley & Sons, New York.

Wisotzki, L. et al. 1996, A&AS, 115, 227

Astronomical Data Analysis Software and Systems VII
ASP Conference Series, Vol. 145, 1998
R. Albrecht, R. N. Hook and H. A. Bushouse, eds.

Information Mining in Astronomical Literature with Tetralogie

D. Egret

CDS, Strasbourg, France, E-mail: egret@astro.u-strasbg.fr

J. Mothe[1], T. Dkaki[2] and B. Dousset

Research Institute in Computer Sciences, IRIT, SIG, 31062 Toulouse Cedex, France, E-mail: {mothe/dkaki/dousset}@irit.fr

Abstract. A tool derived for the technological watch (Tetralogie[3]) is applied to a dataset of three consecutive years of article abstracts published in the European Journal *Astronomy and Astrophysics*. This tool is based on a two step approach: first, a pretreatment step that extracts elementary items from the raw information (keywords, authors, year of publication, etc.); second, a mining step that analyses the extracted information using statistical methods. It is shown that this approach allows one to qualify and visualize some major trends characterizing the current astronomical literature: multi-author collaborative work, the impact of observational projects and thematic maps of publishing authors.

1. Introduction

Electronic publication has become an essential aspect of the distribution of astronomical results (see Heck 1997). The users who want to exploit this information need efficient information retrieval systems in order to retrieve relevant raw information or to extract hidden information and synthesize thematic trends. The tools providing the former functionalities are based on query and document matching (Salton et al. 1983; Frakes et al. 1992; Eichhorn et al. 1997). The latter functionalities (called data mining functionalities) result from data analysis, data evolution analysis and data correlation principles and allow one to discover a priori unknown knowledge or information (Shapiro et al. 1996). We focus on these latter functionalities.

A knowledge discovery process can be broken down into two steps: first, the data or information selection and pre-treatment; second, the mining of these pieces of information in order to extract hidden information. The main objectives of the mining are to achieve classification (i.e., finding a partition of the

[1] Institut Universitaire de Formation des Maîtres de Toulouse

[2] IUT Strasbourg-Sud, Université Robert Schuman, France

[3] http://atlas.irit.fr

data, using a rule deduced from the data characteristics), association (one tries to find data correlations) and sequences (the objective is to identify and to find the temporal relationships between the data). The information resulting from an analysis have then to be presented to the user in the most synthetic and expressive way, including graphical representation.

Tetralogie[4] is an information mining tool that has been developed at the Institut de Recherche en Informatique de Toulouse (IRIT). It is used for science and technology monitoring (Dousset et al. 1995; Chrisment et al. 1997) from document collections. In this paper it is not possible to present in detail all the system functionalities. We will rather focus on some key features and on an example of the results that can be obtained from astronomical records.

2. Information Mining using "Tétralogie"

The information mining process is widely based on statistical methods and more precisely on data analysis methods. First, the raw information have to be selected and pre-treated in order to extract the relevant elements of information and to store them in an appropriate form.

2.1. Information Harvesting and Pre-treatment

This step includes the selection of relevant raw information according to the user's needs. It is generally achieved by querying a specific database or a set of databases. Once the raw information has been selected, the next phase is to extract the relevant pieces of information : e.g., authors, year of publication, affiliations, keywords or main topics of the paper, etc. This is achieved using the "rewriting rule principle". In addition, the feature values can be filtered (e.g., publications written by authors from selected countries) and semantically treated (dictionaries are used to solve synonymy problems). These relevant feature values are stored in contingency and disjunctive tables.

Different kinds of crossing tables can be performed according to the kind of information one wants to discover, for example:

Kind of crossing	Expected discovering
(authors name, authors name)	Multi-author collaborative work
(authors name, document topics)	Thematic map of publishing authors
(document topics, authors affiliation)	Geographic map of the topics

2.2. Data Analysis and Graphical Representation

The preprocessed data from the previous step are directly usable by the implemented mining methods. These methods are founded on statistical fundamentals (see e.g., Benzecri 1973; Murtagh & Heck 1989) and their aim is either to represent the pre-treated information in a reduced space, or to classify them.

The different mining functions used are described in Chrisment et al. (1997). They include: Principal component analysis (PCA), Correspondence Factorial

[4]This project is supported by the French defense ministry and the Conseil Régional de la Haute Garonne.

Analysis (CFA), Hierarchical Ascendant Classification, Classification by Partition and Procustean Analysis.

The Result Visualization The information mining result is a set of points in a reduced space. Tools are proposed for displaying this information in a four dimensional space, and for changing the visualized space, or the point of view (zoom, scanning of the set of points).

The User Role In addition to be a real actor in the information harvesting phase, the user has to intervene in the mining process itself: elimination of some irrelevant data or already studied data, selection of a data subset to analyze it deeper, choice of a mining function, and so on.

3. Application to a Dataset from the Astronomical Literature

3.1. The Information Collection

The information collection used for the analysis was composed of about 3600 abstracts published in the European Journal *Astronomy and Astrophysics* (years 1994 to 96). Note that this dataset is therefore mainly representative of European contributions to astronomy, in the few recent years.

In that abstract sample, it is possible to extract about 1600 different authors, and 200 can be selected as the most prolific. Topics of the documents can be extracted either from the title, from the keywords or from the abstract field. Titles have been considered as too short to be really interesting for the study. In addition, the use of keywords was considered too restrictive, as they belong to a controlled set. Indeed, we preferred to automatically extract the different topics from the words or series of words contained in the abstracts.

3.2. Study of the Collaborative Work

The details concerning the collaborative works can be discovered using (author name / author name) crossing and analyzing it.

The first crossing was done using all the authors. A first view is obtained by sorting the author correlations in order to find the strong connexities. The resulting connexity table shows strongly related groups (about 15 groups appear on the diagonal). They are almost all weakly linked via at least one common author (see the several points above and below the diagonal line, linking the blocks). The isolated groups appear on the bottom right corner. This strong connexity is typical of a scientific domain including large international projects and strong cooperative links.

One can go further and study in depth one of these collaborative groups: a CFA of the (main author / author) crossing shows, for instance, some features of the collaborative work around the Hipparcos project in the years 1994-96 (i.e., before the publication of the final catalogues) as can be viewed by grouping together authors having papers co-authored with M. Perryman (Hipparcos project scientist) and L. Lindegren (leader of one of the scientific consortia). The system allows the extraction of 25 main authors (with more than two publications, and at least one with one of the selected central authors) and the cross-referencing of them with all possible co-authors.

3.3. Thematic Maps of Publishing Authors

The details concerning the thematic map of publishing authors can be discovered using (author name / topic) crossing and analyzing it. The significant words in the abstracts have been automatically extracted during the first stage (see 2.1) and they are crossed with the main authors. That kind of crossing allows one to discover the main topics related to one or several authors ; it can also show what are the keywords that link several authors or that are shared by several authors.

For instance, crossing main authors of the 'Hipparcos' collaboration with topical key words, allowed us to discover the main keywords of the 'peripheral' authors (those who bring specific outside collaborations).

4. Conclusion

In this paper, we have tried to show the usefulness of the TETRALOGIE system for discovering trends in the astronomical literature. We focused on several functionalities of this tool that allow one to find some hidden information such as the teams (through the multi-author collaborative work) or the topical maps of publishing authors. This tool graphically displays the discovered relationships that may exist among the extracted information.

Schulman et al. 1997, using classical statistical approaches, have extracted significant features from an analysis of subsequent years of astronomy literature. In a forthcoming study, we will show how the Tetralogie system can also be used to discover thematic evolutions in the literature over several years.

The Web version[5] contains additional figures for illustration.

References

Benzecri, J.P. 1973, L'analyse de données, Tome 1 et 2, Dunod Edition

Chrisment, C., Dkaki, T., Dousset, & B., Mothe, J. 1997, ISI vol. 5, 3, 367 (ISSN 1247-0317)

Dousset, B., Rommens, M., & Sibue, D. 1995, Symposium International, Omega-3, Lipoprotéines et atherosclerose

Eichhorn, G., et al. 1997, in ASP Conf. Ser., Vol. 125, Astronomical Data Analysis Software and Systems VI, ed. Gareth Hunt & H. E. Payne (San Francisco: ASP), 569

Frakes et al. 1992, Information retrieval, Algorithms and structure (ISBN 0-13-463837-9)

Heck, A. 1997, "Electronic Publishing for Physics and Astronomy", Astrophys. Space Science 247, Kluwer, Dordrecht (ISBN 0-7923-4820-6)

Murtagh, F., & Heck, A. 1989, Knowledge-based systems in astronomy, Lecture Notes in Physics 329, Springer-Verlag, Heidelberg (ISBN 3-540-51044-3)

[5]http://cdsweb.u-strasbg.fr/publi/tetra-1.htx

Salton, G., et al. 1983, Introduction to modern retrieval, McGraw Hill International (ISBN 0-07-66526-5)

Shapiro et al. 1996, Advances in Knowledge discovery and Data Mining, AAAI Press (ISBN 0-262-56097-6)

Schulman, E., et al. 1997, PASP 109, 741

Astronomical Data Analysis Software and Systems VII
ASP Conference Series, Vol. 145, 1998
R. Albrecht, R. N. Hook and H. A. Bushouse, eds.

CDS GLU, a Tool for Managing Heterogeneous Distributed Web Services

P. Fernique, F. Ochsenbein and M. Wenger

Centre de Données astronomiques de Strasbourg, 11 rue de l'université, 67000 STRASBOURG - FRANCE, Email: question@simbad.u-strasbg.fr

Abstract. The Web development has been a very important milestone in unifying the access to remote databases : a unique user interface, a unique network protocol, and a hypertext link mechanism (via URLs - Uniform Resource Links). But if creating URLs is easy, maintaining them is the real challenge for database managers.

In this context, the CDS (Centre de Données astronomiques de Strasbourg) has developed the GLU system (Générateur de Liens Uniformes). This tool allows all managers participating in the system to describe and easily distribute the URL definitions required to access their databases. The GLU system ensures that all GLU members receive these definitions and will receive all future modifications. So, the GLU system is particularly adapted to design cooperative Web services, allowing one to generate links to other services which remains always up-to-date.

1. Introduction

The World Wide Web has allowed significant progress towards astronomical service interoperability : the Web is understood through all the Internet, it is very easy to interface it with databases, and hypertext is a powerful concept to obtain good interoperability between databases.

However for this last functionality, a keystone, the ability to define a unique access key to a given information location, the URN (Uniform Resource Name), is still missing. A IETF Working Group has been working on it for a few years, but there is presently no implementation of such a system (Berners-Lee 1994) Therefore everybody is still using URLs (Uniform Resource Locators) instead of URNs, with their well known drawbacks: at any time, any component of an URL (the hostname, the directory, the resource name and/or the parameter syntax) can be modified without any possibility of informing potential users.

The real challenge is thus to maintain all these URLs. For this, the Centre de Données astronomiques de Strasbourg (CDS[1]) has developed the GLU (Générateur de Liens Uniformes - Uniform Link Generator) which allows one to use *symbolic names* instead of hard-coded URLs in data.

[1]http://cdsweb.u-strasbg.fr

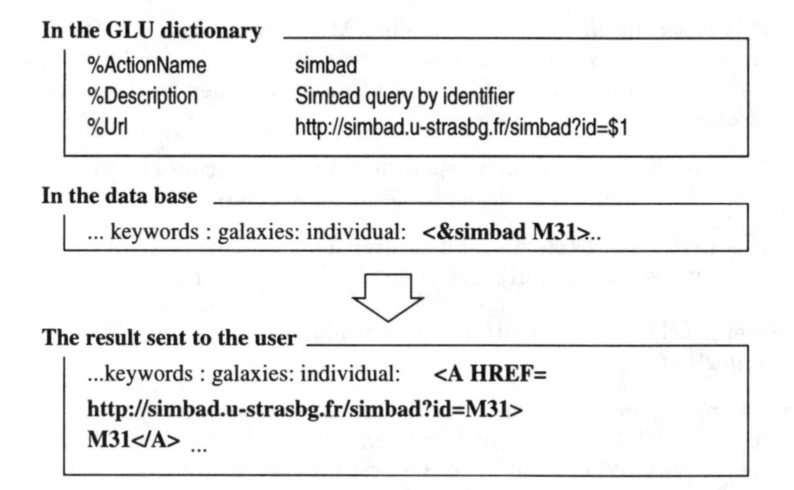

Figure 1. Usage of GLU tag (symbolic name: *simbad*, parameter: *M31*)

2. The GLU System

To fulfill this aim, the GLU implements two concepts:

1. the *GLU dictionary* which is a compilation of symbolic names with their corresponding URLs maintained up-to-date.

2. the *GLU resolver* which replaces symbolic names by the relevant URLs on the fly.

Using the GLU system, data managers can forget the URLs and just use their symbolic names - using *GLU tags* - in all their Web documents. Every time a server sends a database result, the GLU replaces, on the fly, the GLU tags by the corresponding URLs and substitutes the parameters at the proper locations (see the example in Figure 1).

The GLU system differs from other similar tools (such as the PURL system - developed by OCLC Office of Research) in that the GLU allows each of its members to locally resolve the URLs using its own view of the *GLU dictionary.* This is an important choice to ensure the resolution speed and the security of the system. So, for this replacement mechanism, only a simple modification to the Web server configuration is required: specifying that the data streams have to be filtered by the *GLU resolver* during output. For the advanced Web programmer, the GLU has two libraries, one in Perl, the other in C to resolve *GLU tags* directly in CGI (Common Gateway Interface) programs.

An important piece of the GLU system is the mechanism maintaining each view of *GLU dictionary.* This task is performed by a daemon, called *glud,* which has the responsibility to send and to receive the *GLU records* (the entries in the dictionary).

The GLU protocol used by these daemons has the following characteristics:

- It is based on *distribution domains.* Managers choose the *GLU domains* in which they want to distribute their own *GLU records.* In the same way, they choose the *GLU domains* from which they want to receive *GLU records.*

- It uses a hierarchical name space for the *GLU record* identifiers (this ensures their uniqueness through the whole system).

- It is pretty fault tolerant and secure: independent views of the GLU dictionary, mechanism authenticating the update sender, ...

Several other functionalities are also addressed by the GLU system, the most important being:

- The management of clones : in the *GLU dictionary*, it is possible to specify several URLs for a symbolic name. The *GLU daemon* will test each of them regularly and will label the best for the future resolutions (based on connection speed and proximity).

- The capability to use the GLU system as a general macro system : instead of a simple URL, it is possible to specify in the GLU record a full HTML text. In this way, homogeneous HTML pages (for headers, menus, forms,...) are easy to maintain (see the CDS pages).

3. GLU Tools

To take the full benefit of all GLU facilities, three GLU tools have been developed:

1. The **Dictionary Browser**[2]. It is used by the data managers to check the *GLU dictionary* content (see Figure 2).

2. The **Test Sequencer**. Its purpose is to check all the URLs of the GLU dictionary to inform the managers of a disfunctionality.

3. The **Service Browser** (called *AstroGLU*). An "intelligent" service directory using the capability of the GLU to give access to the participating Web services by commonly known data types (see Egret 1997).

4. Conclusion

Presently, two centers use the GLU system : the CDS, for its own Web services, and the CNES, the French Space Agency for the Space Physics database project (CDPP).

The GLU package[3] has been designed for a large cooperation and is opened to other astronomy partners.

---

[2]http://simbad.u-strasbg.fr/demo/cgi-bin/glu_dic.pl

[3]http://simbad.u-strasbg.fr/demo/glu.html

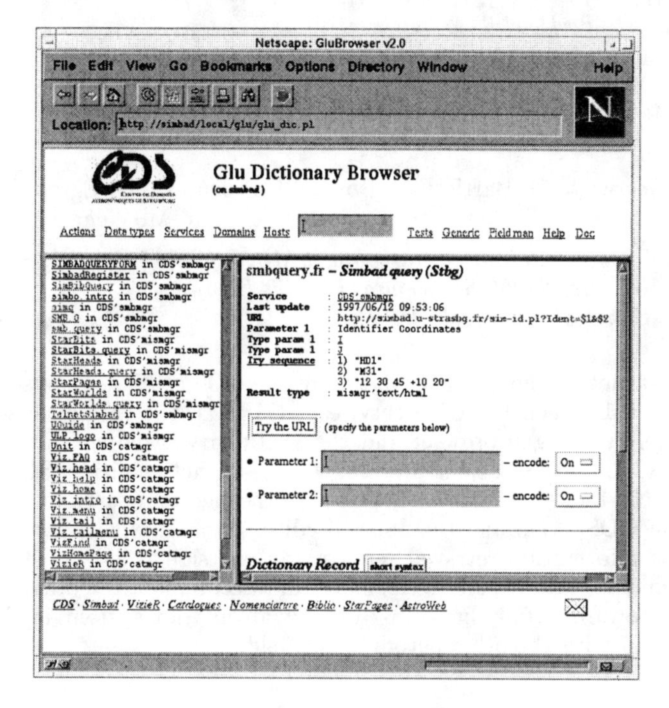

Figure 2. GLU Browser tool

References

Egret, D., et al., 1998, this volume

Wenger, M, 1996 et al., Bull. American Astron. Soc., 189, 602

Berners-Lee, T., 1994, Request For Comments 1630

Astronomical Data Analysis Software and Systems VII
ASP Conference Series, Vol. 145, 1998
R. Albrecht, R. N. Hook and H. A. Bushouse, eds.

The CDS Information Hub

F. Genova, J.G. Bartlett, F. Bonnarel, P. Dubois, D. Egret, P. Fernique, G. Jasniewicz, S. Lesteven, F. Ochsenbein and M. Wenger

CDS, Observatoire de Strasbourg, UMR CNRS 7550, 11 rue de l'Université, 67000 Strasbourg, France, Email: question@astro.u-strasbg.fr

Abstract. The recent evolution of the CDS services is presented. Navigation between the CDS services, SIMBAD, the catalogue service, the VizieR catalogue browser and the Dictionary of nomenclature is developed, as well as graphic and image-based functionalities (SIMBAD, ALADIN). Links are also developed with the electronic astronomical library, from CDS, ADS and the Journal editors, together with innovative 'textual' information retrieval services, on one side, and with data archives from observatories and disciplinary data centers on the other, paving the way towards a fully linked astronomy information system, in close collaboration with the other actors in the field.

The CDS develops information retrieval services which are widely used by the worldwide astronomy community (e.g., Egret et al. 1995; Genova et al. 1996). The two main ones are SIMBAD, the reference database for the identification and bibliography of astronomical objects, which contains nearly 1,550,000 objects, 4,400,000 object names and more than 95,000 references (in November 1997), and the catalogue service, with ftp retrieval of whole catalogues and the VizieR catalogue Browser (Ochsenbein 1997). The CDS WWW service[1] gives access to SIMBAD and the catalogue service/VizieR, and also to other information such as the *Dictionary of Nomenclature of celestial objects*, developed for many years by M.-C. Lortet and her collaborators (Lortet et al. 1994), now fully maintained by CDS in collaboration with the Paris Observatory (DASGAL), the StarPages, maintained by A. Heck (1997), or AstroWeb (Jackson et al. 1994). Moreover, the CDS develops the ALADIN interactive sky atlas, with a dedicated XWindow interface (e.g., Bonnarel et al. 1997).

The CDS services are evolving continuously, taking advantage of new technical possibilities, and taking into account the evolution of astronomy itself, with new domains, new object types, etc, and also new large projects.

From the point of view of the database contents, a particular effort has been made in recent years to improve the multiwavelength coverage of SIMBAD, with for instance the inclusion of the IRAS catalogue and of the IUE log, and the 'cleaning' of the high energy data, from the earlier satellites to Einstein, to be ready for the next generation of X-ray observatories. In parallel, about 90 jour-

[1]http://cdsweb.u-strasbg.fr/CDS.html

nals are regularly scanned, in collaboration with the Institut d'Astrophysique de Paris and the Paris (DASGAL) and Bordeaux Observatories: for instance, more than 70 lists of objects observed by ROSAT published in journals have been included in the database from this bibliography scanning. A large amount of work over several years has also been devoted to the inclusion of new reference stellar catalogues, PPM, CCDM, and the Hipparcos input and final catalogues, in the context of the Hipparcos project, and to prepare the transition to a J2000.0 reference for object positions. In addition, the on–line distribution of the Hipparcos final catalogues was implemented through VizieR, by agreement with ESA, in May 1997.

The evolution towards electronic publication of journals has added an important new function to the catalogue service : it is now the depository of large tables from papers, these tables being very often published in electronic form only and accessible from the data centers (and on CD-ROMs for the *American Astronomical Society* journals). In this context, the CDS builds the electronic tables for *Astronomy and Astrophysics*, as part of the journal publication process, and installs on line the tables published on CD–ROMs by the *American Astronomical Society*, by agreement with the editor. These tables are then shared with the other Data Centers (ADC/GSFC, INASAN/Moscow, NAOJ/Tokyo, the Observatory of Beijing, UICAA/Pune). The key feature for this collaboration between data centers, and with the journal editors, is a common standard description of tables, first proposed by CDS, and now shared with the other data centers and the editors for their electronic tables. This byte–per–byte description of tables, allows an automated check of the quality of the table contents (in addition to verification by the referee), on–line format change (e.g., ASCII to FITS), etc.

From the technical point of view, the major evolution in the last years has been the very rapid development of the WWW, which opens in particular remarkable capabilities for navigation. The CDS has taken advantage of the WWW, first, of course, to develop WWW versions of its services : SIMBAD (1996), the catalogue service (1993), the Dictionary of Nomenclature. VizieR has been directly designed for the WWW, and a detailed description of its recent evolution is given in the companion paper by Ochsenbein (1997).

Navigation between the CDS services, and with external resources, has been implemented. The aim is to complete the evolution from a set of several different CDS services, each with its own contents, user interface and method of access, to an integrated CDS service giving access to the set of CDS information retrieval tools, in a transparent way for the user (keeping however advanced functionalities in dedicated interfaces if necessary).

A few examples:

- from object names in SIMBAD, to the corresponding *Dictionary of nomenclature* entry, which gives information about the origin of the name, access to bibliographic information about the original paper, and to the list in the catalogue service if it is there;

- from bibliography in SIMBAD, to the CDS bibliographic information, with links to the Dictionary of Nomenclature, and/or to the CDS catalogue service, when applicable, to the full electronic paper when available from the journal editor, and also to the ADS services for this reference (e.g., Eichhorn 1997) – reciprocally, ADS implements links to several CDS services, the list of SIMBAD objects for one reference, or the tables in VizieR and in the catalogue service;

- from table to table in the new version of VizieR;

- ...

Another important possibility with the WWW, is to implement relatively easily graphics functionality. For instance, clickable charts are in a prototype phase for SIMBAD, and will be developed in the future for VizieR. On the other hand, a WWW version of ALADIN will progressively be implemented, as a first step to allow the users to get an image of the sky from the other CDS services, SIMBAD, VizieR, and also from the objects cited in texts (e.g., in abstracts), and then with additional functionalities such as SIMBAD or catalogue overlay on the images.

Links to external resources will certainly develop rapidly in the future. On one hand, the first links between the CDS databases and distributed observatory archives have been implemented in September/October 1997 : from SIMBAD to HEASARC, for the objects having a 'high–energy designation', and from VizieR to the FIRST radio database. More links with other data archives will be installed soon. On the other hand, as explained earlier, navigation between the CDS databases and on–line bibliographic resources, the ADS and electronic journals, is already well developed. This has certainly been facilitated by the existence of a *de facto* standard for the description of bibliographic reference, the bibcode, first defined by NED and SIMBAD and now widely used by the ADS (Schmitz et al. 1995). Moreover, the implementation of the European mirror copy of ADS at CDS, thanks to the support of the French Space Agency CNES, together with that of a mirror copy of SIMBAD at CfA, has certainly helped towards better integration of the CDS services and the ADS. In parallel, the European mirror copy of the electronic *Astrophysical Journal* has also been installed at CDS in June 1997. In this domain, innovative services are certainly ahead, for instance new methods to retrieve 'textual' information, such as the Kohonen map method (Lestven et al. 1996), and also more links between the text of articles and databases. For instance, links can be foreseen between object names in the papers, and information about the object in SIMBAD or NED, or images, for instance from the future ALADIN WWW service.

To maintain the set of links between the CDS services, and with external services, a generic tool, the GLU (Fernique 1998), has been developed. This tool is being tested in one of the AstroBrowse prototypes, as discussed in the AstroBrowse BOF during the ADASS'97 meeting.

References

Bonnarel, F., Ziaeepour, H., Bartlett, J. G., Bienaymé, O., Crézé, M., Egret, D., Florsch, J., Genova, F., Ochsenbein, F., Raclot, V., Louys, M., & Paillou,

Ph. 1997, in IAU Symp. 179, New Horizons from Multi-Wavelength Sky Surveys, in press

Egret, D., Crézé, M., Bonnarel, F., Dubois, P., Genova, F., Jasniewicz, G., Heck, A., Lesteven, S., Ochsenbein, F., & Wenger, M. 1995, in Information & On–line Data in Astronomy, ed. D.Egret & M.Albrecht (Kluwer), 163

Genova, F., Bartlett, J. G., Bienaymé, O., Bonnarel, F., Dubois, P., Egret, D., Fernique, P., Jasniewicz, G., Lesteven, S., Monier, R., Ochsenbein, F., & Wenger, M. 1996, Vistas in Astron., 40, 429

Heck, A. 1997, in Electronic Publishing for Physics and Astronomy, ed. A. Heck (Kluwer), 211

Eichhorn, G. 1998, this volume

Fernique, P. 1998, this volume

Jackson, R., Wells, D., Adorf, H. M., Egret, D., Heck, A., Koekemoer, A., & Murtagh, F. 1994, Astron. Astrophys. Suppl., 108, 235

Lesteven, S., Poinçot, P., & Murtagh, F. 1996, Vistas in Astron., 40, 395

Lortet, M.C., Borde, S., & Ochsenbein, F. 1994, Astron. Astrophys. Suppl, 107,193

Ochsenbein, F. 1998, this volume

Schmitz, M., Helou, G., Dubois, P., LaGue, C., Madore, B., Corwin, H.G. Jr, & Lesteven, S. 1995, in Information & On–line Data in Astronomy, ed. D.Egret & M.Albrecht (Kluwer), 259

Astronomical Data Analysis Software and Systems VII
ASP Conference Series, Vol. 145, 1998
R. Albrecht, R. N. Hook and H. A. Bushouse, eds.

Literature and Catalogs in Electronic Form: Questions, Ideas and an Example: the IBVS

A. Holl

Konkoly Observatory, P.O.Box 67, H-1525 Budapest, Hungary, Email: holl@ogyalla.konkoly.hu

Abstract. While transforming astronomical journals and catalogs to electronic form, we should have in sight two questions: making it easier for the human reader to locate and comprehend information. At the same time, some of the text read by humans in the past, will be — or already is — processed by machines, and should be laid down in a different way than formerly. Information should flow more easily, but references to the origin should be kept all the way along. With the same effort, references could be checked automatically. To achieve this goal, appropriate markup should be used. Software technology has applicable ideas for this problem.

In this paper we discuss the problems of transferring old issues of astronomical journals to computerised formats, and designing formats for new material, using the example of the Information Bulletin on Variable Stars, along with experience with other journals — like the AAS CD-ROM and JAD. Some problems with machine-readable catalogs are also investigated, with ideas about improving formats (FITS) and access tools.

1. Introduction

While transforming astronomical journals and catalogs to electronic form, we intend to make them easier to access for the reader, and, at the same time, making them more easily processable by computers.

In this paper we discuss problems of transforming a small astronomical journal, the IAU Comm. 27 & 42 Information Bulletin on Variable Stars (IBVS) to electronic form — including about 15000 pages of previous issues, back to 1961, to the present, computer-typeset ones.

2. The Old Material

Rendering printed textual information to an ASCII computer file is a difficult problem. There are ambiguities in the typesetting (some old typewriters used the same character for zero and capital o, for digit one and the lowercase l character etc.), some places "redundant" characters were spared for economical reasons, math formulae, non-Latin characters and accents are common.

One could re-typeset the text in TeX , for example, but that would be very difficult. We decided to use a format as simple as possible. We have dropped the non-Latin accents; Greek characters, math signs were replaced by their names (as in TEX , but without the leading backslash); for superscripts and subscripts either TEX -like or FORTRAN -like syntax were accepted. We remove hyphenation, for the sake of simple text string searches.

Errors introduced by the Optical Character Recognition (OCR) process make the situation worse. IBVS was published mainly from camera-ready material, therefore pages were extremely heterogeneous. The OCR and primary correction was done by many different persons.

In spite of a three-pass error checking and correction process (including a spell-checker), errors still remain in the text. But what should we do with the errors originally in the text? Our accepted policy was: correct the obvious typos, spelling errors (if found), do not correct foreign spelling, nor semantical errors. We have not checked the references in the papers (except for obvious spelling errors in the names of journals).

In retrospect, we see now that we should have laid down a rule-set for the rendering in advance. It would be desirable to develop a standard, which would produce easily readable, and at the same time, computer-browsable information.

The next question is the format, in which we provide the information to the community. We have chosen plain ASCII text and PostScript images of the pages. We could have devised a simple markup for the ASCII text version, which would have enabled us, for instance, to create tables of contents automatically, or any bibliographical service provider (BSP), like ADS or SIMBAD, to process the references in the papers — we have not done this.

There is one obvious shortcoming of the ASCII text version: the figures are missing. In the final form, we will use a simple markup in the place of the missing figure, adding a brief description, if not available in the caption or in the text (e.g.: [Fig. 1.: V lightcurve for 1973]).

3. The New Material

For the past few years, IBVS has been typeset in LATEX . Source code and PostScript versions are available. Recently, we have introduced a new TEX style file, which uses a simple markup using appropriate macro names, which enables automatic extraction of the title, author name, date information, makes possible the insertion of object, variable type (GCVS standards) and other keywords, and also abstracts. Keywords and abstracts do not appear in print, but they are part of the LATEX source. Macros were designed to enable the extraction of information with very simple and generally available text processing tools (i.e., Unix grep). With these new features, IBVS issues get to the Web automatically, tables of contents, object and author indices are generated automatically too. BSPs could easily process the source text (sometimes using the IBVS-specific markup, otherwise removing all LATEX markup completely).

Here we have to deal with the question of errors. Electronically produced material contains less misspellings, thank to the spell-checkers. But what should we do if we find a mistake in an electronic journal? Should we resist the temptation to correct such a mistake in a paper which has been already available on the Web since some time (after publication)? We adopted the following practice: we do not correct the error, but issue an erratum, which is attached to the end of a new issue (as traditionally), AND gets attached to the end (one might use links in HTML format material) of the original issue too, and the Web-page containing the table of contents would also get a flag, notifying the reader. (This way papers could become more dynamic — one can see a journal publishing comments, discussions of papers — as in conference proceedings — attached to the original paper.)

With the references in the papers we have not done anything so far. Discussing the problem with BSPs, it would be possible to design LaTeX macros in such a way, to help automatic reference processing.

We also have to think about the question of figures. To facilitate indexing the information content of the figures, we suggest moving most textual information from the bitmapped or preferably vector-graphic figure to the caption.

The next point to stop at is the question of tabular material. Tables in the IBVS — and in other journals available in electronic form (like the AAS CD-ROM or the Journal of Astronomical Data, also on CD) are often formatted for the human eye, and would be very difficult to read in by a program (to plot or analyze). Publishers of such journals should take care to provide tables easily processable by programs. A good example is the ADC CD-ROM series. IBVS will make available lengthy tables electronically, in machine readable ASCII text, or FITS ASCII Table form. Those tables are easily readable for humans too. The simple catalog format introduced by STARLINK should be also considered. Besides the widely used graphical or text processing tools, there are specific tools for such tables — like the Fits Table Browser by Lee E. Brotzman for the ASCII FITS Tables. We have just one complaint: FTB is slow. With introducing the notion of unique (for catalog numbers) or ordered (like right ascension for many tables) variables to the FITS standard, those tools could be considerably improved.

4. Formats, Media and Policy

We have decided to put all text (plain ASCII or LaTeX source) on-line, and PostScript format issues for recent material. At the moment we do not expect to introduce PDF format, but, in the future, we might add HTML format with converting the LaTeX sources. Our opinion is, that large volume, static material, which has a well defined user community, who regularly use the information, should be distributed on CD-ROM. So we will put old, digitized IBVS issues to a CD-ROM, in PostScript format — which we do not have storage capacity and bandwidth to provide on-line. The CD-ROM will contain IBVS issues 1-4000 and an HTML interface.

Information, which is dynamic, changeable, which is accessed casually (when a user of a BSP follows a reference), should go on-line. So we serve PostScript versions of the recent issues, and text for all. We must keep in sight that this information should be accessible to the broadest community. In consequence, we serve IBVS with different distribution methods: anonymous ftp and WWW. Readers using a public FTPMAIL server could access IBVS via e-mail too. We intend to use such HTML tags on the Web-pages, which work with all possible browsers.

We want to retain control over the textual material too — so BSPs could have it for indexing, and they could put links to the issues residing on our server for full text. The reasons for this decision are the following: the errors in the old and new material get corrected by us, reader services are provided by us, so the best place for the material is with us. On the other hand, those investing in the project wish to retain full rights over the intellectual property. But with the Web there should be no problem with it — the interested reader, following a link, could get the material promptly, wherever it resides.

5. Conclusions and Remarks

Astronomical literature — old and new alike — gets on-line at a rapid pace. Besides the publishers and readers, third parties: the BSPs are concerned as well. Establishing conventions, standards would be desirable.

One can also envision — similar to software development tools and environments — publication development aids. Such tools, for example, could help check the references, whether they really point to an existing paper or not. The focus of the present FADS session is "the prospects for building customized software systems more or less automatically from existing modules". Would it be possible to build "customized scientific papers", automatically from existing modules? In other words, is component re-use possible for astronomical papers? I think the case of figures, tables and references should be considered.

Acknowledgments. The electronic IBVS project was supported by the Hungarian National Information Infrastructure Development Programme (NIIF).

References

The IBVS homepage, URL: http://www.konkoly.hu/IBVS/IBVS.html

A. Holl: The electronic IBVS, Konkoly Observatory Occasional Technical Notes, No. 5, 1996

Astronomical Data Analysis Software and Systems VII
ASP Conference Series, Vol. 145, 1998
R. Albrecht, R. N. Hook and H. A. Bushouse, eds.

Keeping Bibliographies using ADS

Michael J. Kurtz, Guenther Eichhorn, Alberto Accomazzi, Carolyn Grant and Stephen S. Murray

Smithsonian Astrophysical Observatory, Cambridge, MA 02138, Email: dmink@cfa.harvard.edu

Abstract. Nearly every working astronomer now uses the NASA Astrophysics Data System Abstract Service regularly to access the technical literature. Several advanced features of the system are not very well used, but could be quite useful. Here we describe one of them.

Astronomers can use the ADS to maintain their publication lists; a simple http link can bring a current bibliography at any time. In this paper we show how to form the link, how to include papers which ADS does not currently have, and how to deal with name changes.

The ADS can be reached at: http://adswww.harvard.edu/

1. Introduction

The NASA ADS Abstract Service (Kurtz et al. 1993) is now a standard research tool for the majority of astronomers; more than 10,000 different astronomers use it each month, more than 300,000 queries are made, and more than 30,000 papers are read.

Most astronomers maintain an active list of their publications. Because ADS automatically (from the point of view of a user) obtains nearly all journal articles in astronomy, and many book and conference proceeding articles, it can automatically maintain publication lists for most astronomers.

2. Linking to a Personal Bibliography using ADS

Linking to a personal bibliography is very simple. For example the following links to Martha Hazen's bibliography.

http://adsabs.harvard.edu/cgi-bin/abs_connect?author=hazen,m.
&aut_syn=YES&nr_to_return=all

This sets the author=last, first initial (author), turns on author synonym replacement (aut_syn), and returns all papers in the database (nr_to_return).

Should last name and first initial not be unique, or if other features are desired the query becomes somewhat more complex. If you want to specify the author middle initial in addition to the first initial, use exact author matching (&aut_xct=YES):

http://adsabs.harvard.edu/cgi-bin/abs_connect?author=last,+f.m.
&aut_xct=YES&return_req=no_params&jou_pick=NO

Note that there can not be any spaces in the URL, so the "+" sign replaces spaces. Also, you can search for two different formats of author names by entering two author arguments, separated with a semicolon. This does not include the listing of parameters at the bottom of the page (return_req=no_params). Also it only returns articles from refereed journals (jou_pick=NO).

3. Adding Papers into ADS

While ADS is reasonably complete for astronomy journal articles, it is missing many articles. To have a complete bibliography via ADS one needs but to add the missing articles into ADS. The following link gets the form interface for doing this:

http://adsabs.harvard.edu/submit_abstract.html

Figure 1 shows a completed form.

Note that, while abstracts are most welcome, they are not required for inclusion in the database. If one already has a bibliography, one may reformat it into the simple form shown in:

http://adsabs.harvard.edu/abs_doc/abstract_format.html

then submit the whole list (minus the papers already in ADS) via e-mail to ads@cfa.harvard.edu.

4. Dealing with Name Changes

Many people change their names during the course of their careers. In the example above Martha Hazen was M. Hazen, M. Liller, M. Hazen-Liller, and M. Hazen. ADS tracks these changes. If your name is not properly tracked by ADS, just send e-mail to ads@cfa.harvard.edu and list all the names you have published under. Your bibliography will then work properly, so long as the aut_syn flag is set as YES.

References

Kurtz, M.J., Karakashian, T., Grant, C.S., Eichhorn, G., Murray, S.S., Watson, J.M., Ossorio, P.G., & Stoner, J.L. 1993, in ASP Conf. Ser., Vol. 52, Astronomical Data Analysis Software and Systems II, ed. R. J. Hanisch, R. J. V. Brissenden & Jeannette Barnes (San Francisco: ASP), 121

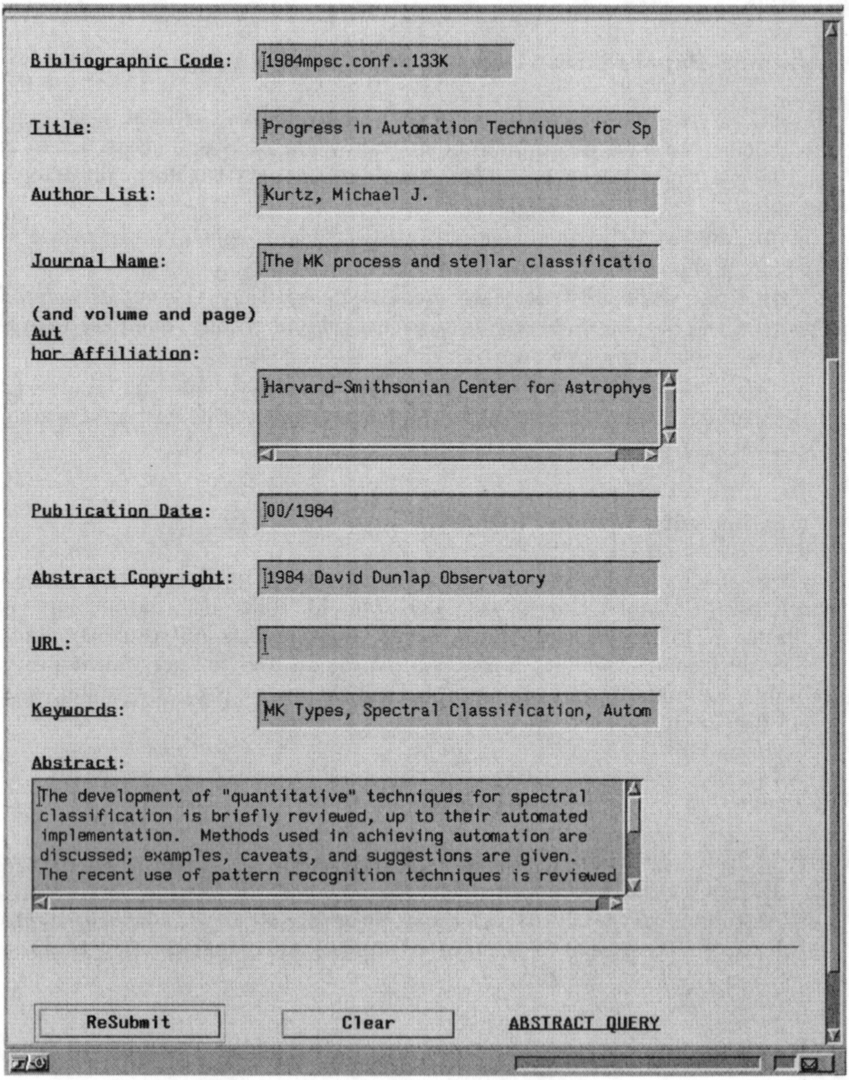

Bibliographic Code: 1984mpsc.conf..133K

Title: Progress in Automation Techniques for Sp

Author List: Kurtz, Michael J.

Journal Name: The MK process and stellar classificatio

(and volume and page)
Aut
hor Affiliation:

Harvard-Smithsonian Center for Astrophys

Publication Date: 00/1984

Abstract Copyright: 1984 David Dunlap Observatory

URL:

Keywords: MK Types, Spectral Classification, Autom

Abstract:

The development of "quantitative" techniques for spectral
classification is briefly reviewed, up to their automated
implementation. Methods used in achieving automation are
discussed; examples, caveats, and suggestions are given.
The recent use of pattern recognition techniques is reviewed

ReSubmit Clear ABSTRACT QUERY

Figure 1. A completed submission form for ADS

Astronomical Data Analysis Software and Systems VII
ASP Conference Series, Vol. 145, 1998
R. Albrecht, R. N. Hook and H. A. Bushouse, eds.

Astrobrowse: A Multi-site, Multi-wavelength Service for Locating Astronomical Resources on the Web

T. McGlynn[1] and N. White

NASA/Goddard Space Flight Center, Greenbelt, MD 20771, Email: tam@silk.gsfc.nasa.gov

Abstract.
We report on the development of a Web agent which allows users to conveniently find information about specific objects or locations from astronomical resources on the World Wide Web. The HEASARC Astrobrowse agent takes a user-specified location and queries up to hundreds of resources on the Web to find information relevant to the given target or position. The current prototype implementation is available through the HEASARC and provides access to resources at the HEASARC, CDS, CADC, STScI, IPAC, ESO and many other institutions. The types of resources the user can get include images, name resolution services, catalog queries or archive indices.

The Astrobrowse effort is rapidly evolving with collaborations ongoing with the CDS and STScI. Later versions of Astrobrowse will use the GLU system developed at CDS to provide a distributable database of astronomy resources. The Astrobrowse agent has been written to be customizable and portable and is freely available to interested parties.

1. Introduction

The myriad astronomical resources now available electronically provide an unprecedented opportunity for astronomers to discover information about sources and regions they are interested in. However, many are intimidated by the very number and diversity of the available sites. We have developed a Web service, Astrobrowse[2], which makes using the Web much easier. The Astrobrowse agent can go and query many other Web sites and provide the user easy access to the results. In the next section we discuss the history and underlying philosophy of our Astrobrowse agent. The subsequent sections address the current implementation, status and future plans.

---

[1]Universities Space Research Association

[2]http://legacy.gsfc.nasa.gov

2. Why Astrobrowse?

Astronomers wishing to use the Web in their research face three distinct problems:

Discovery Given the hundreds of Web sites available it is virtually impossible users to know of all the sites which might have information relevant to a given research project.

Utilization Even when users know the URLs of useful Web sites, each Web site has different formats and requirements for how to get at the underlying resources.

Integration Finally, when users have gotten to the underlying resources, the data are given in a variety of incompatible formats and displays.

As we began to design our Astrobrowse agent to address these problems we factored in several realizations: First, as we looked at the usage of our HEASARC catalogs we found that by about 20 to 1, users simply requested information by asking for data near a specified object or position. The particular ratio may be biased by the data and forms at our site, but clearly being able just to do position based searches would address a major need in the community.

Second, we saw that the CGI protocols are quite restrictive so that regardless of the appearance of the site, essentially all Web sites are queried using a simple keyword=value syntax. This commonality of interface presents a unique opportunity. Earlier X-windows forms that many data providers had created, and emerging technologies like Java do not share this.

Another consideration was that for a system to be successful, it should require only minimal, and preferably no effort, on the part of the data providers. We could not build a successful system if it mandated how other sites use their scarce software development resources.

Finally, and perhaps most important, we recognized that problem of integration is by far the most difficult to solve. Integrating results requires agreement on formats and names to a very low level. This is also an area which can require deep understanding of the resources provided so that it may appropriately be left to the astronomer. We would provide very useful service to users even if we only addressed the issues of discovery and utilization.

With these in mind, the outline of our Astrobrowse system was straightforward: Astrobrowse maintains a database which describes the general characteristics of each resource and detailed CGI `key=value` syntax of the Web page. It takes a given target position, and translates the query into the CGI syntax used at the various sites and stores the results. In current parlance, Astrobrowse is a Web agent which explodes a single position query to all the sites a user selects. Since very many, if not most, astronomy data providers have pages which support positional queries, Astrobrowse can access a very wide range of astronomy sites and services.

3. Implementation

The HEASARC Astrobrowse implementation has three sections: resource selection, where the user chooses the sites to be queried; query exploding where the positional query is sent to all of the selected resources; and results management, where Astrobrowse provides facilities for the user to browse the results from the various sites.

3.1. Resource Selection

Once the total number of resources available to an Astrobrowse agent grows beyond 10-20, it is clear that a user needs to preselect the resources to be queried. The current Astrobrowse implementation provides nearly a thousand resources. Querying all of them all of the time would strain the resources of some of the data providers and would also confuse the user. We currently provide two mechanisms for selecting resources. A tree of resources can be browsed and desired resources selected. Alternatively a user can search for resources by performing Alta-Vista-like queries against the descriptions of those resources. E.g., a user might ask for all queries which have the words 'Guide Star' in their descriptions. The user can then select from among the matching queries.

3.2. Query Exploding

The heart of Astrobrowse is the mechanism by which it takes the position or target specified by the user and then transforms this information into a query against the selected resources. For each resource the Astrobrowse database knows the syntax of the CGI text expected, and especially the format of the positional information, including details like whether sexagesimal or decimal format is used and the equinox expected for the coordinates. The current system uses a simple Perl Web query agent and spawns a separate process for each query.

3.3. Results Management

Astrobrowse takes the text returned from each query and caches it locally. If the query returns HTML text then all relative references in the HTML – which presumably refer to the originating site and thus would not be valid when the file is retrieved from the cache – are transformed into absolute references.

Our Astrobrowse interface uses frames to provide a simple mechanism where the user can easily switch among the pages returned. A number of icons return the status of each request, and allow the user to either delete a page which is no longer of interest, or to display it in the entire browser window.

3.4. The Astrobrowse Database

A database describing Astronomy Web sites is central to the functioning of Astrobrowse. For each resource, a small file describes the CGI parameters and provides some descriptive information about the resource. The file is human-readable and can be generated manually in a few minutes if one has access to the HTML form being emulated. We also provide a page on our Astrobrowse server to automatically build these files so that users can submit new resources to be accessed by our agent.

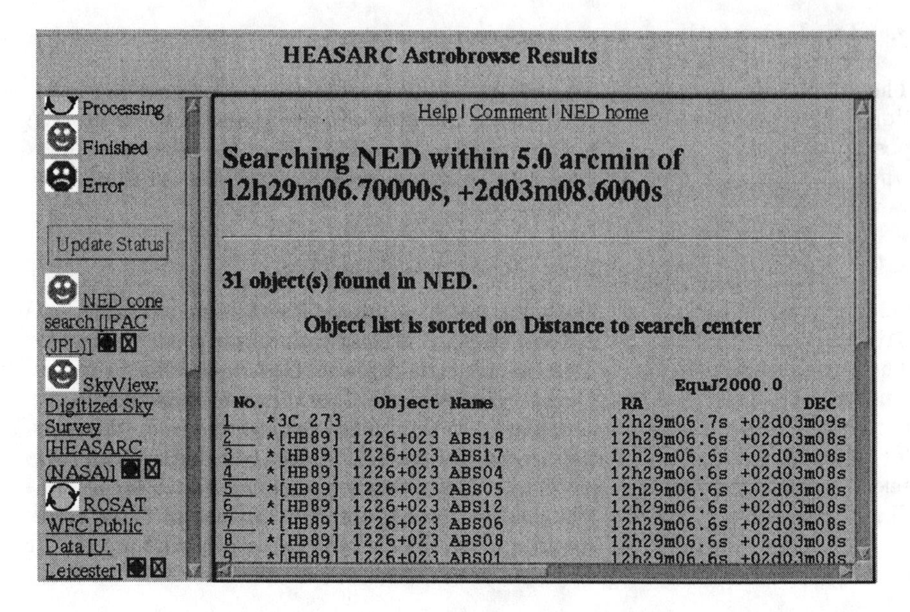

Figure 1. An Astrobrowse Screen Shot.

4. Future Plans

We believe the current Astrobrowse provides a convincing proof-of-concept for an astronomy Web agent and is already a very useful tool but we anticipate many changes in the near term. Among these are:

- Access to virtually all astronomical Web pages supporting positional queries. We expect to support about 3,000 to 10,000 resources compared to the current approximately 1,000 resources.
- Providing a few uniform keys for resources so that users can select data conveniently by waveband and data type.
- Conversion of the database to formats using the CDS's GLU format and use of the GLU system to manage distributed access and updates to the resource database.
- Substantial experimentation in the user interfaces to select resources and display results.

In the longer term we hope that Astrobrowse can be expanded beyond the limits of positional searches for astronomical resources and become the basis for tools to help integrate astronomy, space science and planetary data.

Astronomical Data Analysis Software and Systems VII
ASP Conference Series, Vol. 145, 1998
R. Albrecht, R. N. Hook and H. A. Bushouse, eds.

A Multidimensional Binary Search Tree for Star Catalog Correlations

D. Nguyen, K. DuPrie and P. Zografou

Smithsonian Astrophysical Observatory, Cambridge, MA 02138

Abstract. A multi-dimensional binary search tree, k-d tree, is proposed to support range queries in multi-key data sets. The algorithm can been used to solve a near neighbors problem for objects in a 19 million entries star catalog, and can be used to locate radio sources in cluster candidates for the X-ray/SZ Hubble constant measurement technique. Objects are correlated by proximity in a RA-Dec range and by magnitude. Both RA and Dec have errors that must be taken into account. This paper will present the k-d tree design and its application to the star catalog problem.

1. Problem Description

One of the requirements for creating the Guide and Aspect Star Catalog for the AXAF (Advanced X-ray Astrophysics Facility) project was to calculate a number of spoiler codes for all objects in the catalog. One of these spoiler codes was calculated based on the distance between an object and its nearest neighbor, and on their magnitude difference. Another group of spoiler codes depended on finding the magnitude difference between an object and the brightest object within a specified radius, where the radius ranged from 62.7 arcmin to 376.2 arcmin. These calculations had to be performed on all 19 million objects in the catalog in a reasonable amount of time. Each object had 14 attributes, taking up 75 bytes of space, resulting in a size of approximately 1.5 GB for the entire catalog. In order to solve this problem we needed a way to search the data by RA, Dec and magnitude fast so repeated searches for each object in the database would be possible within our time limitations. Two different options were immediately available none of which however was ideal.

The first option would use the format in which the catalog is distributed and accessed by a number of applications with relatively simple data needs. This is 9537 *grid* FITS files, each covering a small region of the sky. Since all objects within a file are close to each other in RA and Dec, it is possible to do most positional calculations on a file-by-file basis. In our case, where we were concerned with near neighbors, it would be necessary to deal with a number of neighboring files at at time: the central file whose objects we were calculating near neighbors for, and the surrounding files so we could be sure to find all the near neighbors out to the largest radius. By dealing with a small subset of the data at a time it is possible to load the data in memory but one still has to scan all records in memory until a match is found. This is a time consuming process for the amount of data in the catalog.

Another option was to use a relational database which provides a structure much easier to search than a flat file. The limitation here was the number of dimensions in the search. Although one may create a number of indexes on different fields, including multi-dimensional, any given search is really only using the index in one dimension and scans in all other dimensions. These disk based scans were time consuming and resulted in a performance which was unacceptable. It became clear that a preferably memory based true multi-dimensional search structure was needed.

2. Associative Search

An orthogonal range query for records with multiple keys, say k number of keys, such as the problem described above, is commonly known as *associative search.* In its general form, this problem can formally be described as : the data structure of a record is a (k+1)-tuple, (key_1,key_2,..,key_k, I) where key_i is one of the k key components and I is the additional information field of the record. In the problem above, key_1 is RA, key_2 is Dec, key_3 is magnitude and I refers to any other fields that need to be modified as a result of the calculation. An orthogonal range query is to find all objects which satisfy the condition

$$l_i \leq key_i \leq u_i \qquad (1)$$

for i = 1, ... k, where l_i and u_i are the lower and upper ranges of key key_i. A variety of methods have been proposed for solving the multiple keys access problems, unfortunately there is no particular method which is ideal for all applications. The choices for the problem at hand were narrowed down to two : a Multidimensional k-ary Search Tree (MKST) and a k-dimensional (k-d) tree.

2.1. Multidimensional k-ary Search Tree

A MKST is a k-ary search tree generalized for k-dimensional search space in which each non-terminal node has 2^k descendants. Each non-terminal node partitions the records into 2^k subtrees according to the k comparisons of the keys. Note that at each level of the tree, k comparisons must be made and since there are 2^k possible outcomes of the comparisons, each node must have 2^k pointers. The number of child pointers grow exponentially as a function of the dimension of the search space which is a waste of space since many child pointers usually remain unused. An MKST in two (k=2) and three (k=3) dimensional search space is called a quadtree and octree, respectively.

2.2. k-d Tree

A k-d tree is a binary search tree generalized for a k-dimensional search space. Each node of the k-d tree contains only two child pointers, i.e., the size of the node is independent of the number of the dimensional search space. Each non-terminal node of the k-d tree splits its subtree by cycling through the k keys of the k-dimensional search space. For example, the node of the left subtree of the root has a record with value for key key_1 less or equal to value for key key_1 for the root; the node of the right subtree of the root has a record with value for key key_1 greater than the value for key key_1 for the root. The nodes at depth one partition their subtrees depending on the value for the key key_2. In general, nodes at depth h are split according to $key_{h\ mod\ k}$. Note that at each level of the tree only one comparison is necessary to determine the child node. All the attributes of the stars are always known in advanced hence a balanced k-d tree, called an *optimized k-d tree*, can be built. This is done by recursively inserting the median of the existing set of data for the applicable discriminator as the root of the sub tree. An optimized k-d tree can be built in time of $O(n\ log\ n)$, where n is the number of nodes in the tree. A range search in a k-d tree with n nodes takes time O(m + k $n^(1 - 1/k)$) to find m elements in the range.

3. k-d Tree vs k-ary Search Tree

The k-d tree and the MKST do not exhibit any efficient algorithms for maintaining tree balance under dynamic conditions. For the problem at hand, node deletion and tree balancing are not necessary. The k-d tree was chosen over the MKST for the simple reason that it is spatially (2^k vs 2 child pointers per node) and computationally (k vs 1 comparisons at each level of the tree) more efficient to the MKST as the dimension k increases. The advantage of the MKST over the k-d tree is that the code is slightly less complicated to write.

4. Data Volume Limitation

If the k number of keys to be searched for contain RA, Dec and some other attributes then one can minimize the number of nodes in the tree by utilizing only the near neighbor FITS files. If RA and Dec are not a subset of the k number of keys to be searched then the grid FITS files cannot be utilized and all the stars must be inserted in the nodes of the tree. This could potentially be a problem since the size of the catalog is 1.5 GB. The solution of this problem is to store the k number of keys and the tree structure of the internal nodes in main memory subject to the size of available memory, while the information fields of each star can reside on disk. The disk based informational fields can be linked with the tree structure with the *mmap* utility.

Acknowledgments. We acknowledge support from NASA through grant NAGW-3825.

References

Finkel, R. A. & J. L. Bentley, 1974, "Quadtrees: A data structure for retrieval on composite keys," *Acta Informatica* **4**, 1

J. L. Bentley, 1979, "Multidimensional Binary Search Trees Used for Associative Searching," *Communications of the ACM* **19**, 509

J. L. Bentley, 1979, "Multidimensional Binary Search Trees in Database Applications," *IEEE Transactions on Software Engineering* **SE-5**, 333

H. Samet, 1984, "The Quadtree and Related Hierarchical Data Structures," *Computing Surveys **16**, 187*

Astronomical Data Analysis Software and Systems VII
ASP Conference Series, Vol. 145, 1998
R. Albrecht, R. N. Hook and H. A. Bushouse, eds.

Methodology of Time Delay Change Determination for Uneven Data Sets

V.L.Oknyanskij

Sternberg Astronomical Ins., Universitetskij Prospekt 13, Moscow, 119899, Russia

Abstract. At the previous ADASS Conference (Oknyanskij, 1997a) we considered and used a new algorithm for time-delay investigations in the case when the time delay was a linear function of time and the echo response intensity was a power-law function of the time delay. We applied this method to investigate optical-to-radio delay in the double quasar 0957+561 (generally accepted to be a case of gravitational lensing). It was found in this way that the radio variations (5 MHz) followed the optical ones, but that the time delay was a linear function of time with the mean value being about 2370 days and with the rate of increase $\mathbf{V} \approx 110$ days/year.

Here we use Monte-Carlo simulations to estimate the significance of the results. We estimate (with the 95% confidence level) that the probability of getting the same result by chance (if the real $\mathbf{V}$ value was equal to 0 days/year) is less then 5%. We also show that the method can also determine the actual rate of increase $\mathbf{V_a}$ of the time delay in artificial light curves, which have the same data spacing, power spectrum and noise level as real ones.

We briefly consider some other possible fields for using the method.

1. Introduction

At the previous ADASS Conference (Oknyanskij, 1997a, see also Oknyanskij 1997b) we considered and used a new algorithm for time-delay investigations in the case when the time delay was a linear function of time and the echo response intensity was a power-law function of the time delay. We applied this method to investigate optical-to-radio delay in the double quasar 0957+561 (generally accepted to be a case of gravitational lensing). It was found in this way that the radio variations (5 MHz) followed the optical ones, but that the time delay was a linear function of time with the mean value being about 2370 days and with the rate of increase $\mathbf{V} \approx 110$ days/year.

The cross-correlation function for this best fit is shown in Figure 1 together with the cross-correlation function for the data without any corrections for possible variability of the time delay value. The maximum of the cross-correlation for the last case (if $\mathbf{V} = 0$ days/year) is less then 0.5. So we can note that our fit explains the real data significantly better then the simple model with some constant time delay. Meanwhile Monte-Carlo simulations are needed to estimate

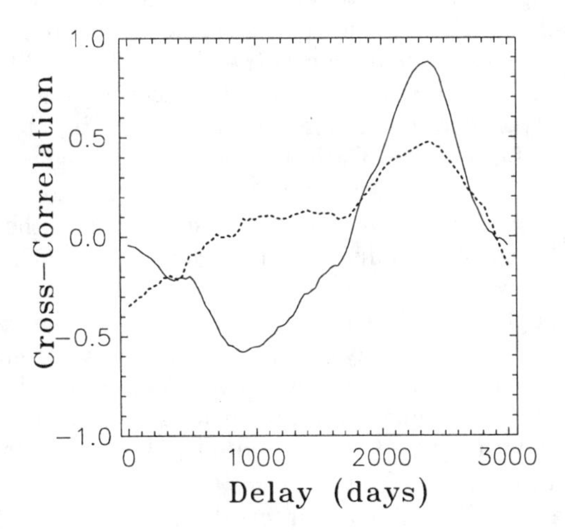

Figure 1. Cross-correlation functions for combined radio and optical light curves. The solid curve shows the cross-correlation function with correction of the data for the best fits parameters: $\mathbf{V} = 110$ days/year and $\alpha = 0.7$ (Oknyanskij, 1997a,b). The dashed curve shows the simple cross-correlation function for the data without any correction i.e., taking $\mathbf{V}$=0 days/year.

the significance of our result. The methodology of these estimations is briefly explained below.

2. Monte-Carlo Simulations

The Monte-Carlo method is a powerful tool that can be used to obtain a distribution-independent test in any hypothesis-testing situation. It is particularly useful for our task, because it is very difficult or practically impossible to use some parametric way to test some null against the alternative hypothesis. We used the Monte-Carlo method to estimate the significance of the result about optical-to-radio delay (Oknyanskij & Beskin, 1993) obtained on the basis of observational data in the years 1979-1990 . With 99% confidence level we rejected the possibility that the high correlation could be obtained if the optical and radio variations were really independent. In a subsequent paper we will estimate the probability of obtaining our result by chance (Oknyanskij 1997a,b) applying the new method to observational data obtained in the years 1979-1994. Here we assume that the optical and radio data are correlated and estimate the significance of the result about possible variability of the time delay value. So here, under H_0, optical and radio data are correlated but there is some constant time delay and the absolute maximum of the correlation coefficient obtained for some $\mathbf{V} \geq \mathbf{V_t} > 0$ days/year reflects random variation. We take $\mathbf{V_t} = 110$ days/year as the test value. When H_1 is true, the time delay is not some constant, but some increasing function of time. We need to calculate the distribution of $\mathbf{V}$ under the null hypothesis. The $p(\mathbf{V})$ value is probability, under H_0, of obtaining some $\mathbf{V}'$ value at least as extreme as (i.e., bigger or equal to) the $\mathbf{V}$. The smaller the $p(\mathbf{V})$ value the more likely we are to reject H_0 and accept H_1. If $p(\mathbf{V_t})$ is less than the typical value of 0.05 then H_0 can be rejected.

The idea of our Monte-Carlo test is the following:

1. We produce m=500 pairs of the simulated light curves which have the same power spectra, time spacing, signal/noise ratio as the real optical and radio data, but with constant value of time delay (τ_{or} = 2370 days) and about the same maximum values of cross-correlation functions.

2. We apply the same method and make all the steps the same as was done for the real data (Oknyanskij, 1997a,b), but for each of m pairs of the simulated light curves. The proportion $\mathbf{p(V)}$ of obtained $\mathbf{V}'$ (see Figure 2) that yield a value bigger or equal to $\mathbf{V}$ provides an estimate of the $p(\mathbf{V})$ value. When $m \geq 100$, standard error of the estimated $p(\mathbf{V})$ value can be approximated by well-known formula $\sigma = [\mathbf{p(1-p)}/m]^{1/2}$ (see Robbins and Van Ryzin 1975). An approximate 95% confidence interval for the true p value can be written as $\mathbf{p} \pm 2\sigma$. As it is seen from Figure 2 - $\mathbf{p(V_t)} \approx 3\%$. The approximate standard error of this value is about 0.8%. We can write the 95% confidence interval for $p(\mathbf{V_t}) = (3 \pm 1.6)\%$ and conclude with a 95% confidence level that $p(\mathbf{V_t}) \leq 5\%$. So the H_0 can be rejected, i.e., time delay is some increasing function of time. For the first step we assume that it is a linear function of time and found $\mathbf{V}$ = 110 days/year, a value approximately equal to the true one.

3. To show that the method has real abilities to determinate a value of of the time delay rate of increase in the light curves we again use Monte-Carlo simulation as it is explained in (1) and (2), but the actual $\mathbf{V_a}$ value is 110 days/year. Then we obtain the histogram (Figure 3), which shows the distribution of obtained $\mathbf{V}'$ values. It is clear that the distribution has some asymmetry, which could be a reason for some small overestimation of $\mathbf{V}$ value, since the mathematical expectation of mean $\mathbf{V}$' is about 114 days/year. Meanwhile we should note that the obtained histogram shows us the ability of the method to get the approximate estimate of the actual $\mathbf{V_a}$ value, since the distribution in the Figure 3 is quite narrow. Using this histogram we approximately estimate the standard error $\sigma(\mathbf{V}) \approx 15$ days/year (for $\mathbf{V}$ value, which has been found for Q 0956+561).

3. Conclusion and Ideas for Possible Applications of the Method

We have found a time delay between the radio and optical flux variations of Q 0957+561 using a new method (Oknyanskij, 1997a,b), which also allowed us to investigate the possibilities that (1) there is some change of time delay that is a linear function of time, and (2) the radio response function has power-law dependence on the time delay value. Here we estimate the statistical significance of the result and the ability of the method to find the actual value of $\mathbf{V}$ as well as its accuracy. We show that with 95% confidence level the probability of getting a value of $\mathbf{V} \geq 110$ days/year (if actual $\mathbf{V_a}$ would be equal to 0 days/year) is less then 5%. We estimate that standard error of the $\mathbf{V}$ value (which has been found for Q 0957+561) is about 15 days/year.

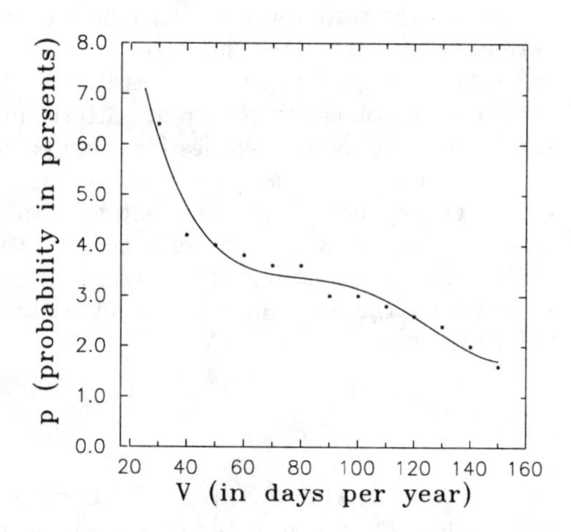

Figure 2. $\mathbf{p(V)}$ is probability to get some value $\mathbf{V'} \geq \mathbf{V}$ if the actual $\mathbf{V_a} = 0$ days/year. All Monte-Carlo test $\mathbf{p}$ values are based on applying our method to 500 pairs of artificial light curves with the actual parameters $\tau_{or}(t_0) = 2370$ days and $\mathbf{V_a} = 0$ days/year.

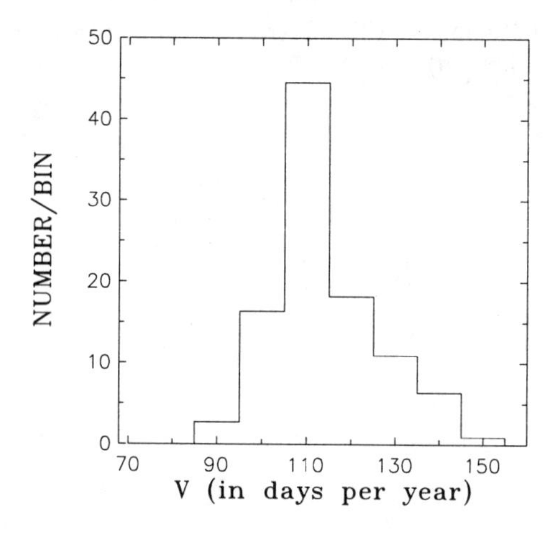

Figure 3. Result of Monte-Carlo test (see Figure 2) with actual values of $V_a = 110$ days/year and $\tau_{or}(t_0) = 2370$ days. The histogram shows distribution of obtained $\mathbf{V'}$ values.

Finally, we can briefly note some other fields where the method may be used:

1. Time delay between continuum and line variability in AGNs may be a function of time as well as the response function possibly being some function of time. So our method can be useful for this case.

2. Recently, it has been suggested by Fernandes et. al (1997) that variability of different AGNs might have coincident recurrent patterns in their light curves. However it has been found that the time-scales for these patterns in NGC 4151 and 5548 are about the same, there are a lot of reasons to expect that the patterns in AGN light curves may be similar, but have different time scales. It is possible to use our method with some enhancements to investigate this possibility. Some Monte-Carlo estimations of the significance would be also very useful. The probability that these common recurrent patterns in AGNs occur by chance should be estimated.

References

Fernandes, R.C., Terlevich R. & Aretxaga I. 1997, MNRAS, 289, 318

Oknyanskij, V.L., & Beskin, G.M. 1993, in: Gravitational Lenses in the Universe: Proceedings of the 31st Liege International Astrophysical Colloquium, eds. J.Surdej *at al.* (Liege, Belgium: Universite de Liege, Institut d'Astrophysique), 65

Oknyanskij, V.L. 1997a, in ASP Conf. Ser., Vol. 125, Astronomical Data Analysis Software and Systems VI, ed. Gareth Hunt & H. E. Payne (San Francisco: ASP), 162

Oknyanskij, V.L. 1997b, Ap&SS, 246, 299

Robbins, H. & Van Ruzin, J. 1975, Introduction to Statistic (Science Research Associates, Chicago), 167

Astronomical Data Analysis Software and Systems VII
ASP Conference Series, Vol. 145, 1998
R. Albrecht, R. N. Hook and H. A. Bushouse, eds.

A Wavelet Parallel Code for Structure Detection

A. Pagliaro

Istituto di Astronomia dell'Università di Catania

U. Becciani, V. Antonuccio and M. Gambera

Osservatorio Astrofisico di Catania

Abstract. We describe a parallel code for 3-D structure detection and morphological analysis. The method is based on a multiscale technique, the wavelet transform and on segmentation analysis. The wavelet transform allows us to find substructures at different scales and the segmentation method allows us to make a quantitative analysis of them. The code is based on the message passing programming paradigm and major speed inprovements are achieved by using a strategy of domain decomposition.

1. Introduction

This paper presents a new parallel code, that allows a rapid structure detection and morphological analysis in a 3-D set of data points. The code is described in greater detail in Becciani & Pagliaro, 1997. A serial version of this code has been successfully used in the analysis of the Coma cluster (Gambera et al., 1997). However, possible applications of this code are not limited to astrophysics but may also benefit several other fields of science.

Our method of structure detection is based on the wavelet transform (see Grossmann & Morlet 1984, 1987) evaluated at several scales and on segmentation analysis (see also Escalera & Mazure 1992, Lega 1994, Lega et al. 1995).

2. Method Overview

The detection method can be divided into three main steps.

1. Computation of the wavelet matrices on all the scales investigated. These are computed, by means of the "á trous" algorithm, for the data to be analyzed and on a random distribution in the same region of space and on the same grid as the real data. On these latter matrices we calculate the threshold corresponding to a fixed confidence level in the structure detection.

2. Segmentation analysis. The aim of this analysis is to have all the connected pixels with a wavelet coefficients greater than the threshold labeled with an integer number different for every single structure.

3. Computation of a morphological parameter for every structure singled out and of a mean morphological parameter for each scale.

3. The Implementation

Our strategy has been to develop a parallel code that can run both on multiprocessors or MPP systems and on clusters of workstations; the latter are easily available at low cost without a specific investment in supercomputing. Hence, we have adopted a **message passing technique** and subdivided the computational domain into subdomains, assigning the wavelet and segmentation analysis of the subdomain to different processors: each processor executes in a parallel way the analysis with memory usage and computational load decreasing as the number of working processors grow. For the development of our code we have choose to use **PVM** (Parallel Virtual Machine), a set of C and FORTRAN function written by the OACK Ridge National Laboratory.

4. The Parallel Code

The computational domain is made of the region of space that comprises all the data points from the simulation or the catalogue to be analyzed. Recognition of the substructures happens by means of the wavelet coefficient computation and by the segmentation analysis. Our code is based on the programming paradigm **MASTER/SLAVE**. The computational domain is subdivided into M subdomains (proportionally to the weights of the hosts), along a split axis chosen as the longest one. Each subdomain is assigned to a host of the virtual machine. On the wave slaves both a fault tolerance mechanism and a dynamical load balance one have been implemented. All the quantities computed are written on a shared area of the disk, accessible to all the tasks. This is useful for an efficient fault tolerance mechanism.

The label slaves are spawned at the end of the previous jobs. Each slave executes the segmentation analysis on the subdomain assigned to it by the master and recognizes and numbers the substructures inside it. The labels found by each slave are written on a shared area.

The last part of the code is executed only by the master, that reads the results written on the shared area by the slaves and rearranges them. The master reads the labels that each slaves has assigned to the substructures found in its subdomain and makes an analysis of the borders between the subdomains. The task is to recognize those structures that different adjacent hosts have singled out and that are really only one substructure across the border. Finally, the morphological analysis is executed by the master.

References

Becciani, U., Pagliaro, A., Comp. Phys. Comms, submitted

Escalera, E., Mazure, A., 1992, ApJ, 388, 23

Gambera, M., Pagliaro, A., Antonuccio-Delogu, V., Becciani, U., 1997, ApJ, 488, 136

Grossmann, A., Morlet, J., 1984, SIAM J. Math., 15, 723

Grossmann, A., Morlet, J., 1987, Math. & Phys., Lectures on recent results, ed. L.Streit, World Scientific

Lega, E., 1994, These de Doctorat, Université de Nice (L94)

Lega, E., Scholl, H., Alimi, J.-M., Bijaoui, A., Bury, P., 1995, Parallel Computing, 21, 265

Astronomical Data Analysis Software and Systems VII
ASP Conference Series, Vol. 145, 1998
R. Albrecht, R. N. Hook and H. A. Bushouse, eds.

Positive Iterative Deconvolution in Comparison to Richardson-Lucy Like Algorithms

Matthias Pruksch and Frank Fleischmann

OES-Optische und elektronische Systeme GmbH, Dr.Neumeyer-Str.240, D-91349 Egloffstein, F.R. Germany

Abstract. Positive iterative deconvolution is an algorithm that applies non-linear constraints, conserves energy, and delivers stable results at high noise-levels. This is also true for Richardson-Lucy like algorithms which follow a statistical approach to the deconvolution problem.

In two-dimensional computer experiments, star-like and planet-like objects are convolved with band-limited point-spread functions. Photon noise and read-out noise are applied to these images as well as to the point-spread functions being used for deconvolution. Why Richardson-Lucy like algorithms favor star-like objects and the difference in computational efforts are discussed.

1. Introduction

The increasing availability of computing power fuels the rising interest in iterative algorithms to reconstruct an unknown object distribution $O(x)$, that was blurred by a linear system's point-spread function $P(x)$. The measured image distribution $I(x)$ is then known to be

$$I(x) = P(x) * O(x), \tag{1}$$

with $*$ denoting convolution.

In the case of band-limited point-spread functions or point-spread functions with incomplete coverage of the Fourier domain (interferometry), information is lost and therefore, deconvolution is not possible. Instead of one unique solution, a space of distributions solves (1) (Lannes, Roques & Casanove 1987). This raises the question of how the algorithms choose their reconstruction out of the space of possible solutions.

2. Algorithms

Richardson-Lucy like algorithms use a statistical model for image formation and are based on the Bayes formula

$$p(P * O|I) = \frac{p(I|P * O)\; p(P * O)}{p(I)}, \tag{2}$$

where $p(P*O|I)$ denotes the probability of an event at $(P*O)(x)$, if an event at $I(x)$ occured. Müller (1997) optimized (2) for $O(x)$ by functional variation. For

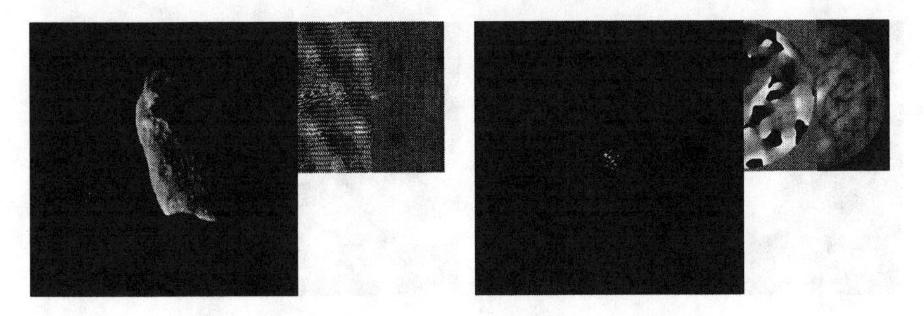

Figure 1. Simulated object on the left, and point-spread function that simulates aberrations by atmospheric turbulence of a ground-based telescope on the right (short exposure).

the case of setting the probability $p(O(x))$ to Poisson statistic the optimization leads to the algorithm invented by Richardson (1972) and Lucy (1974)

$$\hat{O}_{i+1}(x) = \hat{O}_i(x) \left[P(-x) * \frac{I(x)}{P(x) * \hat{O}_i(x)} \right], \tag{3}$$

with $\hat{O}_i(x)$ denoting the iterated object distribution for reconstruction. Setting the probability distribution to have Gauss statistics (Müller 1997) leads to

$$\hat{O}_{i+1}(x) = \hat{O}_i(x) \frac{I(x) * P(-x)}{[P(x) * \hat{O}_i(x)] * P(-x)}. \tag{4}$$

Both algorithms intrinsically apply the positivity constraint and conserve energy.

Positive iterative deconvolution expands the deconvolution to a converging sum. In between iteration-steps the positivity constraint is applied such that the equation

$$I(x) = \hat{O}_i(x) * P(x) + R_i(x) \tag{5}$$

is always satisfied, with $R_i(x)$ denoting the residual of the sum. (Pruksch & Fleischmann 1997)

3. Simulation

In order to test the algorithms for astronomical objects, the simulated object distribution $O(x)$ consists of one planet-like object, and five star-like objects as shown in Figure 1 on the left. The planet-like object in the center is a picture of minor planet Ida[1]. The star-cluster on the lower left has brightness ratios of 1:2:3:4 (right:left:upper:center). The star in the upper right is ten times brighter than the weakest star in the cluster and allows to determine the point-spread function.

---

[1]http://www.jpl.nasa.gov/galileo/messenger/oldmess/Ida2.gif

Figure 2. Convolved image with noise consistent to 10^8 photons per image on the left, and reconstruction by Richardson-Lucy algorithm on the right.

All pictures consist of 256 by 256 pixels and show Ida in full contrast (gamma correction 0.7). The smaller picture on the right side of each picture shows phase (no gamma correction) and modulus (gamma correction 0.25) of the corresponding spectrum. The spectrum is shown for positive horizontal frequencies only, since the values of the negative frequencies are the complex conjugate of the positive ones. The origin of the phase is located on the left center and the origin of the modulus in the center of the small picture.

Different band-limited point-spread functions have been applied to the object, but only one is discussed and illustrated in Figure 1 on the right. The point-spread function simulates wavefront aberrations of approximately $\pm 3\pi$. This is typical for a ground-based telescope that suffers from atmospheric turbulence. The band-limit is visible in the spectrum, since any value beyond the circle is zero (grey for phase and black for modulus).

The peaks in the point-spread function are best seen in Figure 2 on the left which shows the convolved image $I(x)$. In order to avoid spurious information due to round-off errors, the point-spread function was calculated and convolved with the object in the Fourier domain. The contributions beyond the band-limit in Figure 2 on the left originate from noise that was added to the image after convolution.

4. Results

The number of iterations needed by the algorithms to deconvolve the images was in the range of 20 to 100. The computational effort per iteration is dominated by the number of Fourier transforms. The algorithm by Müller is the fastest by performing only two Fourier transforms per iteration, followed by positive iterative deconvolution needing three and Richardson-Lucy with four. To measure convergence, another two Fourier transforms are needed for the Richardson-Lucy like algorithms. If convergence is tested for every iteration-step then positive iterative deconvolution is the fastest algorithm.

All algorithms resolve the objects, as shown by the reconstructions by Richardson-Lucy in Figure 2 on the right, the algorithm by Müller in Figure 3 on the left and positive iterative deconvolution in Figure 3 on the right.

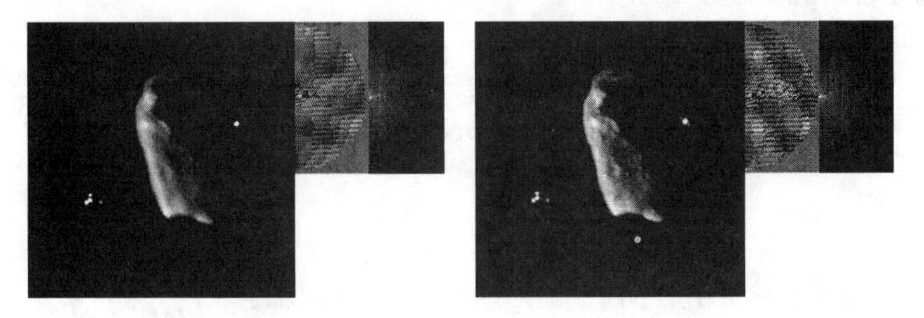

Figure 3. Reconstruction by Müller algorithm on the left, and reconstruction by positive iterative deconvolution on the right.

The Richardson-Lucy like algorithms seem to perform better in the reconstruction of stars, whereas positive iterative deconvolution shows more details for the planet-like object. This is due to the fact that positive iterative deconvolution uses only the positivity constraint, whereas the other algorithms allow statistical deviations. This is clearly seen by comparing the phase distribution of the reconstructions with that of the original object. Since the phase is the most important information, positive iterative deconvolution delivers in this sense the best solution.

5. Summary

Richardson-Lucy like algorithms lead to satisfactory reconstructions. Especially in situations with bad signal-to-noise ratio, the results are very stable. Due to the statistical process of optimization, a different phase distribution is reconstructed, which is favors star-like objects.

Positive iterative deconvolution reconstructs the original data as far as it is available in the measured image and adds information only consistent to that data and the positivity constraint. Therefore, it shows equal quality for all objects. Since positive iterative deconvolution is also the fastest algorithm it is the algorithm of choice, if the reconstructions have to be consistent with measurements.

Acknowledgments. The work reported here has received financial support from the OES-Optische und elektronische Systeme GmbH.

References

Lannes, A., Roques, S., & Casanove, M. J. 1987, J.Mod.Opt., 34, 161

Lucy, L. B. 1974, AJ, 79, 745

Mller, R. 1997, private communication

Pruksch, M. & Fleischmann, F. 1997, accepted by Comput.Phys.

Richardson, W. H. 1972, J.Opt.Soc.Am., 62, 55

Astronomical Data Analysis Software and Systems VII
ASP Conference Series, Vol. 145, 1998
R. Albrecht, R. N. Hook and H. A. Bushouse, eds.

Structure Detection in Low Intensity X-Ray Images using the Wavelet Transform Applied to Galaxy Cluster Cores Analysis

Jean-Luc Starck and Marguerite Pierre

CEA/DSM/DAPNIA F-91191 Gif-sur-Yvette cedex

Abstract. In the context of assessing and characterizing structures in X-ray images, we compare different approaches. The intensity level is often very low and necessitates a special treatment of Poisson statistics. The method based on wavelet function histogram is shown to be the most reliable one. Multi-resolution filtering methods based on the wavelet coefficients detection are also discussed. Finally, using a set of ROSAT HRI deep pointings, the presence of small-scale structures in the central regions of clusters of galaxies is investigated.

1. Wavelet Coefficient Detection

The ability to detect structures in X-ray images of celestial objects is crucial, but the task is highly complicated due to the low photon flux, typically from 0.1 to a few photons per pixel. Point sources detection can be done by fitting the Point Spread Function, but this method does not allow extended sources detection. One way of detecting extended features in a image is to convolve it by a Gaussian. This increases the signal to noise ratio, but at the same time, the resolution is degraded. The VTP method (Scharf et al. 1997) allows detection of extended objects, but it is not adapted to the detection of substructures. Furthermore, in some cases, an extended object can be detected as a set of point sources (Scharf et al. 1997). The wavelet transform (WT) has been introduced (Slezak et al. 1990) and presents considerable advantages compared to traditional methods. The key point is that the wavelet transform is able to discriminate structures as a function of scale, and thus is well suited to detect small scale structures embedded within larger scale features. Hence, WT has been used for clusters and subclusters analysis (Slezak et al. 1994; Grebenev et al. 1995; Rosati et al. 1995; Biviano et al. 1996), and has also allowed the discovery of a long, linear filamentary feature extending over approximately 1 Mpc from the Coma cluster toward NGC 4911 (Vikhlinin et al. 1996). In the first analyses of images by the wavelet transform, the Mexican hat was used. More recently the *à trous* wavelet transform algorithm has been used because it allows an easy reconstruction (Slezak et al. 1994; Vikhlinin et al. 1996). By this algorithm, an image $I(x,y)$ can be decomposed into a set $(w_1, ..., w_n, c_n)$,

$$I(x,y) = c_n(x,y) + \sum_{j=1}^{n} w_j(x,y) \tag{1}$$

Several statistical models have been used in order to say whether an X-ray wavelet coefficient $w_j(x,y)$ is significant, i.e., not due to the noise. In Viklinin et al. (1996), the detection level at a given scale is obtained by an hypothesis that the local noise is Gaussian. In Slezak et al. (1994), the Anscombe transform was used to transform an image with Poisson noise into an image with Gaussian noise. Other approaches have also been proposed using k sigma clipping on the wavelet scales (Bijaoui & Giudicelli 1991), simulations (Slezak et al. 1990, Escalera & Mazure 1992, Grebenev et al. 1995), a background estimation (Damiani et al. 1996; Freeman et al. 1996), or the histogram of the wavelet function (Slezak et al. 1993; Bury 1995). Simulations have shown (Starck and Pierre, 1997) that the best filtering approach for images containing Poisson noise with few events is the method based on histogram autoconvolutions. This method allows one to give a probability that a wavelet coefficient is due to noise. No background model is needed, and simulations with different background levels have shown the reliability and the robustness of the method. Other noise models in the wavelet space lead to the problem of the significance of the wavelet coefficient.

This approach consists of considering that, if a wavelet coefficient $w_j(x,y)$ is due to the noise, it can be considered as a realization of the sum $\sum_{k \in K} n_k$ of independent random variables with the same distribution as that of the wavelet function (n_k being the number of photons or events used for the calculation of $w_j(x,y)$). Then we compare the wavelet coefficients of the data to the values which can taken by the sum of n independent variables. The distribution of one event in the wavelet space is directly given by the histogram H_1 of the wavelet ψ. Since independent events are considered, the distribution of the random variable W_n (to be associated with a wavelet coefficient) related to n events is given by n autoconvolutions of H_1: $H_n = H_1 \otimes H_1 \otimes ... \otimes H_1$

For a large number of events, H_n converges to a Gaussian. Knowing the distribution function of $w_j(x,y)$, a detection level can be easily computed in order to define (with a given confidence) whether the wavelet coefficient is significant or not (i.e not due to the noise).

Significant wavelet coefficients can be grouped into structures (a structure is defined as a set of connected wavelet coefficients at a given scale), and each structure can be analyzed independently. Interesting information which can be easily extracted from an individual structure includes the first and second order moments, the angle, the perimeter, the surface, and the deviation of shape from sphericity (i.e., $4\pi \frac{Surface}{Perimeter^2}$). From a given scale, it is also interesting to count the number structures, and the mean deviation of shape from sphericity.

2. Image Filtering

In the previous section, we have shown how to detect significant structures in the wavelet scales. A simple filtering can be achieved by thresholding the non-significant wavelet coefficients, and by reconstructing the filtered image by the inverse wavelet transform. In the case of the *à trous* wavelet transform algorithm, the reconstruction is obtained by a simple addition of the wavelet scales and the last smoothed array. The solution S is:

$S(x,y) = c_p^{(I)}(x,y) + \sum_{j=1}^{p} M(j,x,y) w_j^{(I)}(x,y)$

where $w_j^{(I)}$ are the wavelet coefficients of the input data, and M is the multiresolution support ($M(j, x, y) = 1$, the wavelet coefficient at scale j and at position (x, y) is significant). A simple thresholding generally provides poor results. Artifacts appear around the structures, and the flux is not preserved. The multiresolution support filtering (see Starck et al (1995)) requires only a few iterations, and preserves the flux. The use of the adjoint wavelet transform operator (Bijaoui et Rué, 1995) instead of the simple coaddition of the wavelet scale for the reconstruction suppresses the artifacts which may appear around objects. Partial restoration can also be considered. Indeed, we may want to restore an image which is background free, objects which appears between two given scales, or one object in particular. Then, the restoration must be performed without the last smoothed array for a background free restoration, and only from a subset of the wavelet coefficients for the restoration of a set of objects (Bijaoui et Rué 1995).

3. Galaxy Cluster Cores Analysis

Cluster cores are thought to be the place where virialisation first occurs and thus in this respect, should present an overall smooth distribution of the X-ray emitting gas. However, in cooling flows (CF) - and most probably in the whole ICM - the presence of small scale inhomogeneities is expected as a result of the development of thermal instability (e.g., Nulsen 1986). (Peculiar emission from individual galaxies may be also observed, although at the redshifts of interest in the present paper (≥ 0.04) - and S/N - such a positive detection would be most certainly due to an AGN.) It is thus of prime interest to statistically investigate at the finest possible resolution, the very center of a representative sample of clusters, in terms of luminosity, redshift and strength of the cooling flow.

Using a set of ROSAT HRI deep pointings, the shape of cluster cores, their relation to the rest of the cluster and the presence of small scale structures have been investigated (Pierre & Starck, 1997). The sample comprises 23 objects up to z=0.32, 13 of them known to host a cooling flow. Structures are detected and characterized using the wavelet analysis described in section 1.

We can summarize our findings in the following way:
- In terms of shape of the smallest central scale, we find no significant difference between, CF and non CF clusters, low and high z clusters.
- In terms of isophote orientation and centroid shift, two distinct regions appear and seem to co-exist: the central inner 50-100 kpc and the rest of the cluster. We find a clear trend for less relaxation with increasing z.
- In general, very few isolated "filaments" or clumps are detected above 3.7σ in the cluster central region out to a radius of ~ 200 kpc. Peculiar central features have been found in a few high z clusters.

This study, down to the limiting instrumental resolution, enables us to isolate - in terms of dynamical and physical state - central regions down to a scale comparable to that of the cluster dominant galaxy. However it was not possible to infer firm connections between central morphologies and cooling flow rates or redshift. Our results allow us to witness for the first time at the cluster center, the competition with the relaxation processes which should here be well advanced and local phenomena due to the presence of the cD galaxy. Forthcoming AXAF and XMM observations at much higher sensitivity, over a wider spectral range and with a better spatial resolution may considerably improve our understanding of the multi-phase plasma and of its inter-connections with the interstellar medium.

References

Bijaoui, A., & Rué, F., 1995, Signal Processing, 46, 345

Biviano, A., Durret, F., Gerbal, D., Le Fèvre, O., Lobo, C., Mazure, A., & Slezak, E., 1996, A&A311, 95

Bury, P., 1995, Thesis, University of Nice-Sophia Antipolis.

Damiani, F., Maggio, A., Micela, G., Sciortino, S, 1996, in ASP Conf. Ser., Vol. 101, Astronomical Data Analysis Software and Systems V, ed. George H. Jacoby & Jeannette Barnes (San Francisco: ASP), 143

Escalera, E., Mazure, A., 1992, ApJ, 388, 23

Freeman, P.E., Kashyap, V., Rosner, R., Nichol, R., Holden, B, & Lamb, D.Q., 1996, in ASP Conf. Ser., Vol. 101, Astronomical Data Analysis Software and Systems V, ed. George H. Jacoby & Jeannette Barnes (San Francisco: ASP), 163

Grebenev, S.A., Forman, W., Jones, C., & Murray, S., 1995, ApJ, 445, 607

Nulsen P. E. J., 1986, MNRAS, 221, 377

Pierre, M., & Starck, J.L., to appear in Astronomy and Astrophysics, (astro-ph/9707302).

Rosati, P., Della Ceca, R., Burg, R., Norman, C., & Giacconi, R., 1995, Apj, 445, L11

Scharf, C.A., Jones, L.R., Ebeling, H., Perlman, E., Malkan, M., & Wegner, G., 1997, ApJ, 477, 79

Slezak, E., Bijaoui, A., & Mars, G., 1990, Astronomy and Astrophysics, 227, 301

Slezak, E., de Lapparent, V. & Bijaoui, A. 1993, ApJ, 409, 517

Slezak, E., Durret, F. & Gerbal, D., 1994, AJ, 108, 1996.

Starck, J.L., & Pierre, M., to appear in Astronomy and Astrophysics, (astro-ph/9707305).

Starck, J.L., Bijaoui, A., & Murtagh, F., 1995, in CVIP: Graphical Models and Image Processing, 57, 5, 420

Vikhlinin, A., Forman, W., and Jones, C., 1996, astro-ph/9610151.

Astronomical Data Analysis Software and Systems VII
ASP Conference Series, Vol. 145, 1998
R. Albrecht, R. N. Hook and H. A. Bushouse, eds.

An Optimal Data Loss Compression Technique for Remote Surface Multiwavelength Mapping

Sergei V. Vasilyev

Solar-Environmental Research Center, P.O. Box 30, Kharkiv, 310052, Ukraine

Abstract. This paper discusses the application of principal component analysis (PCA) to compress multispectral images taking advantage of spectral correlations between the bands. It also exploits the PCA's high generalizing ability to implement a simple learning algorithm for sequential compression of remote sensing imagery. An example of compressing a ground-based multiwavelength image of a lunar region is given in view of its potential application in space-borne imaging spectroscopy.

1. Introduction

Growing interest in lossy compression techniques for space-borne data acquisition is explained by the rapid progress in sensors, which combine high spatial resolution and fine spectral resolution. As the astronomical community becomes aware of multispectral and hyperspectral imaging opportunities, the airborne land sensing heritage (e.g., Birk & McCord 1994; Rao & Bhargava 1996; Roger & Cavenor 1996) can also valuably contribute to future development of the space related instrumentation and data processing methods. On the other hand the on-board storage capacity and throughput of the space-to-ground link are limited, especially for distant and costly astronomical missions. This constrains direct data transmission and the application of traditional image compression algorithms requiring computationally unreasonable expense and inputs, which are not often available on space observatories. In addition, to meet rational operating of the data transmission channels the raw data need on-board compression prior to their down-linking and an option to compress the data flow bit by bit as it is generated onbord without preliminary image store.

Intuitively, when dealing with the hyperspectral/multispectral imagery, one can expect potential benefits from sharing some information between the spectral bands. As a matter of fact, in the case of high spectral correlations between the bands much of information is redundant, that affords a good opportunity for application of lossless compression.

This paper focuses on the ability of principal component analysis (PCA) to reduce dimensions of a given set of correlated patterns and suppress the data noise through data representation with new independent parameters (Vasilyev 1997). It also implements a simple algorithm, which involves some a priori information about the spectral behavior of the object being studied through preliminary learning of the principal components for their further application for practically lossless image compression in a computationally inexpensive way.

2. Essential Concepts

The starting-point for our approach and for the application of the PCA compression to the multispectral/hyperspectral imagery is a spectral decorrelation transformation at each pixel of the frame on the scanning line across the bands. In other words, let us represent a multispectral image set taken in m different wavelengths as the assemblage of n spectra, where n corresponds to the number of pixels in each single-band picture. Then, let A be the matrix composed of the signal values A_{ij} of the i spectrum in the j band.

Obviously, all the data, namely each signal value A_{ij} can be easily restored in the m-dimensional basis simply by the linear combinations of $l \leq m$ eigenvectors (principal components) obtained for the covariance matrix AA': $A_{ij} = \sum_{k=1}^{l} \lambda_{ik} V_{jk}$, where k is the principal component number, V_{jk} is the j-th element of the k-th eigenvector and λ_{ik} is the corresponding eigenvalue at the i-th spectrum.

Generally, m eigenvectors are needed to reproduce A_{ij} exactly. Nevertheless, PCA possesses an amazing feature for the eigenvalues sorted in value-descending order: the eigenvectors corresponding to the first, largest eigenvalues bear the most physical information on the data, while the rest account for the noise and can be neglected from further consideration (Genderon & Goddard 1966; Malinowski 1977). Thus, utilizing $l \ll m$ eigenvectors yields a significant compression of the lossy type and allows the compression rate to be adjusted according to the specific tasks.

Another important feature of PCA that makes this method suitable for the remote sensing applications lies in its powerful generalizing abilities (Liu-Yue Wang & Oja 1993). This feature allows us to describe, using the principal components obtained on the basis of a relatively small calibration data set, a much larger variety of data of the same nature (e.g., Vasilyev 1996; Vasilyev 1997).

3. The Data Used

The lunar spectral data obtained by L. Ksanfomaliti (Shkuratov et al. 1996; Ksanfomaliti et al. 1995) with the *Svet* high-resolution mapping spectrometer (Ksanfomaliti 1995), which was intended for the Martian surface investigations from the spacecraft *Mars 94/96*, and the 2-meter telescope of the Pic du Midi Observatory were selected and used for testing the PCA compression. The 12-band spectra (wavelengths from $0.36\mu m$ to $0.90\mu m$) were recorded in the "scanning line" mode during the Moon rotation at the phase angle of 45°, so that the data acquisition method was similar to the spacecraft scanning. The resulting 19-pixel-wide images were composed of these spectra separately for each band and put alongside in the gray scale (see Figure 1a), where the lower intensity pixels correspond to the higher signal level.

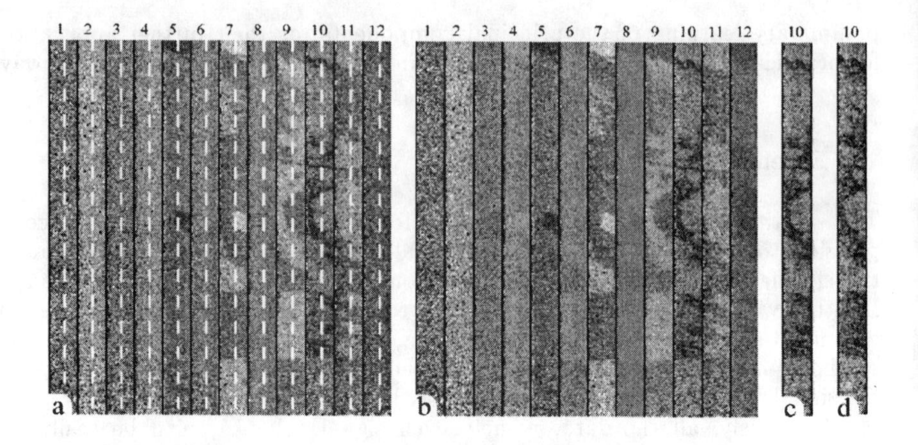

Figure 1. Results of the PCA multiwavelength image compression (a, b) and the whole-band interpolation (c, d). *Data courtesy of Yu. Shkuratov, Astronomical Observatory of Kharkiv University.*

4. Implementation

For the initial training of the principal components a sample multiwavelength scan, indicated with the dashed lines in each of the 12 bands on Figure 1a, was arbitrarily chosen to produce the calibration. The eigenvectors were obtained using this scan containing 250 out of the total 4,750 spectra. Then the entire multispectral image set described above was compressed through encoding the data with the eigenvalues derived from the least-squares fits using the calibration eigenvectors. Each spectrum was processed independently from the others to simulate the real data recording process and without storing all the image in memory.

We found the six principal components providing the compression rate of 1:2, which are able to represent simply by the linear combinations all the features of the original data with the differences not exceeding the *rms* error in each channel. The restored multiwavelength image is shown on Figure 1b.

The calibration eigenvectors allow interpolation for the pixels or even whole spectral bands, which are affected by the impulse noise or other errors. To demonstrate this possibility we have deliberately excluded one channel in the original data set from consideration (shown on Figure 1c) and performed the data encoding with the six principal components. Image restoration made in the ordinary manner proves such a knowledge-based interpolation to be ideal for the remote sensing imagery (see the interpolated single-channel image on Figure 1d).

5. Conclusions

PCA is shown to be applicable to on-board multispectral image compression allowing the incremental compression preceded by the preliminary calibration of the principal components with fairly short learning times. This calibration is stable as it leans on a more varied data library and can be apparently performed on the basis of the laboratory spectra as well.

It is important to note that this type of compression is practically lossless at its reasonably high rates due to the PCA's ability to discard first and foremost the noise and redundant correlations from the data. It should also be noted that compression is better for the higher band numbers where the method retains its robustness and accuracy, our other experiments with hyperspectral land satellite images show that. The compression rate can be further increased by subsequent application of a spatial decorrelating techniques such as JPEG or DCT.

In addition to the reduction of the data dimensions PCA can be used for automatic correction of the impulse noise due to its unique interpolation abilities. The principal components so obtained are characterized by higher informational content than the initial spectra and can be directly used for various data interpretation tasks (see, for example, Shkuratov et al. 1996; Smith et al. 1985; Vasilyev 1996).

Acknowledgments. The author thanks the ADASS VII Organizing Committee for offering him full financial support to attend the Conference.

References

Birk, R. J. & McCord, T. B. 1994, IEEE Aerospace and Electronics Systems Magazine, 9, 26

Genderon, R. G. & Goddard, M.C. 1966, Photogr. Sci. Engng, 10, 77

Ksanfomaliti, L.V. 1995, Solar System Research, 28, 379

Ksanfomaliti, L. V., Petrova, E. V., Chesalin, L. S., et al. 1995, Solar System Research, 28, 525

Liu-Yue Wang & Oja, E. 1993, Proc. 8th Scandinavian Conf. on Image Analysis, 2, 1317

Malinowski, E. R. 1977, Annal. Chem., 49, 612

Rao, A. K. & Bhargava, S. 1996, IEEE Transactions on Geoscience and Remote Sensing, 34, 385

Roger, R. E. & Cavenor, M. C. 1996, IEEE Transactions on Image Processing, 5, 713

Shkuratov, Yu. G., Kreslavskii, M. A., Ksanfomaliti, L. V., et al. 1996, Astron. Vestnik, 30, 165

Smith, M. O., Johnson, P. E. & Adams, J. B. 1985, Proc. Lunar Sci. Conf. XV. Houston. LPI, 797

Vasilyev, S. V. 1996, Ph.D. Thesis, Kharkiv St. Univ., Ukraine

Vasilyev, S. 1997, in ASP Conf. Ser., Vol. 125, Astronomical Data Analysis Software and Systems VI, ed. Gareth Hunt, & H.E. Payne (San Francisco: ASP), 155

Astronomical Data Analysis Software and Systems VII
ASP Conference Series, Vol. 145, 1998
R. Albrecht, R. N. Hook and H. A. Bushouse, eds.

Automated Spectral Classification Using Neural Networks

E. F. Vieira
Laboratorio de Astrofísica Espacial y Física Fundamental
P.O.Box 50727, 28080 Madrid, Spain, Email: efv@laeff.esa.es

J. D. Ponz
GSED/ESA, Villafranca
P.O.Box 50727, 28080 Madrid, Spain, Email: jdp@vilspa.esa.es

Abstract. We have explored two automated classification methods: supervised classification using Artificial Neural Networks (ANN) and unsupervised classification using Self Organized Maps (SOM). These methods are used to classify IUE low-dispersion spectra of normal stars with spectral types ranging from O3 to G5.

1. Introduction

This paper describes the application of automated methods to the problem of the classification of stellar spectra. The availability of the IUE low-dispersion archive (Wamsteker et al. 1989) allows the application of pattern recognition methods to explore the ultraviolet domain. The analysis of this archive is especially interesting, due to the homogeneity of the sample.

The present work has been done within the context of the IUE Final Archive project, to provide an efficient and objective classification procedure to analyze the complete IUE database, based on methods that do not require prior knowledge about the object to be classified. Two methods are compared: a supervised ANN classifier and an unsupervised Self Organized Map (SOM) classifier.

2. The Data Set

The spectra were taken from the IUE Low-Dispersion Reference Atlas of Normal Stars (Heck et al. 1983), covering the wavelength range from 1150 to 3200 Å. The Atlas contains 229 normal stars distributed from the spectral type O3 to K0, that were classified manually, following a classical morphological approach (Jaschek & Jaschek 1984), based on UV criteria alone.

The actual input set was obtained by merging together data from the two IUE cameras, sampled at a uniform wavelength step of 2 Å, after processing with the standard calibration pipeline. Although the spectra are good in quality, there are two aspects that seriously hinder the automated classification: interstellar extinction and contamination with geo-coronal Ly-α emission. Some pre-processing was required to eliminate these effects and to normalize the data.

All spectra were corrected for interstellar extinction by using Seaton's (1979) extinction law. Figure 1 shows original and corrected spectra, corresponding to a O4 star; the wavelength range used in the classification is indicated by the solid line.

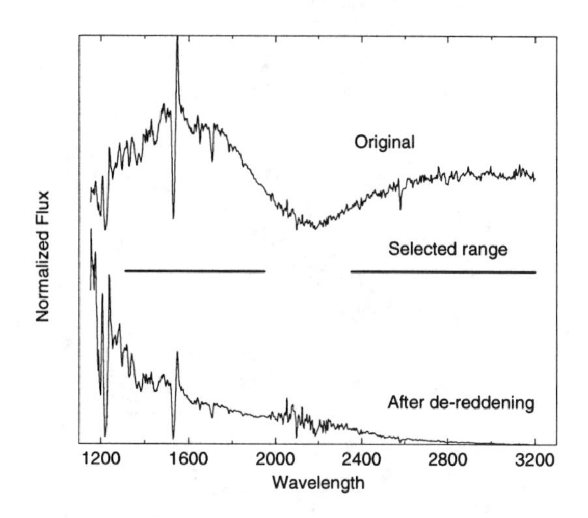

Figure 1. Original and de-reddened spectra.

3. Supervised Classification Using ANN

A supervised classification scheme based on artificial neural networks (ANN) has been used. This technique was originally developed by McCullogh and Pitts (1943) and has been generalized with an algorithm for training networks having multiple layers, known as back-propagation (Rumelhart et al. 1986).

The complete sample in the Atlas was divided into two sets: 64 standard stars, with spectral types from O3 to G5, was used as the training set. The remaining spectra were used as a test to exercise the classification algorithm. The network contains $744 \times 120 \times 120 \times 51$ neurons. The resulting classification error on the test set was 1.1 spectral subclasses. Figure 2 shows the classification diagrams, comparing automatic classification (ANN) with manual (Atlas) and with a simple metric distance algorithm.

4. Unsupervised Classification Using SOM

In the Self Organized Map (SOM) the net organizes the spectra into clusters based on similarities using a metric to define the distance between two spectra. The algorithm used to perform such clustering was developed by Kohonen (1984).

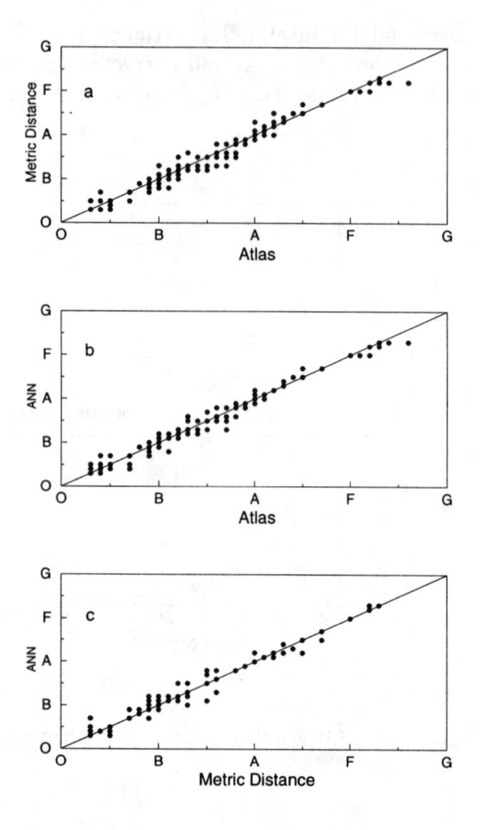

Figure 2. Results of supervised classification.

A 8×8 map with 744 neurons in the input layer was exercised on the same input sample. The training set was used to define the spectral types associated to the elements in the map. This classifier gives an error of 1.62 subclasses when compared with the Atlas, with a correlation of 0.9844. In addition, 27 stars could not be classified according to the classification criterion used in this experiment. Figure 3 shows the classification diagrams, comparing the SOM classifier with ANN and manual classification.

5. Conclusions

Two automated classification algorithms were applied to a well defined sample of spectra with very good results. The error found for supervised algorithm is 1.10 subclasses and 1.62 subclasses for the unsupervised method.

These methods can be directly applied to the set of spectra, without previous analysis of spectral features.

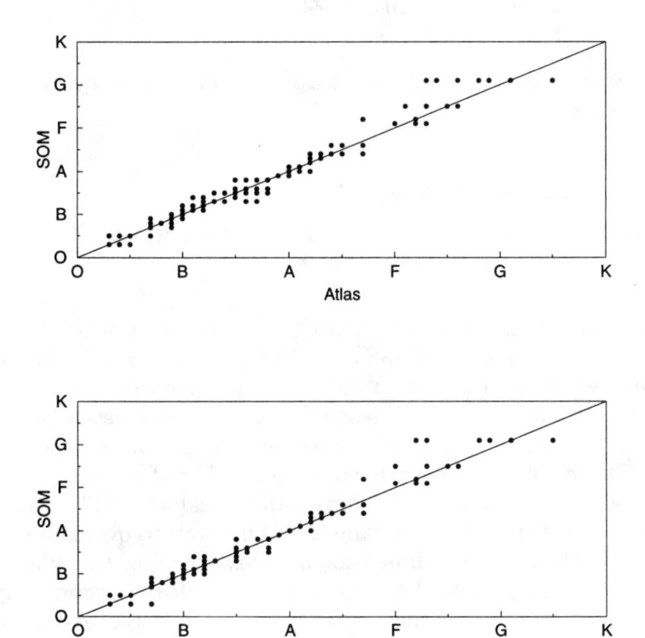

Figure 3. Results of unsupervised classification.

References

Heck, A., Egret, D., Jaschek, M., & Jaschek, C. 1983, (Feb), IUE Low Dispersion Spectra Reference Atlas, SP 1052, ESA, Part1. Normal Stars

Jaschek, M., & Jaschek, C. 1984, in The MK process and Stellar Classification, David Dunlap Observatory, 290

Kohonen, T. 1984, Self-organization and Associative Memory Volume 8 of Springer Series in Information Sciences (Springer Verlag, Nueva York)

McCullogh, W.S., & Pitts, W.H. 1943, Bull. Math. Biophysics, 5, 115

Rumelhart, D.E., Hinton, G.E., & Williams, R.J. 1986, Nature, 323, 533

Seaton, M.J. 1979, MNRAS, 187, 73P

Vieira, E.F., Ponz, J.D. 1995, A&AS, 111, 393

Wamsteker, W., Driessen, C., & Muñoz, J.R. et al. 1989, A&AS, 79, 1

Astronomical Data Analysis Software and Systems VII
ASP Conference Series, Vol. 145, 1998
R. Albrecht, R. N. Hook and H. A. Bushouse, eds.

Methods for Structuring and Searching Very Large Catalogs

A. J. Wicenec and M. Albrecht

European Southern Observatory, Karl-Schwarzschild-Str. 2, D-85748 Garching, Germany ESO

Abstract. Some existing (e.g., USNO-A1.0) and most of the upcoming catalogs (e.g., GSC-II[1] SDSS[2], GAIA[3]) contain data for half a billion up to several billion objects. While the usefulness of the content of these catalogs is undoubted, the feasibility of scientific research and thus the scientific value is very much dependent on the access speed to an arbitrary subset of parameters for a large number of objects. Given the data volume of such catalogs it seems impractical to build indexes for more than two or three of the parameters. One way to overcome this problem is to establish a multi-dimensional index covering the whole parameter space. This implies that the catalog is structured accordingly. The disadvantage of a multidimensional index is that the access to the main index is much slower than it would be using a dedicated structures and indexes. The most commonly used index for astronomical catalogs are the coordinates. The astrophysical importance of coordinates is limited, at least if the coordinates are not suited to the astrophysical question. However for observational purposes coordinates are the primary request key to catalogs. Here we present methods for structuring and searching very large catalogs to provide fast access to the coordinate domain of the parameter space. The principles of these methods can be extended to an arbitrary number of parameters to provide access to the whole parameter space or to another subset of the parameter space.

1. Indexing and Structuring Catalogs

Most of the existing catalogs are physically structured and sorted according to only one parameter. Mostly this parameter is the longitude of the underlying coordinate system (e.g., RA). Some newer catalogs (GSC, Tycho) are structured and sorted by regions, i.e., objects from one region are close together in the catalog. Obviously for accessing large catalogs the first approach is not very well suited. But also the second approach addresses the problem only partly, because the regions have been defined artificially. For very large catalogs the

[1] http://pixela.stsci.edu/gsc/gsc.html

[2] http://www-sdss.fnal.gov:8000

[3] http://astro.estec.esa.nl/SA-general/Projects/GAIA/

region structure should be easily re-producible from basic catalog parameters or, even better, it should be possible to calculate the correct region directly from the main access parameters. Moreover this approach can be hierarchical in the sense that subregions might be accessed by just going one step deeper in the same structure. By generalising such methods it is also possible to produce catalog structures and indexes for accessing a multi dimensional parameter space.

1.1. The Dream

Think of accessing the USNO-A1.0[4] catalog by using a zoom procedure in an online map. The first map shows the whole sky with all objects brighter than a certain limit (e.g., $mag_V \leq 5.0$). In clicking on this map only part of the sky is shown, but the limiting magnitude is now fainter. When getting to the zoom level of about the size of one CCD frame ($\leq$ 100 sq. arcmin.) all objects in this region contained in the catalog are shown. If the upper scenario is possible then think about the implications for astrophysical applications of such a catalog, like e.g., stellar statistics.

1.2. The Reality

Since originally the catalog is structured and sorted by coordinates, any access by magnitudes means reading the whole catalog. Creating a magnitude index means creating a file of about the size of the catalog. Both is unacceptable!

1.3. How to make the Dream Reality

Create small regions of about 100 sq. arcmin sort all objects within these regions by magnitude create an histogram index of these regions, i.e., pointer to first object and number of objects within a magnitude bin. That means less than 30 pointer pairs per region. The last two steps are straightforward but how can the first step be done? There are several ways to create such regions. The HTM or QTM methods (Hierarchical Triangular Mesh, Quaternary Triangular Mesh) is used in the SDSS (Brunner et al. 1994) and GSC-II projects and have already been described at the ADASS (Barrett 1995). In the same paper the 'Bit-Interleaving' method has been proposed for storing astronomical data. Here we describe our experience with 'Bit-Interleaving' and its easy and fast implementation. Bit-Interleaving is just taking the bits of two numbers and creating a new number where every odd bit stems from the first number and every even bit from the second number, i.e., the bits are like the teeth in a zip. The resulting number may be used to sort the catalog in a one-dimensional manner. If the first 14 bits of the bit-interleaved coordinates (in long integer representation) are used as an index, this creates regions of about 71 sq. arcmin. In other words all objects having the same first 14 bits of their bit-interleaved coordinates will be assigned to one region. As already mentioned before the objects in one region may be sorted according to the rest of the bit-interleaved coordinates or, alternatively by magnitudes or any other parameter. The first will produce a hierarchy of subregions, the latter will give the possibility to access the catalog in the magnitude domain. We would like to mention here

---

[4] http://archive.eos.org/skycat/servers/usnoa

the possibility of building real multi-dimensional structures and indexes, like X (Berchtold et al. 1996), R (Brinkhoff et al., 1996), R* (Beckmann et al., 1990) or k-d-trees (various tree structures and comparisons given in the papers of White et al.). However some of these index-structures are hard to build and they slow down the access to the primary key.

2. Testing Bit Interleaving

We have carried out tests using the complete Hipparcos catalog containing about 120.000 objects all over the sky. All code has been written in IDL. The catalog has been accessed through a disk copy (Sun) and directly from the CD (PC). Even in the worst case, a PC with dual-speed CD-ROM drive, it took only about 15 minutes to produce the whole structure.

3. The Future

It is planned to use an advanced catalog structure/sorting for the export version of the GSC-II catalog. This structure might either be a copy of the HTM structure used in the GSC-II database or a structure produced by bit-interleaving. The export catalog will be produced in a collaboration between STScI and ESO. We will also produce a bit-interleaved version of the USNO-A1.0 catalog to test the concepts described above.

References

P. Barrett, 1995, Application of the Linear Quadtree to Astronomical Databases, in ASP Conf. Ser., Vol. 77, Astronomical Data Analysis Software and Systems IV, ed. R. A. Shaw, H. E. Payne & J. J. E. Hayes (San Francisco: ASP), http://www.stsci.edu/stsci/meetings/adassIV/barrettp.html,

David A. White - Shankar Chatterjee - Ramesh Jain, Similarity Indexing for Data Mining Applications http://vision.ucsd.edu/~dwhite/datamine/datamine.html

David A. White - Shankar Chatterjee - Ramesh Jain, Similarity Indexing: Algorithms and Performance, http://vision.ucsd.edu/papers/sindexalg/

R.J.Brunner, K.Ramaiyer, A.Szalay, A.J.Connolly & R.H.Lupton, 1994, An Object Oriented Approach to Astronomical Databases, in ASP Conf. Ser., Vol. 61, Astronomical Data Analysis Software and Systems III, ed. Dennis R. Crabtree, R. J. Hanisch & Jeannette Barnes (San Francisco: ASP), http://tarkus.pha.jhu.edu/database/papers/adass94.ps

Berchtold S., Keim D. A., & Kriegel H.-P., 1996, The X-Tree: An Index Structure for High-Dimensional Data, Proc. 22th Int. Conf. on Very Large Data Bases, Bombay, India, 28

Brinkhoff T., Kriegel H.-P., & Seeger B., 1996, Parallel Processing of Spatial Joins Using R-trees, Proc. 12th Int. Conf. on Data Engineering, New Orleans, LA

Beckmann N., Kriegel H.-P., Schneider R., & Seeger B., The R*-tree: An Efficient and Robust Access Method for Points and Rectangles, 1990, Proc. ACM SIGMOD Int. Conf. on Management of Data, Atlantic City, NJ, 322

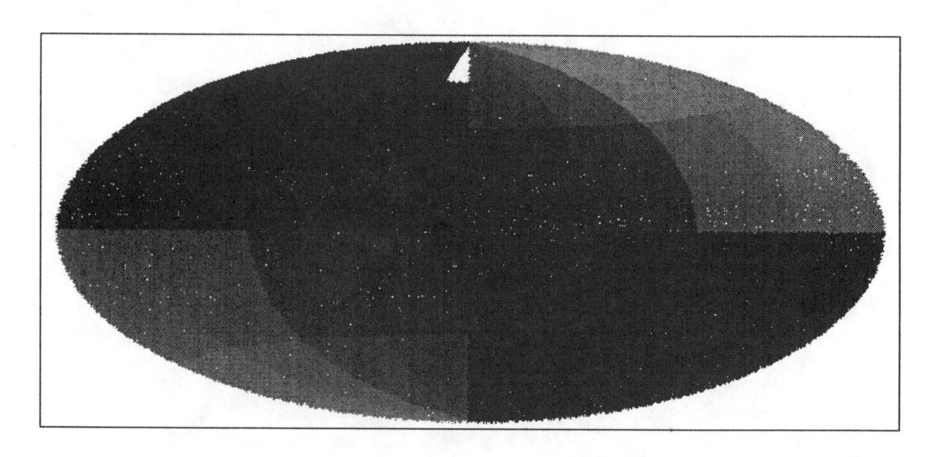

Figure 1. Tessellation of the sky using RA and DEC

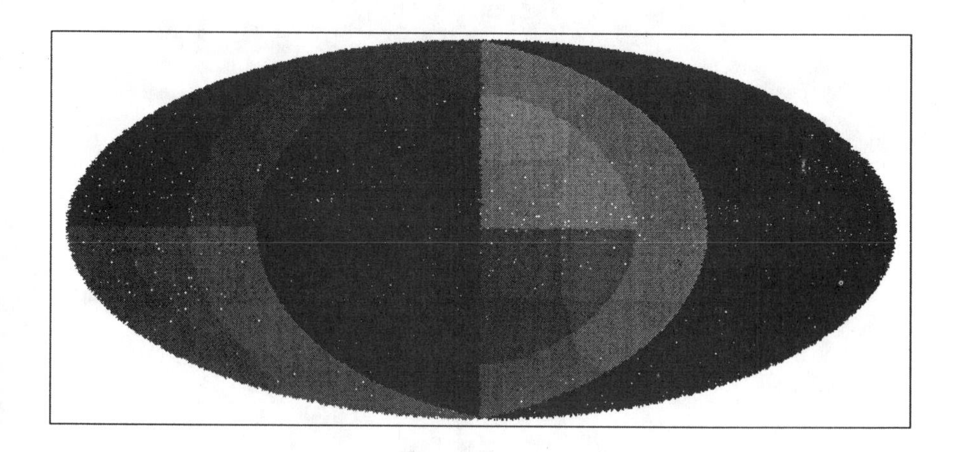

Figure 2. Illustration of a 3-D tessellation using direction cosines

Author Index

Index

Colophon

These proceedings were prepared using many of the tools developed by previous editors of ADASS conferences and we are very grateful for their help. In particular Harry Payne at STScI made available his notes and some software. As before there was a skeleton LaTeX file giving the names of all the contributed papers as chapters which were automatically produced from the contributions submitted electronically by the authors who used the ASP conference style file. This year all manuscripts were submitted either by FTP (the vast majority) or by e-mail. All figures were also deposited by FTP. All manuscripts used LaTeXmarkup and most printed as they were received or with minimal mondifications.

The list of participants was prepared from information held in a Macintosh database by the conference secretary, Britt Sjöberg. The conference photograph was supplied directly as a glossy print.

The proceedings of this conference are once again being made available on the World Wide Web thanks to the permission of the Editors of the Astronomical Society of the Pacific. This is available from the ST-ECF[1] and STScI[2]. The conversion was done by feeding the entire LaTeX document to a modified version of Nikos Drakos' program LaTeX2html[3] v98.1.

[1]http://ecf.hq.eso.org/adass/adassVII

[2]http://www.stsci.edu/stsci/meetings/adassVII

[3]http://cbl.leeds.ac.uk/nikos/tex2html/doc/latex2html/latex2html.html